MW00760766

GEOMETRIC AND TOPOLOGICAL METHODS FOR QUANTUM FIELD THEORY

Aimed at graduate students in physics and mathematics, this book provides an introduction to recent developments in several active topics at the interfaces between algebra, geometry, topology and quantum field theory.

The first part of the book begins with an account of important results in geometric topology. It investigates the differential equation aspects of quantum cohomology, before moving on to noncommutative geometry. This is followed by a further exploration of quantum field theory and gauge theory, describing AdS/CFT correspondence, and the functional renormalization group approach to quantum gravity. The second part covers a wide spectrum of topics on the borderline of mathematics and physics, ranging from orbifolds to quantum indistinguishability and involving a manifold of mathematical tools borrowed from geometry, algebra and analysis.

Each chapter presents introductory material before moving on to more advanced results. The chapters are self-contained and can be read independently of the rest.

HERNÁN OCAMPO is a Professor of Theoretical Physics at the University del Valle, Cali, Colombia.

EDDY PARIGUÁN is an Associate Professor in Mathematics at the University La Javeriana, Bogotà, Colombia.

SYLVIE PAYCHA is a Professor of Mathematics at the University Blaise Pascal, Clermont-Ferrand, France.

GEOMETRIC AND TOPOLOGICAL METHODS FOR QUANTUM FIELD THEORY

Edited by

HERNÁN OCAMPO
University del Valle

EDDY PARIGUÁN
University La Javeriana

SYLVIE PAYCHA
University Blaise Pascal

CAMBRIDGE UNIVERSITY PRESS
Cambridge, New York, Melbourne, Madrid, Cape Town, Singapore, São Paulo, Delhi, Dubai, Tokyo

Cambridge University Press
The Edinburgh Building, Cambridge CB2 8RU, UK

Published in the United States of America by Cambridge University Press, New York

www.cambridge.org
Information on this title: www.cambridge.org/9780521764827

First published 2010

Printed in the United Kingdom at the University Press, Cambridge

A catalogue record for this publication is available from the British Library.

Library of Congress Cataloguing in Publication data
Geometric and topological methods for quantum field theory / edited by Hernán Ocampo,
Eddy Pariguán, Sylvie Paycha.
p. cm.
Includes bibliographical references and index.
ISBN 978-0-521-76482-7 (hardback)
1. Geometry, Differential. 2. Quantum theory. I. Ocampo, Hernán. II. Pariguán, Eddy.
III. Paycha, Sylvie.
QC20.7.D52G46 2010
530.14′301516 – dc22 2009043283

ISBN 978-0-521-76482-7 Hardback

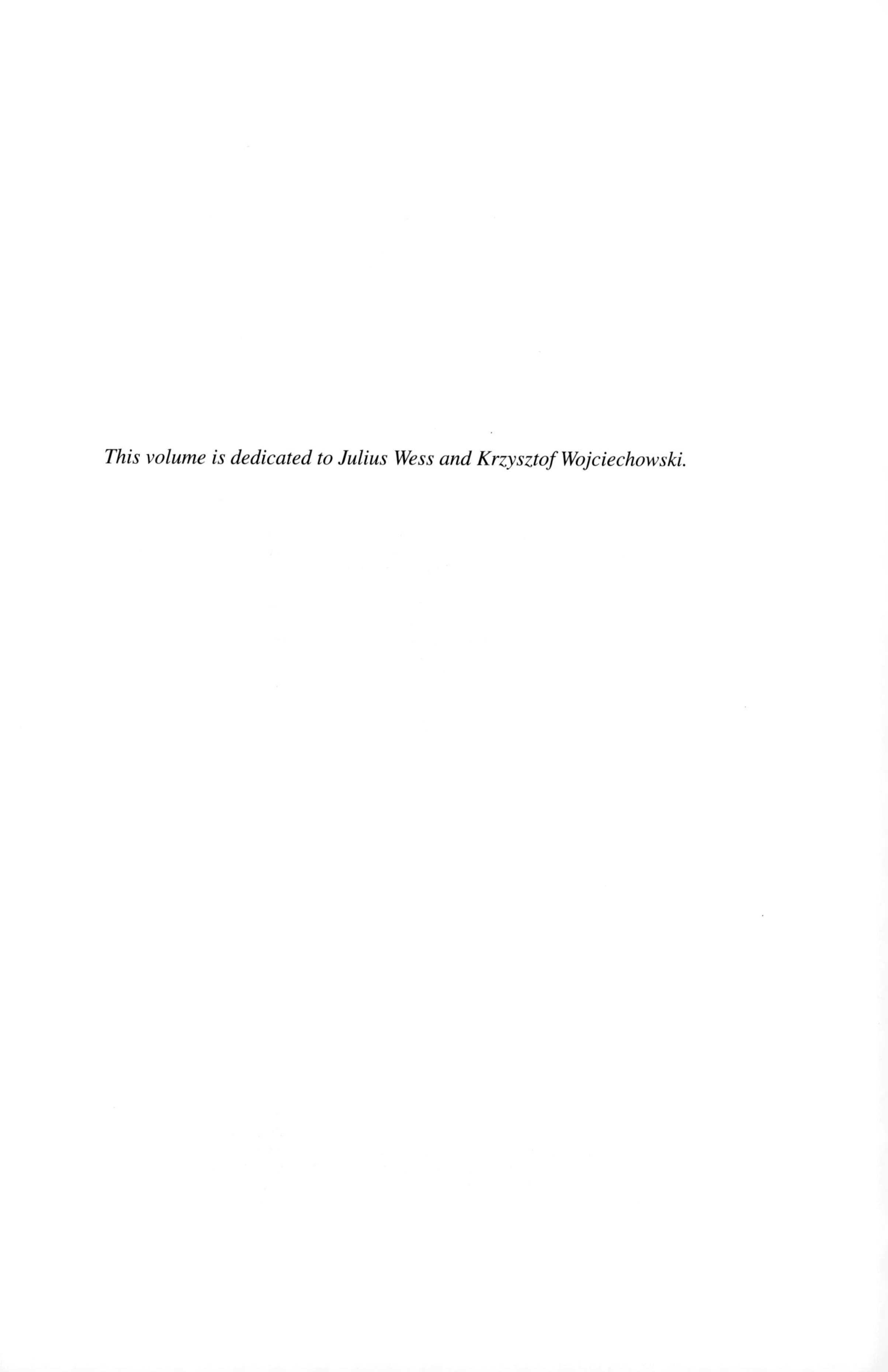

This volume is dedicated to Julius Wess and Krzysztof Wojciechowski.

Contents

List of contributors

ANDRÉS ANGEL, Department of Mathematics, Stanford University, USA

CARLOS BENAVIDES, Departamento de Física, Universidad de los Andes, Colombia

DAVID BERENSTEIN, Department of Physics, University of California Santa Barbara, USA

JAIME R. CAMACARO, Departamento de Matemàticas puras y aplicadas, Universidad Simón Bolívar, Venezuela

CLAIRE DEBORD, Laboratoire de Mathématiques, Universit Blaise Pascal – Clermont II, France

ALESSANDRA FRABETTI, Université de Lyon, France

JOSÉ M. GRACIA-BONDÍA, Departamento de Física Teórica, Universidad de Zaragoza, Spain

MARTIN A. GUEST, Department of Mathematics and Information Sciences, Tokyo Metropolitan University, Japan

PAUL KIRK, Mathematics Department, Indiana University, USA

JEAN-MARIE LESCURE, Laboratoire de Mathématiques, Universit Blaise Pascal – Clermont II, France

JUAN CARLOS MORENO, Departamento de Matemàticas puras y aplicadas, Universidad Simón Bolívar, Venezuela

MARIE-FRANÇOISE OUEDRAOGO, Laboratoire de Mathématiques, Université Blaise Pascal, France and Département de Mathématiques, Université de Ouagadougou, Africa

MARLIO PAREDES, Department of Mathematics, Universidad del Turabo, USA

SOFIA PINZÓN, Escuela de Matemáticas, Universidad Industrial de Santander, Colombia

JORGE PLAZAS, University of Utrecht, The Netherlands

MARTIN REUTER, Institute of Physics, University of Mainz, Germany
ANDRÉS REYES-LEGA, Departamento de Física, Universidad de los Andes, Colombia
FRANK SAUERESSIG, Institute of Theoretical Physics and Spinoza Institute, Utrecht University, The Netherlands

Introduction

This volume offers an introduction to some recent developments in several active topics at the interfaces between algebra, geometry, topology and quantum field theory. It is based on lectures and short communications delivered during the summer school 'Geometric and Topological Methods for Quantum Field Theory' held in Villa de Leyva, Colombia, in July 2007.

The invited lectures, aimed at graduate students in physics or mathematics, start with introductory material before presenting more advanced results. Each lecture is self-contained and can be read independently of the rest.

The volume begins with an introductory course by Paul Kirk on the history and problems of geometric topology, which explains how ideas coming from physics have had an impact on low-dimensional topology in the last 20 years. In the second lecture, Martin Guest discusses differential equation aspects of quantum cohomology, as part of a framework which accommodates the KdV equations and other well-known integrable systems.

We are then led into the realm of noncommutative geometry with a lecture by Claire Debord and Jean-Marie Lescure, who present a proof of Atiyah and Singer's index theorem using groupoids and *KK*-theory, which they then generalize to the case of conical pseudomanifolds.

The remaining lectures take us to the world of quantum field theory, starting with a lecture by Alessandra Frabetti, who presents the Connes–Kreimer algebra for renormalization and its associated proalgebraic group of formal series after having reviewed the Dyson–Schwinger equations for Green's functions and the renormalization procedure for graphs. We then step into gauge theory with José Gracia-Bondía's lecture, which sheds light on BRS invariance of gauge theories using Utiyama's general gauge theory. David Berenstein then gives a short but gentle introduction to the rather sophisticated ideas of AdS/CFT correspondence. In the last lecture, Martin Reuter and Frank Saueressig survey the background material underlying the functional renormalization group approach to quantum gravity.

The invited lectures are followed by six short communications on a wide spectrum of topics on the borderline between mathematics and physics, ranging from orbifolds to quantum indistinguishability and involving a multitude of mathematical tools borrowed from geometry, algebra and analysis.

We hope that these contributions will give – as much as the school itself seems to have given – young students the desire to pursue what might be their first acquaintance with some of the problems on the boundary between mathematics and physics presented here. On the other hand, we hope that the more advanced reader will find some pleasure in reading about different outlooks on related topics and seeing how the well-known geometric tools prove to be useful in some areas of quantum field theory.

We are indebted to various organizations for their financial support for this school. Let us first of all thank the Universidad de los Andes, which has been supporting this and many other schools of this kind that we have been organizing in Colombia since 1999. We are also deeply grateful to the ICTP in Trieste, for its constant financial support over the years and specifically for this school. We also thank the IMU for its support. We are also greatly indebted to other organizations – such as CLAF in Brazil, and Colciencias, ICETEX and ICFES in Colombia – which also contributed in a substantial way to the financial support needed for this school.

Special thanks to Sergio Adarve, Alexander Cardona and Andrés Reyes (Universidad de los Andes), coorganizers of the school, who dedicated time and energy to make this school possible in a country like Colombia where many difficulties are bound to arise along the way due to social, political and economic problems.

We are also very grateful to Marta Kovacsics, Ana Cristina García and Melissa Caro for their help in various essential tasks needed for the successful development of the school. Without the people named here, all of whom helped in the organization in some way or another, before, during and after the school, this scientific event would not have left such vivid memories in the lecturers' and participants' minds. Last but not least, thanks to all the participants who gave us all, lecturers and editors, the impulse to prepare this volume through the enthusiasm they showed during the school, and special thanks to all the contributors and referees for their participation in the realization of these proceedings.

Hernán Ocampo, Eddy Pariguán and Sylvie Paycha

1

The impact of QFT on low-dimensional topology

PAUL KIRK*

Abstract

In this chapter I discuss some of the history and problems of geometric topology and how ideas coming from physics have had an impact on low-dimensional topology in the last 20 years. The ideas are presented largely in simplified (and morally but not necessarily rigorously correct) form to give students an overview of the topics unencumbered by the many technical issues required to put the results on a firm theoretical footing.

The goal of this chapter is to provide theoretical physics students with an introduction to the impact of modern physics on mathematics, as well as to provide for mathematics students a gentle but broad introduction to some of the developments in topology inspired by quantum field theory. No prerequisites are needed besides the usual mathematical maturity, but the astute student will recognize the large role that Morse theory plays. Thus, some familiarity with Morse theory is likely to be useful.

1.1 Geometric topology: a brief history

Geometric topology refers to the study of (usually compact) manifolds.

Let $\mathbb{R}^n_{\geq} = \{(x_1, \ldots, x_n) | x_n \geq 0\}$. An *n-dimensional manifold* is a topological space M equipped with a maximal collection of *charts*

$$\{(U_i, \phi_i) \, | U_i \subset M \text{ open}, \, \phi_i : U_i \to \mathbb{R}^n_{\geq}, \quad \phi_i \text{ a homeomorphism onto an open subset}\}.$$

The set of points mapped to $\{x_n = 0\}$ by the charts ϕ_i is called the *boundary of M*, denoted ∂M. If M is compact and ∂M is empty, we call M *closed*.

There are many different notions of manifold. Manifolds can have many different kinds of extra structures or restrictions (and corresponding equivalences),

* Dedicated to the memory of my friend K. P. Wojciechowski.

such as an orientation (orientation-preserving homeomorphism), a smooth structure (diffeomorphism), a PL structure (PL isomorphism), a spin structure (spin diffeomorphism), a $Spin^c$ structure, an almost complex structure, a symplectic structure (symplectomorphism), a Riemannian or Lorentzian structure (isometry), a flat or spherical or hyperbolic structure, a holomorphic structure (biholomorphism), a Kähler structure, a framing, a trivial fundamental group, a contractible universal cover, etc.

In geometric topology the focus is on *structures such that the corresponding set of equivalence classes is discrete*, and the goal of geometric topology can usually be stated as follows:

Distinguish all equivalence classes of manifolds with a given structure.

1.1.1 Examples

To a compact, connected 2-manifold one can associate its *Euler characteristic* χ (alternating sum of numbers of n-simplices in a triangulation), the number b of boundary circles, and $o \in \{0, 1\}$ keeping track of whether or not the manifold is orientable. Then a classical theorem of topology states that *two compact 2-manifolds have the same triple* (χ, b, o) *if and only if they are homeomorphic.* Thus the class of compact, connected 2-manifolds is classified up to homeomorphism by $(\chi, b, o) \in \mathbb{Z} \times \mathbb{Z}_{\geq 0} \times \mathbb{Z}/2$, and it is simple to determine which triples occur.

Another example is provided by a consequence of Smale's h-cobordism theorem [48]: *Every closed manifold homotopy equivalent to an n-sphere* $S^n = \{x \in \mathbb{R}^{n+1} | \|x\| = 1\}$ *is homeomorphic to an n-sphere, if* $n > 4$. (Freedman [15] proved this for $n = 4$, and Perelman for $n = 3$.)

Two topological spaces X, Y are *homotopy equivalent* if there exists continuous maps $f : X \to Y$ and $g : Y \to X$ and homotopies $H : X \times [0, 1] \to X$ and $K : Y \times I \to Y$ such that $H(x, 0) = x$, $H(x, 1) = g(f(x))$, $K(y, 0) = y$, $K(y, 1) = f(g(y))$.

More interestingly, two *smooth* closed n-manifolds homotopy-equivalent to S^n need not be diffeomorphic. But a consequence of Smale's theorem is that if M^n is a smooth homotopy sphere, $n \geq 5$, then M is obtained from a pair of hemispheres D^n_+ and D^n_- (with $D^n_\pm = \{x \in \mathbb{R}^n \mid \|x\| \leq 1\}$) by gluing their boundaries using a nontrivial diffeomorphism $f : \partial D^n_+ \cong \partial D^n_-$.

Gluing, or *pasting*, topological spaces X, Y along subsets $A \subset X$ and $B \subset Y$ using a *gluing map* $f : A \to B$ refers to forming the quotient space $X \cup Y / \sim$, where $x \sim y$ if $x \in A$, $y \in B$, and $f(x) = y$. Gluing n-manifolds using a homeomorphism f of their boundaries results in an n-manifold. If M is an n-manifold and $N \subset M$ is an $(n-1)$-submanifold with $\partial N = N \cap \partial M$, then *cutting M along N* means forming the manifold (with nonempty boundary) obtained by taking the closure of $M - N$. There are technical issues to worry about, and often one uses instead the complement $M - nbd(N)$ of a small tubular neighborhood of N. Notice that gluing and cutting are inverse operations.

An earlier example is provided by Thom's cobordism theorem [53]: *two closed manifolds M, N are cobordant if and only if they have the same Stiefel–Whitney numbers.* Thom determined which numbers occur as Stiefel–Whitney numbers.

Closed n manifolds M and N are *cobordant* if there exists a compact $n+1$-manifold W with ∂W the disjoint union of M and N. The *Stiefel–Whitney numbers* of a manifold are a collection of numbers $w_I \in \mathbb{Z}/2$; one for each multi-index $I = (i_1, i_2, \ldots, i_n)$, $i_1 + 2i_2 + \cdots + ni_n = n$.

Yet another example is given by Freedman's theorem [15]: *two simply connected (see the following) closed 4-manifolds are homeomorphic if and only if they have isomorphic cohomology rings (see the next section for an introduction to cohomology) and the same Kirby–Seibenmann invariant $KS \in \mathbb{Z}/2$.* One new twist here is that there remains one unsolved case of the classification of possible ring structures; namely, the full classification of unimodular quadratic forms over $\mathbb{Z}$ (which is determined by and determines the cohomology ring of a simply connected 4-manifold) is not known.

Thus one might say that the homeomorphism classification of simply connected 4-manifolds is *reduced* to an algebra problem. In the case of Thom's theorem, Thom first reduced the cobordism classification to a homotopy theory problem, then he solved the homotopy theory problem. For Freedman's theorem, the classification reduces to an algebra problem which is largely solved, but not completely. This is typical.

As a negative example, it is simple to prove that any finitely presented group is the fundamental group of a closed n-manifold for any $n \geq 4$. Logicians tell us the problem of determining whether two group presentations give isomorphic groups is not solvable (no algorithm exists to determine if two presentations determine isomorphic groups). Thus there cannot be an algorithmic (e.g., a finite set of invariants) homeomorphism classification of n-manifolds for $n \geq 4$.

The *fundamental group* $\pi_1(X, x)$ of a topological space X with a distinguished base point is the set of based homotopy classes of loops $\alpha : [0, 1] \to X, \alpha(0) = \alpha(1) = x$. So $\alpha \sim \beta$ if there is a map $H : [0, 1] \times [0, 1] \to X$ so that $H(t, 0) = \alpha(t)$, $H(t, 1) = \beta(t)$, $H(0, u) = x = H(1, u)$. The group structure is given by following one loop, then the next. Concisely, it is the group of path components of the based loop space on X. A continuous map $f : X \to Y$ induces a homomorphism $\pi_1(X) \to \pi_1(Y)$. A connected space X is called *simply connected* if $\pi_1(X, x)$ is the trivial group.

A good 3-dimensional theory is provided by Waldhausen's results on Haken 3-manifolds [56]. A closed, oriented 3-manifold M is called *Haken* if it contains a closed oriented 2-manifold $F \subset M$ (with $F \neq S^2$) such that $\pi_1(F) \to \pi_1(M)$ is injective and such that every sphere $S^2 \subset M$ cuts M into two pieces, one of which is D^3. Then Waldhausen's theorem says: *if M and N are Haken and $\pi_1(M)$ is isomorphic to $\pi_1(N)$, then M and N are homeomorphic (and diffeomorphic).* Thus the fundamental group "classifies" Haken manifolds. Many (but not all) 3-manifolds are Haken, or can be cut into Haken pieces which can be analyzed. Later Thurston proved [54] that *a closed Haken 3-manifold which does not contain a π_1-injective* torus *admits a hyperbolic structure.* Thus this class of 3-manifolds is classified by identifying its fundamental group with a Kleinian group.

An important family of examples comes from considering pairs (M, N) where M is an m-dimensional manifold and N is an n-dimensional submanifold (with $n < m$). One can ask for a *relative* homeomorphism classification, i.e., assuming M_0 is homeomorphic to M_1 and N_0 is homeomorphic to N_1, does there exist a homeomorphism $h : M_0 \cong M_1$ such that $h(N_0) = N_1$? Other interesting questions include the *concordance problem*, where one assumes $M_0 = M_1 = M$, and sets $(M, N_0) \sim (M, N_1)$ if there exists an embedding of $N \times [0, 1] \subset M \times [0, 1]$ which restricts to (M, N_0) and (M, N_1) at the ends $M \times \{0\}$ and $M \times \{1\}$.

The most interesting case is when $n = m - 2$, the *codimension 2 embedding problem*. This topic is generally known as *knot theory*, especially when $M = S^m$ and $N \cong S^{m-2}$. The further specialization when $n = 3$, i.e., the study of embeddings $S^1 \subset S^3$, is usually called *classical knot theory* (and was first systematically studied by the physicist Lord Kelvin, who theorized that tiny knots in the "æther" might explain the subatomic properties of nature).

1.1.2 Invariants

The preceding examples show that classifying manifolds in a certain class is a subtle problem. One should not expect as clean an answer as for 2-manifolds, or even Thom's cobordism theorem. There are several questions to consider: What class of manifolds do we study? Under what equivalence relation? What kind of methods can we use? What is considered progress?

The standard approach to a classification problem in geometric topology is to

(i) find a geometrically meaningful and computable set of *invariants*,
(ii) find a collection of manifolds in the class that *realize* all the invariants (or determine their range),
(iii) prove these invariants classify.

The term *invariants* refers to some way of associating some object (e.g., in a category, or set, or group) to each manifold in the chosen class so that equivalent manifolds are given the same invariant. "Geometrically meaningful" is a vague term, but ideally the invariants should tell us something interesting about the geometric problem. As a negative example, consider the invariant of closed n-manifolds I defined by $I(M) = 0$ if M is homeomorphic to a sphere and $I(M) = 1$ otherwise. Then I is an invariant which partially classifies n manifolds, but it is useless, because its definition reveals nothing about the underlying geometric question.

By contrast, the Euler characteristic of a closed, orientable 2-manifold classifies up to homeomorphism, but in addition it can be defined for any compact space, and it can be computed in many ways (e.g., from a triangulation, by computing homology, from a Morse function, or geometrically by the Gauss–Bonnet theorem). Moreover it has many nice properties (multiplicativity under covers, independence from the triangulation, etc.). Thus producing new invariants is not by itself progress (despite frequent claims made to the contrary).

An important requirement is that invariants be computable. This is also a vague requirement, but to the extent that there are cut-and-paste constructions to produce new manifolds from old in the given class, a good interpretation of this requirement is that it is desirable to be able to compute how the invariant changes under specific cut-and-paste operations.

1.1.3 High and low dimensions

Geometric topology is divided into two distinct topics: *high-dimensional topology*, i.e., the study of manifolds of dimension 5 and higher, and *low-dimensional topology*, i.e., manifolds of dimensions 2, 3, and 4. The reason for this dichotomy is technical, but boils down to the slogan "there is more room to move in high dimensions." A beautiful construction due to Whitney [59], called the *Whitney trick*, uses 2-dimensional disks as guides for various geometric deformations. In n-dimensional topology with $n > 4$, because $2 + 2 < n$ it follows that it is easy to fit the 2-dimensional disk into a manifold in such a way that it does not interfere with itself and other disks in the manifold (in the same way that circles in 3-space can be moved off each other by arbitrarily small perturbations, because $1 + 1 < 3$).

The upshot of this is that the Whitney trick allows one to prove injectivity of invariants produced by counting intersections of submanifolds in high dimensions. This was exploited in the golden era of geometric topology (1955–1980) by many mathematicians, including Smale, Milnor, Wall, Browder, Sullivan, and Novikov, by combining cutting and pasting constructions (surgery) with related algebra (algebraic K-theory) and homotopy theory (bordism theory). These techniques are combined in a powerful machine called *surgery theory*.

It is not correct to say that all classification problems of high-dimensional topology are solved, but it is fair to say that many important ones have been, and that surgery theory provides a starting point for investigating high-dimensional problems which is often powerful enough to solve the problem, or at least to reduce it to purely algebraic or discrete problems.

In low-dimensional topology, the situation is less clear. In dimension 4, there is a chasm between the topological theory and the smooth theory (i.e., between classification up to homeomorphism and up to diffeomorphism). The homeomorphism problem was treated satisfactorily by Freedman. The result is that many of the techniques of surgery theory extend to dimension 4, albeit with much more intricate proofs using infinite processes in point-set topology. Interestingly, the homotopy classification of non-simply-connected 4-manifolds is not well understood. Freedman's results treat the gap between the homotopy problem and the homeomorphism problem for many fundamental groups.

The diffeomorphism problem is quite different, and a breakthrough came at about the same time as Freedman's theorem when Donaldson used ideas from physics (gauge theory) to produce invariants of differentiable 4-manifolds [5]. The development of the ideas pioneered by Donaldson has been the focus of most of the work in the last 20 years in 4-dimensional topology. A major simplification came a few years later with the introduction of Seiberg–Witten theory [28, 47]. However, it is fair to say that any kind of diffeomorphism classification in dimension 4 is a distant goal. As R. Stern puts it, in smooth 4-dimensional topology, "the more we learn the more we realize how little we know."

In 3-dimensional topology (and 2-dimensional topology) there is no difference between homeomorphism and diffeomorphism questions. Depending on one's point of view (and the time of day), 3-manifolds are either well understood or mysterious. One feature of 3-dimensional topology is that there are many structure theorems, notably the existence and uniqueness of decompositions along 2-spheres (the *connected sum* decomposition theorem [30]) and then along tori (the *Jaco–Shalen–Johannsen torus* decomposition theorem [22, 23]). The results of Waldhausen have already been mentioned: these build on many previous results, but notably on Papakyriakopoulos's proofs of Dehn's lemma and the sphere theorem [42].

The next major step forward occurred when Thurston proved his hyperbolization theorem, which inserted the beautiful techniques of Kleinian groups into the study of 3-manifolds. The recent stunning results of Perelman on Thurston's geometrization conjecture can be considered as a continuation of this perspective in 3-dimensional topology.

Fig. 1.1. A projection of the unknotted circle

Fig. 1.2. A framed trefoil knot

1.1.4 Links, Reidemeister moves, and Kirby's theorem

We introduce a few notions from the theory of knots and links, a subject that is both of intrinsic interest in topology and also a useful tool in the construction of manifolds.

A *link* in S^3 is an embedding of a finite disjoint union of circles in S^3,

$$\bigsqcup_{i=1}^{n} S_i^1 \subset S^3.$$

A link with one path component (i.e., $n = 1$) is called a *knot*.

A *projection* of a knot or link is a picture of a generically immersed curve in $\mathbb{R}^2$, with "over and under" data given at each double point, to specify a knot or link in $\mathbb{R}^3 = S^3 - \{p\}$. See Figure 1.1.

A *framed* link in S^3 is an embedding of a finite disjoint union of solid tori in S^3,

$$\bigsqcup_{i=1}^{n} (S^1 \times D^2)_i \subset S^3.$$

See Figure 1.2.

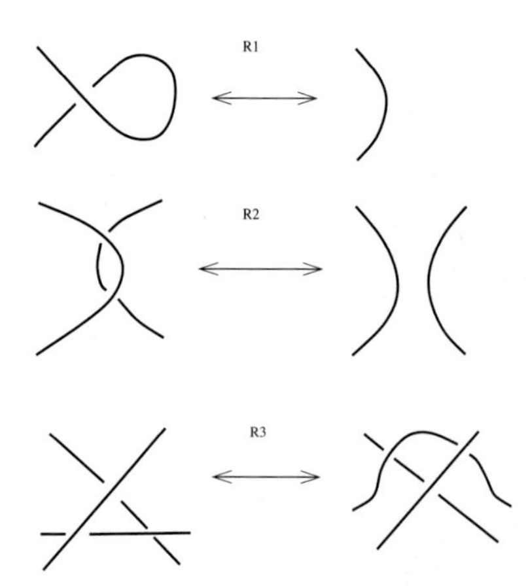

Fig. 1.3. The three Reidemeister moves

The result of *surgery on a framed link* is the 3-manifold, M, obtained by cutting out each $(S^1 \times D^2)$ and gluing in $(D^2 \times S^1)$:

$$M = \left(S^3 - \bigsqcup_i (S^1 \times D^2)_i \right) \cup \bigsqcup_i (D^2 \times S^1)_i .$$

Isotopy of links can be described using projections in terms of the three *Reidemeister moves*, illustrated in Figure 1.3. In fact, Reidemeister proved that these three moves classify links in S^3, in the sense that two *projections* (i.e., pictures like those in Figures 1.1 and 1.2) of links correspond to equivalent links if and only if one can get from one projection to the other by a sequence of Reidemeister moves R1, R2, and R3.

In a different direction, a classification theorem of sorts for 3-manifolds was proven by Kirby in [26].

It has been known since the 1950s that any 3-dimensional manifold is obtained by surgery on a framed link in S^3. However, many different framed links yield the same manifold.

There are geometric moves on the set of framed links: isotopy, stabilization (adding a small, appropriately framed knot away from the rest, as in Figure 1.4), and addition, or sliding (adding a parallel copy of one component to another component; see Figure 1.5).

Kirby's theorem says *two framed links give diffeomorphic 3-manifolds if and only if the framed links are related by these moves*. (The stabilization and sliding moves in this dimension are often called *Kirby moves*.) In other words, 3-manifolds are classified by identifying them with equivalence classes of framed links in S^3 (note that any framed link can be visualized as circles in $\mathbb{R}^3$ with numbers attached to them).

This suggests a strategy to approach the classification problem for 3-manifolds: construct a function which assigns a complex number (or an element in an abelian

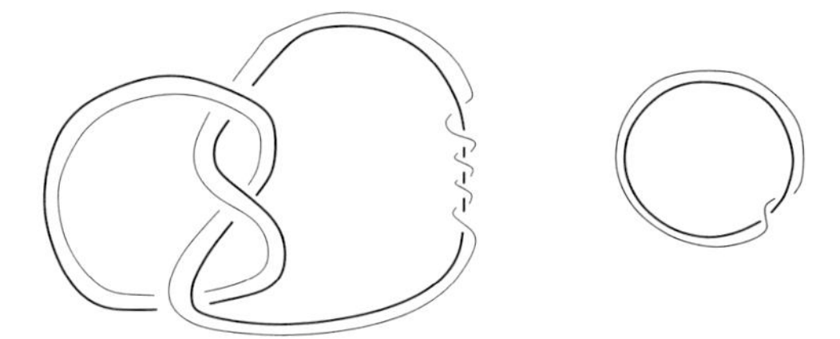

Fig. 1.4. A (-1) stabilization of a framed trefoil knot

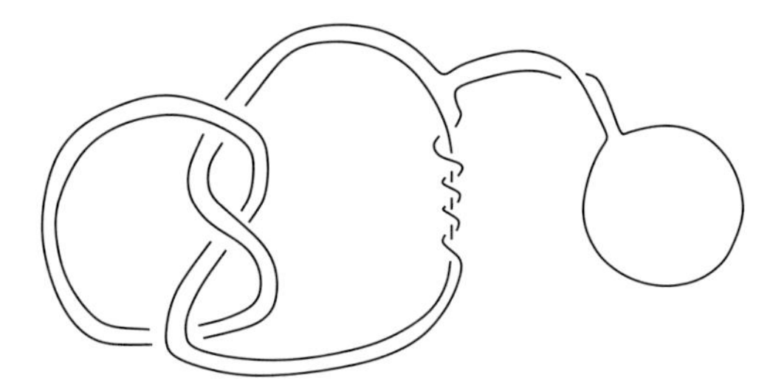

Fig. 1.5. Sliding one component over another (framings omitted)

group) to each link in S^3 (respectively, framed link in S^3) in such a way that links related by Reidemeister moves are assigned the same number (respectively, framed links related by Reidemeister and Kirby moves are assigned the same number). This might be a good strategy because one can draw links and framed links. It may be hard to prove directly that two link projections correspond to different equivalence classes, but straightforward to show that certain functions are preserved by Reidemeister or Kirby moves.

This is one place where QFT has had an impact on low-dimensional topology: The use of Feynman path integrals and the strategy by which physicists compute them have led to the invention of the new mathematical notion of TQFT. Perhaps more importantly, the input from physics has led to the formulation of a set of axioms similar to, but in an important sense fundamentally different from, the axioms of a homology theory.

1.2 Homology theories

We first review axioms for cohomology theories, which reshaped mathematical thinking in the second half of the twentieth century. We will take the point of view suited to algebraic topology, but there are many other points of view, in geometry, algebra, analysis, etc. For reasons of exposition we discuss cohomology theories, but each cohomology theory corresponds to a unique homology theory.

A *category* $\mathcal{C}$ is a class of *objects* $\text{Ob}(\mathcal{C})$, together with a class of disjoint sets $\text{Hom}(A, B)$, called *morphisms*, one for each pair $A, B \in \text{Ob}(\mathcal{C})$. Moreover, we require that for each triple A, B, C of objects there exist a *composition* $\text{Hom}(A, B) \times \text{Hom}(B, C) \to \text{Hom}(A, C)$ denoted $(f, g) \mapsto g \circ f$ satisfying

(i) *(Associativity)* $(f \circ g) \circ h = f \circ (g \circ h)$,
(ii) *(Identity)* for each $A \in \text{Ob}(\mathcal{C})$ there exists a $1_A \in \text{Hom}(A, A)$ such that for each B, we have $1_A \circ f = f$ for $f \in \text{Hom}(B, A)$ and $g \circ 1_A = g$ for $g \in \text{Hom}(A, B)$.

A *covariant functor* $F : \mathcal{C} \to \mathcal{D}$ is one that assigns to each object $A \in \text{Ob}(\mathcal{C})$ an object $F(A) \in \text{Ob}(\mathcal{D})$ and to each morphism $h \in \text{Hom}(A, B)$ a morphism $F(h) \in \text{Hom}(F(A), F(B))$ so that compositions and identity are preserved. A *contravariant functor* is defined in a similar way, except that if $h \in \text{Hom}(A, B)$, then $F(h) \in \text{Hom}(F(B), F(A))$, i.e., the arrows are reversed.

A *natural transformation* between two functors is, loosely speaking, a functor of functors. More precisely, if $F, G : \mathcal{C} \to \mathcal{D}$ are two covariant functors, then a natural transformation $n : F \to G$ assigns to each $A \in \text{Ob}(\mathcal{C})$ a morphism $n(A) \in \text{Hom}(F(A), G(A))$ such that if $f \in \text{Hom}(A, B)$ is a morphism in $\mathcal{C}$, the diagram

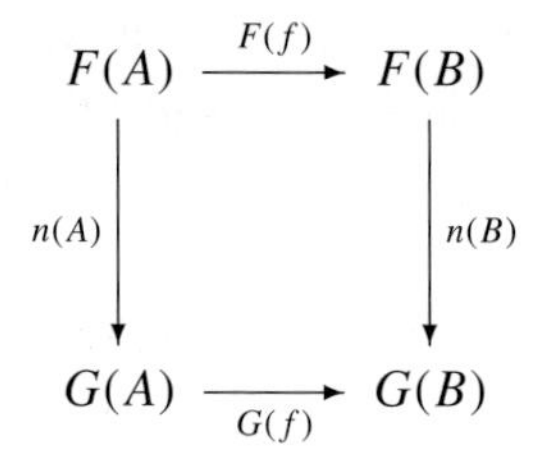

commutes. A similar definition works for contravariant functors.

Definition 1.2.1 Let $\mathcal{T}$ denote the category of topological spaces with base points (or, to be safe, CW complexes) and continuous maps, and $\mathcal{A}$ the category of (graded) abelian groups. Let $S : \mathcal{T} \to \mathcal{T}$ be the *suspension functor*, i.e., $SX = X \times [0, 1]/\sim$, where $X \times \{0, 1\} \cup \{p\} \times [0, 1]$ is collapsed to a point.

A (reduced) *cohomology theory* is a contravariant functor $h : \mathcal{T} \to \mathcal{A}$ together with a degree 1 natural transformation $e : h \circ S \to h$ satisfying the following axioms:

(i) *(Homotopy)* If $f_0, f_1 : X \to Y$ are (based) homotopic, then
$$h(f_0) = h(f_1) : h(Y) \to h(X).$$
(ii) *(Exactness)* If $X \subset Y$, then the sequence
$$h(X/Y) \to h(Y) \to h(X)$$
is an exact sequence.
(iii) *(Suspension)* For each X, the natural transformation
$$e(X) : h(SX) \to h(X)$$
is an isomorphism.

Typically one would also want the cohomology theory to be $\mathbb{Z}$-*graded*, i.e., there is a natural decomposition $h = \bigoplus_{i \in \mathbb{Z}} h^i$ and e has degree -1, i.e., $e(X) : h^n(SX) \cong h^{n-1}(X)$.

The group $h(S^0)$ associated to the zero-dimensional sphere is called the *coefficients* of the cohomology theory. Eilenberg and Steenrod [10] proved (with some technical conditions, always satisfied when working with manifolds, for example) that any two cohomology theories with the same coefficients are isomorphic. More precisely, any natural transformation of cohomology theories which induces an isomorphism on the coefficients induces an isomorphism for all spaces.

The philosophy that underpins this result is the following. Cohomology had been constructed in different ways (singular, simplicial, de Rham), and basic properties (suspension, homotopy, exactness) proved. But in the use of cohomology, typically the construction of the theory is not important: only its existence and its basic properties are used for calculations. The Eilenberg–Steenrod theorem confirms this. One has a list of properties one wishes to have, they are declared to be axioms, and if one can prove that a unique functor exists which has these properties, then one expects to be able to calculate using only the axioms. As modest as these three axioms appear, they form the basis of the vast and beautiful subject of stable homotopy theory.

From the perspective of geometric topology, several remarks are in order. First, the homotopy axiom, which gives cohomology its flexibility, implies that homology theories (which exist in abundance) are not set up to directly investigate problems unique to manifolds, in particular to attack homeomorphism or diffeomorphism questions. Second, the exactness axiom implies that the cohomology of a finite union of spaces is the direct sum of the homology of the pieces.

In contrast, as we will see, the TQFT axioms are built from the start by considering the category of manifolds rather than all spaces. Moreover, the theory is multiplicative, in the sense that unions are assigned the tensor product rather than direct sum. These are two properties which emerge naturally from the physical considerations.

There is an important way in which homotopy functors like cohomology theories are used in geometric topology. Given an n-dimensional manifold M, it may be the boundary of a manifold of dimension $n+1$: $M = \partial W$. Assuming it is, one can try to define *geometric (i.e., homeomorphism) invariants of* M by taking *cohomology invariants of* W.

This is a successful strategy. We will give one example, which is based on an additional axiom which some, but not all, homology theories satisfy, namely the existence of a ring structure.

(iv) *(Ring structure)* There exists a natural ring structure $h(X) \times h(X) \to h(X)$.

For example, the ordinary cohomology $h = H^*$ with integer coefficients (i.e., the unique cohomology theory h such that $h^0(S^0) = \mathbb{Z}$ and $h^i(S^0) = 0$ for $i \neq 0$) has a ring structure called the *cup product* (in terms of differential forms, this is just the wedge product of forms). The *Poincaré duality theorem* implies that if Z is a compact oriented manifold of dimension $4k$, there is an isomorphism $H^{4k}(Z) \to \mathbb{Z}$ which leads to a symmetric pairing

$$H^{2k}(Z) \times H^{2k}(Z) \to \mathbb{Z}. \tag{1.1}$$

Poincaré duality is the main homotopy-theoretical property that distinguishes manifolds from arbitrary spaces. It says that the ring structure on integral cohomology induces an isomorphism modulo torsion $H^p(Z) \cong H^{n-p}(Z, \partial Z)$ for any compact oriented n-manifold Z and index p. This implies, among other things, that the pairing (1.1) is unimodular (determinant ± 1) when Z is closed. Indeed, much of the theory of manifolds can be (and is) studied in the *Poincaré category*, that is, for general spaces (or CW complexes) satisfying Poincaré duality.

Choosing a basis for the free part of $H^{2k}(Z)$, we can think of the pairing (1.1) as a symmetric integral (or real) matrix. The *signature*, $\sigma(Z)$, of this matrix is the number $b^+(Z)$ of positive eigenvalues minus the number $b^-(Z)$ of negative eigenvalues.

When $Z = \partial W$ with W compact oriented, it is an elementary exercise to show that $\sigma(Z) = 0$. In other words the *geometric information* ($Z = \partial W$) has the *cohomology (homotopy-theoretical)* implication $\sigma = 0$. Hence if $\sigma(Z) \neq 0$, then Z is not the boundary of an $n{+}1$-dimensional manifold.

A more subtle version of this line of reasoning goes as follows. Suppose that one knows that a class $\mathcal{M}$ of $4k$ manifolds has the property that whenever $Z \in \mathcal{M}$ is *closed*, then Z has signature that lies in a subgroup $m\mathbb{Z}$ of $\mathbb{Z}$. For example, Rohlin's theorem [44] implies that if Z is a closed spin 4-manifold, then $\sigma(Z) = 16\ell$ for some integer ℓ.

Then given a closed $4k - 1$-manifold M, if we assume that $M = \partial Z$ for some $Z \in \mathcal{M}$, we can try to define an invariant d of M by the definition $d(M) = \sigma(Z)$. This may depend on the choice of Z: perhaps $\sigma(Z') \neq \sigma(Z)$ for some manifold Z' with $\partial Z' = M$. But gluing Z and Z' along W yields a *closed* manifold with σ in this subring. Novikov's additivity of the signature theorem implies that $\sigma(Z) \equiv \sigma(Z')$ mod m. Hence $d(M) = \sigma(Z) \in \mathbb{Z}/m$ is a well-defined invariant of M, independent of Z, called the *signature defect* of M.

A similar reasoning applied to the Chern–Weil definition of the second Chern class exploits the fact that $c_2(P)[M^4] \in \mathbb{Z}$ rather than $\mathbb{R}$ to define the *Chern–Simons invariant* for a connection on a 3-manifold in $\mathbb{R}/\mathbb{Z}$ (see the next section).

Milnor used a similar approach to construct smooth manifolds homeomorphic but not diffeomorphic to S^7 [31]. He constructs a certain 8-manifold Z (in fact,

Z is a D^4 bundle over S^4) whose boundary M is homotopy equivalent to S^7, and hence homeomorphic to S^7 by Smale's theorem [48]. If M were diffeomorphic to S^7, then M would bound the 8-disk D^8, which could then be glued to Z, yielding a *smooth* closed manifold $Z \cup D^8$, which cannot exist, for reasons similar to Rohlin's theorem.

A similar approach was taken to try to find counterexamples to the Poincaré conjecture in dimension 3 using Rohlin's theorem. Any 3-manifold M with the homology of S^3 admits a unique spin structure, and is the boundary of a spin 4 manifold W. The signature defect invariant $\mu(M) = \sigma(W) \bmod 16$ is well defined (i.e., depends only on M), and it was hoped that one could find a simply connected example M with $\mu(M) \neq 0$; this would provide a counterexample to the Poincaré conjecture.

In 1985 Casson [1] showed that this approach was doomed to failure: $\mu(M) = 0$ for any simply connected 3-manifold. However, his methods were far reaching, and provided another deep link between gauge theory and topology.

Remark. The two fundamental invariants of a closed 4-dimensional manifold M are:

(i) Its *signature* $\sigma(M)$, defined as the difference $b^+(M) - b^-(M)$, where $b^\pm(M)$ is the dimension of a maximal positive/negative definite subspace of $H^2(M;\mathbb{R})$ with respect to the *intersection form*, i.e., the symmetric unimodular bilinear pairing of Equation (1.1).

(ii) Its *Euler characteristic*, defined as $\chi(M) = \sum_n (-1)^n \mathrm{rank}(H^n(M))$. It can also be defined as the alternating sum of the number of n-simplexes in a triangulation of M.

Both of these invariants have the property that they are multiplicative with respect to finite covering spaces $\tilde{M} \to M$.

1.3 Axioms derived from QFT

In this section I explain how considerations of Feynman integration led to the formulation of a new set of axioms for a mathematical theory built to study manifolds. My first exposure to this axiomatic point of view (the "gluing laws") came from Witten's influential article [60]. It seems that the idea was already present in many forms before that, including work of Segal [46] on conformal field theory, and Witten himself had interpreted Donaldson's 4-manifold invariants from this perspective in an earlier article [61]. These ideas have been floating around topology in some form or another since Feynman introduced the Feynman integral, but they gained traction in geometric topology with the appearance of Jones's work on new knot polynomials [24] and Witten's interpretation and "extension" in [60]. The axiomatic point of view was popularized in Atiyah's book [2].

A few words about mathematical rigor are in order here. Since presumably the goal of physics is to describe the nature of the physical universe, and the goal of mathematics is to describe the nature of the logical universe, it might be unfair for mathematicians to expect physicists to justify their claims to a mathematician's satisfaction. After all, if a model uses mathematical language to accurately predict physical phenomena, whether or not it makes any mathematical sense need not concern a physicist. However, for a while it became common for some physicists, following Witten's lead, to make mathematical claims (in knot theory for example) that were not justified mathematically (and presumably had limited physical interest). In addition, some mathematicians began to introduce jargon and even reasoning from theoretical physics into their work, in an attempt to ride the wave that formed from this new relationship between the disciplines. This led to an interesting and lively (and ongoing) debate, part of which can be found in the AMS bulletin articles [3, 18] (see also Kauffman's review of these articles [62]).

The positive side of the issue is that because speculation in theoretical physics need not be constrained by mathematical rigor, physicists have been able to make deep mathematical conjectures which would not have manifested themselves in the presence of the timidity prescribed by the methods of mathematics. Some of these conjectures have been subsequently proven by mathematical methods, and it is uncontroversial to claim that the influence of theoretical physics on mathematics has been overwhelmingly positive in the last 20 years.

The various examples in use of TQFTs satisfy slightly different axioms, operate on different categories of manifolds, etc. In some cases the axioms are used in the construction, in others the axioms serve as guides for desired theorems, and in yet others the axioms serve only as a loose framework to organize the theory. Ultimately the goal of a good theory is to provide a systematic way to prove theorems and construct examples. Adherence to a predetermined collection of axioms is useful, but secondary.

The exposition I give here is derived (inspired?) from the clear work of Kevin Walker, who wrote an influential manuscript in 1991 [57], and more recently has posted a draft of a book on TQFTs [58]. The reader interested in a careful exposition of this topic is encouraged to read these two sources, as well as many other useful approaches, such as that taken in Turaev's book [55] and the many references cited in that book.

1.3.1 Formal properties of functional integrals

I give an outline of how formal properties of functional integrals from theoretical physics (also called *path integrals* or *Feynman integrals*) motivate the introduction of axioms. The reader is to keep in mind that none of what follows is supposed to be rigorous mathematics, nor for that matter precise. It is meant purely speculatively. Many details are ignored, and in fact more careful treatments make more careful constructions. The exposition is meant only to show how the the functional-integral formalism meshes with the geometric topology goal of understanding how cut-and-paste constructions of manifolds lead to a list of desirable axioms or properties. Rigorous mathematical examples will be described later.

First, we will use the word *manifold* to mean a compact oriented manifold. It may have a boundary, and typically we will assume that is has a smooth structure. It is also important to allow *corners*: a corner is a codimension 2 submanifold which is itself a codimension 1 submanifold of the boundary; it is defined by relaxing the charts so that they map to open sets in the quadrant $\{(x_1, x_2, \ldots, x_n) \in \mathbb{R}^n \mid x_1 \geq 0, x_2 \geq 0\}$. The corners then correspond to the points mapped to $(0, 0, x_3, \ldots, x_n)$ by the charts. Sometimes it is even necessary to go further and define higher-codimensional corners. The motivation comes from the observation that if one cuts a manifold along a union of transversely intersecting hypersurfaces, the pieces in the decomposition will be manifolds with corners.

Assume to any compact oriented manifold M of dimension less than or equal to $n + 1$ we have a way to assign some kind of spaces or sets of functions $\mathcal{F}(M)$, called *fields*. Suppose that we have a well-defined way to restrict fields to compact submanifolds (usually of codimension 0 or 1) $\Sigma \subset M$, $r_\Sigma : \mathcal{F}(M) \to \mathcal{F}(\Sigma)$ (we also write $f|_\Sigma$ for $r_\Sigma(f)$). For each $f \in \mathcal{F}(\Sigma)$ denote by $\mathcal{F}(M, f)$ the preimage $r_\Sigma^{-1}(f)$. Assume that all these restrictions commute. Moreover, we assume that the homeomorphism (or diffeomorphism) group of M acts on $\mathcal{F}(M)$, and the action is compatible with restrictions in the sense that if $h : M \to M$ is a homeomorphism which preserves a submanifold Σ, then the restriction map r_Σ satisfies $r_\Sigma \circ h_* = (h|_\Sigma)_* \circ r_\Sigma$. This is obvious when $\mathcal{F}$ denotes the set of maps into some space B.

Example. Take $n = 2$. For each 3-manifold M, define $\mathcal{F}(M)$ to be the space $\mathcal{A}_M$ of $su(2)$ connections on M, which we can identify with the vector space of $su(2)$-valued differential 1-forms on M, $\Omega^2_M \otimes su(2)$. Restricting connections yields a map $\mathcal{A}_M \to \mathcal{A}_\Sigma$.

Next, assume that for each $n + 1$-dimensional manifold M, we have a function $S_M : \mathcal{F}(M) \to \mathbb{R}$ (called the *action*) which is local in the sense that if M is the union of two codimension zero manifolds M_1 and M_2 along a hypersurface Σ, $M = M_1 \cup_\Sigma M_2$, then

$$S_M(f) = S_{M_1}(f|_{M_1}) + S_{M_2}(f|_{M_2}).$$

Here we mean that M is formed by gluing M_1 to M_2 along embeddings $e_i : \Sigma \subset \partial M_i$, or equivalently, that M contains Σ and $M_1 \sqcup M_2$ is obtained by cutting M along Σ.

Example continued. For each $su(2)$ connection $a \in \mathcal{A}_M$ and integer k, define the *Chern–Simons action* $S_M : \mathcal{A}_M \to \mathbb{R}$ to be

$$S_M(a) = \frac{k}{8\pi}\int_M \mathrm{Tr}(da \wedge a + \tfrac{2}{3} a \wedge a \wedge a).$$

Here the "$\wedge$" is a combination of wedge product on forms and matrix multiplication on the coefficients. "Tr" refers to the usual trace, and the result is an ordinary real-valued (rather than complex-valued) differential 3-form on M, for the coefficients are taken to be $su(2)$. This 3-form can be integrated on the 3-manifold M, yielding a real number.

If $M = M_1 \cup_\Sigma M_2$, then because integrating forms is a local operation,

$$\int_M \mathrm{Tr}(da \wedge a + \tfrac{2}{3} a \wedge a \wedge a) = \int_{M_1} \mathrm{Tr}(da \wedge a + \tfrac{2}{3} a \wedge a \wedge a) + \int_{M_2} \mathrm{Tr}(da \wedge a + \tfrac{2}{3} a \wedge a \wedge a),$$

and so $S_M(a) = S_{M_1}(a|_{M_1}) + S_{M_2}(a|_{M_2})$. Notice that no Riemannian metric is used to define S_M, and hence one says the action is "topological."

Next we assume that there exists some kind of measure on $\mathcal{F}(M, f)$ for each M and $f \in \mathcal{F}(\partial M)$, so that the circle-valued function $e^{iS_M} : \mathcal{F}(M, f) \to S^1$ is integrable. Then define

$$Z(M, f) = \int_{\mathcal{F}(M,f)} e^{iS_M} \in \mathbb{C}. \tag{1.2}$$

The main point of this expression is that the result depends only on M and f. Hence we think of $Z(M, -)$ as a complex-valued function on the set of fields $\mathcal{F}(\partial M)$ on ∂M. In the case that ∂M is empty, we write

$$Z(M) = \int_{\mathcal{F}(M)} e^{iS_M} \in \mathbb{C}. \tag{1.3}$$

We assume that all the structure (including the measure on $\mathcal{F}(M)$) is equivariant with respect to diffeomorphisms.

Example continued. For the Chern–Simons action this gives Witten's "invariant"

$$Z(M) = \int_{\mathcal{A}_M} e^{iS_M}.$$

This is not a mathematical definition, because the integral is not defined; in fact no appropriate measure exists.

Suppose that M is the disjoint union of two manifolds, $M = M_1 \sqcup M_2$. We have restriction maps $\mathcal{F}(M) \to \mathcal{F}(M_i)$, $i = 1, 2$, which combine to give a product map $\mathcal{F}(M) \to \mathcal{F}(M_1) \times \mathcal{F}(M_2)$, $f \mapsto (f|_{M_1}, f|_{M_2})$. We assume this map is a natural bijection, i.e., we assume that a field is determined by its two restrictions, and that any two restrictions determine a unique field. Similar comments apply to a disjoint union $\Sigma = \Sigma_1 \sqcup \Sigma_2$.

Since $S_M = S_{M_1} + S_{M_2}$, assuming Fubini's theorem holds, it follows that

$$Z(M) = \int_{\mathcal{F}(M_1)\times\mathcal{F}(M_2)} e^{iS_{M_1}} e^{iS_{M_2}} = Z(M_1)\cdot Z(M_2) \tag{1.4}$$

and, if the M_i have boundaries,

$$Z(M, f) = Z(M_1, f|_{\partial M_1}) \cdot Z(M_2, f|_{\partial M_2}). \tag{1.5}$$

Next we examine what happens when a manifold is cut. Suppose that the closed $n+1$-manifold M contains a hypersurface Σ. Cutting M along Σ yields a manifold M_{cut} with two boundary components: $\partial M_{cut} = \Sigma \sqcup -\Sigma$. Notice that cutting a manifold along a hypersurface need not disconnect it, and so M_{cut} might be connected. We assume that fields on M correspond (bijectively) to fields on M_{cut} which agree on Σ and $-\Sigma$, i.e.,

$$\mathcal{F}(M) = \bigcup_{a\in\mathcal{F}(\Sigma)} \mathcal{F}(M_{cut}, a, a).$$

Thus it is natural to assume that

$$Z(M) = \int_{a\in\mathcal{F}(\Sigma)} \int_{\mathcal{F}(M_{cut},a,a)} e^{iS_M} = \int_{a\in\mathcal{F}(\Sigma)} Z(M_{cut}, a, a). \tag{1.6}$$

If M is not closed, assume that Σ lies in the interior of M. Thus $\partial M_{cut} = \partial M \sqcup \Sigma \sqcup -\Sigma$. We interpret Equation (1.6) as an equation of functions on $\mathcal{F}(\partial M)$, i.e., for $f \in \mathcal{F}(\partial M)$,

$$Z(M, f) = \int_{a\in\mathcal{F}(\Sigma)} \int_{\mathcal{F}(M_{cut},a,a)} e^{iS_M} = \int_{a\in\mathcal{F}(\Sigma)} Z(M_{cut}, f, a, a). \tag{1.7}$$

A special case is when $M = \Sigma \times [0, 1]$. We can cut M along $\Sigma \times \{\frac{1}{2}\}$, for each of the two pieces of M_{cut} is homeomorphic to M; Equation (1.7) implies that

$$Z(\Sigma \times I, f_0, f_1) = \int_{a\in\mathcal{F}(\Sigma)} Z(\Sigma \times I, f_0, a) Z(\Sigma \times I, a, f_1). \tag{1.8}$$

More generally, every manifold with boundary has a collar $\partial M \times [0, 1] \subset M$ with $\partial M = \partial M \times \{0\}$. Cutting M along $\Sigma \times \{1\}$, one sees that

$$Z(M, f) = \int_{a\in\mathcal{F}(\partial M)} Z(\partial M \times I, f, a) Z(M, a). \tag{1.9}$$

Equation (1.9) restricts the class of functions $\mathcal{F}(\Sigma) \to \mathbb{C}$ that can arise as functions of the form $Z(M, -)$: if we define $\pi_\Sigma : \text{Funct}(\mathcal{F}(\Sigma), \mathbb{C}) \to \text{Funct}(\mathcal{F}(\Sigma), \mathbb{C})$ by

$$(\pi_\Sigma(\Phi))(b) = \int_{a\in\mathcal{F}(\Sigma)} Z(\Sigma \times I, a, b)\Phi(a), \tag{1.10}$$

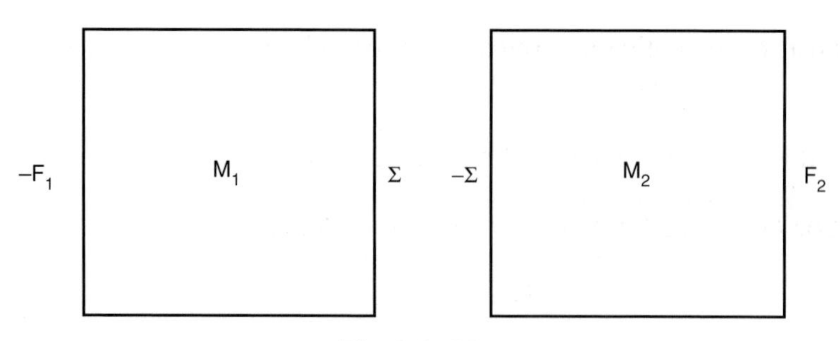

Fig. 1.6. M_{cut}

then Equation (1.8) shows that π_Σ is a projection ($\pi_\Sigma^2 = \pi_\Sigma$), and Equation (1.9) shows that $\pi_\Sigma(Z(M, -)) = Z(M, -)$.

Notice that if M_{cut} is the union of two manifolds, $M_{cut} = M_1 \sqcup M_2$ with $\partial M_1 = -F_1 \sqcup \Sigma$ and $\partial M_2 = -\Sigma \sqcup F_2$ (see Figure 1.6), then Equations (1.7) and (10.7) yield

$$Z(M, f_1, f_2) = \int_{a \in \mathcal{F}(\Sigma)} Z(M_1, f_1, a) \cdot Z(M_2, a, f_2) \tag{1.11}$$

for fixed $f_1 \in \mathcal{F}(F_1)$ and $f_2 \in \mathcal{F}(F_2)$. This can be interpreted as an L^2 inner product of the two functions $\overline{Z(M_1, f_1, -)} : \mathcal{F}(\Sigma) \to \mathbb{C}$ and $Z(M_2, -, f_2) : \mathcal{F}(\Sigma) \to \mathbb{C}$.

Working formally (and fixing f_1, f_2), we rewrite Equation (1.11) as

$$Z(M) = \langle Z(M_1, -), \overline{Z(M_2, -)} \rangle_{L^2(\mathcal{F}(\Sigma))}.$$

This suggests that one might be able to find a subspace

$$V(\Sigma) \subset \mathrm{Funct}(\mathcal{F}(\Sigma), \mathbb{C}),$$

namely, the image of the projection π_Σ (or a subspace of this image, which contains all the functions $Z(M, -)$ for any M with $\partial M \supset \Sigma$ and in which these functions span a dense subspace), and an identification

$$\bar{\ }: V(-\Sigma) \cong V(\Sigma)^* \tag{1.12}$$

compatible with this L^2 inner product in the sense that if $\langle -, - \rangle : V(\Sigma) \times V(\Sigma)^* \to \mathbb{C}$ denotes the usual contraction, then Equation (1.11) reads

$$Z(M, f_1, f_2) = \langle Z(M_1, f_1, -), Z(M_2, -, f_2) \rangle.$$

Moreover, $Z(M_{cut}, f_1, f_2) \in V(\Sigma \sqcup -\Sigma)$, but because $M_{cut} = M_1 \sqcup M_2$, we have $Z(M_{cut}, f_1, f_2, a, a) = Z(M_1, f_1, a) \cdot Z(M_2, a, f_2)$ with $Z(M_1, f_1, -) \in V(\Sigma)$ and $Z(M_2, -, f_2) \in V(-\Sigma)$. So we assume that $V(\Sigma \sqcup -\Sigma) = V(\Sigma) \otimes V(\Sigma)^*$. One can similarly justify

$$V(F_1 \sqcup F_2) = V(F_1) \otimes V(F_2). \tag{1.13}$$

This is most conveniently reinterpreted as a *cobordism* property: if M is a manifold with $\partial M = -F \cup \Sigma$ then

$$Z(M) \in V(-F \sqcup \Sigma) = V(F)^* \otimes V(\Sigma) = \text{Hom}(V(F), V(\Sigma)). \tag{1.14}$$

Equation (1.11) implies that if $M = M_1 \cup_\Sigma M_2$, then

$$Z(M) = Z(M_2) \circ Z(M_1) : V(F_1) \to V(F_2).$$

In more detail: $Z(M_1) \in V(F_1)^* \otimes V(\Sigma)$, $Z(M_2) \in V(-\Sigma)^* \otimes V(F_2)$, so $Z(M_{cut}) \in V(F_1)^* \otimes V(\Sigma) \otimes V(-\Sigma)^* \otimes V(F_2)$. Equation (1.11) shows that contracting out the middle two factors yields $Z(M)$. But this is the same as composing the two linear maps $Z(M_1)$ and $Z(M_2)$.

Since any manifold M can be "cut" along the empty manifold into two pieces $M_1 = M$ and $M_2 = \emptyset$, for the theory to be nontrivial one should assume that $V(\emptyset) = \mathbb{C}$. Then, thinking of M as a cobordism from $\emptyset$ to ∂M, one gets the linear map $Z(M) : \mathbb{C} \to V(\partial M)$, which is the same thing as a vector $Z(M) \in V(\partial M)$.

Example continued. For the Chern–Simons action, if the orientation of a 3-manifold M is reversed, the value of the Chern–Simons function changes sign: $S_{-M}(a) = -S_M(a)$. Hence $e^{iS_{-M}(a)} = \overline{e^{iS_M(a)}}$, and so

$$Z(-M) = \overline{Z(M)}.$$

Finally, we assume that the path integral is an oriented homeomorphism (or diffeomorphism) invariant. If $h : \Sigma \to \Sigma$ is a homeomorphism, then it induces a bijection on fields, $h_* : \mathcal{F}(\Sigma) \to \mathcal{F}(\Sigma)$, and hence an isomorphism h : Funct($\mathcal{F}(\Sigma), \mathbb{C}$) $\to$Funct($\mathcal{F}(\Sigma), \mathbb{C}$). Because any homeomorphism of Σ extends to some manifold, one of whose boundary components is Σ (e.g., $\Sigma \times [0, 1]$), it is reasonable to assume that it preserves the subspace $V(\Sigma)$, i.e., we assume there is an action of Homeo(Σ) on $V(\Sigma)$.

However, the setup allows us to conclude a much stronger *modular invariance* as follows. Suppose that the homeomorphism $h : \Sigma \to \Sigma$ extends to a diffeomorphism $H : \Sigma \times [0, 1] \to \Sigma \times [0, 1]$ such that $H|_{\Sigma\times\{0\}} = Id$ and $H|_{\Sigma\times\{0\}} = h$ (e.g., if h is isotopic to the identity). The existence of the diffeomorphism H and naturality allow us to conclude that the following diagram commutes:

$$\begin{array}{ccc} V(\Sigma) & \xrightarrow{Z(\Sigma\times I)} & V(\Sigma) \\ \downarrow{\scriptstyle Id} & & \downarrow{\scriptstyle h_*} \\ V(\Sigma) & \xrightarrow[Z(\Sigma\times I)]{} & V(\Sigma) \end{array}$$

But it is safe to assume that $Z(\Sigma \times I) : V(\Sigma) \to V(\Sigma)$ equals the identity. This is because $Z(\Sigma \times I) = Z(\Sigma \times I) \circ Z(\Sigma \times I)$, so that $Z(\Sigma \times I)$ is an idempotent. But also, if M is any manifold with boundary Σ, then $Z(\Sigma \times I)Z(M) = Z(M)$, and so $Z(\Sigma \times I)$ acts as the identity on the (dense) span of the $Z(M)$ in $V(\Sigma)$. Hence we conclude that the action of Homeo(Σ) on $V(\Sigma)$ factors through the *mapping class group* Homeo(Σ)/isotopy.

1.3.2 Axioms for a TQFT

Motivated by the (nonrigorous) musings of the previous subsection, as well as some known examples, a picture gradually emerged of a new kind of functor with some superficial similarities to cohomology theory, but suited to the cut-and-paste methods of geometric topology. Many mathematicians and physicists have contributed to the development of these ideas, starting perhaps with Segal's conformal theory article [46] and Witten's Jones polynomial article [60], with Atiyah as a strong advocate for the exploration of these ideas.

The axioms listed in the following are probably best thought of as guideposts for what one would like a TQFT to satisfy, rather than a complete collection. These axioms may not to be strong enough to compute with: extra structure and/or axioms may be needed. For example, Witten argued in his article that for the theory based on the Chern–Simons integral, each 3-manifold must be endowed with a trivialization of its tangent bundle. Moreover, any computable cut-and-paste theory will require cutting along properly embedded manifolds with boundary, leading to the introduction of many more necessary (and technical) axioms, as was made clear in Walker's article [57]. Further technicalities arise when one tries to deduce existence or uniqueness of a theory constructed in a particular way (e.g., starting with a quantum group).

Nevertheless, the perspective has proved extremely useful. Profound theorems have been proven in Donaldson–Floer theory, Gromov–Witten theory, Seiberg–Witten theory, Heegaard–Floer homology, etc., and the TQFT perspective affords a way to organize the results (and provides conjectures to pursue).

A mathematical way to describe what a *theory* is typically takes the following form. One starts with a category $\mathcal{C}$ (or a more general notion) one wishes to study, and a theory will be a functor (or a sequence of functors) to some other (usually algebraic) category $\mathcal{A}$ which satisfies properties that reflect the kinds of operations in $\mathcal{C}$ one wants to understand, in a hopefully simpler way.

So we need to start with an appropriate $\mathcal{C}$. Because our goal is to study compact manifolds, we might want $\mathcal{C}$ to include all compact manifolds of dimension $n+1$. They may be equipped with additional structure, such as an orientation, a smooth structure, a *Spin*^{c} structure, a framing of the tangent bundle, a choice of cohomology class, an equivalence class of triangulations, a parameterization of its boundary, etc. We might also want to study knots or links, so perhaps some extra structure on our $n+1$-manifolds consists of a properly embedded codimension k (typically $k=2$) submanifold, or even a properly embedded graph of some sort.

We wish to study the homeomorphism problem, so we take our morphisms to be homeomorphisms (or diffeomorphisms) which preserve the extra structure, rather than all continuous maps (in marked contrast to the case for cohomology theories). Because we are interested in cut-and-paste along codimension 1 submanifolds, we need to consider n-dimensional manifolds and their homeomorphisms also. The corresponding extra structure should be present.

A good start is the following. Let $\mathcal{M}_{n+1}$ denote the category of oriented compact manifolds of dimension $n+1$ with possibly nonempty boundary, with homeomorphisms as the morphisms. Let $\mathcal{M}_n$ denote the category of closed oriented compact n-manifolds. There is a boundary functor $\partial : \mathcal{M}_{n+1} \to \mathcal{M}_n$ which takes a manifold M to its boundary ∂M. Let $\mathcal{A}$ denote the category of finite-dimensional complex vector spaces and *isomorphisms* (or, more generally, R-modules and isomorphisms for some commutative ring R).

Then, motivated by the discussion of the previous section, we can define an $n+1$-*dimensional TQFT* to be a pair (V, Z) where:

(i) V is a functor

$$V : \mathcal{M}_n \to \mathcal{A}$$

satisfying the following:

(a) There is a natural identification of $V(-\Sigma) \cong V(\Sigma)^*$.

(b) There is a natural identification $V(\Sigma_1 \sqcup \Sigma_2) \cong V(\Sigma_1) \otimes V(\Sigma_2)$.

(c) $V(\emptyset) = \mathbb{C}$.

(ii) Z is a natural transformation from the functor $T : \mathcal{M}_{n+1} \to \mathcal{A}$ taking every $n+1$-manifold to $\mathbb{C}$ and every homeomorphism to the identity to the functor $V \circ \partial$. That is, Z assigns a homomorphism $Z(M) : \mathbb{C} \to V(\partial M)$ (equivalently, the element $Z(M)(1) = Z(M) \in V(\partial M)$) to each $M \in \mathcal{M}_{n+1}$. The assignment Z satisfies:

(a) (Collar) $Z(\Sigma \times I) = \mathrm{Id} \in V(-\Sigma \sqcup \Sigma) = V(\Sigma)^* \otimes V(\Sigma) = \mathrm{Hom}(V(\Sigma), V(\Sigma))$.

(b) (Disjoint union) If $M = M_1 \sqcup M_2$ is a disjoint union, then

$$Z(M) = Z(M_1) \otimes Z(M_2) \in V(\partial M_1) \otimes V(\partial M_2) = V(\partial M_1 \sqcup \partial M_2).$$

(c) (Pasting) If $\partial M = F$, and $\Sigma \subset M$ is a closed surface in the interior of M, so that $\partial M_{cut} = F \sqcup -\Sigma \sqcup \Sigma$, then

$$c(Z(M_{cut})) = Z(M) \in V(F),$$

where $c : V(F \sqcup -\Sigma \sqcup \Sigma) \to V(F)$ is the contraction

$$V(F \sqcup -\Sigma \sqcup \Sigma) = V(F) \otimes V(\Sigma)^* \otimes V(\Sigma) \to V(F) \otimes \mathbb{C} = V(F).$$

To have any hope of being able to compute "from the axioms," one would surely need a more detailed list of axioms, which includes cutting along n-manifolds Σ with nonempty boundary properly embedded in M. This might require $\mathcal{M}_{n+1}$ to be defined as the category of compact $n+1$-manifolds with boundary and corners up to codimension 2, and $\mathcal{M}_n$ to be compact n-manifolds with boundary, and $\mathcal{M}_{n-1}$ to be closed $n-1$-manifolds. (For example, consider cutting the 2-disk $D^2 = \{x \in \mathbb{R}^2 \mid \|x\| \leq 1\}$ along the properly embedded arc $A = [-1, 1] \times \{0\}$.) More sophisticated theories might require looking even deeper, i.e., allowing corners of higher codimension.

As more precise properties are proscribed, one expects a more computable theory, but in any specific example it may be hard (or impossible) to establish the desired properties.

1.4 Donaldson–Floer theory

1.4.1 Outline of the construction and properties

Donaldson–Floer theory is roughly a 3+1 theory. It did not develop from the TQFT perspective, but fits (loosely) into the framework. We outline its construction, but the novice should understand that the precise mathematical definitions and constructions are complicated, particularly in their analytical aspects. Two careful expositions of this topic are [8] and [7].

To begin with, we outline the construction for closed and simply connected 4-manifolds. Given a smooth closed 4-manifold M, a principal $SU(2)$ bundle $\pi : P \to M$ over M is uniquely determined by its second Chern class $c_2(P) \in H^4(M) = \mathbb{Z}$. An $SU(2)$ *connection*, a, on P is a choice of a subspace $H_p \subset T_pP$ at each point $p \in P$ such that the composite $H_p \subset T_pP \xrightarrow{d\pi} T_{\pi(p)}M$ is an isomorphism. The precise definition can be found in any geometry book. The important properties are as follows (here $\Omega^i_M(ad\ P) = C^\infty(\wedge^i T^*(M) \otimes ad\ P)$ denotes the vector space of differential i-forms with values in the Lie algebra vector bundle associated to P):

(i) A connection a gives a way to take directional derivatives of sections of vector bundles E associated to P. The directional derivative is denoted $d_a : C^\infty(E) = \Omega^0_M(E) \to C^\infty(T^*M \otimes E) = \Omega^1_M(E)$.

(ii) A connection a has a *curvature* 2-form $F(a) \in \Omega^2_M(ad\ P)$ which obstructs the commuting of second directional derivatives ($F(a) = d_a^2$).

(iii) The space of all $SU(2)$ connections $\mathcal{A}(P)$ on P forms an infinite-dimensional affine space modeled on $\Omega^1_M(ad\ P)$.

(iv) The infinite-dimensional Lie group $\mathcal{G}(P)$ (with Lie algebra $\Omega^0_M(ad\ P)$) of automorphisms of P acts on $\mathcal{A}$.

The (square of the) L^2 norm of the curvature of a connection defines the *Yang–Mills action*:

$$YM(a) = \int_M \|F(a)\|^2 dvol_M,$$

which is constant on $\mathcal{G}(P)$ orbits, and hence descends to the orbit space $\mathcal{A}(P)/\mathcal{G}(P)$. Chern–Weil theory implies that when $c_2(P) \leq 0$ then $YM(a) \geq -8\pi^2 c_2(P)$. Thus if a connection a satisfies $YM(a) = -8\pi^2 c_2(P)$, it realizes the absolute minimum of $YM(a)$. Such connections are called *instantons*. They are also called *anti-self-dual (ASD)* because they satisfy the equation $*F(a) = -F(a)$. Here $*$ denotes the *Hodge $*$-operator* acting on 2-forms; this operator induces the Poincaré duality isomorphism on harmonic forms. (There are no instantons on P if $c_2(P) > 0$, but changing the orientation of M changes the sign of $c_2(P)$.)

The *moduli space $\mathcal{M}(P)$ of instantons on $P \to M$* is defined to be the quotient

$$\mathcal{M}(P) = YM^{-1}(-8\pi^2 c_2(P))/\mathcal{G}(P). \tag{1.15}$$

After perturbing the action YM slightly if needed, and with the additional topological hypotheses $b^+(M) \geq 2$ on M, $\mathcal{M}(P)$ is a smooth orientable (typically noncompact) manifold of dimension

$$\dim \mathcal{M}(P) = -8c_2(E) - \tfrac{3}{2}(\chi(M) + \sigma(M)).$$

(See Equation (1.1) and the last paragraph of Section 1.2) A choice of orientation of a maximal positive definite subspace of $H_2(M; \mathbb{R})$ determines an orientation of $\mathcal{M}(P)$.

What one does next is technical, but the following comment may help. Given a manifold B, one way to distinguish submanifolds of B is to pair them with cohomology classes of B. In other words, if M_1 and M_2 are n-dimensional (closed, oriented) submanifolds of B, then one can ask whether $\alpha(M_1) \neq \alpha(M_2)$ for some cohomology class $\alpha \in H^n(B)$. What follows is an application of this idea when $B = \mathcal{A}(P)/\mathcal{G}(P)$ and $M_1 = \mathcal{M}(P)$.

There is a map $\mu : H_2(M) \to H^2(\mathcal{M}(P))$ which (essentially) takes an embedded smooth 2-manifold $F \subset M$ representing $\alpha \in H_2(M)$ to the first Chern class of the line bundle $L_F \to \mathcal{M}(P)$. Here L_F is the (Quillen) determinant line bundle over $\mathcal{M}(P)$, a bundle whose fiber over $a \in \mathcal{M}(P)$ is the complex line $(\wedge^{top} \ker \not{\partial}_a)^* \otimes (\wedge^{top} \operatorname{coker} \not{\partial}_a)$ for an appropriate Dirac operator $\not{\partial}$ on F coupled to the (restriction to F of) a.

With some hard work due to many people, including Donaldson, Taubes, and Uhlenbeck, one can define integer-valued diffeomorphism invariants of M for M a 4-manifold with the following extra structure:

(i) M is closed,
(ii) M is oriented,

(iii) M is simply connected,
(iv) $b^+(M) \geq 2$,
(v) a positive integer k is given such that $8k - \frac{3}{2}(\chi(M) + \sigma(M))$ is even, say $2n_k$,
(vi) n_k classes $x_1, x_2, \ldots, x_{n_k}$ in $H_2(M)$ are specified,
(vii) an orientation of a maximal positive definite subspace of $H^2(M; \mathbb{R})$ is specified.

Note that a diffeomorphism $\phi : M \to M'$ preserves (or pulls back) all this structure.

To this data, associate

$$D(M) = \big(\mu(x_1) \cup \cdots \cup \mu(x_{n_k})\big) \in H_c^{2n_k}(\mathcal{M}(P)) = \mathbb{Z},$$

where $H_c^{2n_k}(\mathcal{M}(P))$ denotes the compactly supported cohomology of $\mathcal{M}(P)$. Equivalently, count the number of points in the intersection of the vanishing loci of sections of each line bundle L_{F_i}, where $F_i \subset M$ represents x_i. This is most easily understood in the special case when $8k - \frac{3}{2}(\chi(M) + \sigma(M)) = 0$, in which case there are no x_i and $D(M)$ is just the count (with signs) of the (compact, hence finite) 0-dimensional space $\mathcal{M}(P)$, i.e., the algebraic count of instantons on $P \to M$.

No doubt this is a massively complicated and technical definition, and the proof that it is well defined, independent of all the auxiliary choices (such as a Riemannian metric) used to define it, and is a diffeomorphism invariant of M with this extra data is difficult. However in the 25 years that this subject has existed, the path one follows in proving such facts has become clear, if not simple. Indeed, Seiberg–Witten (SW) theory has largely replaced Donaldson's instanton theory because on the one hand the gauge theory framework had become mainstream by the time SW came along, but also because many of the analytic features of SW theory are simpler than those of Donaldson theory.

Nevertheless, for a decade, Donaldson theory was the only game in town for smooth 4-manifold topology. At the end of this section I will list some of the many important theorems proven using this technology.

So far I have not made much progress towards describing any TQFT-like structure: what has been accomplished is the definition of an invariant for closed 4-manifolds (with all the extra structure). We still need a theory of 3-manifolds, and this is provided by Floer's instanton homology.

Some hints on how to proceed have already appeared. First, the Yang–Mills action (given by the L^2 norm of the curvature of a connection) is borrowed from physics; indeed, this action was studied by particle physicists interested in phenomena arising on considering non-abelian structure groups, like $SU(2)$. Second, it is known in physics that functional integrals of the type

$$\int_{\mathcal{F}} e^{kiS}$$

can sometimes be understood (approximated or asymptotically computed as $k \to \infty$) in terms of data (the signature and determinant of the Hessian) coming from the critical points of the action S (the method of *stationary phase*). Since the set of absolute minima of the Yang–Mills action is the space $\mathcal{M}(P)$, one then expects to find a relationship between the path integral and invariants derived from $\mathcal{M}(P)$.

The construction of the corresponding $V(\Sigma)$ for a 3-manifold Σ is intuitively simple. For technical reasons one needs to assume that Σ is a closed 3-manifold with the same homology as S^3. Any principal $SU(2)$ bundle P over Σ is necessarily topologically trivial, $P \cong \Sigma \times SU(2)$. One denotes by $\mathcal{A}(\Sigma)$ the space of $su(2)$ connections on the trivial bundle, and by $\mathcal{G}(\Sigma)$ the corresponding group of bundle automorphisms. Then the Chern–Simons action previously defined gives an $\mathbb{R}/\mathbb{Z} \cong S^1$-valued function:

$$CS : \mathcal{A}(\Sigma)/\mathcal{G}(\Sigma) \to S^1, \qquad a \mapsto \int_\Sigma Tr(da \wedge a + \tfrac{2}{3} a \wedge a \wedge a). \tag{1.16}$$

Building on work of Taubes [49], Floer [14] viewed this as a Morse function on the infinite-dimensional singular manifold $\mathcal{A}(\Sigma)/\mathcal{G}(\Sigma)$ and constructed the corresponding chain complex.

Digression on Morse theory

Recall that Morse showed how a generic function $f : M \to \mathbb{R}$ on a closed (finite-dimensional) manifold gives rise to a chain complex.

Morse proved every smooth function $f : M \to \mathbb{R}$ is arbitrarily close (in the C^r topology) to a function for which every critical point $m \in M$ (a *critical point of* f is a point $m \in M$ satisfying $df_m = 0$; other points are called *regular points*) has a parameterized neighborhood $(U, p) \cong (\mathbb{R}^n, 0)$ such that in these coordinates,

$$f(x_1, \ldots, x_n) = -x_1^2 - \cdots - x_i^2 + x_{i+1}^2 + \cdots + x_n^2.$$

Thus critical points are isolated, and hence on a compact manifold finite.

The number i is called the *Morse index of f at p*. The implicit function theorem implies that at a regular point coordinates can be chosen so that $f(x_1, \ldots, x_n) = x_1$. A *gradient flow line* for f, defined in terms of a fixed Riemannian metric on M, is a smooth path $\alpha : \mathbb{R} \to M$ such that $\langle \alpha'(t), X\rangle = df_{\alpha(t)}(X)$ for all t and $X \in T_{\alpha(t)}M$.

Morse takes the free abelian group $C_i(M)$ generated by this set of critical points of index i. Then a differential $\partial : C_i(M) \to C_{i-1}(M)$ is defined by setting

$$\partial p = \sum_q [p, q] q,$$

where the integer $[p, q]$ is defined to be the number of gradient flow lines $\alpha : \mathbb{R} \to M$ of f (counted with sign, i.e., taking a certain orientation into account) limiting to p as $t \to -\infty$ and to q as $t \to \infty$. (The number of flow lines from x to y is finite if the Morse index of y is one less than the Morse index of x.)

Then $\partial^2 = 0$, and the homology of the chain complex $(C_*(M), \partial)$

$$H_i(M) = \frac{\ker \partial : C_i(M) \to C_{i-1}(M)}{\text{image } \partial : C_{i+1}(M) \to C_i(M)}$$

is isomorphic to the ordinary (e.g., singular) homology of M. An exposition of this topic can be found in Milnor's classic text [32].

Floer applied these notions to the infinite-dimensional manifold $\mathcal{A}(\Sigma)/\mathcal{G}(\Sigma)$, viewing the Chern–Simons function CS as a Morse function. Floer showed that $\partial^2 = 0$ and that the homology is independent of the (appropriate) perturbation of CS and the choice of Riemannian metric on Σ.

This produces the *instanton chain complex* $IC_*(\Sigma)$, and its homology, called *instanton homology*, produces the abelian group $IH_*(\Sigma)$. This group plays the role of $V(\Sigma)$, i.e., we set

$$V(\Sigma) = IH_*(\Sigma).$$

In fact $IH_*(\Sigma)$ is a topological invariant of Σ. The fact that CS is a circle-valued function, rather than a real-valued function, introduces some subtleties; for example, the Morse index is only well defined modulo 8, and so $IH_*(\Sigma)$ is $\mathbb{Z}/8$ graded. (We are allowing $V(\Sigma)$ to be an abelian group rather than a complex vector space.)

The generators of $IC_*(\Sigma)$ are the critical points of CS. A straightforward calculation shows that these critical points are precisely the equivalence classes of connections a whose curvature $F(a)$ equals zero, the *flat* $su(2)$ *connections on* Σ. It is well known that the space $\chi_{SU(2)}(\Sigma)$ of gauge equivalence classes of $SU(2)$ flat connections on Σ (or G connections for any compact Lie group) is a real-algebraic variety homeomorphic to the conjugacy classes of representations of the fundamental group $\pi_1(\Sigma)$:

$$\chi_{SU(2)}(\Sigma) = \{a \in \mathcal{A}(\Sigma) \mid F(a) = 0\}/\mathcal{G}(\Sigma) \cong \text{Hom}(\pi_1(\Sigma), SU(2))/\text{conjugation}.$$

The fundamental group is the most important invariant of a 3-manifold, and so the critical points of CS can be understood in terms of this important algebraic topological invariant. Indeed, an algebraic count of the points in $\chi_{SU(2)}(\Sigma)$ gives Casson's invariant [1, 49], a manifestly topological invariant whose construction solved several important problems in low-dimensional topology and motivated the enormous growth in the study of combinatorial invariants of 3-manifolds. (From the current perspective, Casson's invariant is just the alternating sum of the ranks of the instanton homology groups.)

For disconnected manifolds $\Sigma = \Sigma_1 \sqcup \Sigma_2$, one defines inductively $IH_*(\Sigma)$ to be the tensor product $IH_*(\Sigma_1) \otimes IH_*(\Sigma_2)$. It is a bit tricky to justify $V(\emptyset) = \mathbb{Z}$, for the empty manifold admits no connections (and hence no flat connections). If one tries to avoid this problem by working with the three-sphere S^3, one runs into some difficulties in that the only flat connection on S^3 is the trivial connection up to the $\mathcal{G}$ action (the only representation of the trivial group is trivial). Moreover, one really needs to remove the trivial connection as a generator of $IC_*(\Sigma)$, because the quotient $\mathcal{A}(\Sigma)/\mathcal{G}(\Sigma)$ is not a manifold near the trivial connection. These problems are serious enough to prevent viewing Donaldson–Floer theory as a bona fide TQFT.

The differential $\partial : IC_i(\Sigma) \to IC_{i-1}(\Sigma)$ provides the connection between instanton homology and the Donaldson invariants, and in fact explains the name "instanton homology." This works as follows. A gradient flow line for the Chern–Simons functional CS is a path a_t, $t \in \mathbb{R}$ such that

$$dCS_{a_t}(X) = \langle a_t', X\rangle \qquad \text{for all } t \text{ and } X \in T_{a_t}(\mathcal{A}(\Sigma)/\mathcal{G}(\Sigma)). \tag{1.17}$$

To make sense of this, we first need to explain in what sense the infinite-dimensional manifold $\mathcal{A}(\Sigma)/\mathcal{G}(\Sigma)$ is a Riemannian manifold. This is done using the L^2 inner product on $\mathcal{A}(\Sigma)$. Now, $\mathcal{A}(\Sigma)$ is an affine space with linear model the space of $su(2)$-valued 1-forms $\Omega^1_\Sigma \otimes su(2)$ on Σ; thus, given any $a \in \mathcal{A}(\Sigma)$, the tangent space $T_a\mathcal{A}(\Sigma)$ is canonically identified with the vector space $\Omega^1_\Sigma \otimes su(2)$. Thus we define the Riemannian metric on $\mathcal{A}(\Sigma)$ as an L^2 inner product:

$$\langle X, Y\rangle = \int_\Sigma (X, Y)_s dvol_\Sigma \qquad \text{for } X, Y \in \Omega^1_\Sigma \otimes su(2), \tag{1.18}$$

where $(X, Y)_s$ denotes the inner product in the fiber over $s \in \Sigma$ of the bundle of $su(2)$-valued 1-forms (explicitly, $(X, Y)_s = tr(X_s \wedge *Y_s)$, where $*$ denotes the Hodge $*$ operator on Σ and tr the trace on 2×2 matrices). The inner product is preserved by the $\mathcal{G}(\Sigma)$ action and descends to an appropriate Riemannian metric on the quotient $\mathcal{A}(\Sigma)/\mathcal{G}(\Sigma)$. This explains the inner product in Equation (1.17).

A straightforward calculation using Equations (1.16), (1.17), and (1.18) and the formula for the curvature in terms of 1-forms, $F(a) = da + \frac{1}{2}a \wedge a$, shows that the gradient flow equation can be rewritten in the form

$$\frac{da_t}{dt} = *F(a_t). \tag{1.19}$$

A path of functions $f_t : \mathbb{R} \to \text{Maps}(A, B)$ can be viewed as a function $\mathbb{R} \times A \to B$ by the rule $(t, a) \mapsto f_t(a)$. In the same way, a path of connections a_t on Σ can be viewed as a single connection (which we denote **a**) on the 4-manifold $\mathbb{R} \times \Sigma$. A simple calculation shows that on giving $\mathbb{R} \times \Sigma$ the product Riemannian metric, Equation (1.19) is converted into the instanton equation

$$*_4F(\mathbf{a}) = -F(\mathbf{a})$$

on $\mathbb{R} \times \Sigma$. Up to the action of $\mathcal{G}$ this works in reverse as well: an instanton $\mathbf{a}$ on $\mathbb{R} \times \Sigma$ (with finite L^2 energy) can be gauge transformed to the form $\mathbf{a} = a_t$ for some path of connections on Σ that is a gradient flow line for the Chern–Simons action; in particular, the limits as $t \to \pm\infty$ are critical points of CS, i.e., flat connections on Σ.

Recall that in the formula $\partial p = \sum_q [p, q]q$, the expression $[p, q]$ denotes the algebraic count of gradient flow lines between critical points p and q. In our context this is just the count of instantons on $\mathbb{R} \times \Sigma$. This establishes the link between instanton homology and Donaldson's invariant, which counts instantons on closed 4-manifolds.

Now to continue building in the TQFT apparatus, one would like to define $D(M) \in IH_*(\Sigma)$ when M is a 4-manifold with nonempty boundary (with extra structure as in the closed case), and then to prove a gluing theorem. The details are even more daunting than what has come so far, so I only give the barest of outlines.

Given a 4-manifold M with boundary Σ and a principal $su(2)$ bundle P over M, add a collar $[0, \infty) \times \Sigma$ to M (call the result M_∞), and for each fixed critical point a of CS on Σ, consider the instanton moduli space

$$\mathcal{M}(P, a) = \{A \in \mathcal{A}(P) \mid *F(A) = -F(A), \|F(A)\|_{L^2} < \infty, \lim_{t \to \infty} A|_{t \times \Sigma} = a\}/\mathcal{G}(P)$$

– in other words, the moduli space of finite Yang–Mills energy instantons over M_∞ which limit to the flat connection a on Σ. As before, use cohomology classes $x_i \in H_2(M)$ and the map $\mu : H_2(M) \to H^2(\mathcal{M}(P, a))$ to cut down this moduli space until it is a 0-dimensional, compact oriented (using an orientation of $H_2(M)$ as in the closed case) manifold, i.e., a finite number of signed points. The sum of these signs gives an integer n_a. Then take

$$D(M) = \sum_{a \in \mathrm{crit} CS} n_a \cdot a \in IH_*(\Sigma).$$

This leaves the problem of gluing, which is analytically the greatest challenge in this subject and, although I didn't mention it before, has already come up at every stage in the construction. The question is: How does one glue instantons on one 4-manifold to instantons on another 4-manifold which share a common boundary to produce an instanton on their union?

Changing the orientation of Σ does not change the critical points of CS, and so the generators of the chain complexes $IC_*(\Sigma)$ and $IC_*(-\Sigma)$ are the same, although their Morse indices are different. One defines an inner product $IC_*(\Sigma) \times IC_*(-\Sigma) \to \mathbb{Z}$ using instantons again: given a generator of $IC_*(\Sigma)$ (i.e., a flat connection $a \in IC_*(\Sigma)$) and similarly a generator $b \in IC_*(-\Sigma)$, define $\langle a, b\rangle$ to be the signed count of the finite-energy instantons on $\mathbb{R} \times \Sigma$ that limit to a as $t \to -\infty$ and b as $t \to \infty$. This descends to an inner product $IH_*(\Sigma) \times IH_*(-\Sigma) \to \mathbb{Z}$.

Suppose then that M_1 and M_2 are 4-manifolds, with $\partial M_1 = \Sigma$ and $\partial M_2 = -\Sigma$. We let $M = M_1 \cup_\Sigma M_2$ be the closed manifold obtained by gluing them along their boundaries; equivalently $M_{cut} = M_1 \sqcup M_2$. Now $D(M_1) \in IH_*(\Sigma)$ keeps track of how many instantons on $M_1 \cup ([0, \infty) \times \Sigma)$ limit to each flat connection a on Σ. (We suppress any mention of the x_i, which may be needed to cut down the moduli space. Note, however, that the Mayer–Vietoris sequence allows us to relate $H_2(M)$ with $H_2(M_1) \oplus H_2(M_2)$.)

Similarly, $D(M_2) \in IH_*(-\Sigma)$ keeps track of how many instantons on $M_2 \cup ((-\infty, 0] \times \Sigma)$ limit to each flat connection b on $-\Sigma$. The proof of a gluing formula of the form

$$D(M) = \langle D(M_1), D(M_2)\rangle$$

is typically carried out by a "stretch the neck" argument. One makes use of the metric independence of all the invariants to find a sequence of metrics which allows one to identify an exhausting subset of $\mathcal{M}(M)$ as the union of $\mathcal{M}(M_1)$, $\mathcal{M}(M_2)$, and $\mathcal{M}(\mathbb{R} \times \Sigma)$. The sequence of metrics is obtained by identifying a tubular neighborhood of Σ in M with $(-\epsilon, \epsilon) \times \Sigma$, and then stretching in the normal direction, i.e., replacing $(-\epsilon, \epsilon)$ by $(-R, R)$ where $R \to \infty$. Conversely, given a triple of instantons in $\mathcal{M}(M_1)$, $\mathcal{M}(M_2)$, and $\mathcal{M}(\mathbb{R} \times \Sigma)$, one uses cutoff functions to glue them together to get a near-instanton on M, which, by PDE methods, is then shown to lie close to an instanton.

1.4.2 Some results proven using Donaldson–Floer theory

The results obtained using the theory of instantons on 4-manifolds were spectacular breakthroughs in geometric topology. To fully appreciate them it helps to know a little bit of the high-dimensional manifold theory (i.e., surgery theory), in that perhaps the main lesson learned is that smooth 4-dimensional topology is completely unlike higher dimensions (dim ≥ 5) or lower dimensions (dim ≤ 3). Moreover, the contrast with *topological* 4-manifold theory sharpens this distinction. The important results of Freedman [15] showing that, if one ignores smooth structures, 4-manifolds behave similarly to high-dimensional manifolds, appeared at roughly the same time as Donaldson's first breakthrough.

Donaldson proved [5] that a simply connected smooth 4-manifold M with $b^-(M) = 0$ (i.e., with positive definite intersection form) has a *standard* intersection form, i.e., there exists a basis $\{x_i\}$ for the free abelian group $H^2(M)$ such that $x_i \cup x_j = \delta_{i,j}$. Using Freedman's theorem, the conclusion reads that M is homeomorphic to a connected sum of $\mathbb{C}P^2$s.

Shortly after, building on ideas of Freedman, Donaldson, and Gompf, Taubes proved [51] that $\mathbb{R}^4$ admitted uncountably many distinct (i.e., nondiffeomorphic) smooth structures.

One important result [6] is that the cut-and-paste machinery described can be used to show that if the closed 4-manifold M is the connected sum of two manifolds $M = M_1 \# M_2$, then $D(M) = 0$. This provides a way to prove that manifolds are homeomorphic but not not diffeomorphic to connected sums.

Donaldson theory is well suited to study Kähler surfaces, and in particular algebraic surfaces over $\mathbb{C}$. This is because the instanton equations can be perturbed on such manifolds to take a special form whose solutions can be understood using the tools of algebraic geometry. Namely, the instanton moduli spaces on a Kähler manifold M can be identified with the moduli space of stable holomorphic rank 2 bundles over M. In particular, it can be shown in many cases that $D(M)$ is nonzero when M is a Kähler surface. As a consequence, Donaldson showed that there exist smooth complex surfaces that are homeomorphic but not diffeomorphic.

This approach was pushed much farther by many mathematicians, notably Friedman and Morgan [16]. For example, they proved that up to holomorphic deformation, there are only a finite number of algebraic surfaces within a given diffeomorphism type.

Further results include calculations by Kronheimer and Mrowka [63], Fintushel and Stern [11, 12], and others on how the invariants $D(M)$ behave under various cut-and-paste operations.

There have also been important contributions to 3-dimensional problems. An early consequence concerns the following problem about knots $K : S^1 \subset S^3$: when is such a knot the boundary of a smoothly or continuously embedded disk $D^2 \subset D^4$? If yes, the knot K is called *smoothly or topologically slice*. The set of knots modulo slice knots forms an important but mysterious group. Many examples of knots which are topologically but not smoothly slice have been constructed (see, e.g., [20]).

Other results include the work of Fintushel and Stern [11], who computed the instanton homology of a class of 3-manifolds (the *Seifert-fibered* homology spheres) and proved that certain infinite families of Seifert-fibered homology spheres are linearly independent in the (abelian) *homology cobordism group*. (This is the group of all oriented 3-manifolds Σ homology-equivalent to S^3 modulo the equivalence relation that identifies Σ_1 and Σ_2 if there is a 4-manifold M with $\partial M = \Sigma_1 \sqcup -\Sigma_2$ such that M has the homology of $S^3 \times [0, 1]$.) The calculation of this group is needed to solve the problem of deciding whether all manifolds (in all dimensions) are triangulable.

1.5 Seiberg–Witten theory

In 1994 Seiberg and Witten published the article [47] which introduced a new PDE, the *Seiberg–Witten (SW) equations*, into the study of 4-manifolds. By this point the framework of gauge theory coming from Donaldson theory had become part of the arsenal of 4-dimensional topology. Applying the Donaldson–Floer machine is difficult for several reasons, including the fact that the structure group $SU(2)$ is non-abelian, but most seriously because the instanton moduli spaces $\mathcal{M}(P)$ are noncompact, and their natural compactifications tend to be quite complicated.

By contrast, although the SW equations are a bit more complicated algebraically than the instanton equations, they are based on the structure group $U(1)$, and, most importantly, the SW moduli space (solutions of the SW equation modulo bundle automorphisms) *is compact*. This was quickly recognized to be a fundamental simplification, and it is fair to say that SW theory immediately replaced Donaldson theory as the tool of choice to study smooth 4-manifolds.

Since the introduction of SW invariants for closed manifolds, cut-and-paste constructions have figured prominently in the applications. Constructing a 3+1-dimensional TQFT from SW theory is an ongoing project. I will outline how SW invariants are constructed, but much more briefly than I did for the Donaldson theory, hopefully to illustrate some similarities and highlight the differences.

1.5.1 *Spin^c structures*

As with Donaldson invariants, the SW invariants of a closed 4-manifold M are defined from a moduli space of solutions to a nonlinear PDE on M. The equations depend on a choice of *Spin^c structure*, which we now define.

The group $SO(n)$ has fundamental group $\mathbb{Z}/2$ for $n > 2$, and therefore has a connected 2-fold cover $Spin(n) \to SO(n)$. When $n = 2$, $SO(2) = U(1)$ and has a 2-fold self-cover $U(1) \to U(1)$, $z \mapsto z^2$. The group $Spin(n) \times U(1)$ thereby inherits a diagonal $\mathbb{Z}/2$ action, and one defines the quotient group

$$Spin^c(n) = \big(Spin(n) \times U(1)\big)/\mathbb{Z}/2.$$

The 2-fold covers give two Lie group homomorphisms

$$Spin^c(n) \to SO(n), \quad [s,u] \mapsto [s], \quad \text{and} \quad Spin^c(n) \to U(1), \quad [s,u] \mapsto u^2. \tag{1.20}$$

An oriented n-manifold M is a manifold whose tangent bundle admits transition functions to $SO(n)$. Thus we define a $Spin^c(n)$ structure to be (an equivalence class) of lifts of these transition functions to $Spin^c(n)$ using the first projection of Equation (1.20). Equivalently, a $Spin^c$ structure is a principal $Spin^c$ bundle $P_s \to M$

which lifts the principal $SO(n)$ frame bundle $FM \to M$ of M with respect the first projection.

Dold and Whitney proved that all oriented 4-manifolds admit $Spin^c(4)$ structures [4]. Because oriented 3-manifolds have trivial tangent bundles, they admit $Spin^c(3)$ structures. We denote by $\mathcal{S}(M)$ the set of $Spin^c$ structures on M. Elementary obstruction theory shows that $\mathcal{S}(M)$ is in (noncanonical) bijective correspondence with $H^2(M;\mathbb{Z})$.

Given a $Spin^c(n)$ structure s on an n-manifold, the first projection of Equation (1.20) recovers the orientation of M, and the second projection determines a $U(1)$ bundle over $L_s \to M$. Denote by $c_1(s) \in H^2(M;\mathbb{Z})$ the first Chern class of the $U(1)$ bundle L_s. Another way to explain this is as follows. Let P_s denote the corresponding principal $Spin^c(n)$ bundle, $FM \to M$ the $SO(4)$ frame bundle M, and L_s the $U(1)$ bundle. Then P_s fibers over FM and also over L_s, corresponding to these two Lie group homomorphisms.

The representation theory of $Spin^c(n)$ is known, and has special properties which allow the construction of a complex vector bundle $S \to M$ called the *spinor bundle*. Moreover, given a connection A on the $U(1)$ bundle L_s, one can construct a self-adjoint operator $D_A : C^\infty(S) \to C^\infty(S)$ called the *Dirac operator coupled to A*.

1.5.2 Outline of the construction of SW invariants for closed 4-manifolds

On a 4-manifold M with $Spin^c(4)$ structure $s \in \mathcal{S}(M)$, the spinor bundle splits as a sum of two bundles, $S = S^+ \oplus S^-$, and the Dirac operator splits accordingly: $D_A^\pm : C^\infty(S^\pm) \to C^\infty(S^\mp)$. Set $\mathcal{A}$ to be the product of $C^\infty(S^+)$ and the space $\mathcal{C}(P)$ of $U(1)$ connections on L_s,

$$\mathcal{A} = C^\infty(S^+) \times \mathcal{C}(P).$$

The group $\mathcal{G} = \text{Maps}(M, U(1))$ acts on $\mathcal{A}$. One thinks of $\mathcal{G}$ as the subgroup of automorphisms of the principal $Spin^c(4)$ bundle P_s which cover the identity on the frame bundle FM.

The SW equations on M associated to the $Spin^c(4)$ structure are defined for $(\psi, A) \in \mathcal{A}$ by

$$D_A(\psi) = 0,$$

$$F(A) + *F(A) = q(\psi),$$

where $F(A) \in \Omega^2_M \otimes \mathbb{C}$ denotes the curvature of A, and $q : C^\infty(S^+) \to \Omega^2_M \otimes \mathbb{C}$ is a certain quadratic (and algebraic) function. In other words, (ψ, A) satisfies the SW equations if it is sent to zero by the function

$$SW : C^\infty(S^+) \times \mathcal{C}(P) \to C^\infty(S^-) \times (\Omega^2_M \otimes \mathbb{C}),$$

$$(\psi, A) \mapsto (D_A(\psi), F(A) + *F(A) - q(\psi)).$$

The action of $\mathcal{G}$ preserves solutions, and so the *Seiberg–Witten moduli space associated to the structure* $s \in \mathcal{S}(M)$ is defined to be

$$\mathcal{M}(M, s) = SW^{-1}(0, 0)/\mathcal{G}.$$

The basic facts about $\mathcal{M}(M, s)$ are:

(i) $\mathcal{M}(M, s)$ is a closed manifold of dimension

$$d = (c_1(s)^2 - 2\chi(M) - 3\sigma(M))/4$$

(for generic small perturbations of the SW equations, and if $b^+(M) > 1$).

(ii) Orientations of $H_1(M; \mathbb{R})$ and a maximal positive definite subspace of $H^2(M; \mathbb{R})$ determine an orientation of the moduli space $\mathcal{M}(M, s)$.

(iii) There is a canonical element $\mu = c_1(\mathcal{L}) \in H^2(\mathcal{M}(M, s); \mathbb{Z})$. Here $\mathcal{L} \to \mathcal{M}(M, s)$ is the $U(1)$ bundle corresponding to the basepoint fibration. That is, let $\mathcal{G}_0 \subset \mathcal{G}$ denote the subgroup of Maps$(M, U(1))$ sending the basepoint to the identity. Then $\mathcal{L} = SW^{-1}(0)/\mathcal{G}_0$.

Except for compactness of $\mathcal{M}(M, s)$, these facts follow much in the same way they did in Donaldson's theory. Compactness is a consequence of a-priori bounds that are known to hold because of the special properties of Dirac operators, and this is the main property that distinguishes SW from Donaldson theory.

One can now define the SW invariants of a closed 4-manifold M, together with a choice of $Spin^c(4)$ structure and an orientation of its homology, to be

$$SW(M, s) = \mu^{d/2}([\mathcal{M}(M, s)])$$

when d is even, and 0 otherwise. In other words, view μ as a 2-form on $\mathcal{M}(M, s)$, and take the integral

$$SW(M, s) = \int_{\mathcal{M}(M,s)} \mu \wedge \mu \wedge \cdots \wedge \mu.$$

The simplest case occurs when $d = 0$: $SW(M, s)$ is then just a signed count of a finite number of points. It is conjectured that if $d > 0$, then $SW(M, s) = 0$, in other words, that one need only consider the 0-dimensional case.

Changing the orientation of the homology of M (item (ii) in the preceding list) at most changes the sign of $SW(M, s)$, and so SW is typically considered as a function

$$SW_M : \mathcal{S}(M) \to \mathbb{Z}.$$

In other words, the partition function for SW theory is defined on the category of closed 4-manifolds with $Spin^c$ structure, and $Spin^c$-preserving diffeomorphisms preserve SW invariants.

I will omit the discussion of how one constructs a TQFT, including the delicate issues of gluing formulae. Much deep work has been done by Taubes, Mrowka, Kronheimer, Ozsváth, Szabó, Morgan, and many others. The book [27] by Kronheimer and Mrowka treats the formal construction of a TQFT, taking the point of view of Morse theory of the *Chern–Simons–Dirac* functional on a 3-manifold defined by

$$S(\psi, A) = -\tfrac{1}{8}\int_\Sigma (A - A_0)\wedge(F(A) - F(A_0)) + \tfrac{1}{2}\int_\Sigma \langle D_A(\psi), \psi\rangle dvol_M.$$

In this definition, a $Spin^c(3)$ structure s on a 3-manifold Σ is fixed. Then ψ is a section of the corresponding spinor bundle $S \to \Sigma$, A a connection on the $U(1)$ bundle associated to s, and D_A the corresponding Dirac operator.

This is used to construct chain complexes and abelian groups $\widehat{HM}_*(\Sigma, s)$ and $\check{HM}_*(\Sigma, s)$, called *monopole Floer homology groups*, for each 3-manifold Σ and $Spin^c(3)$ structure s on Σ. Then 4-dimensional cobordisms induce homomorphisms on the monopole Floer homology groups.

1.5.3 Some results obtained from Seiberg–Witten theory

Like Donaldson invariants, SW invariants vanish for connected sums. This gives a method to show that certain manifolds are *irreducible*, i.e., not diffeomorphic to connected sums, provided one can compute their SW invariants. For Kähler surfaces this is straightforward, and led to much quicker proofs of irreducibility of Kähler surfaces. But Taubes proved that closed simply connected *symplectic* 4-manifolds (M, ω) satisfy $SW(M, s_\omega) = \pm 1$ [52]. Here s_ω is a $Spin^c$ structure canonically associated to the symplectic form ω.

A *symplectic manifold* (M, ω) is a smooth manifold M of even dimension with a 2-form $\omega \in \Omega^2_M$ which is *closed* ($d\omega = 0$) and *nondegenerate* ($\omega(X, Y) = 0$ for all Y implies $X = 0$). Every Kähler manifold (and hence smooth projective complex variety) is symplectic. But the notion is much less rigid than the Kähler condition. Small (closed) perturbations of a symplectic form remain symplectic, because the nondegeneracy condition is open. This permits one to make many cut-and-paste constructions of symplectic manifolds which have no counterpart in Kähler geometry. See Gompf's article [21].

In particular, one can now show that many 4-manifolds known to admit symplectic structures cannot be diffeomorphic to connected sums, even though they are homeomorphic to connected sums. This result, combined with the results of [21], spurred an explosion in the study of symplectic 4-manifolds and their 3-dimensional analogues, *contact 3-manifolds*. In another direction, one can use Taubes's result to prove that there exist many smooth manifolds homeomorphic but not diffeomorphic to symplectic manifolds.

Symplectic topology is interesting to geometric topologists because, on the one hand, the existence of a symplectic (or tight contact) structure guarantees the nontriviality of the SW invariants. On the other hand, these structures are much more flexible than Riemannian, holomorphic, or algebraic structures, and thereby amenable to cut-and-paste operations. Moreover, contact structures on 3-manifolds have a close link to the theory of foliations of 3-manifolds, a subject extensively studied by Thurston, Gabai, and others in the 1980s, and a continuation of the classical 3-manifold theory of Dehn, Haken, Waldhausen, and others. Thus this development has formed an important bridge between gauge theory and classical 3-manifold theory.

Another important set of problems that were attacked using SW theory are the *minimal-genus problems*. In a 4-manifold M every class $\alpha \in H_2(M)$ is *represented* by an embedded oriented surface $F \subset M$ in the sense that a triangulation of F represents the cycle α. The problem is to determine, for a given M and α, what is the minimal genus smoothly embedded surface F representing α. The problem manifests itself in many contexts, perhaps most notably knot theory.

For example, for $\mathbb{C}P^2$, we have $H_2(\mathbb{C}P^2) = \mathbb{Z}$, generated by a hyperplane $\mathbb{C}P^1 \subset \mathbb{C}P^2$. It is known that any smooth algebraic curve in $\mathbb{C}P^2$ described as the zero locus of a degree n polynomial has genus $\frac{1}{2}(n-1)(n-2)$ and represents the class $n \in H_2(\mathbb{C}P^2)$. The *Thom conjecture* asserts that this cannot be improved by considering smooth surfaces rather than algebraic curves. More precisely, if $F \subset \mathbb{C}P^2$ is a smoothly embedded surface that represents $n \in H_2(\mathbb{C}P^2)$, then the genus of F is at least $\frac{1}{2}(n-1)(n-2)$. This was proven by Kronheimer and Mrowka in [28]. (The Thom conjecture is false *topologically*, i.e., for continuously but not smoothly embedded surfaces [29].)

A bit later Ozsváth and Szabó [35] vastly generalized this result by showing that an embedded symplectic surface in a closed symplectic 4-manifold is genus minimizing. In fact they proved an *adjunction inequality* for surfaces in manifolds with simple type (a class of smooth 4-manifolds that includes all symplectic 4-manifolds and is conjectured to include *all* smooth 4-manifolds). This inequality says that if $SW(M,s) \neq 0$ and $\alpha \in H_2(M)$ is represented by a smoothly embedded surface $F \subset M$, then

$$2\ \mathrm{genus}(F) - 2 \geq \alpha \cdot \alpha + |c_1(s)(\alpha)|.$$

These and many other theorems, including determining the effect of *blowing up* (taking a connected sum with $-\mathbb{C}P^2$) [12], or cutting and pasting along T^3 [33, 35, 50], on the SW invariants are obtained by proving delicate gluing theorems for solutions to the SW equations. This has led to new results about the nonuniqueness of smooth structures on 4-manifolds. Fintushel and Stern's knot surgery method [13] shows that most smooth closed compact 4-manifolds admit infinitely many nondiffeomorphic smooth structures (given a knot $K \in S^3$ and a simply connected smooth 4-manifold M containing a suitable torus, they show how to construct a smooth 4-manifold M_K whose SW invariants differ from those of M by the Alexander polynomial of K.) Fintushel and Stern have conjectured that *every* 4-manifold which admits one smooth structure admits infinitely many nondiffeomorphic smooth structures.

Another important set of results concerns inequalities that the signature $\sigma(M)$ and Betti numbers $b_i(M) = \text{rank}(H_i(M))$ of a smooth manifold must satisfy. Furuta [17] proved that if M is simply connected and admits a *Spin*(4) structure, then

$$b_2(M) \geq \tfrac{5}{4}|\sigma(M)| + 2.$$

(The *11/8 conjecture* asserts that $b_2(M) \geq \frac{11}{8}|\sigma(M)|$ for simply connected spin 4-manifolds.) Taubes proved that if M admits a symplectic structure and is simply connected, then $2\chi(M) + 3\sigma(M) \geq 0$.

These are strong restrictions, which fail for 4-manifolds which do not admit smooth structures.

There are many more results, including the existence and nonexistence of complex, or symplectic, or smooth structures on 4-manifolds, and further results in knot theory. It is fair to say that SW theory has revealed that the study of 4-manifolds is an extremely rich and (at least at the present) mysterious subject.

1.6 Heegaard–Floer homology

The final $3+1$ TQFT I will describe is the *Heegaard–Floer theory* introduced by Oszváth and Szabó in 2001 [36]. Since its introduction there has been an explosion of results in all aspects of 3- and 4-dimensional topology. Most of the theorems proven using Donaldson and SW theory have alternative proofs using Heegaard–Floer theory. Many new results in knot theory and contact 3-manifold theory have been obtained, and a robust bridge to classical 3-manifold theory has been produced. Moreover, the TQFT philosophy is central in the construction of Heegaard–Floer theory. Finally, although the construction of this machine requires some of the hard analysis of moduli spaces, in a more recent development new *combinatorial* methods have been discovered, i.e., constructions and calculations of Heegaard–Floer invariants that do not require analytic methods at all. For "old-school" topologists, this is a welcome development.

The survey articles [37, 38] by Oszváth and Szabó outline the constructions and some results, and our exposition draws from these articles.

1.6.1 Lagrangian Floer homology

To begin with, before Floer defined instanton homology, inspired by ideas of Gromov, he defined a different homology, called *Lagrangian Floer homology*,

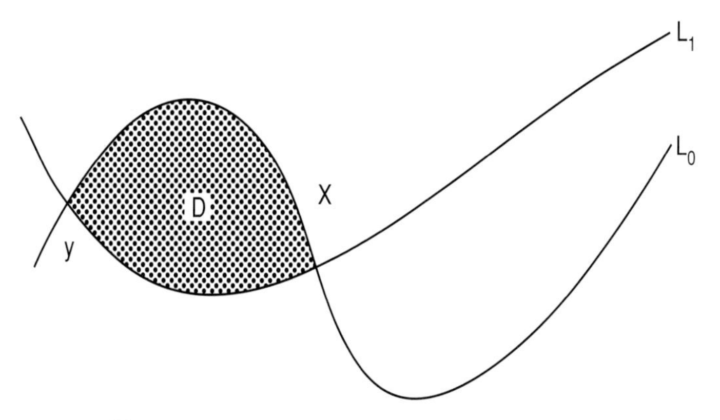

Fig. 1.7. A J-holomorphic disk D from x to y

associated to a symplectic manifold (M, ω) and a pair of Lagrangian submanifolds $L_0, L_1 \subset M$.

A *Lagrangian submanifold* $L \subset M$ of a symplectic manifold is a submanifold of half dimension, $\dim(L) = \frac{1}{2}\dim(M)$, such that ω restricts to the zero form on L, i.e., $\omega(a, b) = 0$ for all $a, b \in T_\ell L$, for all $\ell \in L$.

Assuming L_0 and L_1 are transverse (and M, L_0, L_1 are compact), Floer constructed a chain complex with chain groups CF the free abelian groups generated by the intersection points of L_0 and L_1. The differentials are defined by the formula

$$\partial x = \sum_y [x, y] y,$$

where the integer index $[x, y]$ is a signed count of the J *holomorphic* disks from x to y.

If (M, ω) is a symplectic manifold, then there exists a bundle isomorphism $J : T_* M \to T_* M$ such that $J^2 = -Id$ and such that the inner product $\langle x, y \rangle := \omega(x, Jy)$ is a Riemannian metric on M. This is called a *compatible almost complex structure*. Each tangent space $T_m M$ admits a complex structure such that multiplication by i is given by $J_m : T_m M \to T_m M$.

With D the unit disk in $\mathbb{C}$, a J*-holomorphic disk* is a map $f : D \to M$ such that $df(i\mathbf{v}) = J(df(\mathbf{v}))$ for $\mathbf{v} \in T_* M$. If L_0, L_1 are transverse submanifolds of M and $x, y \in L_0 \cap L_1$, then a J-holomorphic disk *from* x *to* y is a J-holomorphic disk $f : D \to M$ such that

$$f(1) = x, \qquad f(-1) = y,$$
$$f(e^{it}) \in L_0 \quad \text{if } t \in [0, \pi], \quad \text{and} \quad f(e^{it}) \in L_1 \quad \text{if } t \in [\pi, 2\pi]$$

(see Figure 1.7).

Gromov showed that in such contexts the space of J-holomorphic disks (modulo the $\mathbb{R}$ action of holomorphic reparameterizations of D fixing ± 1) is typically a finite-dimensional manifold, and in suitable contexts a compact oriented manifold.

Thus one defines $[x, y]$ to be the signed number of points in the space of J holomorphic disks modulo parameterization whenever this number is finite, and sets it equal to zero otherwise. Floer showed that in certain situations $\partial^2 = 0$. In fact, he also set up a corresponding Morse theory by studying a certain action functional on the space of paths in M that start on L_0 and end at L_1, such that the critical points are constant paths, i.e., intersections of L_0 and L_1, and gradient flow lines correspond to J-holomorphic disks.

The resulting homology group $HF(M, L_0, L_1)$ contains information about how the Lagrangian submanifolds are situated symplectically in M. For example, it is unchanged by Lagrangian isotopies of L_0 and L_1.

Similar ideas and constructions occur in Gromov and Gromov–Witten theory, quantum cohomology, and other tools used to study symplectic manifolds. Ozsváth and Szabó had the insightful idea of applying this to a context arising from a particular description of 3-manifolds, namely in terms of their *Heegaard splittings*.

1.6.2 Heegaard splittings of 3-manifolds

Morse theory (again!) shows that every closed oriented 3-manifold Σ admits a Morse function $f : \Sigma \to \mathbb{R}$ with one minimum (i.e., Morse index 0) point m satisfying $f(m) = 0$, with g Morse index 1 critical points p_i satisfying $f(p_i) = 1$, with g Morse index 2 critical points q_i satisfying $f(q_i) = 2$, and with one maximum n satisfying $f(n) = 3$ (this is called a *self-indexing Morse function*).

The closed, oriented 2-manifold $F = f^{-1}(\frac{3}{2}) \subset \Sigma$ is called a *Heegaard surface*, and splits M into two manifolds

$$H_1 = f^{-1}([0, \tfrac{3}{2}]), \qquad H_2 = f^{-1}([\tfrac{3}{2}, 3]).$$

Thus Σ has a decomposition

$$\Sigma = H_1 \cup_F H_2.$$

The manifolds H_1 and H_2 are simple 3-manifolds with boundary, called *handlebodies of genus g*. These manifolds are easily described as the solid bounded by the standard embedding of a genus g surface in $\mathbb{R}^3$; see Figure 1.8.

One can find disjointly embedded curves $\alpha_1, \ldots, \alpha_g$ in $F = \partial H_1$ that bound disjoint disks $D_1, \ldots, D_g$ in H_1 in such a way that cutting H_1 along these disks yields a 3-ball D^3 (see Figure 1.9). Similarly one can find disjointly embedded curves $\beta_1, \ldots, \beta_g$ in $F = \partial H_2$ that bound disjoint disks $D_1, \ldots, D_g$ in H_2 in such a way that cutting H_2 along these disks yields a 3-ball D^3.

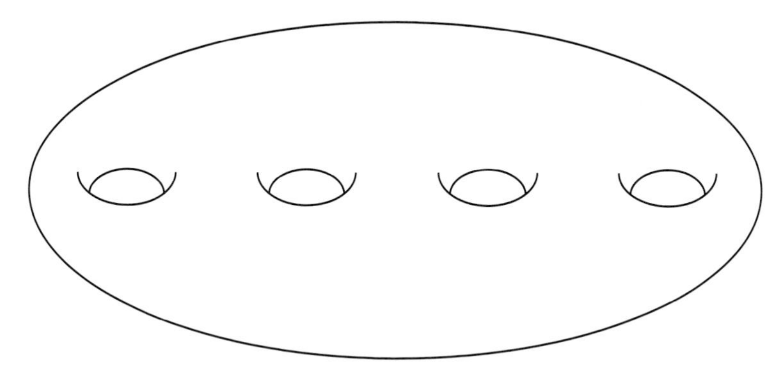

Fig. 1.8. A solid genus 4 handlebody

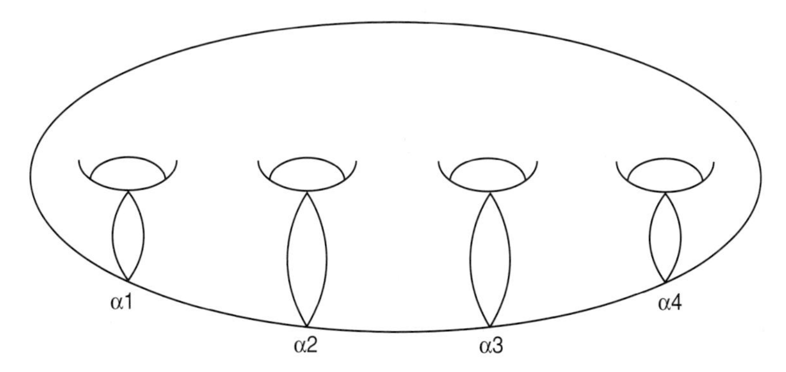

Fig. 1.9. The α-curves

Consider the following data, called a *Heegaard splitting of* Σ:

(i) a surface F of genus g,
(ii) g disjointly embedded curves $\alpha_1, \ldots, \alpha_g$ that cut F into a punctured 2-sphere (the *α-curves*),
(iii) g disjointly embedded curves $\beta_1, \ldots, \beta_g$ that cut F into a punctured 2-sphere (the *β-curves*).

It is not hard to show that this data determines the 3-manifold Σ up to diffeomorphism. However, Σ may have many different Heegaard splittings, just as it may have many Morse functions.

1.6.3 Heegaard–Floer homology

Let $Sym^g(F)$ denote the g-fold symmetric product of the Heegaard surface F. Officially,

$$Sym^g(F) = \underbrace{F \times F \times \cdots \times F}_{g} / S_g,$$

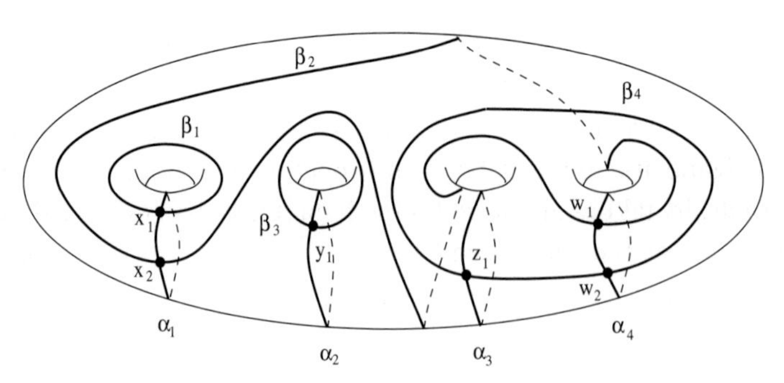

Fig. 1.10. A Heegaard splitting

where the symmetric group S_g acts on $F \times F \times \cdots \times F$ by permuting coordinates. More concretely, $Sym^g(F)$ is the set of *unordered* g-tuples of points on F, with repetition allowed.

Because F is a 2-manifold, $Sym^g(F)$ is a smooth manifold of dimension $2g$. Moreover, a choice of complex structure on F induces a Kähler, and hence symplectic, structure on $Sym^g(F)$.

We have a symplectic manifold, so we need a pair of Lagrangian submanifolds to apply Floer's construction. These are provided by the α and β curves as follows. View each α_i as a smooth embedding of a circle $\alpha_i : S^1 \to F$. The product map

$$\alpha_1 \times \cdots \times \alpha_g : S^1 \times \cdots \times S^1 \to F \times \cdots \times F$$

is an embedding. Moreover, because the α_i have disjoint images, the composite

$$\alpha_1 \times \cdots \times \alpha_g : S^1 \times \cdots \times S^1 \to F \times \cdots \times F \to Sym^g(F)$$

is also an embedding. Its image is a Lagrangian g-torus denoted T_α. Similarly the β_i define a Lagrangian submanifold $T_\beta \subset Sym^g(F)$.

The intersection points of T_α and T_β are transverse (because we assume the α_i and β_i are transverse). One should think of each intersection point as an unordered g tuple of intersections of the α_i and β_i. So for example, for the Heegaard splitting of Figure 1.10 the set $\{x_1, y_1, z_1, w_1\}$ denotes one intersection point of T_α and T_β, and $\{x_1, y_1, z_1, w_2\}$ denotes a different intersection point. For this example, T_α and T_β intersect in four points.

Ozsváth and Szabó's idea is to take (a properly modified version of) the Lagrangian Floer homology of the triple $(Sym^g(F), T_\alpha, T_\beta)$. The intersection points of T_α and T_β keep track of subtle information about the Heegaard splitting. Moreover, J-holomorphic disks $D \to Sym^g(F)$ contain information about the combinatorics of the complementary regions obtained by cutting F along the union of the α_i and β_i, and how they fit together. Thus the idea of assigning the abelian group $HF(Sym^g(F), T_\alpha, T_\beta)$ to the 3-manifold Σ has potential for organizing and providing deep information about Σ.

This is indeed what is done to define Heegaard–Floer homology. Actually Ozsváth and Szabó use more refined information to produce a more sophisticated set of tools, by taking into account the homotopy class of the J-holomorphic disks, a $Spin^c(3)$ structure on Σ, and the *Maslov index*, a certain integer-valued symplectic invariant of loops on Lagrangian submanifolds. The construction yields the following.

Theorem 1.6.1

(i) *To any closed 3-manifold Σ and $Spin^c$ structure $s \in \mathcal{S}(\Sigma)$ one can assign the graded abelian groups $\widehat{HF}_*(\Sigma, s)$, $HF_*^+(\Sigma, s)$, $HF_*^-(\Sigma, s)$, and $HF_*^\infty(\Sigma, s)$. These are diffeomorphism invariants, in particular independent of the choice of Heegaard splitting of Σ.*

(ii) *(Duality) For $\circ = +, -,$ or ∞, there is a natural bilinear pairing*

$$HF^\circ(\Sigma, s) \otimes HF^\circ(-\Sigma, s) \to \mathbb{Z}.$$

(iii) *These groups fit in long exact sequences; in particular,*

$$\cdots \to \widehat{HF}_*(\Sigma, s) \to HF_*^+(\Sigma, s) \to HF_*^+(\Sigma, s) \to \cdots$$

and

$$\cdots \to HF_*^-(\Sigma, s) \to HF_*^\infty(\Sigma, s) \to HF_*^+(\Sigma, s) \to \cdots$$

are exact. These groups are graded by $\mathbb{Q}$, although for each (Σ, s), $HF_k^*(\Sigma, s)$ is nonzero only for $k \in \mathbb{Z} + q$ for some $q \in \mathbb{Q}$ (which depends on (Σ, s)).

(iv) *(Cobordism property) Given any 4-manifold M with boundary $-\Sigma_1 \sqcup \Sigma_2$, and s a $Spin^c(4)$ structure on M (restricting to s_1 and s_2 on Σ_1 and Σ_2), then M induces a homomorphism $F_{M,s}^\circ : HF_*^\circ(\Sigma_1, s_1) \to HF_*^\circ(\Sigma_2, s_2)$ (for $\circ = +, -,$ or ∞) which induces maps of the long exact sequences. The map $F_{M,s}$ shifts grading by $(c_1(s)^2 - 2\chi(M) - 3\sigma(M))/4$.*

(v) *(Gluing) Given a cobordism (M_1, s) from (Σ_1, s_1) to (Σ, s_2) and a cobordism (M_2, t) from $(\Sigma_2, t_2 = s_2)$ to (Σ_3, t_3), let $M = M_1 \cup_{\Sigma_2} M_2$. Then for $\circ = +, -,$ or ∞,*

$$F_{M_2,t}^\circ \circ F_{M_1,s}^\circ = \sum_{\{u \in \mathcal{S}(M) \,|\, u|_{M_1} = s,\, u|_{M_2} = t\}} F_{M,u}.$$

The groups $\widehat{HF}(\Sigma, s)$, $HF^\pm(\Sigma, s)$, $HF^\infty(\Sigma, s)$ are conjectured to be equal to their analogues in SW theory constructed by Kronheimer and Mrowka, and this has been proved in many cases.

When M is a simply connected closed 4-manifold with $b^+(M) > 0$, one can remove two small 4-balls from M and consider the result X as a cobordism from S^3 to itself. The corresponding homomorphism $F_{X,s} : HF^-(S^3) \to HF^-(S^3)$, together with the known calculation $HF^-_*(S^3) = \mathbb{Z}[U]$ (polynomial ring), allows one to interpret $F_{X,s}$ as a polynomial. Its coefficients then give integer invariants of M [39]. These are conjectured to equal the SW invariants of M, and in fact agree with them (suitably interpreted) in all known calculations.

Virtually all the main results of Donaldson–Floer and Seiberg–Witten theory have corresponding proofs in Heegaard–Floer (HF) theory. Typically these proofs are much simpler technically in HF theory. Moreover, the moduli space of J-holomorphic disks can often be understood, at least well enough to compute some of the differentials in the HF chain complex.

More interestingly, the technically simpler nature of HF theory allows other constructions to be made, most notably of a functor HF for knots in 3-manifolds. This has led to many new discoveries in knot theory, especially the minimal-genus problem.

For example, through a remarkable chain of ideas and deep theorems of Ozsváth and Szabó, Thurston and Eliashberg, Gabai, and others, it has been proved that the *genus of a knot* [40] as well as whether a knot is *fibered* is determined by its HF invariants [19, 34]. Moreover, HF theory has been used to construct new invariants that bound the *4-ball genus* of knots [41].

The *genus of a knot* $K \subset S^3$, $g(K)$, is the minimal genus of any oriented surface embedded in S^3 with boundary the knot (such surfaces always exist and are called Seifert surfaces). A knot is *fibered* if there is a (smooth) fiber bundle $(S^3 - K) \to S^1$. The *4-ball genus* of K, $g_4(K)$, is the minimal genus of a (properly embedded) surface in the 4-ball D^4 whose boundary is K. Because one can push a surface in S^3 into D^4, one has $g(K) \leq g_4(K)$.

A recent development, following discoveries of Sarkar and Wang [45], is the realization that many of the HF invariants can be computed (and in some cases defined) combinatorially, i.e., without reference to or calculations of J-holomorphic disks.

1.7 TQFTs in dimension 2+1

As a last topic in our introduction to TQFTs in low-dimensional topology, we consider aspects of the 2+1-dimensional TQFTs. In contrast to the 3+1 examples previously discussed, there are "axiomatically complete" examples in this dimension. Briefly, the article [57] lists 10 (precise) axioms that a TQFT must satisfy. In [55], Turaev shows how to any *modular tensor category* (essentially the category of representations of some quantum group) one can construct a corresponding 2+1 TQFT. Using the *Kauffmann bracket*, he constructs infinitely many modular tensor categories, and hence infinitely many 2+1 TQFTs.

What follows is a brief description of this subject, which has blossomed into a huge mathematical enterprise spanning not just topology, but also representation theory, operator theory, quantum physics, combinatorics, and many other areas.

Many mathematicians have made important contributions to this topic, and should be mentioned here. Instead I refer to the bibliography of Turaev's book. In contrast to the previous examples, I will be rather explicit (although there are some technical categorical issues that I sweep under the rug).

1.7.1 Modular tensor categories

A modular tensor category is an abstract algebraic construct arrived at from several directions by many authors, including Drinfeld, Reshetikhin and Turaev, Jimbo, and others. The definition is technical, but ultimately useful and as concise as necessary to relate topology and algebra. We follow Turaev's book [55].

A *tensor product* in a category $\mathcal{C}$ is a covariant functor $\otimes : \mathcal{C} \times \mathcal{C} \to \mathcal{C}$. For example, the category of representations of a fixed group G has a tensor product. If $\mathcal{C}$ contains an object $\mathbf{1}$ such that $V \otimes \mathbf{1} = V = \mathbf{1} \otimes V$, and if in addition the tensor product is associative, then $\mathcal{C}$ is called a *strict monoidal category*.

The category $\mathcal{V}$ of vector spaces over $\mathbb{C}$ is a strict monoidal category, with $\mathbf{1} = \mathbb{C}$. In this category there is a canonical isomorphism $V \otimes W$ with $W \otimes V$. This isomorphism has many nice properties, some of which (but not all) we would like to axiomatize.

A *braiding* in a strict monoidal category $\mathcal{C}$ is a natural family $\{c_{V,W} : V \otimes W \to W \otimes V\}$ of isomorphisms, one for each pair $V, W \in \mathcal{C}$, compatible with the tensor product in the sense that for $U, V, W \in \mathcal{C}$,

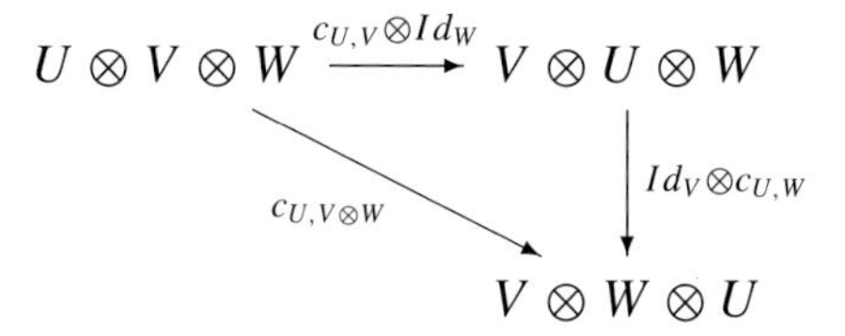

commutes, as well as a similar diagram for $c_{U \otimes V,W}$. Figure 1.11 provides both a mnemonic device to remember $c_{V,W}$ and a description of a map F to be defined. The left picture represents $c_{V,W}$, and the right $c^{-1}_{V,W}$.

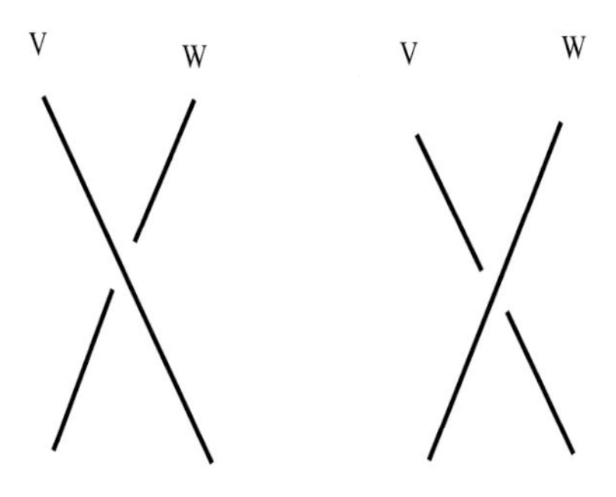

Fig. 1.11.

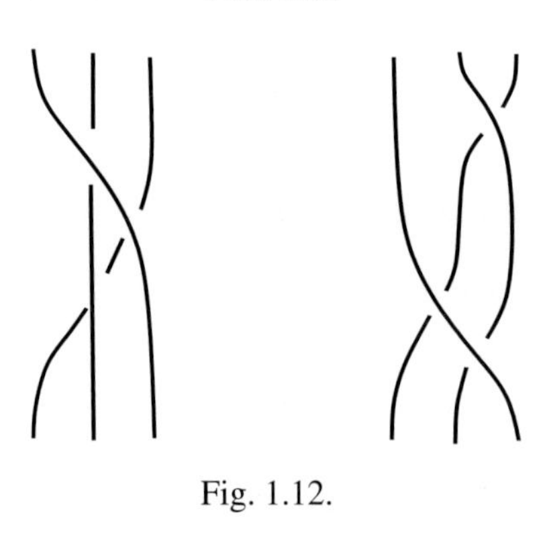

Fig. 1.12.

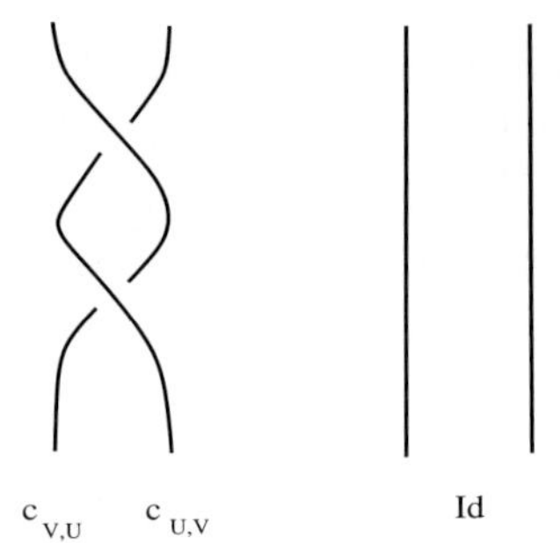

Fig. 1.13. $(c_{V,U})(c_{U,V}) \neq Id$

Figure 1.12 motivates, explains, and justifies the *Yang–Baxter equation*

$$(c_{V,W} \otimes Id_U)(Id_V \otimes c_{U,W})(c_{U,V} \otimes Id_W)$$
$$= (Id_W \otimes c_{U,V})(c_{U,W} \otimes Id_V)(Id_U \otimes c_{V,W}).$$

A critical difference between the category of vector spaces and a general braided monoidal category is that we do not want to assume that $c_{U,V}c_{V,U} = Id_{V\otimes U}$. To require this ultimately would lead to trivial invariants of 3-manifolds. This is hinted at by Figure 1.13: intuitively we see that if we were to require $c_{U,V}c_{V,U} = Id_{V\otimes U}$, then we could change crossings in any knot or link, a process which can be used to trivialize any link.

But some constraints are needed to keep control. These are provided by a *twist*. This is a natural family $\{\theta_V : V \to V\}$ of isomorphisms for each V. See Figure 1.14.

One requires that for any two objects V, W the diagram

$$\begin{array}{ccccc} V \otimes W & \xrightarrow{\theta_V \otimes \theta_W} & V \otimes W & \xrightarrow{c_{V,W}} & W \otimes V \\ & \searrow^{\theta_{V\otimes W}} & & \swarrow_{c_{W,V}} & \\ & & V \otimes W & & \end{array} \tag{1.21}$$

Fig. 1.14. θ_V

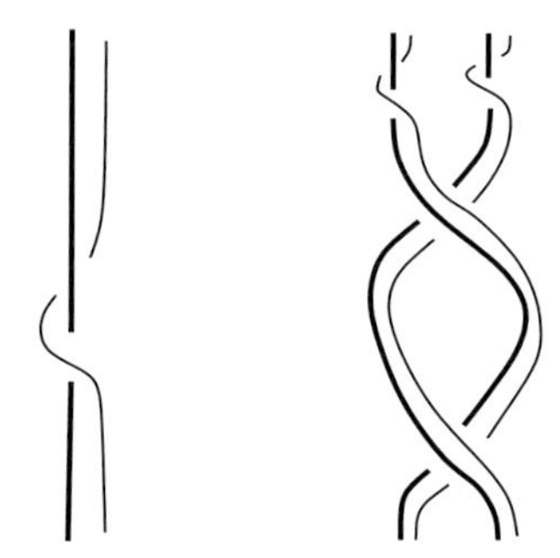

Fig. 1.15. Commutativity of the diagram (1.21)

commute. This is explained in Figure 1.15, in which the dark strand on the left is to be labeled $V \otimes W$ and the two dark strands on the right are to be labeled V and W.

Next, we need to axiomatize the concept of duality on the category of vector spaces. To this end, we assume that to each object $V \in \mathcal{V}$ we can assign another object V^*, and two maps $b_V : \mathbf{1} \to V \otimes V^*$ and $d_V : V^* \otimes V \to \mathbf{1}$. These should be such that

$$V = \mathbf{1} \otimes V \xrightarrow{b_V \otimes Id_V} V \otimes V^* \otimes V \xrightarrow{Id_V \otimes d_V} V \otimes 1 = V \tag{1.22}$$

equals Id_V, and similarly for the other obvious composition. The motivating figures for b_V and d_V are given in Figure 1.16. The reader should draw the picture that corresponds to Equation (1.22).

Compatibility with $c_{U,V}$ and θ_V is required; this is expressed by insisting that the equation

$$(\theta_V \otimes Id_{V^*})b_V = (Id_V \otimes \theta_V^*)b_V$$

hold. (What is the picture?) It can be shown that $(V^*)^* = V$.

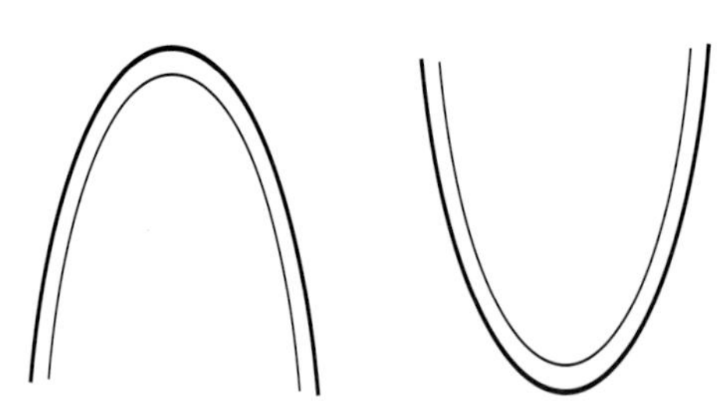

Fig. 1.16. b_V and d_V

A monoidal category $\mathcal{C}$ with a braiding $c_{*,*}$, twist θ_*, and compatible duality $(*, b_*, d_*)$ is called a *ribbon category*. The definition is technical, but the mental picture provided by braids, and more precisely ribbons, or *framed braids*, helps to understand the axioms. The reader should think about Reidemeister moves (with ribbons) and write down the corresponding equations.

We still have a little way to go to finish defining a modular tensor category. Next we assume that $\mathcal{C}$ has the property that its morphism sets form abelian groups: for each $V, W \in \mathcal{C}$ we require $\text{Hom}(V, W)$ to be an abelian group. One can show that this implies $\text{Hom}(\mathbf{1}, \mathbf{1})$ is a commutative ring, which we denote by $K_{\mathcal{C}}$. Moreover, $\text{Hom}(V, W)$ becomes a left $K_{\mathcal{C}}$ module for any objects V, W. Call an object V *simple* if $\text{Hom}(V, V)$ is a free rank 1 $K_{\mathcal{C}}$ module. (For example, in the category of complex representations of a group G, $\mathbf{1} = \mathbb{C}$, $K_{\mathcal{C}} = \mathbb{C}$, and a representation V is simple if it is irreducible.)

One then defines a *modular tensor category* to be a pair $(\mathcal{C}, I)$ where $\mathcal{C}$ is a ribbon category whose morphism sets are abelian groups, and $I = \{V_0, \ldots, V_m\}$ is a finite set of simple objects in $\mathcal{C}$ such that $V_0 = \mathbf{1}$ and which is closed under the duality $V \mapsto V^*$, such that the objects in I "generate" $\mathcal{C}$ in the sense that for each object V the images of the compositions

$$\text{Hom}(V, V_i) \otimes_{K_{\mathcal{C}}} \text{Hom}(V_i, V) \to \text{Hom}(V, V)$$

generate $\text{Hom}(V, V)$ as an abelian group. (This is analogous to the idea in representation theory that all representations are sums of irreducible representations). There is one further *nondegeneracy* axiom required, which ensures that the set I is as small as possible: the morphism $S_{i,j} := c_{V_j, V_i} c_{V_i, V_j} \in \text{Hom}(V_i, V_i) = K_{\mathcal{C}}$ determines a square matrix which is invertible over $K_{\mathcal{C}}$.

1.7.2 Construction of a TQFT from a modular tensor category

Recall from Section 1.1.4 that Kirby proved that every closed 3-manifold is obtained by surgery on a framed link in S^3. To put this in a setting convenient for our present

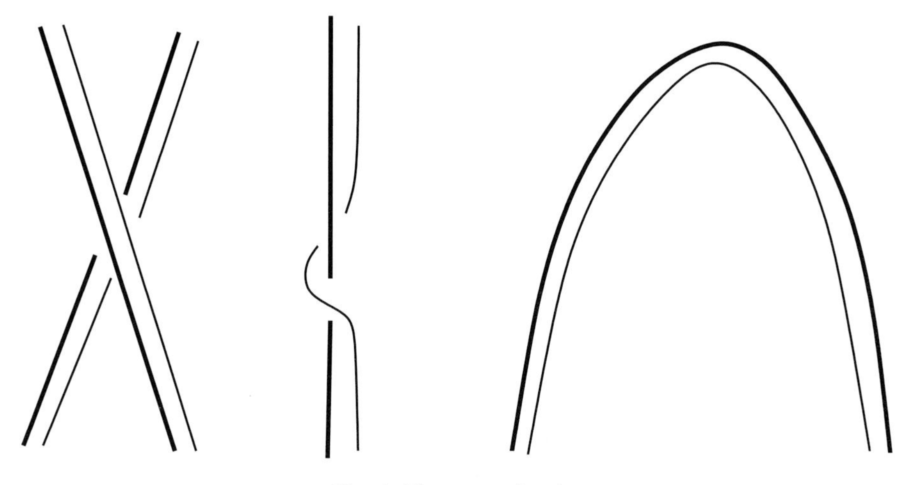

Fig. 1.17. $c_{V_i,V_j}, \theta_{V_i}, b_{V_i}$

purposes, define an n-component *ribbon link* to be an embedding of a disjoint union

$$L : S_1^1 \times [0,\epsilon] \sqcup \cdots \sqcup S_n^1 \times [0,\epsilon] \subset \mathbb{R}^3 \subset S^3.$$

It is clear how this uniquely determines a framed link.

Given a modular tensor category $(\mathcal{C}, I)$, define a *coloring* of the ribbon link to be an assignment of one of the simple objects $V_i \in I$ to each link component; thus if I has m objects and L has n components, there are m^n colorings of L.

Take a *projection* of L (its image under a projection to $\mathbb{R}^2$, with crossing data). After a small isotopy if needed, one may find finitely many horizontal lines in $\mathbb{R}^2$ such that between any two, only one change occurs. It is harder to explain this precisely than to understand it, but consideration of the the local pictures makes it clear.

Consider Figure 1.17. In the left picture, if the dark strand from top left to bottom right is colored V_i and the dark strand from the top right to bottom left is colored V_j, assign the morphism $\cdots \otimes c_{V_i,V_j} \otimes \cdots$, where the "$\cdots$" refers to strands not illustrated. In the middle picture, if the dark strand is colored V_i, assign $\cdots \otimes \theta_{V_i} \otimes \cdots$. In the right picture, if the dark strand is colored V_i, assign $\cdots \otimes b_{V_i} \otimes \cdots$. Turning the right picture upside down yields a picture which is assigned $\cdots \otimes d_{V_i} \otimes \cdots$.

One starts at the top of the projection of a colored ribbon link, and by these rules one produces a morphism in $\mathcal{C}$ from $\mathbf{1}$ to $\mathbf{1}$. For example, for the framed link of Figure 1.18, if the left component is colored V_1 and the right V_2, one obtains the

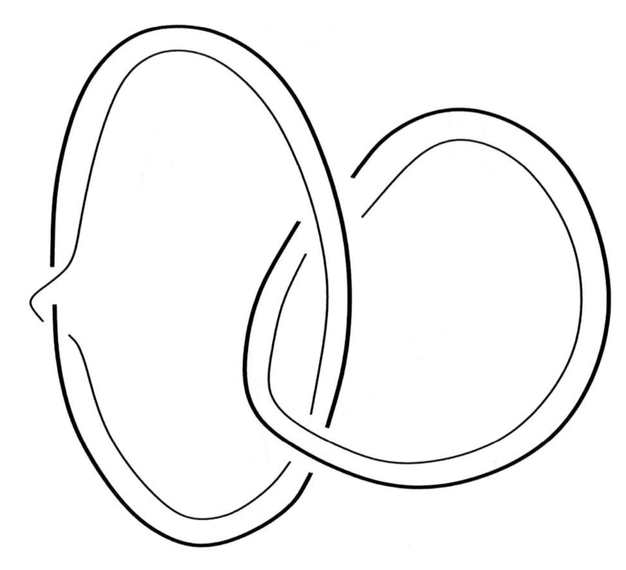

Fig. 1.18.

composite (reading from top to bottom)

$$\mathbf{1} \xrightarrow{b_{V_1}} V_1 \otimes V_1^* \xrightarrow{Id \otimes b_{V_2}} V_1 \otimes V_1^* \otimes V_2 \otimes V_2^*$$

$$\xrightarrow{Id \otimes c_{V_1^*, V_2} \otimes Id} V_1 \otimes V_2 \otimes V_1^* \otimes V_2^* \xrightarrow{\theta_{V_1}^{-1} \otimes Id} V_1 \otimes V_2 \otimes V_1^* \otimes V_2$$

$$\xrightarrow{Id \otimes c_{V_2, V_1^*} \otimes Id} V_1^* \otimes V_2 \otimes V_2^* \xrightarrow{Id \otimes d_{V_2^*}} V_1 \otimes V_1^* \xrightarrow{d_{V_1^*}} \mathbf{1}.$$

Note that we used the inverse of θ_{V_1}, because the twist in the framing is opposite to that in Figure 1.14.

It should be clear what to do in general: given a modular tensor category $(\mathcal{C}, I)$, this procedure assigns, to a projection of a colored framed link L, an endomorphism $F(L)$ of $\mathbf{1}$, i.e., an element of $K_{\mathcal{C}}$.

What we want to do to get a topological invariant of a manifold M is sum $F(L, \lambda)$ over all colorings of a framed link L which yields M by surgery (this may be thought of as analogous to integrating all fields in a path integral). This is almost correct, but some slight modification, taking into account the need for a framing of the 3-manifold, is required.

The *dimension of* V is defined to be $\dim(V) = d_V c_{V,V^*}(\theta_V \otimes Id_{V^*}) b_V$, e.g., $\dim(V_i) = F(L, \lambda)$ for (L, λ) the ribbon link in Figure 1.19 colored with V_i.

Before we can write down the formula for a closed 3-manifold invariant, we must observe that because $\theta_{V_i} \in \mathrm{Hom}(V_i, V_i)$ and $\mathrm{Hom}(V_i, V_i)$ is free of rank 1 over $K_{\mathcal{C}}$ for $i \in I$, we have $\theta_{V_i} = v_i Id_{V_i}$ for some $v_i \in K_{\mathcal{C}}$. Moreover, one needs to assume that the modular tensor category $\mathcal{C}$ has the property that the element $\sum_{i \in I} (\dim(V_i))^2$ has a square root $\mathcal{D} \in K_{\mathcal{C}}$.

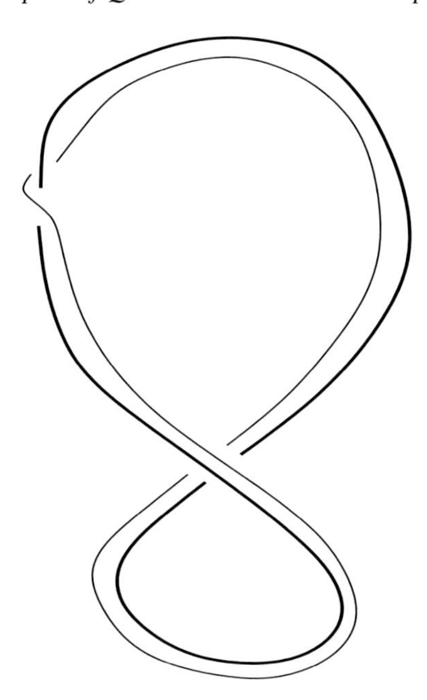

Fig. 1.19. $\dim(V)$

Definition. Given a generic projection of a framed (i.e., ribbon) link L in S^3 of ℓ components and a modular tensor category $(\mathcal{C}, I)$, let $col(L)$ denote the set of all colorings λ of L (a set of cardinality $\ell^{|I|}$). Let $\sigma(L) \in \mathbb{Z}$ denote the signature of the linking matrix of L. Let M denote the closed 3-manifold obtained by surgery on the framed link L. Define

$$Z(M) = \left(\sum_{i \in I} v_i^{-1}(\dim(V_i))^2\right)^{\sigma(L)} \mathcal{D}^{-\sigma(L)-\ell-1} \sum_{\lambda \in col(L)} \left(\prod_{n=1}^{m} \dim(V_{\lambda(n)})\right) F(L, \lambda).$$

The following theorem is proved in [55].

Theorem 1.7.1 *The invariant $Z(M) \in K_{\mathcal{C}}$ is a well-defined diffeomorphism invariant of M, i.e., independent of the choice of the framed link L presenting M and of the projection of L to $\mathbb{R}^2$. Moreover, this invariant is the partition function of a 2+1-dimensional TQFT which satisfies 10 precisely stated axioms.*

Defining the $K_{\mathcal{C}}$ modules $V(\Sigma)$ for a 2-manifold Σ takes more work. However, the reader should notice that the map $F(L, \lambda)$ makes sense for any *colored ribbon tangle*, that is, any union of ribbons (each homeomorphic to $I \times I$ or $S^1 \times I$) properly embedded in $\mathbb{R}^2 \times [0, 1]$ with its path components colored by I. Then $F(L, \lambda)$ is a homomorphism between a tensor product of the form $V_{i_1} \otimes V_{i_2} \otimes \cdots \otimes V_{i_k}$ corresponding to $\mathbb{R}^2 \times \{0\}$ and a similar tensor product corresponding to $\mathbb{R}^2 \times \{1\}$. One sees that the cobordism property of TQFTs is built into the construction. I refer the interested reader to [55, 57].

The important point is that in this situation a TQFT is precisely constructed. The work of Walker addresses to what extent the axioms determine the theory.

Of course, we have said nothing about whether any modular tensor categories exist. In fact they do, in abundance. There are two sources of examples: One is the representation theory of *quantum groups*. These are certain Hopf algebras introduced by physicists studying lattice models and statistical mechanics. The ideas were picked up and formalized by Drinfel'd [9] and refined by Reshetikhin and Turaev in a long series of joint and individual articles starting with [43]. In a certain sense one can show that any modular tensor category arises as the representation category of a quantum group. The other method of producing modular tensor categories is more geometric and is based on the skein theory of the Jones polynomial [24] and its generalization, the Kauffman bracket [25].

Thus 2+1-dimensional TQFTs exist. These are combinatorial in the sense that they begin with a projection of a framed link, and a modular tensor category, which can be described ultimately in terms of an *R-matrix* which determines solutions to the Yang–Baxter equations. No PDE need be solved, no Sobolev spaces introduced, no holomorphic disks are used, and no path integration is invoked. The axioms are precisely stated and satisfied. In this sense a 2+1-dimensional TQFT is a completely satisfactory realization of the objective of constructing a rigorous mathematical theory from the notions coming from quantum field theory and functional integral methods.

Nevertheless, explicit computation is not practical in most cases. Despite its depth and the many connections between 2+1-dimensional TQFTs and different disciplines in mathematics (and in contrast to the Donaldson–Floer, Seiberg–Witten, and Heegaard–Floer theories), from the perspective of this author few problems of independent interest to geometric topologists (classification theorems of manifolds and their submanifolds) have been solved by the 2+1-dimensional theory. Its success stems more from its value as a source of rich and beautiful mathematical ideas than from its use as a machine to solve old problems.

Bibliography

[1] Akbulut, Selman; McCarthy, John D. "Casson's invariant for oriented homology 3-spheres. An exposition." Mathematical Notes, 36. Princeton University Press, Princeton, NJ, 1990. xviii+182 pp.

[2] Atiyah, Michael. "The geometry and physics of knots." Lezioni Lincee [Lincei Lectures]. Cambridge University Press, Cambridge, UK, 1990. x+78 pp.

[3] Atiyah, Michael, et al. *Responses to: A. Jaffe and F. Quinn, "Theoretical mathematics: toward a cultural synthesis of mathematics and theoretical physics."* Bull. Amer. Math. Soc. (N.S.) 30 (1994), no. 2, 178–207.

[4] Dold, A.; Whitney, H. *Classification of oriented sphere bundles over a* 4*-complex.* Ann. of Math. (2) 69 (1959), 667–677.
[5] Donaldson, S. K. *An application of gauge theory to four-dimensional topology.* J. Differential Geom. 18 (1983), no. 2, 279–315.
[6] Donaldson, S. K. *Polynomial invariants for smooth four-manifolds.* Topology 29 (1990), no. 3, 257–315.
[7] Donaldson, S. K. "Floer homology groups in Yang–Mills theory." With the assistance of M. Furuta and D. Kotschick. Cambridge Tracts in Mathematics, 147. Cambridge University Press, Cambridge, UK, 2002. viii+236 pp.
[8] Donaldson, S. K.; Kronheimer, P. B. "The geometry of four-manifolds." Oxford Mathematical Monographs. Oxford Science Publications. The Clarendon Press, Oxford University Press, New York, 1990. x+440 pp.
[9] Drinfel'd, V. G. *Hopf algebras and the quantum Yang–Baxter equation.* Dokl. Akad. Nauk SSSR 283 (1985), no. 5, 1060–1064.
[10] Eilenberg, Samuel; Steenrod, Norman. "Foundations of algebraic topology." Princeton University Press, Princeton, New Jersey, 1952. xv+328 pp.
[11] Fintushel, Ronald; Stern, Ronald J. *Instanton homology of Seifert fibred homology three spheres.* Proc. London Math. Soc. (3) 61 (1990), no. 1, 109–137.
[12] Fintushel, Ronald; Stern, Ronald J. *Immersed spheres in* 4*-manifolds and the immersed Thom conjecture.* Turkish J. Math. 19 (1995), no. 2, 145–157.
[13] Fintushel, Ronald; Stern, Ronald J. *Knots, links, and* 4*-manifolds.* Invent. Math. 134 (1998), no. 2, 363–400.
[14] Floer, Andreas. *An instanton-invariant for* 3*-manifolds.* Comm. Math. Phys. 118 (1988), no. 2, 215–240.
[15] Freedman, Michael Hartley. *The topology of four-dimensional manifolds.* J. Differential Geom. 17 (1982), no. 3, 357–453.
[16] Friedman, Robert; Morgan, John W. "Smooth four-manifolds and complex surfaces." Ergebnisse der Mathematik und ihrer Grenzgebiete (3) [Results in Mathematics and Related Areas (3)], 27. Springer-Verlag, Berlin, 1994. x+520 pp.
[17] Furuta, M. *Monopole equation and the* $\frac{11}{8}$*-conjecture.* Math. Res. Lett. 8 (2001), no. 3, 279–291.
[18] Jaffe, Arthur; Quinn, Frank. *"Theoretical mathematics": toward a cultural synthesis of mathematics and theoretical physics.* Bull. Amer. Math. Soc. (N.S.) 29 (1993), no. 1, 1–13.
[19] Ghiggini, Paolo. *Knot Floer homology detects genus-one fibred knots.* Preprint math. GT/0603445.
[20] Gompf, Robert E. *Periodic ends and knot concordance.* Proceedings of the 1987 Georgia Topology Conference (Athens, GA, 1987). Topology Appl. 32 (1989), no. 2, 141–148.
[21] Gompf, Robert E. *A new construction of symplectic manifolds.* Ann. of Math. (2) 142 (1995), no. 3, 527–595.
[22] Jaco, William H.; Shalen, Peter B. "Seifert fibered spaces in 3-manifolds." Mem. Amer. Math. Soc. 21 (1979), no. 220, viii+192 pp.
[23] Johannson, Klaus. "Homotopy equivalences of 3-manifolds with boundaries." Lecture Notes in Mathematics, 761. Springer, Berlin, 1979. ii+303 pp.
[24] Jones, Vaughan F. R. *A polynomial invariant for knots via von Neumann algebras.* Bull. Amer. Math. Soc. (N.S.) 12 (1985), no. 1, 103–111.
[25] Kauffman, Louis H. *State models and the Jones polynomial.* Topology 26 (1987), no. 3, 395–407.

[26] Kirby, Robion. *A calculus for framed links in* $S\mathfrak{sp}3$. Invent. Math. 45 (1978), no. 1, 35–56.

[27] Kronheimer, Peter; Mrowka, Tomasz. *Monopoles and three-manifolds*. New Mathematical Monographs, 10. Cambridge University Press, Cambridge, UK, 2007. xii+796 pp.

[28] Kronheimer, P. B.; Mrowka, T. S. *The genus of embedded surfaces in the projective plane.* Math. Res. Lett. 1 (1994), no. 6, 797–808.

[29] Lee, Ronnie; Wilczyński, Dariusz M. *Locally flat 2-spheres in simply connected 4-manifolds.* Comment. Math. Helv. 65 (1990), no. 3, 388–412.

[30] Milnor, John. *A unique decomposition theorem for 3-manifolds.* Amer. J. Math. 84 (1962), 1–7.

[31] Milnor, John. *On manifolds homeomorphic to the 7-sphere.* Ann. of Math. (2) 64 (1956), 399–405.

[32] Milnor, John. “Lectures on the h-cobordism theorem.” Notes by L. Siebenmann and J. Sondow. Princeton University Press, Princeton, NJ, 1965. v+116 pp.

[33] Morgan, John W.; Mrowka, Tomasz S.; Szabó, Zoltán. *Product formulas along* $T\mathfrak{sp}3$ *for Seiberg–Witten invariants.* Math. Res. Lett. 4 (1997), no. 6, 915–929.

[34] Ni, Yi. *Knot Floer homology detects fibred knots.* Invent. Math. 170 (2007), no. 3, 577–608.

[35] Ozsváth, Peter; Szabó, Zoltán. *The symplectic Thom conjecture.* Ann. of Math. (2) 151 (2000), no. 1, 93–124.

[36] Ozsváth, Peter; Szabó, Zoltán. *Holomorphic discs and topological invariants for closed 3-manifolds.* Ann. of Math. (2) 159 (2004), no. 3, 1027–1158.

[37] Ozsváth, Peter; Szabó, Zoltán. *Heegaard diagrams and holomorphic discs.* Different faces of geometry, 301–348, Int. Math. Ser. (N.Y.), 3, Kluwer/Plenum, New York, 2004.

[38] Ozsváth, Peter; Szabó, Zoltán. *Heegaard diagrams and Floer homology.* International Congress of Mathematicians. Vol. II, 1083–1099, Eur. Math. Soc., Zürich, 2006.

[39] Ozsváth, Peter; Szabó, Zoltán. *Holomorphic triangles and invariants for smooth four-manifolds.* Adv. Math. 202 (2006), no. 2, 326–400.

[40] Ozsváth, Peter; Szabó, Zoltán. *Holomorphic disks and genus bounds.* Geom. Topol. 8 (2004), 311–334.

[41] Ozsváth, Peter; Szabó, Zoltán. *Knot Floer homology and the four-ball genus.* Geom. Topol. 7 (2003), 615–639.

[42] Papakyriakopoulos, C. D. *On Dehn's lemma and the asphericity of knots.* Proc. Nat. Acad. Sci. U.S.A. 43 (1957), 169–172.

[43] Reshetikhin, N. Yu.; Turaev, V. G. *Ribbon graphs and their invariants derived from quantum groups.* Comm. Math. Phys. 127 (1990), no. 1, 1–26.

[44] Rohlin, V. A. *New results in the theory of four-dimensional manifolds.* Doklady Akad. Nauk SSSR (N.S.) 84 (1952), 221–224.

[45] Sarkar, Sucharit; Wang, Jiajun. *An algorithm for computing some Heegaard Floer homologies.* Preprint math. GT/0607777.

[46] Segal, G. B. *The definition of conformal field theory.* “Differential geometrical methods in theoretical physics (Como, 1987),” 165–171, NATO Adv. Sci. Inst. Ser. C Math. Phys. Sci., 250, Kluwer Acad. Publ., Dordrecht, 1988.

[47] Seiberg, N.; Witten, E. *Electric–magnetic duality, monopole condensation, and confinement in* $N = 2$ *supersymmetric Yang–Mills theory.* Nuclear Phys. B 426 (1994), no. 1, 19–52.

[48] Smale, Stephen. *Generalized Poincaré's conjecture in dimensions greater than four.* Ann. of Math. (2) 74 (1961), 391–406.
[49] Taubes, Clifford Henry. *Casson's invariant and gauge theory.* J. Differential Geom. 31 (1990), no. 2, 547–599.
[50] Taubes, Clifford Henry. *The Seiberg–Witten invariants and 4-manifolds with essential tori.* Geom. Topol. 5 (2001), 441–519.
[51] Taubes, Clifford Henry. *Gauge theory on asymptotically periodic* 4*-manifolds.* J. Differential Geom. 25 (1987), no. 3, 363–430.
[52] Taubes, Clifford Henry. *The Seiberg–Witten invariants and symplectic forms.* Math. Res. Lett. 1 (1994), no. 6, 809–822.
[53] Thom, René. *Quelques propriétés globales des variétés différentiables.* Comment. Math. Helv. 28 (1954), 17–86.
[54] Thurston, William P. *Three-dimensional manifolds, Kleinian groups and hyperbolic geometry.* Bull. Amer. Math. Soc. (N.S.) 6 (1982), no. 3, 357–381.
[55] Turaev, V. G. "Quantum invariants of knots and 3-manifolds." de Gruyter Studies in Mathematics, 18. Walter de Gruyter & Co., Berlin, 1994. x+588 pp.
[56] Waldhausen, Friedhelm. *On irreducible* 3*-manifolds which are sufficiently large.* Ann. of Math. (2) 87 (1968), 56–88.
[57] Walker, Kevin. "On Witten's 3-manifold invariants." Available at http://canyon23.net/math/.
[58] Walker, Kevin. "TQFTs." Available at http://canyon23.net/math/.
[59] Whitney, Hassler. *The self-intersections of a smooth n-manifold in $2n$-space.* Ann. of Math. (2) 45 (1944), 220–246.
[60] Witten, Edward. *Quantum field theory and the Jones polynomial.* Comm. Math. Phys. 121 (1989), no. 3, 351–399.
[61] Witten, Edward. *Topological quantum field theory.* Comm. Math. Phys. 117 (1988), no. 3, 353–386.
[62] Kauffiman, L. Review of Jaffe, Arthur, and Quinn, Frank, *Theoretical mathematics: toward a cultural synthesis of mathematics and theoritical physics* and *Response to: "Responses to: A. Jaffe and F. Quinn, 'Theoretical mathematics: toward a cultural synthesis of mathematics and theoretical physics'."* Math. Rev., MR1202292 and MR1254077.
[63] Kronheimer, P.; Mrowka, T. *Gauage theory for embedded surfaces. I.* Topology 32 (1993), no. 4, 773–826.

2

Differential equations aspects of quantum cohomology

MARTIN A. GUEST*

Abstract

The quantum differential equations can be regarded as examples of equations with certain universal properties, which are of interest beyond quantum cohomology itself. We present this point of view as part of a framework which accommodates the KdV equation and other well-known integrable systems. In the case of quantum cohomology, the theory is remarkably effective in packaging geometric information, as will be illustrated with reference to simple examples of Gromov–Witten invariants, variations of Hodge structure, the Reconstruction Theorem and the Crepant Resolution Conjecture.

The concept of quantum cohomology arose in string theory around 20 years ago. Its mathematical foundations were established around 10 years ago, based on the theory of Gromov–Witten invariants. There are two approaches to Gromov–Witten invariants, via symplectic geometry and via algebraic geometry. Both approaches give the same results for the three-point Gromov–Witten invariants of familiar manifolds M like Grassmannians and flag manifolds, and these invariants may be viewed as the structure constants of the quantum cohomology algebra QH^*M, a modification of the ordinary cohomology algebra H^*M.

However, the name 'quantum cohomology' may be misleading. On the one hand, the 'quantum' and 'cohomology' aspects are somewhat removed from the standard ideas of quantum physics and cohomology theory. On the other hand, there are strong relations between quantum cohomology and several other areas

* I am grateful to the organizers of the summer school for their invitation to give these lectures and for their careful planning, which resulted in an effective and pleasant environment. I thank Ramiro Carrillo-Catalán for preparing a Spanish language version of the original lecture notes. I am grateful to PIMS at the University of British Columbia for its hospitality in August 2007, where some of this material was written. I thank the referee for refreshingly frank comments and for urging me to write a much better article, which I hope to do some day. This research was supported by a grant from the JSPS.

of mathematics: symplectic geometry and algebraic geometry, of course, but also differential geometry, the theory of integrable systems (soliton equations) and even number theory.

In this chapter we shall focus on the *quantum differential equations* as the fundamental concept (due to Alexander Givental [13–15]), which encapsulates many aspects of quantum cohomology. This is 'more elementary' than the definition of Gromov–Witten invariants, in the same way that de Rham cohomology is 'more elementary' than the definition of simplicial or singular cohomology. In addition, it is essential for understanding the relation between quantum cohomology and the theory of integrable systems, which is becoming increasingly important. The language of D-module theory is convenient for this purpose. It provides a unified way to think about (a) classical integrable systems, such as the KdV equation, (b) integrable systems in differential geometry, such as harmonic maps and (c) quantum cohomology.

The abstract theory of D-modules is well developed but not widely used by nonspecialists. One goal of this chapter is to give some motivation for D-module theory (Sections 2.1, 2.3), and to advertise some of its uses (Sections 2.2, 2.4). A second goal is to explain in simple terms how quantum cohomology is related to other parts of mathematics. These links have deep origins and are still evolving. It can be difficult to grasp them from articles which use haphazardly technical language from symplectic geometry, algebraic geometry and singularity theory, especially when some of the links are conjectural.

The lecture series "From quantum cohomology to integrable systems" (based on the book [19]) traced a path from ordinary cohomology theory to the quantum differential equations and their role in the theory of integrable systems. The introductory lectures on cohomology and quantum cohomology do not appear here (they are the subject of an earlier survey article [18]). The differential equations aspects of the lectures have been expanded slightly, and the applications have been gathered together in Section 2.4. The lectures contained various concrete examples from [19], which have been omitted here to save space.

2.1 Linear differential equations and D-modules

Consider the linear ordinary differential equation

$$(\partial^{s+1} + a_s\partial^s + \cdots + a_1\partial + a_0)y = 0$$

for $y = y(z)$, where $\partial = \partial_z = \frac{d}{dz}$ and the coefficients $a_0, \ldots, a_s$ are functions of the complex variable z. We assume that $a_0, \ldots, a_s$ belong to the ring $\mathcal{H} = \mathcal{H}_z$

of holomorphic functions on an open disk $N = N_z$ in $\mathbb{C}$. We shall write $T = \partial^{s+1} + a_s\partial^s + \cdots + a_1\partial + a_0$.

Almost all textbooks on differential equations contain a proof of the following basic result:

Theorem 2.1.1 *For any point $z_0 \in N$ and any values $c_0, \dots, c_s \in \mathbb{C}$, there is a unique solution $y \in \mathcal{H}$ of $Ty = 0$ which satisfies the initial conditions $y(z_0) = c_0, y'(z_0) = c_1, \dots, y^{(s)}(z_0) = c_s$.*

Corollary 2.1.2 *The set of all holomorphic solutions (on N) of the o.d.e. $Ty = 0$ is a vector space of dimension $s + 1$.*

Introducing new variables

$$y_0 = y, \qquad y_1 = \partial y = y', \dots, \qquad y_s = \partial^s y = y^{(s)},$$

we may convert the preceding scalar equation of order $s + 1$ to an equivalent system of $s + 1$ first-order equations of the form $\partial Y = AY$:

$$\partial \begin{pmatrix} y_0 \\ \vdots \\ y_{s-1} \\ y_s \end{pmatrix} = \begin{pmatrix} 0 & 1 & & \\ & \ddots & \ddots & \\ & & 0 & 1 \\ -a_0 & \cdots & -a_{s-1} & -a_s \end{pmatrix} \begin{pmatrix} y_0 \\ \vdots \\ y_{s-1} \\ y_s \end{pmatrix}.$$

Corollary 2.1.3 *The set of all holomorphic solutions (on N) of the system $\partial Y = AY$ is a vector space of dimension $s + 1$.*

Let us choose a basis $y_{(0)}, \dots, y_{(s)}$ of solutions of the scalar equation. The corresponding vector functions

$$Y_{(i)} = \begin{pmatrix} y_{(i)} \\ \partial y_{(i)} \\ \vdots \\ \partial^s y_{(i)} \end{pmatrix}, \qquad 0 \le i \le s,$$

constitute a basis of solutions of $\partial Y = AY$. The fundamental solution matrix

$$H = \begin{pmatrix} | & & | \\ Y_{(0)} & \cdots & Y_{(s)} \\ | & & | \end{pmatrix}$$

satisfies $\partial H = AH$, and it takes values in the group $\mathrm{GL}_{s+1}(\mathbb{C})$ of invertible $(s + 1) \times (s + 1)$ complex matrices. This 'matrix o.d.e.' is a third incarnation of the original equation. Its solutions correspond to initial conditions $H(z_0) = C \in \mathrm{GL}_{s+1}(\mathbb{C})$.

The matrix A depends on the definition of $y_0, \ldots, y_s$. Instead of using the successive derivatives of y, let us set

$$y_0 = P_0 y, \qquad y_1 = P_1 y, \ldots, \qquad y_s = P_s y,$$

where $P_0, \ldots, P_s$ are differential operators. This leads to an equivalent system of first-order equations if the equivalence classes $[P_0], \ldots, [P_s]$ form a basis (over $\mathcal{H}$) of the *D-module*

$$\mathcal{M} = D/(T),$$

where D denotes the ring of differential operators (polynomials in ∂ with coefficients in $\mathcal{H}$), and (T) denotes the left ideal generated by T. The ring operations on D are addition and composition of differential operators.

Each such choice of basis corresponds to a way of converting the scalar equation to a matrix equation $\partial Y = AY$; the matrix A is given explicitly by[1] $\partial[P_j] = [\partial P_j] = \sum_{k=0}^{s} A_{jk}[P_k]$. It should be noted that $\mathcal{M}$ is an infinite-dimensional complex vector space; indeed, it can be identified with the space $\mathrm{Map}(N, \mathbb{C}^{s+1})$ of (holomorphic) maps from N to $\mathbb{C}^{s+1}$.

This discussion can be generalized to partial differential equations. We shall be concerned only with 'overdetermined' linear systems of p.d.e., which share many common features with linear o.d.e., in particular finite-dimensionality of the solution space. Let $N = N_{z_1,\ldots,z_r}$ be a fixed open polydisk in $\mathbb{C}^r$. Writing $\partial_1 = \partial/\partial z_1, \ldots, \partial_r = \partial/\partial z_r$, we consider a system of p.d.e.

$$T_1 y = 0, \ldots, \qquad T_u y = 0$$

for a scalar function $y(z_1, \ldots, z_r)$ on N. The T_i are differential operators, that is, polynomials in $\partial_1, \ldots, \partial_r$ with coefficients in the ring $\mathcal{H} = \mathcal{H}_{z_1,\ldots,z_r}$ (functions of $z_1, \ldots, z_r$ which are holomorphic in N).

In contrast to the o.d.e. case, it is not at all clear whether the solution space is finite-dimensional (or what the dimension is). The concept of D-module and the closely related concept of flat connection are essential at this point. We shall just give a brief discussion, referring to [27] for the general theory. Let $D = D_{z_1,\ldots,z_r}$ be the ring of differential operators generated by $\partial_1, \ldots, \partial_r$ with coefficients in the ring $\mathcal{H}$ of holomorphic functions on N. Let

$$\mathcal{M} = D/(T_1, \ldots, T_u),$$

where $(T_1, \ldots, T_u)$ means the left ideal generated by the differential operators $T_1, \ldots, T_u$.

[1] The notation ∂P_j here means composition of differential operators; this conflicts with our earlier usage of ∂f to mean $\partial f/\partial z$. We shall just rely on the context to distinguish these: ∂f means the function $\partial f/\partial z$ when used in a differential equation, whereas in a D-module computation it is the same as the operator $f\partial + \partial f/\partial z$.

Assumption: $\mathcal{M}$ is a free module of rank $s+1$ over $\mathcal{H}$.

This is a strong assumption, but the following proposition (whose proof consists merely of unravelling the definitions) allows us to conclude that the solution space of the original system has dimension $s+1$:

Proposition 2.1.4 *The map*

$$\begin{aligned} \theta : \{y \mid T_i y = 0, 1 \leq i \leq u\} &\rightarrow \mathrm{Hom}_D(\mathcal{M}, \mathcal{H}), \\ y &\mapsto ([X] \mapsto Xy) \end{aligned}$$

is an isomorphism of complex vector spaces.

As in the o.d.e. case, the D-module point of view allows us to make clear the relation between scalar and matrix equations, under the preceding assumption. However, there is an important new ingredient, a certain flat connection, which leads to the relation with integrable systems. We shall therefore review the whole procedure carefully, taking the opportunity to introduce some further notation.

How to convert a scalar system to a matrix system. Let $[P_0], \dots, [P_s]$ be a basis of $\mathcal{M}$ over $\mathcal{H}$. We define $(s+1) \times (s+1)$ matrix functions $\Omega_1, \dots, \Omega_r$ by

$$\partial_i [P_j] = [\partial_i P_j] = \sum_{k=0}^{s} (\Omega_i)_{kj} [P_k].$$

We set $\Omega = \sum_{i=1}^{r} \Omega_i dz_i$. Then $\nabla = d + \Omega$ defines a connection in the trivial vector bundle $N \times \mathbb{C}^{s+1}$, whose space of sections is $\mathrm{Map}(N, \mathbb{C}^{s+1})$. Namely,

$$\begin{aligned} \nabla_{\partial_i} (\textstyle\sum_{j=0}^{s} f_j [P_j]) &= \sum_{j=0}^{s} \partial f_j / \partial z_i [P_j] + \textstyle\sum_{j=0}^{s} f_j \nabla_{\partial_i} [P_j] \\ &= \textstyle\sum_{i=0}^{s} \partial f_j / \partial z_i [P_j] + \sum_{j,k=0}^{s} f_j (\Omega_i)_{kj} [P_k]. \end{aligned}$$

Let us now recall a well-known fact about such connections (see Section 4.5 of [19]).

Theorem 2.1.5 *The following statements are equivalent:*

(1) The connection $d + \Omega$ is flat (i.e., has zero curvature).
(2) $d\Omega + \Omega \wedge \Omega = 0$.
(3) $[\partial_i + \Omega_i, \partial_j + \Omega_j] = 0$ for all i, j.
(4) $\Omega = L^{-1} dL$ for some $L : N \rightarrow GL_{s+1}\mathbb{C}$ (for this it is essential that N is simply connected).

Using this we obtain:

Proposition 2.1.6 *The connection $\nabla = d + \Omega$ obtained from $\mathcal{M}$ is flat.*

Proof Because $\partial_i\partial_j = \partial_j\partial_i$, we have $\partial_i(\partial_j[P]) = \partial_j(\partial_i[P])$ for any $[P] \in \mathcal{M}$; hence $(\partial_i + \Omega_i)(\partial_j + \Omega_j)f = (\partial_j + \Omega_j)(\partial_i + \Omega_i)f$ for any f, i.e. $[\partial_i + \Omega_i, \partial_j + \Omega_j] = 0$. □

The map L^t in Theorem 2.1.5(4) can be regarded as a fundamental solution matrix for the system $(\partial_i - \Omega_i^t)Y = 0$, $1 \le i \le r$. If we introduce $A_i = \Omega_i^t$, we obtain a matrix system

$$\partial_i Y = A_i Y, \quad 1 \le i \le r$$

of the required form. It should be noted that $(\partial_i - \Omega_i^t)Y = 0$ is the equation for parallel (covariant constant) sections with respect to the *dual* connection $\nabla^* = d - \Omega^t$, rather than $\nabla = d + \Omega$ itself. The identification of the solution space of the system with Ker ∇^* may be regarded as a matrix version of Proposition 2.1.4.

To summarize, we can say that the choice of basis $[P_0], \ldots, [P_s]$ produces a matrix system from a scalar system in the following way:

Proposition 2.1.7 *The map* $y \longmapsto Y = (P_0 y, \ldots, P_s y)^t$ *from the solution space*

$$\{y \mid T_j y = 0, 1 \le j \le u\} \cong \operatorname{Hom}_D(\mathcal{M}, \mathcal{H})$$

to the solution space

$$\{Y \mid \partial_i Y = A_i Y, 1 \le i \le r\} = \operatorname{Ker} \nabla^*$$

is an isomorphism of $(s+1)$*-dimensional vector spaces.*

The appearance of the dual connection ∇^* and the dual D-module $\mathcal{M}^* = \operatorname{Hom}_{\mathcal{H}}(\mathcal{M}, \mathcal{H})$ (of which $\operatorname{Hom}_D(\mathcal{M}, \mathcal{H})$ is a subspace) is an important feature of the construction. We shall make essential use of this in describing the reverse construction, next.

How to convert a matrix system to a scalar system. Given a system

$$\partial_i Y = A_i Y, \quad 1 \le i \le r$$

of first-order matrix p.d.e. whose coefficients are holomorphic on N, it is possible to construct a system of higher-order scalar p.d.e., provided the connection $d - A$ corresponding to the matrix system is flat.

Step 1: We begin with the D-module $\mathcal{N} = \operatorname{Map}(N, \mathbb{C}^{s+1}) = \mathcal{H}^{s+1}$, where the action of ∂_i is given by $\partial_i \cdot Y = (\partial_i - A_i)Y$. (This extends to an action of D because the flatness condition – i.e., (3) of Theorem 2.1.5 – ensures that $\partial_i \cdot (\partial_j \cdot Y) = \partial_j \cdot (\partial_i \cdot Y)$.)

Step 2: The dual D-module $\mathcal{N}^*$ is defined by $\mathcal{N}^* = \operatorname{Hom}_{\mathcal{H}}(\mathcal{N}, \mathcal{H})$ with action of ∂_i given by $\partial_i \cdot p(Y) = -p(\partial_i \cdot Y) + \partial p(Y)/\partial z_i$.

Step 3: Choose[2] a cyclic element of $\mathcal{N}^*$, namely an element p_{cyclic} such that $D \cdot p_{\text{cyclic}} = \mathcal{N}^*$.

Step 4: It follows that $\mathcal{N}^* \cong D/I$, where I is the (left) ideal of operators which annihilate p_{cyclic}.

Step 5: Choose[3] generators $T_1, \ldots, T_u$ for the ideal I. Then a suitable scalar system (not unique) is $T_1 y = 0, \ldots, T_u y = 0$.

We illustrate the procedure with the following (artificial) example. In situations which arise from geometry, cyclic elements (step 3) and generators for ideals (step 5) often arise naturally.

Example 2.1.8 Consider the matrix system

$$\partial \begin{pmatrix} y_0 \\ y_1 \end{pmatrix} = \begin{pmatrix} 0 & u \\ v & 0 \end{pmatrix} \begin{pmatrix} y_0 \\ y_1 \end{pmatrix},$$

where u, v are given functions of z, holomorphic on N. As a candidate for a cyclic element of the dual D-module we try p_0, defined by $p_0(Y) = y_0$. Because

$$\begin{aligned} \partial \cdot p_0(Y) &= -p_0(\partial \cdot Y) + p_0(Y)' \\ &= -p_0 \left(\begin{pmatrix} y_0 \\ y_1 \end{pmatrix}' - \begin{pmatrix} 0 & u \\ v & 0 \end{pmatrix} \begin{pmatrix} y_0 \\ y_1 \end{pmatrix} \right) + y_0' \\ &= -(y_0' - u y_1) + y_0' \\ &= u y_1 \\ &= u p_1(Y), \end{aligned}$$

we have $\partial \cdot p_0 = u p_1$ (where p_1 is defined by $p_1(Y) = y_1$). If u is never zero on N, then p_0 and $\partial \cdot p_0$ span the D-module over $\mathcal{H}$, so p_0 is a cyclic element. A similar calculation gives $\partial^2 \cdot p_0 = uv p_0 + u' p_1 = uv p_0 + (u'/u)\partial \cdot p_0$. We obtain $(\partial^2 - (u'/u)\partial - uv) \cdot p_0 = 0$, so the scalar o.d.e. is $(\partial^2 - (u'/u)\partial - uv)y = 0$. (This computation amounts to 'declaring that $y = y_0$' and computing the scalar system for y from the matrix system for Y.) If $1/u$ does not belong to $\mathcal{H}$, we must either enlarge $\mathcal{H}$ or try another candidate for a cyclic element.

Although the D-module $\mathcal{M} = D/(T_1, \ldots, T_u)$ is the fundamental object, we can regard $\mathcal{H}^{s+1}$ (with D-module structure given by $d + \Omega$) as a concrete representation.

[2] To guarantee the existence of a cyclic vector, it is necessary to enlarge $\mathcal{H}$ in step 1, for example, to the field of meromorphic functions on N. A proof can be found in [29] (Proposition 2.9 and Lemma D.5). If the coefficients are polynomial functions, it suffices to replace $\mathcal{H}$ by the algebra of polynomial functions (Theorem 8.18 and Corollary 8.19 of [5]).

[3] To guarantee that D is Noetherian (so that a finite set of generators of I always exists), it is necessary to replace $\mathcal{H}$ by, for example, the ring of holomorphic functions on a closed polydisk (Section 4 of [27]) or the ring of polynomial functions (Section 3 of the Introduction of [5]).

This representation is often useful for calculations. Moreover, H can be regarded as a gauge transformation which converts $d+\Omega$ to the trivial connection d. To express these correspondences it is convenient to introduce the following J*-function* (a name introduced by Givental in the context of quantum cohomology).

Notation 2.1.9 Let $J=(y_{(0)},\dots,y_{(s)})$, where $y_{(0)},\dots,y_{(s)}$ is any basis of solutions of the scalar system $T_j y=0$, $1\le j\le u$.

We obtain a basis $Y_{(0)},\dots,Y_{(s)}$ of solutions of the matrix system, whose fundamental solution matrix can be written

$$H=\begin{pmatrix} | & & | \\ Y_{(0)} & \cdots & Y_{(s)} \\ | & & | \end{pmatrix}=\begin{pmatrix} - & P_0 J & - \\ & \vdots & \\ - & P_s J & - \end{pmatrix}.$$

It is usually possible to take $P_0=1$, in which case the top row of the last matrix is just J.

The identifications just described are as follows:

$$\begin{array}{ccccc} \partial_i & & \partial_i+\Omega_i & & \partial_i \\ \mathcal{M} & \xrightarrow{[P_0],\dots,[P_s]} & \mathcal{H}^{s+1} & \xrightarrow{H} & \mathcal{H}^{s+1}. \\ P=\sum f_i P_i & & (f_0,\dots,f_s)^t & & PJ \end{array}$$

In each column, the operator (top) acts on elements (bottom) of the D-module (middle) in the natural way.

2.2 The quantum differential equations

We begin by summarizing briefly the notation from cohomology theory that we shall use.

Let M be a connected simply connected compact Kähler manifold, of complex dimension n. *For simplicity we assume that the nonzero integral cohomology groups of M are even-dimensional and torsion-free.* We generally use lower-case letters $a,b,c,\dots\in H^*(M;\mathbb{Z})$ for cohomology classes, and upper-case letters $A=\mathrm{PD}(a)$, $B=\mathrm{PD}(b)$, $C=\mathrm{PD}(c)$, $\dots\in H_*(M;\mathbb{Z})$ for their Poincaré dual homology classes. We often refer to $A,B,C,\dots$ as though they were submanifolds or subvarieties of M (rather than equivalence classes of cycles). We often regard $a,b,c,\dots$ as differential forms on M.

Let $\langle\ ,\ \rangle : H^*(M;\mathbb{Z})\times H_*(M;\mathbb{Z})\to\mathbb{Z}$ denote the natural nondegenerate pairing. In de Rham notation, $\langle a,B\rangle=\int_B a$. The intersection pairing is defined by

$$(\ ,\ ): H^*(M;\mathbb{Z})\times H^*(M;\mathbb{Z})\to\mathbb{Z},\quad (a,b)=\langle ab,M\rangle=\int_M a\wedge b.$$

We have $\langle ab, M\rangle = \langle a, B\rangle = \langle b, A\rangle$. This is a nondegenerate symmetric bilinear form.

We are interested primarily in the cup product operation on cohomology, and its generalization to the quantum product. It is convenient for this purpose to specify the cup product by giving its 'structure constants' with respect to a basis. We can choose bases as follows:

$$H_*(M;\mathbb{Z}) = \bigoplus_{i=0}^{s} \mathbb{Z}A_i, \quad H^*(M;\mathbb{Z}) = \bigoplus_{i=0}^{s} \mathbb{Z}a_i$$

and then define dual cohomology classes $b_0, \ldots, b_s$ by $(a_i, b_j) = \delta_{ij}$. Then for any i, j we have

$$a_i a_j = \sum_{i,j,k} \lambda_{ijk} b_k,$$

where

$$\lambda_{ijk} = \langle a_i a_j a_k, M\rangle = \int_M a_i \wedge a_j \wedge a_k = \#\, A_i \cap A_j \cap A_k.$$

Note that the intersection form itself can be specified in a similar way by the integers

$$(a_i, a_j) = \int_M a_i \wedge a_j = \#\, A_i \cap A_j.$$

It is modern practice to regard a cohomology theory as a functor from a certain category of topological spaces to the category of groups which satisfies the Eilenberg–Steenrod axioms. However, from the point of view of quantum cohomology, which we shall consider next, it is preferable to regard the cohomology of M as a collection of numbers

$$\#\, A_i \cap A_j, \quad \#\, A_i \cap A_j \cap A_k.$$

This primitive viewpoint is necessary because quantum cohomology does not (at present) have a functorial characterization.

If we denote $\#\, A_i \cap A_j \cap A_k$ by $\langle A_i | A_j | A_k \rangle_0$, the quantum product is obtained by extending the mentioned collection of numbers to an infinite sequence of *Gromov–Witten invariants* $\langle A_i | A_j | A_k \rangle_D$ for any $D \in \pi_2(M) \cong H_2(M;\mathbb{Z})$, as follows. Let p, q, r be three distinct points in $\mathbb{C}P^1$. Informally, the definition is

$$\langle A|B|C\rangle_D = \#\ \mathrm{Hol}_D^{A,p} \cap \mathrm{Hol}_D^{B,q} \cap \mathrm{Hol}_D^{C,r},$$

where

$$\mathrm{Hol}_D^{A,p} = \{\text{holomorphic maps } f : \mathbb{C}P^1 \to M \mid f(p) \in A \text{ and } [f] = D\},$$

and $[f]$ is the homotopy class of f.

As explained in [9] (for example), $\langle A|B|C\rangle_D$ can be defined rigorously under very general conditions. The definition has the form

$$\langle A|B|C\rangle_D = \int_{[\overline{M}(D)]^{\mathrm{virt}}} ev_1^* a \wedge ev_2^* b \wedge ev_3^* c,$$

where $M(D)$ is a certain moduli space of 'curves', $\overline{M}(D)$ is a compactification of $M(D)$, obtained by adding suitable 'boundary components', and $[\overline{M}(D)]^{\mathrm{virt}}$ denotes the 'virtual fundamental class' over which integration is carried out. The evaluation map $ev_i : \overline{M}(D) \to M$ assigns to a curve its value at a given ith base-point ($i = 1, 2, 3$).

To define $a \circ_t b$ for $a, b \in H^*M$ and $t \in H^2M$, it suffices to define $\langle a \circ_t b, C\rangle$ for all $C \in H_*M$. The definition is:

Definition 2.2.1 Assume that M is a Fano manifold. Then the quantum product $a \circ_t b$ of two cohomology classes $a, b \in H^*M$ is defined by

$$\langle a \circ_t b, C\rangle = \sum_{D \in H_2(M;\mathbb{Z})} \langle A|B|C\rangle_D \, e^{\langle t, D\rangle}.$$

The Fano condition ensures that the sum is finite, and in this case one has the following nontrivial theorem (see Section 8.1 of [9]):

Theorem 2.2.2 *For each $t \in H^2M$, $\circ_t$ is a commutative, associative product operation on H^*M.*

In general the quantum product is supercommutative, but it is commutative here, as we are assuming that the odd-dimensional cohomology of M is zero. We denote the algebra $(H^*M, \circ_t)$ (more precisely, family of algebras) by QH^*M, and refer to it as the quantum cohomology algebra of M.

Because the second cohomology group plays a prominent role in quantum cohomology, we shall assume that the basis $A_0, \dots, A_s$ has been chosen so that $A_1, \dots, A_r$ span H_2M and $b_1, \dots, b_r$ span H^2M. A general element of H^2M will be written $t = \sum_1^r t_i b_i \in H^2M$; the Poincaré dual homology class is then $T = \sum_1^r t_i A_i$. It is conventional to introduce the notation $q_i = e^{t_i}$. However, ∂_i always denotes the derivative with respect to t_i; thus $\partial_i = q_i \partial/\partial q_i$.

Example 2.2.3 The standard basis of $H^*\mathbb{C}P^n$ is $1, b, b^2, \dots, b^n$, where b is the (Poincaré dual of the) hyperplane class. We take this basis as $b_0, \dots, b_n$. A well-known calculation (see Section 8.1 of [9]) gives the quantum products

$$b_i \circ_t b_j = \begin{cases} b_{i+j} & \text{if } 0 \le i + j \le n, \\ b_{i+j-(n+1)} q & \text{if } n + 1 \le i + j \le 2n. \end{cases}$$

In particular we obtain the presentation $QH^*\mathbb{C}P^n \cong \mathbb{C}[b,q]/(b^{n+1}-q)$, in which q is regarded as a formal parameter, rather than the number e^t. *In this article we shall switch between these two versions of quantum cohomology without further comment.*

Example 2.2.4 Let $M = M_N^k$ be a nonsingular complex hypersurface of degree k in $\mathbb{C}P^{N-1}$. All such hypersurfaces have the same cohomology algebra. The Lefschetz theorems show that $H_i M_N^k \cong H_i\mathbb{C}P^{N-1}$ for $0 \le i \le 2N-4$ except possibly for the middle dimension $i = N-2$, and that the subalgebra $H^\sharp M_N^k$ generated by $H^2 M_N^k$ has additive generators represented by cycles of the form $M_N^k \cap \mathbb{C}P^j$. To avoid odd-dimensional cohomology and make use of these cycles, we shall restrict attention to the subalgebra $H^\sharp M_N^k$ and its quantum version $QH^\sharp M_N^k$.

Let us write $b = b_1$ for the hyperplane class, i.e., the cohomology class Poincaré dual to $M_N^k \cap \mathbb{C}P^{N-2}$. We have $c_1(TM_N^k) = (N-k)b$. It follows that M_N^k is Fano if and only if $1 \le k \le N-1$. The classes $1, b, \dots, b^{N-2}$ are an additive basis (over $\mathbb{C}$) for $H^\sharp M_N^k = H^\sharp(M_N^k;\mathbb{C})$, and the intersection form is given by $(b^i, b^j) = k\delta_{i+j,N-2}$.

As a concrete example, we shall just give the quantum products for M_5^3. All quantum products in this case follow from

$$
\begin{aligned}
b \circ_t 1 &= b,\\
b \circ_t b &= b^2 + 6q,\\
b \circ_t b^2 &= b^3 + 15qb,\\
b \circ_t b^3 &= 6qb^2 + 36q^2
\end{aligned}
$$

(see [8,24]). In particular $b \circ_t b \circ_t b = b^3 + 21qb$ and $b \circ_t b \circ_t b \circ_t b = 27qb \circ_t b$. We deduce that $QH^\sharp M_5^3 \cong \mathbb{C}[b,q]/(b^4 - 27qb^2)$.

As we have mentioned, the construction $M \mapsto QH^*M$ is, unfortunately, not functorial. This is perhaps not surprising in view of the fact that the Gromov–Witten invariants $\langle A|B|C\rangle_D$ contain much more information than the isomorphism class of the algebra QH^*M. Therefore we are led to consider other objects constructed from the Gromov–Witten invariants, and the most prominent of these is the quantum D-module $\mathcal{M}$ (see [13, 14]).

Let us consider the space of sections of the trivial vector bundle

$$H^2M \times H^*M \to H^2M$$

or, more generally, the space of sections over an open subset N of H^2M. This is just the vector space consisting of all H^*M-valued functions on N. The quantum product $\circ_t$ on H^*M gives a way of multiplying sections. Thus the space of sections becomes an algebra over $\mathcal{H}_t$.

Next, we introduce the action of a ring of differential operators on sections, i.e., a D-module structure. We do this by defining a connection ∇, called the Dubrovin connection or Givental connection. The definition is: $\nabla_{\partial_i} = \partial_i + \frac{1}{\hbar} b_i \circ_t$ for $1 \le i \le r$, where $\hbar$ is a parameter. If ω_i is the matrix of quantum multiplication by b_i with respect to the basis $b_0, \ldots, b_s$, then we can also write $\nabla_{\partial_i} = \partial_i + \frac{1}{\hbar}\omega_i$. The D-module structure is specified by saying that ∂_i acts as ∇_{∂_i}. This extends to an action of the ring of all differential operators if the identity $\nabla_{\partial_i}\nabla_{\partial_j} = \nabla_{\partial_j}\nabla_{\partial_i}$ holds for all i, j, and this identity does hold because the connection is flat – a consequence of the properties of the quantum product (see Section 8.5 of [9]).

It is convenient to incorporate the parameter $\hbar$ into the ring of differential operators. Thus, we shall take as ring of differential operators the ring $D^\hbar$ which is generated by $\hbar\partial_1, \ldots, \hbar\partial_r$ and whose elements have coefficients which are holomorphic in $t \in N$ and holomorphic in $\hbar$ in a neighbourhood of $\hbar = 0$. This acts on the enlarged space of sections, in which the sections are allowed also to depend on $\hbar$ (holomorphically, in a neighbourhood of $\hbar = 0$). The quantum D-module $\mathcal{M}$ is defined to be this enlarged space of sections. (We use the generic term 'quantum D-module', rather than 'quantum $D^\hbar$-module', for simplicity, but $\mathcal{M}$ is of course a module over the ring $D^\hbar$.)

The most important property of the quantum D-module is its close relation with the quantum cohomology algebra QH^*M. We shall discuss the relation in this section *under the assumption that H^2M generates H^*M as an algebra and M is a Fano manifold.* These hypotheses imply that QH^*M has a presentation

$$QH^*M = \mathbb{C}[b_1, \ldots, b_r, q_1, \ldots, q_r]/(\mathcal{R}_1, \ldots, \mathcal{R}_u)$$

and H^*M has a presentation

$$H^*(M; \mathbb{C}) = \mathbb{C}[b_1, \ldots, b_r]/(R_1, \ldots, R_u),$$

where $\mathcal{R}_i|_{q=0} = R_i$. However, there is a more precise connection between QH^*M and H^*M, which generalizes to a precise connection between $\mathcal{M}$ and QH^*M, so let us review this.

First, for any polynomial c in 'abstract variables' $b_1, \ldots, b_r, q_1, \ldots, q_r$, let us denote by $[c]$ the corresponding element of QH^*M, and by $[[c]]$ the corresponding element of $H^*M \otimes \mathbb{C}[q_1, \ldots, q_r]$. We claim that there exist suitable polynomials $c_0, \ldots, c_s$ such that

$$[b_i][c_j] = \sum_{k=0}^{s} (\omega_i)_{kj} [c_k],$$

$$[[b_i]] \circ_t [[c_j|_{q=0}]] = \sum_{k=0}^{s} (\omega_i)_{kj} [[c_k|_{q=0}]]$$

for $1 \le i \le r$. In other words, if we identify QH^*M with $H^*M \otimes \mathbb{C}[q_1, \dots, q_r]$ via the bases given by $c_0, \dots, c_s$ and $c_0|_{q=0}, \dots, c_s|_{q=0}$, then quantum multiplication in H^*M corresponds to the natural multiplication in QH^*M. This follows from the observation (Theorem 2.2 of [30]) that 'any quantum polynomial may be written as the same classical polynomial plus lower classical terms, and vice versa'. Namely, if we regard b_j as a polynomial (with respect to the cup product) in $b_1, \dots, b_r$, then the polynomial c_j is obtained by expressing b_j as a polynomial with respect to the quantum product in $b_1, \dots, b_r$. The polynomials c_j satisfy $c_j|_{q=0} = b_j$.

Exactly the same method gives the analogous result for $\mathcal{M}$ in the next theorem, because any polynomial in the operators $\hbar\partial_1 + \omega_1, \dots, \hbar\partial_r + \omega_r$ can be expressed as the same polynomial in $\hbar\partial_1, \dots, \hbar\partial_r$ plus terms of lower order. Moreover, because the lower-order terms contain additional powers of $\hbar$, if we replace $\hbar\partial_i$ by b_i (for each i) and then set $\hbar$ equal to 0, these lower-order terms all vanish and we are left with the original polynomial expressed in terms of the variables $b_1, \dots, b_r$.

Theorem 2.2.5 *The quantum D-module is isomorphic to a D-module of the form $D^\hbar/(D_1, \dots, D_u)$, where $D_1, \dots, D_u$ are converted to $\mathcal{R}_1, \dots, \mathcal{R}_u$ when $\hbar\partial_i$ is replaced by b_i (for each i) and then $\hbar$ is set equal to 0. Furthermore, there exists a basis $[P_0], \dots, [P_s]$ of $D^\hbar/(D_1, \dots, D_u)$, with respect to which the (connection) matrix of $\hbar\partial_i$ is ω_i. This basis is converted to $[c_0], \dots, [c_s]$ when $\hbar\partial_i$ is replaced by b_i (for each i) and $\hbar$ is then set equal to 0.*

The basis $[P_0], \dots, [P_s]$ gives a correspondence between a scalar system and a matrix system (cf. Section 2.1), both of which are referred to as the quantum differential equations. A particular choice of J-function (see Section 5.2 of [19] and the original paper [15]) has the remarkable property that it can be written explicitly in terms of Gromov–Witten invariants.

Example 2.2.6 Let us continue Example 2.2.4 by finding a relation D_1 and a basis $[P_0], [P_1], [P_2], [P_3]$ as predicted by the preceding theorem. (The computation of D_1 is analogous to steps 1–5 in Section 2.1; it would be possible to obtain a version of Theorem 2.2.5 for more general modules over $D^\hbar$ this way.) From the quantum products, the D-module structure is given by

$$\hbar\partial \cdot \begin{pmatrix} f_0 \\ f_1 \\ f_2 \\ f_3 \end{pmatrix} = (\hbar\partial + \omega) \begin{pmatrix} f_0 \\ f_1 \\ f_2 \\ f_3 \end{pmatrix}, \qquad \omega = \begin{pmatrix} & 6q & & 36q^2 \\ 1 & & 15q & \\ & 1 & & 6q \\ & & 1 & \end{pmatrix}.$$

Writing e_0, e_1, e_2, e_3 for the standard basis of column vectors, repeated application of $\hbar\partial$ gives

$$(\hbar\partial)^1 \cdot e_0 = e_1,$$
$$(\hbar\partial)^2 \cdot e_0 = 6qe_0 + e_2,$$
$$(\hbar\partial)^3 \cdot e_0 = 6\hbar qe_0 + 21qe_1 + e_3.$$

This shows that e_0 is a cyclic element. Now we 'solve' for e_0, e_1, e_2, e_3, to obtain

$$e_1 = \hbar\partial \cdot e_0,$$
$$e_2 = \big((\hbar\partial)^2 - 6q\big) \cdot e_0,$$
$$e_3 = \big((\hbar\partial)^3 - 21q\hbar\partial - 6\hbar q\big) \cdot e_0.$$

It follows that the matrix of $\hbar\partial$ with respect to $[1]$, $[\hbar\partial]$, $[(\hbar\partial)^2 - 6q]$, $[(\hbar\partial)^3 - 21q\hbar\partial - 6\hbar q]$ is just the matrix ω, so this is the required basis.

To obtain a relation for the D-module, i.e., a differential operator, which annihilates the cyclic element, we differentiate once more:

$$(\hbar\partial)^4 \cdot e_0 = 162q^2e_0 + 27qe_2 + 27\hbar qe_1 + 6\hbar^2qe_0.$$

Substituting for e_1, e_2, e_3, we obtain

$$\big((\hbar\partial)^4 - 27q(\hbar\partial)^2 - 27\hbar q(\hbar\partial) - 6\hbar^2q\big) \cdot e_0 = 0.$$

We conclude that $\mathcal{M}^{M_5^3} = D^\hbar / \big((\hbar\partial)^4 - 27q(\hbar\partial)^2 - 27\hbar q(\hbar\partial) - 6\hbar^2q\big)$.

At the commutative level, i.e., in the quantum cohomology algebra, the analogous calculation would give

$$b^1 \cdot 1 = b,$$
$$b^2 \cdot 1 = 6q + b^2,$$
$$b^3 \cdot 1 = 21qb$$

(these were stated at the end of Example 2.2.4; the notation $b^i\cdot$ here indicates the i-fold iteration of $b\circ_t$). Then

$$b^1 = b \cdot 1,$$
$$b^2 = \big(b^2 - 6q\big) \cdot 1,$$
$$b^3 = \big(b^3 - 21qb\big) \cdot 1.$$

Thus we obtain $c_0 = 1$, $c_1 = b$, $c_2 = b^2 - 6q$, $c_3 = b^3 - 21qb$. Applying b again leads to the relation $b^4 - 27qb^2$, as expected.

The replacement of QH^*M by $\mathcal{M}$ is not unlike the process of quantization in physics. In the preceding example the relation $b^4 - 27qb^2$ is 'quantized' to the relation $(\hbar\partial)^4 - 27q(\hbar\partial)^2 - 27\hbar q(\hbar\partial) - 6\hbar^2 q$. For $\mathbb{C}P^n$ the same argument shows that the relation $b^{n+1} - q$ is converted to the naive quantization $(\hbar\partial)^{n+1} - q$. However, the naive quantization does not always work: there are examples where the naive 'quantization' is not a quantization at all, because it gives a D-module of the wrong rank. (For the preceding example, the naive quantization $(\hbar\partial)^4 - 27q(\hbar\partial)^2$ gives the correct rank, but it gives the wrong quantum products. We shall return to this point in Section 2.5.)

In general, the parameter $\hbar$ keeps track of the difference between the commutative and noncommutative multiplications. The incompatibility of the commutative and noncommutative situations reveals the key property of the quantum D-module. Namely, the action of $\hbar\partial_i$ on the quantum D-module matches exactly the action of b_i on the quantum cohomology algebra – both are given by the same matrix. However, to accomplish this, a careful choice of basis is necessary in each case, and these bases (like the relations) do not match exactly, only 'mod $\hbar$'. If the bases did match exactly (e.g., $(\hbar\partial)^i$ and b^i), then the matrices of $\hbar\partial_i$ and b_i would not in general be the same.

In the preceding example, the basis $[P_0], [P_1], [P_2], [P_3]$ was produced by modifying the monomial basis $[1], [\hbar\partial], [(\hbar\partial)^2], [\hbar\partial)^3]$, and this involved solving a system of linear equations – by Gaussian elimination. Gaussian elimination may be described as the process of finding a lower-triangular–upper-triangular factorization of matrices. It turns out that such a modification is always possible (under our assumption that H^2M generates H^*M and M is Fano), by using a suitable factorization which takes account of the parameter $\hbar$, known as the Birkhoff factorization, from [28]. This says that "almost every" loop $\gamma \in \Lambda\mathrm{GL}_{s+1}(\mathbb{C})$ (i.e., almost every smooth map $\gamma : S^1 \to \mathrm{GL}_{s+1}\mathbb{C})$ may be factorized in the form

$$\gamma(\hbar) = \underbrace{(a_0 + \tfrac{1}{\hbar}a_1 + \tfrac{1}{\hbar^2}a_2 + \cdots)}_{\gamma_-(\hbar)}\underbrace{(b_0 + \hbar b_1 + \hbar^2 b_2 + \cdots)}_{\gamma_+(\hbar)}.$$

The subgroup of $\Lambda\mathrm{GL}_{s+1}(\mathbb{C})$ consisting of 'negative' ('positive') loops is denoted $\Lambda_-\mathrm{GL}_{s+1}(\mathbb{C})$ $(\Lambda_+\mathrm{GL}_{s+1}(\mathbb{C}))$. The meaning of 'almost every' is that the product set $\Lambda_-\mathrm{GL}_{s+1}(\mathbb{C})\,\Lambda_+\mathrm{GL}_{s+1}(\mathbb{C})$ is open and dense in $\Lambda\mathrm{GL}_{s+1}(\mathbb{C})$.

Theorem 2.2.7 *Assume that M is Fano, and that H^2M generates H^*M. Let L^{m} be any solution of $(L^{\mathrm{m}})^{-1}dL^{\mathrm{m}} = \Omega^{\mathrm{m}}$, where Ω^{m}_i is the matrix of $\hbar\partial_i$ with respect to a monomial[4] basis $[P^{\mathrm{m}}_0], \dots, [P^{\mathrm{m}}_s]$. Let $L^{\mathrm{m}} = L^{\mathrm{m}}_- L^{\mathrm{m}}_+$ be the[5] Birkhoff factorization.*

[4] If $\dim H^2M = 1$ and $\dim H^*M = s+1$, a monomial basis is $[1], [\hbar\partial], \dots, [(\hbar\partial)^s]$. The meaning of monomial basis in general is explained in Section 6.6 of [19].

[5] It can be shown that this factorization is possible in a punctured neighbourhood of $q = 0$, if L^{m}_- is allowed to be multiple-valued. See Sections 5.3 and 5.4 of [19].

Then the Dubrovin–Givental connection is given by $\Omega = L^{-1}dL$, *where* $L = L^{\mathrm{m}}_{-}$. *A basis* $[P_0], \dots, [P_s]$ *of the type predicted in Theorem 2.2.5 is given by* $(L^{\mathrm{m}}_{+})^{-1} \cdot P^{\mathrm{m}}_0, (L^{\mathrm{m}}_{+})^{-1} \cdot P^{\mathrm{m}}_1, \dots, (L^{\mathrm{m}}_{+})^{-1} \cdot P^{\mathrm{m}}_s$, *where* $(L^{\mathrm{m}}_{+})^{-1} \cdot P^{\mathrm{m}}_i$ *means* $\sum_{j=0}^{s} (L^{\mathrm{m}}_{+})^{-1}_{ji} P^{\mathrm{m}}_j$.

For this we refer to [17] and Section 6.6 of [19], where it is also shown that L^{m}_{+} is of the form $L^{\mathrm{m}}_{+} = Q_0(I + \hbar Q_1 + \hbar^2 Q_2 + \cdots + \hbar^N Q_N)$, i.e., a finite series in $\hbar$, and that the coefficient matrices $Q_0, Q_1, \dots, Q_N$ may be found by a simple algorithm. The advantage of $\mathcal{M}$ over QH^*M is that it contains all the Gromov–Witten invariants, and this algorithm shows how to extract them.

Example 2.2.8 We examine the results of applying this algorithm to Example 2.2.6, i.e., $QH^\sharp M^3_5$. First, with respect to the basis $[1], [\hbar\partial], [(\hbar\partial)^2], [(\hbar\partial)^3]$ the connection matrix is

$$\Omega^{\mathrm{m}} = \frac{1}{\hbar} \begin{pmatrix} & & & 6q\hbar^2 \\ 1 & & & 27q\hbar \\ & 1 & & 27q \\ & & 1 & \end{pmatrix}.$$

Then it turns out that $L^{\mathrm{m}}_{+} = Q_0(I + \hbar Q_1)$, with

$$Q_0 = \begin{pmatrix} 1 & & 6q & \\ & 1 & & 21q \\ & & 1 & \\ & & & 1 \end{pmatrix}, \qquad Q_1 = \begin{pmatrix} & & & 6q \\ & & & \\ & & & \\ & & & \end{pmatrix}.$$

Computing $(L^{\mathrm{m}}_{+})^{-1} \cdot (\hbar\partial)^i$ gives the answer $P_0 = 1$, $P_1 = \hbar\partial$, $P_2 = (\hbar\partial)^2 - 6q$, $P_3 = (\hbar\partial)^3 - 21q\hbar\partial - 6\hbar q$ that we obtained in Example 2.2.6.

2.3 A D-module construction of integrable systems

An *integrable p.d.e.* is a p.d.e. which can be written as a zero-curvature condition $d\Omega + \Omega \wedge \Omega = 0$, where Ω is given in terms of some auxiliary function(s) $u = u(z_1, \dots, z_r)$. This concept is somewhat related to the 'explicit solvability' of the p.d.e., and closely related to the concept of 'integrable system'. It is easy to write down connection forms Ω which depend on auxiliary functions, and then compute the condition $d\Omega + \Omega \wedge \Omega = 0$. However, it is not easy to produce *nontrivial* examples this way. In terms of D-modules, a random choice of ideal I generally leads to a D-module D/I of rank infinity or rank zero.

We have seen that quantum cohomology leads to D-modules of finite rank. It is natural to ask whether the quantum cohomology of a particular space can be regarded as a solution of an integrable p.d.e., and whether more general quantum-cohomology-like finite-rank D-modules can be constructed. Let us begin with two simple examples. In this section, D denotes $D_{z_1,\dots,z_r}$; we omit $z_1, \dots, z_r$ when there is no danger of confusion.

Example 2.3.1 Let

$$T_1 = \partial_1 + f, \quad T_2 = \partial_2 + g,$$

where f and g are functions of z_1, z_2. Clearly the rank of $D/(T_1, T_2)$ is either 1 or 0, for we have $[\partial_1] = -f[1], [\partial_2] = -g[1]$. The only question is whether $[1] = [0]$ in the D-module, and this depends on f and g. Alternatively, the rank is 1 if and only if the solution space of the linear system $(\partial_1 + f)y = 0$, $(\partial_2 + g)y = 0$ has dimension 1. The condition for this is $\partial f/\partial z_2 = \partial g/\partial z_1$, which can be regarded as a p.d.e. for the functions f, g.

Example 2.3.2 Let

$$T_1 = \partial_1^2 + u, \quad T_2 = \partial_2 - (\tfrac{1}{2}u_{z_1} - u\partial_1),$$

where u is a given function of z_1, z_2. It is clear that the rank of $D/(T_1, T_2)$ is at most 2, because ∂_2 is expressed in terms of ∂_1, and T_1 is quadratic in ∂_1. Whether the rank is exactly 2 depends on whether $[1]$ and $[\partial_1]$ are independent, and this depends on the nature of u. It can be shown that the rank is 2 if and only if u satisfies the condition $u_{z_2} = 3uu_{z_1} + \frac{1}{2}u_{z_1z_1z_1}$, which is the KdV equation.

Both of these examples arise in the following way: first, we fix a value of z_2, and consider the single variable D-module $D/(T_1)$, whose rank is obvious (the order of T_1); then, we attempt to 'extend' to a two-variable D-module of the same rank by adding a relation of the form $T_2 = \partial_2 - P$.

Let us make this into a general procedure. We call the variables x and t, as the procedure can be interpreted as producing a t-flow of the original D-module in x. Thus, we start with a D-module $D_x/(T)$ of rank $s + 1$, where

$$T = \partial_x^{s+1} + a_s(x)\partial_x^s + \cdots + a_1(x)\partial_x + a_0(x).$$

We wish to extend this to a D-module $D_{x,t}/(T_1, T_2)$ of rank $s + 1$ by extending T to a t-family T_1 (with $T_1|_{t=0} = T$) and adjoining a further partial differential operator T_2. If we take T_2 of the form $T_2 = \partial_t - P$, where P does not involve ∂_t, then it is obvious that

$$\text{rank } D_{x,t}/(T_1, T_2) \leq s + 1,$$

for T_2 may be used to eliminate ∂_t.

Proposition 2.3.3 *Let* $T_2 = \partial_t - P$. *Then* rank $D_{x,t}/(T_1, T_2) = s + 1$ *if and only if any of the following equivalent conditions hold:*

(a) $[T_2,(T_1)] \subseteq (T_1)$,
(b) $[T_2,T_1] \equiv 0 \mod T_1$,
(c) $(T_1)_t \equiv [P,T_1] \mod T_1$
(where $(T_1)_t$ means the result of differentiating the coefficients of T_1 with respect to t).

Proof We just sketch the proof, which hinges on the construction of a certain connection in the trivial bundle $N_{x,t} \times \mathbb{C}^{s+1}$. We shall define the connection form with respect to the local basis $[1], [\partial_x], \dots, [\partial_x^s]$. First of all, the t-family of D-modules $D_x/(T_1)$ gives a connection ∇_{∂_x} in the x-direction, namely $\nabla_{\partial_x}[\partial_x^i] = [\partial_x^{i+1}]$. Next we define ∇_{∂_t} by $\nabla_{\partial_t}[\partial_x^i] = [\partial_x^i P]$ for $0 \le i \le s$. We claim that the resulting connection ∇ is flat if and only if condition (c) holds. Details can be found in Section 4.4 of [19]. □

The proof suggests a useful computational method: first write down the connection form $\Omega = \Omega_1 dx + \Omega_2 dt$ with respect to $[1], [\partial_x], \dots, [\partial_x^s]$; then calculate $d\Omega + \Omega \wedge \Omega$. Let us apply this to the two examples given earlier.

For Example 2.3.1 the connection form is just $\Omega = (-f)dz_1 + (-g)dz_2$. Because we are dealing with 1×1 matrices, we have $\Omega \wedge \Omega = 0$, so the flatness condition is $d\Omega = 0$, i.e., $f_{z_2} = g_{z_1}$. For Example 2.3.2, let us consider a more general relation $T_2 = \partial_t - P$ where $P = f + g\partial_x$ (keeping $T_1 = \partial_x^2 + u$). To find the connection matrix of ∇_{∂_x}, we compute

$$\partial_x[1] = [\partial_x] = 0[1] + 1[\partial_x],$$

$$\partial_x[\partial_x] = [\partial_x^2] = [-u] = -u[1] + 0[\partial_x],$$

so

$$\Omega_1 = \begin{pmatrix} 0 & -u \\ 1 & 0 \end{pmatrix}.$$

Similarly, from

$$\partial_t[1] = [\partial_t] = [P] = f[1] + g[\partial_x],$$
$$\partial_t[\partial_x] = [\partial_x P] = [\partial_x(f + g\partial_x)] = (f_x - ug)[1] + (f + g_x)[\partial_x]$$

we obtain

$$\Omega_2 = \begin{pmatrix} f & f_x - ug \\ g & f + g_x \end{pmatrix}.$$

Now, the zero-curvature condition $d\Omega + \Omega \wedge \Omega = 0$ reduces to

$$-u_t = f_{xx} - u_x g - 2ug_x,$$

$$0 = 2f_x + g_{xx};$$

hence

$$u_t = \tfrac{1}{2}g_{xxx} + gu_x + 2g_x u.$$

To obtain an evolution equation such as the KdV equation it is natural to take f and g to be differential polynomials in u (it suffices to choose g, as without loss of generality $f = -\frac{1}{2}g_x$). The choice $g = -u$ (giving $P = \frac{1}{2}u_x - u\partial_x$) produces the KdV equation of Example 2.3.2. There are many other choices, and the possibilities multiply further when we start with a general operator T_1 instead of $T_1 = \partial_x^2 + u$. Thus our construction produces a vast number of examples of 'integrable p.d.e'.

It is necessary to make a remark here about the special role of the KdV equation, which is more commonly viewed as the Lax eqation $(T_1)_t = [P, T_1]$ with $P = \partial_x^3 + \frac{3}{2}u\partial_x + \frac{3}{4}u_x$. (In the D-module we have $P \equiv -\frac{1}{4}u_x + \frac{1}{2}u\partial_x$, and the Lax equation implies that $(T_1)_t \equiv [-\frac{1}{4}u_x + \frac{1}{2}u\partial_x, T_1]$, so we obtain the same KdV equation from this P.) The condition $[\partial_t - P, T_1] \equiv 0 \bmod T_1$ can be regarded as the intrinsic scalar version of the matrix zero curvature condition $d\Omega + \Omega \wedge \Omega = 0$, but, of course, it is weaker than the condition $[\partial_t - P, T_1] = 0$, in general. The KdV equation is very special, as in this case the scalar version *can* be written in the form $[\partial_t - P, T_1] = 0$.

The proof of Proposition 2.3.3 easily generalizes in one direction (see Section 4.4 of [19]):

Proposition 2.3.4 *Let T_i be a t-family of differential operators in the variables $z_1, \ldots, z_r$ such that the D-module $\mathcal{M} = D_{z_1,\ldots,z_r}/(T_1, \ldots, T_u)$ has rank $s+1$ for each value of t. Let P be a t-family of differential operators in $z_1, \ldots, z_r$ such that $[\partial_t - P, I] \subseteq I$. Then the extended D-module $D_{z_1,\ldots,z_r,t}/(T_1, \ldots, T_u, \partial_t - P)$ also has rank $s+1$.*

This can be used inductively to construct 'hierarchies' of integrable p.d.e., including the well-known KdV hierarchy.

Our extension procedure appears to produce very special D-modules, but it is in fact rather general. Namely, in a 'generic' D-module of rank $s+1$ of the form $D_{x,t}/I$, the elements

$$[1],\ \partial_x[1],\ \partial_x^2[1], \ldots,\ \partial_x^s[1]$$

will be independent. They necessarily satisfy a relation of the form $T = \partial_x^{s+1} + a_s\partial_x^s + \cdots + a_0$, i.e., $T[1] = 0$. The element $[\partial_t]$ can be expressed as a linear combination of the preceding basis vectors, i.e., $(\partial_t - P)[1] = 0$ for some polynomial P in ∂_x. Hence the D-module is of the type constructed in this section.

We conclude with a brief comment on the 'spectral parameter'. It is easy to write down a connection matrix Ω with a sprinkling of λs, then obtain an 'integrable p.d.e.

with spectral parameter' $d\Omega + \Omega \wedge \Omega = 0$, but, just as when λ is absent, it is not easy to produce nontrivial examples. However, such a parameter appears naturally in many integrable systems. For example, the Lax form of the KdV equation is often written as $[\partial_t - P, T_1 - \lambda] = 0$, rather than $[\partial_t - P, T_1] = 0$. These are equivalent, but the parameter λ (eigenfunction of the Schrödinger operator T_1) plays an important role in describing the *solutions* of the KdV equation. For the quantum differential equations we have a natural parameter $\lambda = \hbar$ from the start. In such cases, the D-module treatment can be modified by incorporating the spectral parameter into the ring of differential operators, although some care is needed, as the nature of the λ-dependence of the operators plays a crucial role.

2.4 Applications

The main justification for the D-module language of Sections 2.1 and 2.3 is that it provides a unified approach to various kinds of integrable systems with quite different geometrical interpretations. Superficially the geometry arises from flat connections, but there are deeper undercurrents flowing between differential geometry, symplectic geometry and algebraic geometry that produce these connections.

In the case of quantum cohomology, we have already seen (Theorem 2.2.7, Example 2.2.8) how Gromov–Witten invariants are packaged efficiently by the quantum D-module. It is natural to expect that properties of quantum cohomology will correspond to properties of D-modules. We shall examine several examples in this direction, all of which make contact with current research.

It is also natural to expect benefits from thinking of quantum cohomology in terms of integrable systems, and, conversely, developing a theory of integrable systems which resemble quantum cohomology in some way. Two key examples are the direct relation between 'higher-genus' quantum cohomology and the KdV hierarchy discovered by E. Witten and M. Kontsevich, and the classification of certain integrable systems developed by B. Dubrovin and Y. Zhang. We do not discuss these here, as they primarily involve infinite hierarchies and D-modules of infinite rank.

2.4.1 The WDVV equation and reconstruction of big quantum cohomology

It is time to address the question: 'Of which integrable system is the quantum cohomology (of a given space) a solution?' There are two main candidates, and each of them involves a considerable digression.

The first candidate is the WDVV equation. This applies to *big quantum cohomology* rather than the *small quantum cohomology* that we have seen so far, but the former may be 'reconstructed' from the latter and this is where the D-module extension procedure of Section 2.3 is relevant.

Let us briefly give the definition of big quantum cohomology and the WDVV equation. First, the Gromov–Witten potential of a manifold M is the generating function

$$\mathcal{F}^M(t) = \sum_{l\geq 3, D} \frac{1}{l!} \langle \underbrace{T|\ldots|T}_{l} \rangle_D$$

for the Gromov–Witten invariants. (We assume that this function converges for t in some open subset of H^*M, i.e., for T in some open subset of H_*M.) It follows from this and the definition of the small quantum product that

$$\partial_i \partial_j \partial_k \mathcal{F}^M|_{H^2M}(t) = (b_i \circ_t b_j, b_k).$$

It is natural to define a new product, called the big quantum product, as follows:

Definition 2.4.1 For any $t \in H^*M$ such that $\mathcal{F}^M(t)$ converges, we define $\circ_t$ on H^*M by $(b_i \circ_t b_j, b_k) = \partial_i \partial_j \partial_k \mathcal{F}^M(t)$ for all $i, j, k \in \{0, \ldots, s\}$.

It can be proved (see Section 8.2 of [9]) that this big quantum product is commutative and associative, and has the same identity element 1 as the small quantum product. The most difficult part of this is the associativity.

For *any* (smooth or analytic) $\mathbb{C}$-valued function $\mathcal{F}$ on (an open subset of) H^*M, we can define a product operation $*_t$ in the same way:

$$(b_i *_t b_j, b_k) = \partial_i \partial_j \partial_k \mathcal{F}(t) \overset{\text{def}}{=} \mathcal{F}_{ijk}(t).$$

Whether this product is associative is a nontrivial condition on $\mathcal{F}$. Commutativity is obvious.

Definition 2.4.2 The WDVV equation is the system of third-order nonlinear partial differential equations for $\mathcal{F}$ given by the associativity conditions $(b_i *_t b_j) *_t b_k = b_i *_t (b_j *_t b_k)$.

In general, solutions of the WDVV equation correspond to *Frobenius manifolds*, a generalization of quantum cohomology (see Section 8.4 of [9] and the references there).

Let us see how this leads to an integrable p.d.e. which admits the big quantum cohomology of $\mathbb{C}P^2$ as a distinguished solution. Then we shall return to the matter of reconstructing the big quantum cohomology from the small quantum cohomology. This famous example is taken from [26].

We consider the product operation defined in this way by a function $\mathcal{F}$ on the three-dimensional complex vector space $H^*\mathbb{C}P^2 = \mathbb{C}1 \oplus \mathbb{C}b \oplus \mathbb{C}b^2$. We assume that 1 is the identity element; commutativity is automatic. It follows that

$b *_t 1 = b = 1 *_t b$, $b *_t b = b^2 + \mathcal{F}_{111} b + \mathcal{F}_{112} 1$, and $b *_t b^2 = b^2 *_t b = \mathcal{F}_{121} b + \mathcal{F}_{222} 1$, $b^2 *_t b^2 = \mathcal{F}_{221} b + \mathcal{F}_{222} 1$. There is just one nontrivial associativity condition in this example, namely $(b *_t b) *_t b^2 = b *_t (b *_t b^2)$. In terms of $\mathcal{F}$ this condition is $\mathcal{F}_{222} + \mathcal{F}_{111}\mathcal{F}_{122} = \mathcal{F}_{112}^2$, which is by definition the WDVV equation.

Now, it turns out that the associativity condition is equivalent to the flatness of the connection $d + \frac{1}{\hbar}\omega$ (this connection is defined in the same way as for small quantum cohomology). From the mentioned products, we see that the connection form is given explicitly by

$$\omega = \begin{pmatrix} 1 & 0 & 0 \\ 0 & 1 & 0 \\ 0 & 0 & 1 \end{pmatrix} dt_0 + \begin{pmatrix} 0 & \mathcal{F}_{112} & \mathcal{F}_{122} \\ 1 & \mathcal{F}_{111} & \mathcal{F}_{121} \\ 0 & 1 & 0 \end{pmatrix} dt_1 + \begin{pmatrix} 0 & \mathcal{F}_{212} & \mathcal{F}_{222} \\ 0 & \mathcal{F}_{211} & \mathcal{F}_{221} \\ 1 & 0 & 0 \end{pmatrix} dt_2.$$

This exhibits the WDVV equation as an integrable p.d.e. with spectral parameter $\hbar$ (cf. the end of Section 2.3) for the function $\mathcal{F}$. The particular solution given by the Gromov–Witten potential of $\mathbb{C}P^2$ turns out to be

$$\mathcal{F}^{\mathbb{C}P^2}(t_0, t_1, t_2) = \tfrac{1}{2}(t_0 t_1^2 + t_0^2 t_2) + \sum_{d \geq 1} N_d \, e^{d t_1} \frac{t_2^{3d-1}}{(3d-1)!},$$

where the N_d are determined recursively by $N_1 = 1$ and

$$N_d = \sum_{i+j=d} \left(\binom{3d-4}{3i-2} i^2 j^2 - i^3 j \binom{3d-4}{3i-1} \right) N_i N_j.$$

The positive integer N_d can be interpreted as the number of rational curves of degree d in $\mathbb{C}P^2$ which hit $3d - 1$ generic points. As a function of the variables $t_0, q_1 = e^{t_1}, t_2$ the series for $\mathcal{F}^{\mathbb{C}P^2}$ converges in a neighbourhood of the point $(t_0, q_1, t_2) = (0, 0, 0)$ (see Section 2 of [10]).

The *Reconstruction Theorem* of [26] says that all of this highly nontrivial information may be 'reconstructed' from the (much simpler) small quantum cohomology of $\mathbb{C}P^2$. More precisely, any flat connection of the form

$$\omega = \begin{pmatrix} 1 & 0 & 0 \\ 0 & 1 & 0 \\ 0 & 0 & 1 \end{pmatrix} dt_0 + \begin{pmatrix} 0 & * & * \\ 1 & * & * \\ 0 & 1 & 0 \end{pmatrix} dt_1 + \begin{pmatrix} 0 & * & * \\ 0 & * & * \\ 1 & 0 & 0 \end{pmatrix} dt_2$$

which satisfies the initial condition

$$\omega|_{t_0=t_2=0} = \begin{pmatrix} 1 & 0 & 0 \\ 0 & 1 & 0 \\ 0 & 0 & 1 \end{pmatrix} dt_0 + \begin{pmatrix} 0 & 0 & e^{t_1} \\ 1 & 0 & 0 \\ 0 & 1 & 0 \end{pmatrix} dt_1 + \begin{pmatrix} 0 & e^{t_1} & 0 \\ 0 & 0 & e^{t_1} \\ 1 & 0 & 0 \end{pmatrix} dt_2$$

must be of the previous form for some $\mathcal{F}$, and, furthermore, $\mathcal{F}$ is essentially unique. An elementary discussion of this can be found in [10]. More sophisticated and more general versions of this argument have been given, starting with [21].

In terms of our extension procedure, this example can be formulated as follows. If the D-module basis giving rise to the connection $d + \frac{1}{\hbar}\omega$ is $[P_0] = [1], [P_1], [P_2]$, then the component $\frac{1}{\hbar}\omega_1 dt_1$ shows that $P_1 \equiv \hbar\partial_1$ and $P_2 \equiv (\hbar\partial_1)^2 - \mathcal{F}_{111}\hbar\partial_1 - \mathcal{F}_{112}$. Computing $\hbar\partial_1[P_2]$ gives a third-order relation

$$T_1 = (\hbar\partial_1)^3 - \mathcal{F}_{111}(\hbar\partial_1)^2 - (2\mathcal{F}_{112} + \hbar\mathcal{F}_{1111})\hbar\partial_1 - (\mathcal{F}_{122} + \hbar\mathcal{F}_{1112}).$$

Similarly, the component $\frac{1}{\hbar}\omega_2 dt_2$ gives $\hbar\partial_2[P_0] = [P_2] = [(\hbar\partial_1)^2 - \mathcal{F}_{111}\hbar\partial_1 - \mathcal{F}_{112}]$; hence

$$T_2 = \hbar\partial_2 - P, \qquad \text{where } P = (\hbar\partial_1)^2 - \mathcal{F}_{111}\hbar\partial_1 - \mathcal{F}_{112},$$

is also a relation. These two relations generate the ideal of relations of the D-module. This is an example of the situation of Proposition 2.3.3.

2.4.2 Crepant resolutions

In [6], two examples were given to illustrate a general principle known as the *Crepant Resolution Conjecture*. The simpler of the two relates the quantum cohomology of the Hirzebruch surface $\mathbb{F}_2 = \mathbb{P}(\mathcal{O}(0) \oplus \mathcal{O}(-2))$ (a $\mathbb{C}P^1$-bundle over $\mathbb{C}P^1$; the fibrewise one-point compactification of $T\mathbb{C}P^1$) to the quantum cohomology of the weighted projective space $\mathbb{P}(1, 1, 2)$ (the one-point compactification of $T\mathbb{C}P^1$). The natural map $\mathbb{F}_2 \to \mathbb{P}(1, 1, 2)$ is biholomorphic away from the singular point $[0, 0, 1]$ of $\mathbb{P}(1, 1, 2)$. It is a crepant resolution, and the crepant resolution conjecture predicts that the (orbifold) quantum cohomology of $\mathbb{P}(1, 1, 2)$ can be obtained by specializing the quantum parameters q_1, q_2 of $\mathbb{F}_2$ to certain values. Coates et al. [6] confirm the conjecture in this case by comparing the D-modules of each space and carefully matching up their J-functions after analytic continuation. Such examples are valuable as a guide to finding the most appropriate formulation of the crepant resolution conjecture, and more generally to understanding the functorial properties of quantum cohomology under birational maps.

We shall explain this example very simply, using the method of Section 2.3. As this does not involve direct geometric arguments, it suggests the possibility of a purely D-module-theoretic formulation of the conjecture.

First, we express the quantum D-modules of each space, which are well-known. Because $\mathbb{F}_2$ is a $\mathbb{C}P^1$-bundle over $\mathbb{C}P^1$, $H^2\mathbb{F}_2$ has two additive generators which we call b_1, b_2. Geometrically their Poincaré duals may be represented by a fibre and the infinity section of the bundle, respectively. With respect to this basis, it can be shown (Section 5 of [17] or Chapter 11 of [9]) that

$$\mathcal{M}^{\mathbb{F}_2} = D^{\hbar}_{t_1,t_2}/(F_1, F_2),$$

where $F_1 = (\hbar\partial_1)^2 - q_1q_2$, $F_2 = \hbar\partial_2(\hbar\partial_2 - 2\hbar\partial_1) - q_2(1-q_1)$. This is a 'quantization' of $QH^*\mathbb{F}_2 = \mathbb{C}[b_1, b_2, q_1, q_2]/(b_1^2 - q_1q_2, b_2(b_2 - 2b_1) - q_2(1-q_1))$.

For $\mathbb{P}(1,1,2)$, the (orbifold) quantum cohomology D-module was calculated in [7]. The (orbifold) cohomology group $H^2_{\text{orbi}}\mathbb{P}(1,1,2)$ contains an obvious 'hyperplane class' b. With respect to this, one has

$$\mathcal{M}^{\mathbb{P}(1,1,2)} = D_t^{\hbar}/(P),$$

where $P = (\hbar\partial)^4 - \frac{1}{2}\hbar(\hbar\partial)^3 - \frac{1}{4}q$.

Now, $H^2_{\text{orbi}}\mathbb{P}(1,1,2)$ has rank two; it has another additive generator called $\mathbf{1}_{\frac{1}{2}}$, which arises from the orbifold structure at the singular point $[0,0,1]$. The definitions of orbifold cohomology and orbifold quantum cohomology are substantial generalizations of the non-orbifold case, and we shall not discuss them here. However, the available evidence suggests that the orbifold quantum differential equations behave in a similar way to those in the non-orbifold case. In particular, the orbifold Gromov–Witten invariants of weighted projective space may be extracted by the method of Section 2.2 – see [20]. The canonical[6] bases of the quantum D-modules $\mathcal{M}^{\mathbb{F}_2}$, $\mathcal{M}^{\mathbb{P}(1,1,2)}$, with their corresponding cohomology bases, are as follows:

$$\begin{array}{c|l} 1 & 1 \\ b_1 & \hbar\partial_1 \\ b_2 & \hbar\partial_2 \\ b_1b_2 & \hbar\partial_1\hbar\partial_2 - q_1q_2 \end{array} \qquad \begin{array}{c|l} 1 & 1 \\ b & \hbar\partial \\ b^2 & (\hbar\partial)^2 \\ \mathbf{1}_{\frac{1}{2}} & 2q^{-1/2}(\hbar\partial)^3. \end{array}$$

Theorem 2.4.3 [6] *The orbifold quantum D-module $\mathcal{M}^{\mathbb{P}(1,1,2)}$ is obtained from the quantum D-module $\mathcal{M}^{\mathbb{F}_2}$ by setting $(q_1, q_2) = (-1, iq^{1/2})$. This is a natural identification in which the basis $1, b_1, b_2, b_1b_2$ of $H^2\mathbb{F}_2$ corresponds to the basis $1, b - i\mathbf{1}_{\frac{1}{2}}, 2b, 2b^2$ of $H^2\mathbb{P}(1,1,2)$.*

We can derive this easily (with hindsight) by expressing the D-module in the form given at the end of Section 2.3. We begin by computing expressions for the powers of $\hbar\partial_2$ by differentiating the relations F_1, F_2:

(a) $(\hbar\partial_2)^2 \equiv 2\hbar\partial_1\hbar\partial_2 + q_2(1-q_1)$,

(b) $(\hbar\partial_2)^3 \equiv (3q_1q_2 + q_2)\hbar\partial_2 + 2q_2(1-q_1)\hbar\partial_1 + \hbar q_2(1+q_1)$,

(c) $(\hbar\partial_2)^4 \equiv 2q_2(1+q_1)(\hbar\partial_2)^2 + \hbar(\hbar\partial_2)^3 + \hbar q_2(1+q_1)\hbar\partial_2 - q_2^2(1-q_1)^2$.

From (a) and (b) we see that $[1]$, $[\hbar\partial_2]$, $[(\hbar\partial_2)^2]$, $[(\hbar\partial_2)^3]$ are linearly independent; they form a basis of $\mathcal{M}^{\mathbb{F}_2}$. The fourth-order relation

$$T_1 = (\hbar\partial_2)^4 - 2q_2(1+q_1)(\hbar\partial_2)^2 - \hbar(\hbar\partial_2)^3 - \hbar q_2(1+q_1)\hbar\partial_2 + q_2^2(1-q_1)^2$$

[6] Canonical basis means a basis constructed from a monomial basis by the canonical procedure of Theorem 2.2.7.

is given by (c). From (b) we obtain

$$\hbar\partial_1 \equiv \frac{1}{2q_2(1-q_1)}\left((\hbar\partial_2)^3 - (3q_1q_2 + q_2)\hbar\partial_2 - \hbar q_2(1+q_1)\right) = P \text{ (say)}.$$

This gives a relation $T_2 = \hbar\partial_1 - P$. For dimensional reasons we must have

$$\mathcal{M}^{\mathbb{F}_2} = D^{\hbar}_{t_1,t_2}/(F_1, F_2) = D^{\hbar}_{t_1,t_2}/(T_1, T_2).$$

Let us now put $(q_1, q_2) = (-1, iq^{1/2})$. From $q_2 = iq^{1/2}$, we see that the operator ∂_2 restricts to 2∂, so the operator T_1 restricts to

$$16\left((\hbar\partial)^4 - \tfrac{1}{2}\hbar(\hbar\partial)^3 - \tfrac{1}{4}q\right),$$

which is the quantum differential operator of $\mathcal{M}^{\mathbb{P}(1,1,2)}$, up to a scalar multiple. Thus, we have exhibited $\mathcal{M}^{\mathbb{F}_2}$ as a t_1-extension of $\mathcal{M}^{\mathbb{P}(1,1,2)}$, in the manner of Section 2.3.

It remains to extract the relation between the canonical bases from this description. Under the specialization of variables, we have already seen that $\hbar\partial_2$ restricts to $2\hbar\partial$. Next, formula (b) shows that the operator $\hbar\partial_1$ restricts to

$$\frac{1}{4iq^{1/2}}\left((2\hbar\partial)^3 + 4iq^{1/2}\hbar\partial\right) = -2iq^{-1/2}(\hbar\partial)^3 + \hbar\partial.$$

Finally, because $(\hbar\partial_2)^2$ restricts to $4(\hbar\partial)^2$, we see from the relation F_2 that $\hbar\partial_1\hbar\partial_2$ restricts to $2(\hbar\partial)^2 - iq^{1/2}$, and hence $\hbar\partial_1\hbar\partial_2 - q_1q_2$ restricts to $2(\hbar\partial)^2$. From the preceding table, we read off that the basis elements $1, b_1, b_2, b_1b_2$ correspond to $1, b - i\mathbf{1}_{\frac{1}{2}}, 2b, 2b^2$, as required.

The correspondence between the bases may be justified geometrically, by examining the map $\mathbb{F}_2 \to \mathbb{P}(1,1,2)$, but it is remarkable that the quantum D-module contains this information implicitly – along with all the Gromov–Witten invariants of both spaces. For a recent update on the conjecture we refer to [23].

2.4.3 Harmonic maps and mirror symmetry

The second candidate for an integrable system whose solutions include quantum cohomology is the harmonic (or pluriharmonic) map equation. Small quantum cohomology is sufficient for this, but, as in the case of the WDVV equation, an entirely new direction – this time toward mirror symmetry – is required. Further details of the following discussion may be found in chapter 10 of [19].

The harmonic map equation. The equation for a harmonic map $\phi : \mathbb{R}^2 = \mathbb{C} \to G$, where G is a (compact or noncompact) Lie group, is

$$\partial_x(\phi^{-1}\partial_x\phi) + \partial_y(\phi^{-1}\partial_y\phi) = 0.$$

Writing $z = x + iy$ and $\partial = \partial/\partial z = \frac{1}{2}(\partial/\partial x - i\partial/\partial y)$, $\bar{\partial} = \partial/\partial \bar{z} = \frac{1}{2}(\partial/\partial x + i\partial/\partial y)$, the equation becomes

$$\partial(\phi^{-1}\bar{\partial}\phi) + \bar{\partial}(\phi^{-1}\partial\phi) = 0.$$

This notation assumes that G is a matrix group. If $G^{\mathbb{C}}$ is the complexification of G, and $C : G^{\mathbb{C}} \to G^{\mathbb{C}}$ is the natural conjugation[7] map, and $c : \mathfrak{g}^{\mathbb{C}} \to \mathfrak{g}^{\mathbb{C}}$ is the induced conjugation map of Lie algebras, then $\phi^{-1}\bar{\partial}\phi, \phi^{-1}\partial\phi$ take values in $\mathfrak{g}^{\mathbb{C}}$ and satisfy $c(\phi^{-1}\partial\phi) = \phi^{-1}\bar{\partial}\phi$.

The harmonic map equation can be represented as an integrable p.d.e. with spectral parameter if we introduce the $\mathfrak{g}^{\mathbb{C}}$-valued 1-form

$$\alpha = \tfrac{1}{2}(1 - \tfrac{1}{\lambda})(\phi^{-1}\partial_1\phi)dz_1 + \tfrac{1}{2}(1 - \lambda)(\phi^{-1}\partial_2\phi)dz_2,$$

where λ is a complex parameter. Namely, the connection $d + \alpha$ is flat for every (nonzero) value of λ if and only if ϕ satisfies the harmonic map equation. In fact, it is wellknown and easy to prove (see Sections 4.3 and 7.3 of [19]) that the following more general statement holds:

Proposition 2.4.4 *Let* $\alpha = \frac{1}{2}(1 - \frac{1}{\lambda})\alpha_1 dz_1 + \frac{1}{2}(1 - \lambda)\alpha_2 dz_2$ *be a* $\mathfrak{g}^{\mathbb{C}}$*-valued 1-form on* $\mathbb{C}^2$ *(or a simply connected open subset of* $\mathbb{C}^2$*). If* $d + \alpha$ *is flat for every (nonzero) value of* λ*, then there exists a map* $\phi : \mathbb{C}^2 \to G^{\mathbb{C}}$ *such that* $\alpha_1 = \phi^{-1}\partial_1\phi, \alpha_2 = \phi^{-1}\partial_2\phi$*, and this map satisfies the equation*

$$\partial_1(\phi^{-1}\partial_2\phi) + \partial_2(\phi^{-1}\partial_1\phi) = 0.$$

Conversely, let $\phi : \mathbb{C}^2 \to G^{\mathbb{C}}$ *be a map which satisfies the equation* $\partial_1(\phi^{-1}\partial_2\phi) + \partial_2(\phi^{-1}\partial_1\phi) = 0$*. Then* $\alpha = \frac{1}{2}(1 - \frac{1}{\lambda})(\phi^{-1}\partial_1\phi)dz_1 + \frac{1}{2}(1 - \lambda)(\phi^{-1}\partial_2\phi)dz_2$ *defines a flat connection* $d + \alpha$*.*

This remains true when the reality conditions $z_1 = z$, $z_2 = \bar{z}$, $c(\alpha_1) = \alpha_2$ *are imposed, giving the harmonic map equation (on* $\mathbb{C}$*, or a simply connected open subset of* $\mathbb{C}$*).*

The spectral parameter here plays a crucial role, because α may be regarded as a 1-form taking values in the based loop algebra $\Omega\mathfrak{g}$, and the flatness condition implies (Theorem 2.1.5) that $\alpha = \Phi^{-1}d\Phi$ for some map $\Phi : N \to \Omega G$ (on a simply connected open subset N of $\mathbb{C}$). This Φ is called an 'extended solution', or 'extended harmonic map'. Moreover, the shape of α implies that Φ is holomorphic with respect to the natural complex structure of the based loop group ΩG. Because this complex structure may be obtained from an identification of ΩG with an open subset[8] of $\Lambda G^{\mathbb{C}}/\Lambda_+ G^{\mathbb{C}}$, it follows that $\Phi = [L]$ for some holomorphic map

[7] If $G = U_n$, then $G^{\mathbb{C}} = GL_n(\mathbb{C})$, and $C : G^{\mathbb{C}} \to G^{\mathbb{C}}$, $c : \mathfrak{g}^{\mathbb{C}} \to \mathfrak{g}^{\mathbb{C}}$ are given respectively by $C(A) = A^{*-1}$, $c(A) = -A^*$.

[8] See Section 8.8 of [19]. If G is compact, ΩG may be identified with $\Lambda G^{\mathbb{C}}/\Lambda_+ G^{\mathbb{C}}$.

$L : N \to \Lambda G^{\mathbb{C}}$, i.e., $L = \Phi B$ for some (smooth) $B : N \to \Lambda_+ G^{\mathbb{C}}$. This L is of course not unique, but there is a canonical choice for it, obtained from the Birkhoff factorization $\Phi = \Phi_- \Phi_+$ and taking $L = \Phi_-$, $B = \Phi_+^{-1}$. It can then be shown (Section 7.3 of [19]) that L satisfies an equation of the form

$$L^{-1} dL = \tfrac{1}{\lambda} \omega$$

where ω is a holomorphic $\mathfrak{g}^{\mathbb{C}}$-valued 1-form on N. Conversely, if ω is *any* holomorphic $\mathfrak{g}^{\mathbb{C}}$-valued 1-form, then $\frac{1}{\lambda}\omega$ is of the form $L^{-1}dL$ (because ω depends only on a single variable z, and is therefore flat). From the Iwasawa factorization $L = L_{\mathbb{R}} L_+$, it is easy to show that the map given by $\Phi = L_{\mathbb{R}}$ is an extended harmonic map. This correspondence

$$\phi \longleftrightarrow \omega$$

between harmonic maps ϕ and 'unrestricted holomorphic data' ω is known as the DPW correspondence, or generalized Weierstrass representation. Further details can be found in Section 7.3 of [19] or the original paper [12].

The D-module. The harmonic map equation in the form $L^{-1}dL = \frac{1}{\lambda}\omega$ can be rewritten in an illuminating way if we make use of the Grassmannian model of ΩG. This is an identification

$$\Omega G \cong \mathrm{Gr}^{\mathfrak{g}}$$

of ΩG with a certain infinite-dimensional Grassmannian manifold (Section 8.6 of [28]). The holomorphic map $\Phi = [L] : N \to \Omega G$ corresponds to a holomorphic map $W : N \to \mathrm{Gr}^{\mathfrak{g}}$, and hence to a holomorphic vector bundle $W^* \mathcal{T}$ where $\mathcal{T}$ is the tautologous vector bundle on $\mathrm{Gr}^{\mathfrak{g}}$. The harmonic map equation $L^{-1}dL = \frac{1}{\lambda}\omega$ can then be written as

$$\lambda \partial_z \, \Gamma W^* \mathcal{T} \subset \Gamma W^* \mathcal{T},$$

where Γ denotes the space of holomorphic sections. This says that $\Gamma W^* \mathcal{T}$ has a D-module structure: it is acted upon by the ring D_z^λ of differential operators which is generated by $\lambda \partial_z$ and whose elements have coefficients which are holomorphic in $z \in N$ and holomorphic in λ in a neighbourhood of $\lambda = 0$. In the case $G = \mathrm{U}_n$ this is explained in detail in Section 8.2 of [19].

This D-module does not generally have a distinguished cyclic generator, although some examples with distinguished cyclic generators can be found in [16] (Proposition 2.3 and Theorem 2.4). We shall focus on one particular kind of harmonic map which arises from quantum cohomology, where the D-module may be identified with the quantum D-module.

Quantum cohomology as a (pluri)harmonic map. To explain the link with quantum cohomology, two extensions are needed (Sections 7.4 and 7.5 of [19]).

First, the theory applies also to pluriharmonic maps $\phi : \mathbb{C}^r \to G$, whose equations have a similar zero-curvature form. However, when $r > 1$, the holomorphic data $\omega = \sum_{i=1}^r \omega_i dz_i$ is no longer 'unrestricted'; it is subject to the nontrivial flatness condition $d\omega = \omega \wedge \omega = 0$. Second, the theory applies to harmonic (or pluriharmonic) maps $\phi : \mathbb{C} \to G/K$ where G/K is a symmetric space. Here the 1-form α looks simpler, as it can be written

$$\alpha = (\alpha_1^{\mathfrak{k}} + \tfrac{1}{\lambda}\alpha_1^{\mathfrak{m}})dz + (\alpha_2^{\mathfrak{k}} + \lambda\alpha_2^{\mathfrak{m}})d\bar{z},$$

where $\alpha_i = \alpha_i^{\mathfrak{k}} + \alpha_i^{\mathfrak{m}}$ denotes the eigenspace decomposition of the involution $\sigma : \mathfrak{g}^{\mathbb{C}} \to \mathfrak{g}^{\mathbb{C}}$ which defines the symmetric space.

The first main observation is that the map $L : N \to \mathrm{GL}_{s+1}(\mathbb{C})$ of quantum cohomology (where N is an open subset of $H^2M \cong \mathbb{C}^r$) is exactly of the mentioned form, that is, it satisfies $L^{-1}dL = \frac{1}{\lambda}\omega$ where ω is given by the quantum products. Moreover, ω takes values in $\mathfrak{m}^{\mathbb{C}}$, the (-1)-eigenspace of a certain natural involution σ on $\mathfrak{g}^{\mathbb{C}} = \mathfrak{gl}_{s+1}(\mathbb{C})$ (this fact corresponds to the Frobenius property of the quantum product). By the preceding general theory, it follows that the quantum cohomology of M defines a pluriharmonic map into a symmetric space G/K, where G is any real form of $\mathrm{GL}_{s+1}(\mathbb{C})$. One natural real form[9] is $\mathrm{GL}_{s+1}(\mathbb{R})$, corresponding to the cohomology with real coefficients $H^*(M;\mathbb{R})$, and this gives the symmetric space $\mathrm{GL}_{s+1}(\mathbb{R})/\mathrm{O}_{s+1}$.

The second main observation concerning quantum cohomology – and the link with mirror symmetry – is that in certain situations this (pluri)harmonic map has an independent geometrical interpretation, as the period map for a variation of Hodge structure. The most famous example is the quintic threefold M in $\mathbb{C}P^4$. The harmonic map obtained from the quantum cohomology of M can be described simply as follows: For a certain holomorphic $\mathbb{C}^4$-valued function u, consider the holomorphic map

$$U = \mathrm{Span}\{u\} \subseteq \mathrm{Span}\{u, u'\} \subseteq \mathrm{Span}\{u, u', u''\} \subseteq \mathbb{C}^4$$

to the flag manifold $\mathrm{SU}_4/\mathrm{S}(\mathrm{U}_1 \times \mathrm{U}_1 \times \mathrm{U}_1 \times \mathrm{U}_1)$. This flag manifold can be identified with the space of quadruples (L_1, L_2, L_3, L_4) of mutually orthogonal complex lines in the Hermitian space $\mathbb{C}^4$, and it is well-known (cf. Example 8.16 of [19]) that the composition of any map U of the displayed form with the projection map $(L_1, L_2, L_3, L_4) \mapsto L_1 \oplus L_3$ is a harmonic map into the symmetric space $\mathrm{Gr}_2\mathbb{C}^4 = \mathrm{SU}_4/\mathrm{S}(\mathrm{U}_2 \times \mathrm{U}_2)$. In terms of the general theory of Section 2.2, the map u can be identified with the J-function, and the map U with L, if the flag manifold

[9] In Chapter 10 of [19], an (indefinite) unitary group based on $H^*(M;\mathbb{R})$ was used. Any choice gives a pluriharmonic map, so the "best" choice depends on imposing further criteria. Further discussion of this point, in particular the relation with [22], can be found in Section 6 of [11].

is embedded suitably in the loop group $\Omega \mathrm{SU}_4$. (This is consistent with the above choice of symmetric space $\mathrm{GL}_4(\mathbb{R})/\mathrm{O}_4$, because U actually takes values in the smaller symmetric space $\mathrm{Sp}_4(\mathbb{R})/\mathrm{U}_2$, a symplectic Grassmannian.) Mirror symmetry says that U can be identified with the period map

$$H^{3,0} \subseteq H^{3,0} \oplus H^{2,1} \subseteq H^{3,0} \oplus H^{2,1} \oplus H^{1,2} \subseteq H^{3,0} \oplus H^{2,1} \oplus H^{1,2} \oplus H^{0,3},$$

where $H^{i,j} = H^{i,j}\tilde{M}$ for a 'mirror partner' $\tilde{M}$ of M. The domain of this map is (an open subset of) the moduli space of complex structures of $\tilde{M}$, as $H^{i,j}\tilde{M}$ depends on the complex structure.

It is a special feature of Calabi–Yau manifolds (such as the quintic threefold) that the harmonic map can be described in this way, in elementary terms, without using loop group theory. However, for the quantum cohomology of Fano manifolds (such as $\mathbb{C}P^n$), the map L does not factor through a finite-dimensional submanifold of the loop group. It does still have a variation of Hodge structure interpretation, in a generalized sense (due to Barannikov [3,4] and Katzarkov et al. [25]), because of the Grassmannian model of the loop group: instead of U, we use the holomorphic map W (associated to L) which was described above.

The correspondence

$$\phi \text{ (variation of Hodge structure)} \longleftrightarrow \omega \text{ (quantum cohomology)}$$

between the pluriharmonic map ϕ and the holomorphic data ω can be regarded as an expression of mirror symmetry. It is given explicitly by the Birkhoff and Iwasawa loop factorizations:

$$\Phi \text{ (or } \phi) \quad \underset{\xleftarrow{\text{Iwasawa}}}{\xrightarrow{\text{Birkhoff}}} \quad L \text{ (or } \omega)$$

In this way, the quantum D-module contains not only the geometric information consisting of the Gromov–Witten invariants, but also the much less visible geometric information consisting of the variation of Hodge structure of the mirror partner.

2.5 Conclusion

Much remains to be done to clarify the integrable systems aspects of quantum cohomology, but an even more elusive goal is to *characterize* quantum cohomology in purely differential-equation-theoretic terms. The quantum D-module will attain the status of de Rham cohomology (for example) only if those D-modules which occur as quantum D-modules can be described precisely. This goal is probably too optimistic, but one can at least make a start by listing some conditions, such as:

– quantization of a commutative algebra
– regular singular point of maximal unipotent monodromy at $q = 0$
– homogeneity
– self-adjointness.

In Section 2.2 we have focused on the first of these, so let us comment briefly on the others, taking the case of M_5^3 (Examples 2.2.4, 2.2.6 and 2.2.8) as a concrete example. The quantum differential operator here is

$$(\hbar\partial)^4 - 27q(\hbar\partial)^2 - 27\hbar q(\hbar\partial) - 6\hbar^2 q.$$

The second condition has the usual meaning from o.d.e. theory. The third means that the operator is weighted homogeneous, the weights of the symbols $\partial, \hbar, q$ being $0, 2, 4$, respectively. So far, any quantization of the quantum cohomology relation $b^4 - 27qb^2$ of the form $(\hbar\partial)^4 - 27q(\hbar\partial)^2 - \alpha\hbar q(\hbar\partial) - \beta\hbar^2 q$ would have all these properties, where α, β are constants. The fourth condition means that the operator is formally self-adjoint with respect to the involution defined by $\partial^* = -\partial$, $\hbar^* = -\hbar$ (see Section 6.3 of [19]). This condition forces α to be 27. However, only the value $\beta = 6$ gives the correct quantum products (or Gromov–Witten invariants), and our conditions do not pin this down; we need more, and these are not going to be straightforward.

One source of additional conditions is the global behaviour of the associated (pluri)harmonic map, regarded as a generalized period map. There is a positivity condition which generalizes the second Riemann–Hodge bilinear relation, and it is natural to insist on this. Some ideas in this direction can be found in [22] (see also [11] for a particular example). A related source is the arithmetic behaviour of the differential equation. Even in the Calabi–Yau case where the second Riemann–Hodge bilinear relation is known to hold, it is difficult to characterize differential equations whose solutions have the expected integrality properties (related to the integrality or rationality of the Gromov–Witten invariants) — see, e.g., [1] and [2].

Bibliography

[1] G. Almkvist, C. van Enckevort, D. van Straten, and W. Zudilin, *Tables of Calabi–Yau equations*, preprint, math.AG/0507430.
[2] G. Almkvist and W. Zudilin, *Differential equations, mirror maps and zeta values*, Mirror symmetry V, AMS/IP Stud. Adv. Math. 38, Amer. Math. Soc., 2006, pp. 481–515, math.NT/0402386.
[3] S. Barannikov, *Quantum periods. I. Semi-infinite variations of Hodge structures*, Internat. Math. Res. Notices **2001-23** (2001), 1243–1264.
[4] S. Barannikov, *Non-commutative periods and mirror symmetry in higher dimensions*, Comm. Math. Phys. **228** (2002), 281–325.
[5] J. Björk, *Rings of Differential Operators*, North-Holland, 1979.

[6] T. Coates, H. Iritani, and H.-H. Tseng, *Wall-crossings in toric Gromov–Witten theory I: crepant examples*, preprint, math.AG/0611550.
[7] T. Coates, A. Corti, Y.-P. Lee, and H.-H. Tseng, *The quantum orbifold cohomology of weighted projective space*, preprint, math.AG/0608481.
[8] A. Collino and M. Jinzenji, *On the structure of the small quantum cohomology rings of projective hypersurfaces*, Comm. Math. Phys. **206** (1999), 157–183.
[9] D. A. Cox and S. Katz, *Mirror Symmetry and Algebraic Geometry*, Math. Surveys and Monographs 68, Amer. Math. Soc., 1999.
[10] P. Di Francesco and C. Itzykson, *Quantum intersection rings*, The Moduli Space of Curves, Prog. Math. 129, eds. R. H. Dijkgraaf et al., Birkhäuser, 1995, pp. 81–148.
[11] J. Dorfmeister, M. Guest, and W. Rossman, *The tt^* structure of the quantum cohomology of $\mathbb{C}P^1$ from the viewpoint of differential geometry*, preprint, arXiv:0905.3876.
[12] J. Dorfmeister, F. Pedit, and H. Wu, *Weierstrass type representations of harmonic maps into symmetric spaces*, Comm. Anal. Geom. **6** (1998), 633–668.
[13] A. B. Givental, *Homological geometry and mirror symmetry*, Proc. Int. Congress of Math. I, Zürich 1994, ed. S. D. Chatterji, Birkhäuser, 1995, pp. 472–480.
[14] A. B. Givental, *Homological geometry I. Projective hypersurfaces*, Selecta Math. **1** (1995), 325–345.
[15] A. B. Givental, *Equivariant Gromov–Witten invariants*, Internat. Math. Res. Notices **1996-13** (1996), 1–63, alg-geom/9603021.
[16] M. A. Guest, *An update on harmonic maps of finite uniton number, via the zero curvature equation*, Contemp. Math. **309** (2002), 85–113.
[17] M. A. Guest, *Quantum cohomology via D-modules*, Topology **44** (2005), 263–281, math.DG/0206212.
[18] M. A. Guest, *Introduction to homological geometry: I,II*, Integrable Systems, Geometry, and Topology, ed. C.-L. Terng, AMS/IP Studies in Advanced Mathematics 36, Amer. Math. Soc. and International Press, 2006, pp. 83–121, 123–150, math.DG/0104274, math.DG/0105032.
[19] M. A. Guest, *From Quantum Cohomology to Integrable Systems*, Oxford Univ. Press, 2008.
[20] M. A. Guest and H. Sakai, *Orbifold quantum D-modules associated to weighted projective spaces*, preprint, arXiv:0810.4236.
[21] C. Hertling and Y. Manin, *Unfoldings of meromorphic connections and a construction of Frobenius manifolds*, Frobenius manifolds, Quantum cohomology and singularities, Aspects of Math. E 36, eds. C. Hertling et al., Vieweg, 2004, pp. 113–144, math.AG/0207089.
[22] H. Iritani, *Real and integral structures in quantum cohomology I: toric orbifolds*, preprint, arXiv:0712.2204.
[23] H. Iritani, *Ruan's conjecture and integral structures in quantum cohomology*, preprint, arXiv:0809.2749.
[24] M. Jinzenji, *Gauss–Manin system and the virtual structure constants*, Internat. J. Math **13** (2002), 445–477.
[25] L. Katzarkov, M. Kontsevich, and T. Pantev, *Hodge theoretic aspects of mirror symmetry*, preprint, arXiv:0806.0107.
[26] M. Kontsevich and Y. Manin, *Gromov–Witten classes, quantum cohomology, and enumerative geometry*, Commun. Math. Phys. **164** (1994), 525–562, hep-th/9402147.
[27] F. Pham, *Singularités des systèmes différentiels de Gauss-Manin*, Progr. Math. 2, Birkhäuser, 1979.

[28] A. N. Pressley and G. B. Segal, *Loop Groups*, Oxford Univ. Press, 1986.
[29] M. van der Put, and M. F. Singer, *Galois Theory of Linear Differential Equations*, Grundlehren der Mathematischen Wissenschaften 328, Springer, 2003.
[30] B. Siebert and G. Tian, *On quantum cohomology rings of Fano manifolds and a formula of Vafa and Intriligator*, Asian J. Math. **1** (1997), 679–695, alg-geom/9403010.

3

Index theory and groupoids

CLAIRE DEBORD AND JEAN-MARIE LESCURE*

Abstract

This chapter is mainly devoted to a proof, using groupoids and KK-theory, of Atiyah and Singer's index theorem on compact smooth manifolds. We first present an elementary introduction to groupoids, C^*-algebras, KK-theory and pseudodifferential calculus on groupoids. We then show how the point of view adopted here generalizes to the case of conical pseudomanifolds.

3.1 Introduction

This chapter is meant to give the tools involved in our approach to index theory for singular spaces. The global framework adopted here is noncommutative geometry, with a particular focus on groupoids, C^*-algebras and bivariant K-theory.

The idea of using C^*-algebras to study *spaces* may be understood with the help of the Gelfand theorem, which asserts that Hausdorff locally compact spaces are in one-to-one correspondence with commutative C^*-algebras. A starting point in noncommutative geometry is then to think of noncommutative C^*-algebras as corresponding to a wider class of spaces, more singular than Hausdorff locally compact spaces. As a first consequence, given a geometrical or topological object which is badly behaved with respect to classical tools, noncommutative geometry suggests defining a C^*-algebra encoding relevant information carried by the original object.

* We would like to thank Georges Skandalis, who allowed us to use several of his works, in particular the manuscript of one of his courses [48, 49]. We would like to warmly thank Jorge Plazas for having typewritten a part of this chapter during the summer school, and Jérôme Chabert, who carefully read the chapter and corrected several mistakes. We are grateful to all the organizers for their kind invitation to the extremely stimulating summer school held at Villa de Leyva in July 2007, and we particularly thank Sylvie Paycha, both as an organizer and for her valuable comments on this document.

Refining this construction, one may try to define this C^*-algebra as the C^*-algebra of a groupoid [46, 47]. That is, one can try to build a groupoid directly, encoding the original object and regular enough to allow the construction of its C^*-algebra. In the ideal case where the groupoid is smooth, one gets much more than a C^*-algebra, which only reflects topological properties: the groupoid has a geometrical and analytical flavor enabling many applications.

An illuminating example is the study of the space of leaves of a foliated manifold $(M, \mathcal{F})$ [10, 11, 14]. Although this space $M/\mathcal{F}$ is usually very singular, the holonomy groupoid of the foliation leads to a C^*-algebra $C^*(M, \mathcal{F})$ replacing with great success the algebra of continuous functions on the space $M/\mathcal{F}$. Moreover, the holonomy groupoid is smooth and characterizes the original foliation.

Once a C^*-algebra is built for the study of a given problem, one can look for *invariants* attached to it. For ordinary spaces, basic invariants live in the homology or cohomology of the space. When dealing with C^*-algebras, the suitable homology theory is K-theory, or better the KK-theory developed by G. Kasparov [30, 31, 49] (when a smooth subalgebra of the C^*-algebra is specified, which for instance is the case if a smooth groupoid is available, one may also consider cyclic (co)homology, but this theory is beyond the scope of these notes).

There is a fundamental theory which links the previous ideas, namely index theory. In the 1960s, Atiyah and Singer [6] proved their famous index theorem. Roughly speaking, they showed that, given a closed manifold, one can associate to any elliptic operator an integer called the *index*, which can be described in two different ways: one purely analytic and the other purely topological. This result is stated with the help of K-theory of spaces. Hence, using the Swan–Serre theorem, it can be formulated with K-theory of (commutative) C^*-algebras. This point, and the fact that the index theorem can be proved in many ways using K-theoretic methods, leads to the attempt to generalize it to more *singular* situations where appropriate C^*-algebras are available. Noncommutative geometry therefore offers a general framework in which one can try to state and prove index theorems. The case of foliations illustrates this perfectly again: elliptic operators along the leaves, equivariant with respect to the holonomy groupoid, admit an analytical index living in the K-theory of the C^*-algebra $C^*(M, \mathcal{F})$. Moreover, this index can also be described in a topological way, and this is the content of the index theorem for foliations of Connes and Skandalis [14].

Connes [13] also observed the important role played by groupoids in the definition of the index map: in both cases of closed manifolds and foliations, the analytical index map can be described with the use of a groupoid, namely a *deformation groupoid*. This approach has been extended by the authors and Nistor [20], who showed that the topological index of Atiyah and Singer can also be described using deformation groupoids. This leads to a *geometrical proof* of the index theorem of Atiyah and Singer; moreover, this proof is easily applied to a class of singular spaces (namely, pseudomanifolds with isolated singularities).

The content of this chapter is divided into three parts. Let us briefly describe them:

Part I: Groupoids and their C^*-algebras. As mentioned earlier, the first problem in the study of a singular geometrical situation is to associate to it a mathematical object which carries the information one wants to study and which is regular enough to be analyzed in a reasonable way. In noncommutative geometry, answering this question amounts to looking for a good *groupoid* and constructing its *C^*-algebra*. These points will be the subject of Sections 3.2 and 3.4.

Part II: KK-theory. Once the situation is desingularized, say through the construction of a groupoid and its C^*-algebra, one may look for invariants which capture the basic properties. Roughly speaking, the *KK-theory* groups are convenient groups of invariants for C^*-algebras, and KK-theory comes with powerful tools to carry out computations. Kasparov's bivariant K-theory will be the main topic of Sections 3.4 to 3.6.

Part III: Index theorems. We first briefly explain in Section 3.7 the pseudodifferential calculus on groupoids. Then, in Section 3.8, we give a geometrical proof of the Atiyah–Singer index theorem for closed manifolds, using the language of groupoids and KK-theory. Finally we show in the last section how these results can be extended to *conical pseudomanifolds*.

Prerequisites. The reader interested in this course should have background in several domains. Familiarity with basic differential geometry (manifolds, tangent spaces) is needed. The notions of fiber bundle and of K-theory for locally compact spaces should be known. Basic functional analysis, including the analysis of linear operators on Hilbert spaces, should be familiar. The knowledge of pseudodifferential calculus (basic definitions, ellipticity) is necessary. Although it is not absolutely necessary, some familiarity with C^*-algebras is preferable.

I. Groupoids and their C^*-Algebras

This first part will be devoted to the notion of groupoid, specifically that of differentiable groupoid. We provide definitions and consider standard examples. The interested reader may look for example at [12,35]. We then recall the definition of C^*-algebras and see how one can associate a C^*-algebra to a groupoid. The theory of C^*-algebras of groupoids was initiated by Jean Renault [46]. A good reference for the construction of groupoid C^*-algebras is [32], by which the end of Section 3.3.2 is inspired.

3.2 Groupoids

3.2.1 Definitions and basic examples of groupoids

Definition 3.2.1 Let G and $G^{(0)}$ be two sets. A *groupoid* structure on G over $G^{(0)}$ is given by the following homomorphisms:

- An injective map $u : G^{(0)} \to G$. The map u is called the *unit map*. We often identify $G^{(0)}$ with its image in G. The set $G^{(0)}$ is called the *set of units* of the groupoid.
- Two surjective maps: $r, s : G \to G^{(0)}$, which are, respectively, the *range* and *source* maps. They are equal to the identity on the space of units.
- An involution

$$i : G \to G,$$
$$\gamma \mapsto \gamma^{-1},$$

called the *inverse* map. It satisfies $s \circ i = r$.
- A map

$$p : G^{(2)} \to G,$$
$$(\gamma_1, \gamma_2) \mapsto \gamma_1 \cdot \gamma_2,$$

called the *product*, where the set

$$G^{(2)} := \{(\gamma_1, \gamma_2) \in G \times G \mid s(\gamma_1) = r(\gamma_2)\}$$

is the set of *composable pairs*. Moreover, for $(\gamma_1, \gamma_2) \in G^{(2)}$ we have $r(\gamma_1 \cdot \gamma_2) = r(\gamma_1)$ and $s(\gamma_1 \cdot \gamma_2) = s(\gamma_2)$.

The following properties must be fulfilled:

- The product is associative: for any $\gamma_1,\ \gamma_2,\ \gamma_3$ in G such that $s(\gamma_1) = r(\gamma_2)$ and $s(\gamma_2) = r(\gamma_3)$ the following equality holds:

$$(\gamma_1 \cdot \gamma_2) \cdot \gamma_3 = \gamma_1 \cdot (\gamma_2 \cdot \gamma_3).$$

- For any γ in G, one has $r(\gamma) \cdot \gamma = \gamma \cdot s(\gamma) = \gamma$ and $\gamma \cdot \gamma^{-1} = r(\gamma)$.

A groupoid structure on G over $G^{(0)}$ is usually denoted by $G \rightrightarrows G^{(0)}$, where the arrows stand for the source and target maps.

We will often use the following notation:

$$G_A := s^{-1}(A), \qquad G^B = r^{-1}(B) \quad \text{and} \quad G_A^B = G_A \cap G^B.$$

If x belongs to $G^{(0)}$, the s-*fiber* (r-*fiber*) of G over x is $G_x = s^{-1}(x)$ ($G^x = r^{-1}(x)$).

The groupoid is *topological* when G and $G^{(0)}$ are topological spaces with $G^{(0)}$ Hausdorff, the structural homomorphisms are continuous, and i is a homeomorphism. We will often require that our topological groupoids be *locally compact*.

This means that $G \rightrightarrows G^{(0)}$ is a topological groupoid, such that G is second countable, each point γ in G has a compact (Hausdorff) neighborhood, and the map s is open. In this situation the map r is open and the s-fibers of G are Hausdorff.

The groupoid is *smooth* when G and $G^{(0)}$ are second-countable smooth manifolds with $G^{(0)}$ Hausdorff, the structural homomorphisms are smooth, u is an embedding, s is a submersion, and i is a diffeomorphism.

When $G \rightrightarrows G^{(0)}$ is at least topological, we say that G is *s-connected* when for any $x \in G^{(0)}$ the s-fiber of G over x is connected. The s-connected component of a groupoid G is $\bigcup_{x \in G^{(0)}} CG_x$, where CG_x is the connected component of the s-fiber G_x which contains the unit $u(x)$.

Examples

1. A space X is a groupoid over itself with $s = r = u = \mathrm{Id}$.
2. A group $G \rightrightarrows \{e\}$ is a groupoid over its unit e, with the usual product and inverse map.
3. A group bundle : $\pi : E \to X$ is a groupoid $E \rightrightarrows X$ with $r = s = \pi$ and algebraic operations given by the group structure of each fiber E_x, $x \in X$.
4. If $\mathcal{R}$ is an equivalence relation on a space X, then the graph of $\mathcal{R}$,

$$G_{\mathcal{R}} := \{(x, y) \in X \times X \mid x\mathcal{R}y\},$$

admits a structure of groupoid over X, which is given by

$$u(x) = (x, x), \qquad s(x, y) = y, \qquad r(x, y) = x,$$
$$(x, y)^{-1} = (y, x), \qquad (x, y) \cdot (y, z) = (x, z)$$

for $x,\ y,\ z$ in X. When $x\mathcal{R}y$ for any x, y in X, $G_{\mathcal{R}} = X \times X \rightrightarrows X$ is called the *pair groupoid*.

5. If G is a group acting on a space X, the *groupoid of the action* is $G \times X \rightrightarrows X$ with the following structural homomorphisms:

$$u(x) = (e, x), \qquad s(g, x) = x, \qquad r(g, x) = g \cdot x,$$
$$(g, x)^{-1} = (g^{-1}, g \cdot x), \qquad (h, g \cdot x) \cdot (g, x) = (hg, x)$$

for x in X and $g,\ h$ in G.

6. Let X be a topological space. The *homotopy groupoid* of X is

$$\Pi(X) := \{\bar{c} \mid c : [0, 1] \to X \text{ a continuous path}\} \rightrightarrows X,$$

where $\bar{c}$ denotes the homotopy class (with fixed endpoints) of c. We let

$$u(x) = \overline{c_x},$$

where c_x is the constant path equal to x,

$$s(\overline{c}) = c(0), \qquad r(\overline{c}) = c(1), \qquad \overline{c}^{-1} = \overline{c^{-1}},$$

where $c^{-1}(t) = c(1-t)$,

$$\overline{c_1} \cdot \overline{c_2} = \overline{c_1 \cdot c_2},$$

where $c_1 \cdot c_2(t) = c_2(2t)$ for $t \in [0, \frac{1}{2}]$ and $c_1 \cdot c_2(t) = c_1(2t-1)$ for $t \in [\frac{1}{2}, 1]$. When X is a smooth manifold of dimension n, $\Pi(X)$ is naturally endowed with a smooth structure (of dimension $2n$). A neighborhood of $\bar{c}$ is of the form $\{\bar{c_1}\bar{c}\bar{c_0} \mid c_1(0) = c(1),\ c(0) = c_0(1),\ \mathrm{Im}\ c_i \subset U_i,\ i = 0, 1\}$ where U_i is a given neighborhood of $c(i)$ in X.

3.2.2 Homomorphisms and Morita equivalences

3.2.2.1 Homomorphisms

Let $G \rightrightarrows G^{(0)}$ be a groupoid of source s_G and range r_G, and $H \rightrightarrows H^{(0)}$ be a groupoid of source s_H and range r_H. A groupoid *homomorphism* from G to H is given by two maps:

$$f : G \to H \quad \text{and} \quad f^{(0)} : G^{(0)} \to H^{(0)}$$

such that

- $r_H \circ f = f^{(0)} \circ r_G$,
- $f(\gamma)^{-1} = f(\gamma^{-1})$ for any $\gamma \in G$,
- $f(\gamma_1 \cdot \gamma_2) = f(\gamma_1) \cdot f(\gamma_2)$ for $\gamma_1,\ \gamma_2$ in G such that $s_G(\gamma_1) = r_G(\gamma_2)$.

We say that f is a *homomorphism over* $f^{(0)}$. When $G^{(0)} = H^{(0)}$ and $f^{(0)} = \mathrm{Id}$, we say that f is a *homomorphism over the identity*.

The homomorphism f is an *isomorphism* when the maps $f, f^{(0)}$ are bijections and $f^{-1} : H \to G$ is a homomorphism over $(f^{(0)})^{-1}$.

As usual, when dealing with topological groupoids we require that f be continuous and, when dealing with smooth groupoids, that f be smooth.

3.2.2.2 Morita equivalence

In most situations, the right notion of isomorphism of locally compact groupoids is the weaker notion of Morita equivalence.

Definition 3.2.2 Two locally compact groupoids $G \rightrightarrows G^{(0)}$ and $H \rightrightarrows H^{(0)}$ are *Morita equivalent* if there exists a locally compact groupoid $P \rightrightarrows G^{(0)} \sqcup H^{(0)}$ such that

- the restrictions of P over $G^{(0)}$ and $H^{(0)}$ are, respectively, G and H:

$$P_{G^{(0)}}^{G^{(0)}} = G \quad \text{and} \quad P_{H^{(0)}}^{H^{(0)}} = H;$$

- for any $\gamma \in P$ there exists η in $P_{G^{(0)}}^{H^{(0)}} \cup P_{H^{(0)}}^{G^{(0)}}$ such that (γ, η) is a composable pair (i.e., $s(\gamma) = r(\eta)$).

Examples

1. Let $f : G \to H$ be an isomorphism of locally compact groupoid. Then the following groupoid defines a Morita equivalence between H and G:

$$P = G \sqcup \tilde{G} \sqcup \tilde{G}^{-1} \sqcup H \rightrightarrows G^{(0)} \sqcup H^{(0)},$$

where with the obvious notation we have

$$G = \tilde{G} = \tilde{G}^{-1},$$

$$s_P = \begin{cases} s_G & \text{on } G, \\ s_H \circ f & \text{on } \tilde{G}, \\ r_G & \text{on } \tilde{G}^{-1}, \\ s_H & \text{on } H, \end{cases} \qquad r_P = \begin{cases} r_G & \text{on } G \sqcup \tilde{G}, \\ s_H \circ f & \text{on } \tilde{G}^{-1}, \\ r_H & \text{on } H, \end{cases}$$

$$u_P = \begin{cases} u_G & \text{on } G^{(0)}, \\ u_H & \text{on } H^{(0)}, \end{cases} \qquad i_P(\gamma) = \begin{cases} i_G(\gamma) & \text{on } G, \\ i_H(\gamma) & \text{on } H, \\ \gamma \in \tilde{G}^{-1} & \text{on } \tilde{G}, \\ \gamma \in \tilde{G} & \text{on } \tilde{G}^{-1}, \end{cases}$$

$$p_P(\gamma_1, \gamma_2) = \begin{cases} p_G(\gamma_1, \gamma_2) & \text{on } G^{(2)}, \\ p_H(\gamma_1, \gamma_2) & \text{on } H^{(2)}, \\ p_G(\gamma_1, \gamma_2) \in \tilde{G} & \text{for } \gamma_1 \in G,\ \gamma_2 \in \tilde{G}, \\ p_G(\gamma_1, f^{-1}(\gamma_2)) \in \tilde{G} & \text{for } \gamma_1 \in \tilde{G},\ \gamma_2 \in H, \\ p_G(\gamma_1, \gamma_2) \in G & \text{for } \gamma_1 \in \tilde{G},\ \gamma_2 \in \tilde{G}^{-1}, \\ f \circ p_G(\gamma_1, \gamma_2) \in H & \text{for } \gamma_1 \in \tilde{G},\ \gamma_2 \in \tilde{G}^{-1}. \end{cases}$$

2. Suppose that $G \rightrightarrows G^{(0)}$ is a locally compact groupoid and $\varphi : X \to G^{(0)}$ is an open surjective map, where X is a locally compact space. The *pullback groupoid* is the groupoid

$$^*\varphi^*(G) \rightrightarrows X,$$

where

$$^*\varphi^*(G) = \{(x, \gamma, y) \in X \times G \times X \mid \varphi(x) = r(\gamma) \text{ and } \varphi(y) = s(\gamma)\}$$

with $s(x, \gamma, y) = y$, $r(x, \gamma, y) = x$, $(x, \gamma_1, y) \cdot (y, \gamma_2, z) = (x, \gamma_1 \cdot \gamma_2, z)$ and $(x, \gamma, y)^{-1} = (y, \gamma^{-1}, x)$. One can show that this endows $^*\varphi^*(G)$ with a locally compact groupoid structure. Moreover, the groupoids G and $^*\varphi^*(G)$ are Morita equivalent, but not isomorphic in general. To prove this last point, one can put a locally compact groupoid structure on $P = G \sqcup X \times_r G \sqcup G \times_s X \sqcup {}^*\varphi^*(G)$ over $X \sqcup G^{(0)}$, where $X \times_r G = \{(x, \gamma) \in X \times G \mid \varphi(x) = r(\gamma)\}$ and $G \times_s X = \{(\gamma, x) \in G \times X \mid \varphi(x) = s(\gamma)\}$.

3.2.3 The orbits of a groupoid

Suppose that $G \rightrightarrows G^{(0)}$ is a groupoid of source s and range r.

Definition 3.2.3 The *orbit* of G passing trough x is the following subset of $G^{(0)}$:

$$Or_x = r(G_x) = s(G^x).$$

We let $G^{(0)}/G$ or $Or(G)$ be the *space of orbits*.

The *isotropy group* of G at x is G_x^x, which is naturally endowed with a group structure with x as unit. Notice that multiplication induces a free left (right) action of G_x^x on G^x (G_x). Moreover, the orbits space of this action is precisely Or_x and the restriction $s : G^x \to Or_x$ is the quotient map.

Examples and remarks

1. In example 4 in Section 3.2.1, the orbits of $G_{\mathcal{R}}$ correspond exactly to the orbits of the equivalence relation $\mathcal{R}$. In example 5, the orbits of the groupoid of the action are the orbits of the action.
2. The second assertion in the definition of Morita equivalence precisely means that both $G^{(0)}$ and $H^{(0)}$ meet all the orbits of P. Moreover, one can show that the map

$$Or(G) \to Or(H),$$
$$Or(G)_x \mapsto Or(P)_x \cap H^{(0)}$$

is a bijection. In other word, when two groupoids are Morita equivalent, they have the same orbit space.

Groupoids are often used in noncommutative geometry for the study of singular geometrical situations. In many geometrical situations, the topological space which arises is strongly non-Hausdorff, and the standard tools do not apply. Nevertheless, it is sometimes possible to associate to such a space X a relevant C^*-algebra as a substitute for $C_0(X)$. Usually we first associate a groupoid $G \rightrightarrows G^{(0)}$ such that its space of orbits $G^{(0)}/G$ is (equivalent to) X. If the groupoid is regular enough (smooth, for example), then we can associate natural C^*-algebras to G. This point will be discussed later. In other words, we desingularize a singular space by viewing it as coming from the action of a nice groupoid on its space of units. To illustrate this point let us consider two examples.

3.2.4 Groupoids associated to a foliation

Let M be a smooth manifold.

Definition 3.2.4 A (regular) smooth *foliation* $\mathcal{F}$ on M of dimension p is a partition $\{F_i\}_I$ of M where each F_i is an immersed submanifold of dimension p called a

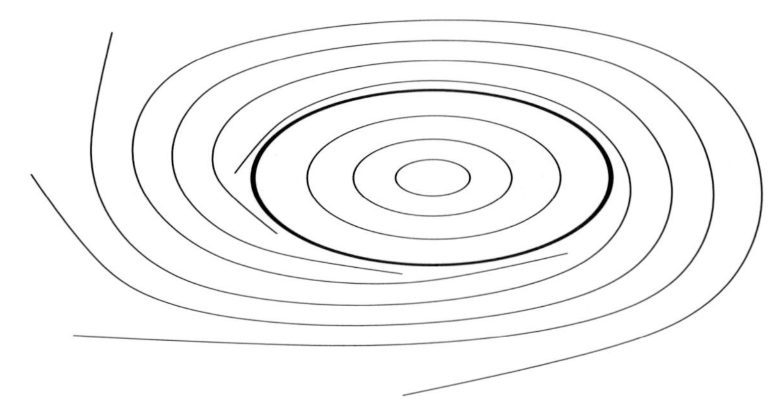

Fig. 3.1.

leaf. Moreover, the manifold M admits charts of the following type:

$$\varphi : U \to \mathbb{R}^p \times \mathbb{R}^q,$$

where U is open in M and such that for any connected component P of $F_i \cap U$ where $i \in I$, there is a $t \in \mathbb{R}^q$ such that $\varphi(P) = \mathbb{R}^p \times \{t\}$.

In this situation the *tangent space to the foliation*, $T\mathcal{F} := \bigcup_I TF_i$, is a subbundle of TM stable under Lie bracket.

The *space of leaves* $M/\mathcal{F}$ is the quotient of M by the equivalence relation of being on the same leaf.

A typical example: Take $M = P \times T$, where P and T are connected smooth manifolds with the partition into leaves given by $\{P \times \{t\}\}_{t \in T}$. Every foliation is locally of this type.

The space of leaves of a foliation is often difficult to study, as appears in the following two examples:

Examples

1. Let $\tilde{\mathcal{F}}_a$ be the foliation on the plane $\mathbb{R}^2$ by lines $\{y = ax + t\}_{t \in \mathbb{R}}$ where a belongs to $\mathbb{R}$. Take the torus $T = \mathbb{R}^2/\mathbb{Z}^2$ to be the quotient of $\mathbb{R}^2$ by translations of $\mathbb{Z}^2$. We denote by $\mathcal{F}_a$ the foliation induced by $\tilde{\mathcal{F}}_a$ on T. When a is rational the space of leaves is a circle, but when a is irrational it is topologically equivalent to a point (i.e., each point is in any neighborhood of any other point).
2. Let $\mathbb{C} \setminus \{(0)\}$ be foliated by

$$\{S_t\}_{t \in]0,1]} \cup \{D_t\}_{t \in]0,2\pi]},$$

where $S_t = \{z \in \mathbb{C} \mid |z| = t\}$ is the circle of radius t, and $D_t = \{z = e^{i(x+t)+x} \mid x \in \mathbb{R}_*^+\}$. (See Figure 3.1.)

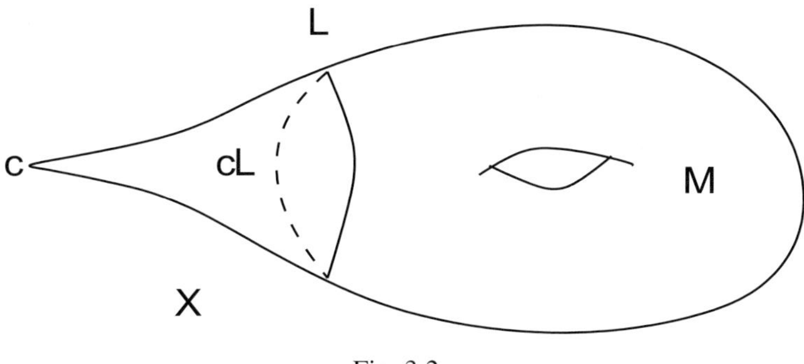

Fig. 3.2.

A *holonomy groupoid* is a smooth groupoid which desingularizes the space of leaves of a foliation. Precisely, if $\mathcal{F}$ is a smooth foliation on a manifold M, its *holonomy groupoid* is the smallest s-connected smooth groupoid $G \rightrightarrows M$ whose orbits are precisely the leaves of the foliation. Here, *smallest* means that if $H \rightrightarrows M$ is another s-connected smooth groupoid whose orbits are the leaves of the foliation, then there is a surjective groupoid homomorphism : $H \to G$ over identity.

The first naive attempt to define such a groupoid is to consider the graph of the equivalence relation defined by being on the same leaf. This does not work: you get a groupoid, but it may not be smooth. This fact can be observed in the preceding example 2. Another idea consists in looking at the *homotopy groupoid*. Let $\Pi(\mathcal{F})$ be the set of homotopy classes of smooth paths lying on leaves of the foliation. It is naturally endowed with a groupoid structure similarly to the homotopy groupoid of Section 3.2.1, example 6. Such a groupoid can be naturally equipped with a smooth structure (of dimension $2p + q$), and the holonomy groupoid is a quotient of this homotopy groupoid. In particular, when the leaves have no homotopy, the holonomy groupoid is the graph of the equivalence relation of being in the same leaf.

3.2.5 The noncommutative tangent space of a conical pseudomanifold

It may happen that the underlying topological space which is under study is a nice compact space which is "almost" smooth. This is the case of pseudomanifolds [24, 36, 53]; for a review on the subject see [9, 28]. In such a situation we can desingularize the tangent space [18, 19]. Let us see how this works in the case of a conical pseudomanifold with one singularity.

Let M be an m-dimensional compact manifold with a compact boundary L. We attach to L the cone $cL = L \times [0, 1]/L \times \{0\}$, using the obvious map $L \times \{1\} \to L \subset \partial M$. The new space $X = cL \cup M$ (see Figure 3.2) is a compact pseudomanifold with a singularity [24]. In general, there is no manifold structure around the vertex c of the cone.

We will use the following notation: $X^\circ = X \setminus \{c\}$ is the *regular part*, X^+ denotes $M \setminus L = X \setminus cL$, $\overline{X_+} = M$ denotes its closure in X, and $X^- = L \times]0, 1[$. If y is a point of the cylindrical part of $X \setminus \{c\}$, we write $y = (y_L, k_y)$, where $y_L \in L$ and $k_y \in]0, 1]$ are the tangential and radial coordinates. The map $y \to k_y$ is extended into a smooth defining function for the boundary of M. In particular, $k^{-1}(1) = L = \partial M$ and $k(M) \subset [1, +\infty[$.

Let us consider $T\overline{X^+}$, the restriction to $\overline{X^+}$ of the tangent bundle of X°. As a C^∞ vector bundle, it is a smooth groupoid with unit space $\overline{X^+}$. We define the groupoid $T^{\mathsf{S}}X$ as the disjoint union

$$T^{\mathsf{S}}X = X^- \times X^- \cup T\overline{X^+} \underset{r}{\overset{s}{\rightrightarrows}} X^\circ,$$

where $X^- \times X^- \rightrightarrows X^-$ is the pair groupoid.

In order to endow $T^{\mathsf{S}}X$ with a smooth structure, compatible with the usual smooth structure on $X^- \times X^-$ and on $T\overline{X^+}$, we have to take care of what happens around points of $T\overline{X^+}|_{\partial\overline{X^+}}$.

Let τ be a smooth positive function on $\mathbb{R}$ such that $\tau^{-1}(\{0\}) = [1, +\infty[$. We let $\tilde{\tau}$ be the smooth map from X° to $\mathbb{R}^+$ given by $\tilde{\tau}(y) = \tau \circ k(y)$.

Let (U, ϕ) be a local chart for X° around $z \in \partial\overline{X^+}$. Setting $U^- = U \cap X^-$ and $\overline{U^+} = U \cap \overline{X^+}$, we define a local chart of $T^{\mathsf{S}}X$ by

$$\tilde{\phi} : U^- \times U^- \cup T\overline{U^+} \longrightarrow \mathbb{R}^m \times \mathbb{R}^m,$$

$$\tilde{\phi}(x, y) = \left(\phi(x), \frac{\phi(y) - \phi(x)}{\tilde{\tau}(x)}\right) \quad \text{if } (x, y) \in U^- \times U^-, \tag{3.1}$$

$$\tilde{\phi}(x, V) = (\phi(x), (\phi)_*(x, V)) \quad \text{elsewhere.}$$

We define in this way a smooth groupoid structure on $T^{\mathsf{S}}X$. Note that at the topological level, the space of orbits of $T^{\mathsf{S}}X$ is equivalent to X. there is a canonical isomorphism between the algebras $C(X)$ and $C(X^\circ/T^{\mathsf{S}}X)$.

The smooth groupoid $T^{\mathsf{S}}X \rightrightarrows X^\circ$ is called the *noncommutative tangent space* of X.

3.2.6 Lie theory for smooth groupoids

Let us go into the more specific world of smooth groupoids. Similarly to Lie groups which admit Lie algebras, any smooth groupoid has a *Lie algebroid* [42, 43].

Definition 3.2.5 A *Lie algebroid* $\mathcal{A} = (p : \mathcal{A} \to TM, [\ ,\ ]_{\mathcal{A}})$ on a smooth manifold M is a vector bundle $\mathcal{A} \to M$ equipped with a bracket $[\ ,\ ]_{\mathcal{A}} : \Gamma(\mathcal{A}) \times \Gamma(\mathcal{A}) \to \Gamma(\mathcal{A})$ on the module of sections of $\mathcal{A}$ together with a homomorphism of fiber bundle $p : \mathcal{A} \to TM$ from $\mathcal{A}$ to the tangent bundle TM of M called the *anchor*, such that:

(i) the bracket $[\ ,\ ]_{\mathcal{A}}$ is $\mathbb{R}$-bilinear, is antisymmetric and satisfies the Jacobi identity,
(ii) $[X, fY]_{\mathcal{A}} = f[X, Y]_{\mathcal{A}} + p(X)(f)Y$ for all $X,\ Y \in \Gamma(\mathcal{A})$ and f a smooth function of M,
(iii) $p([X, Y]_{\mathcal{A}}) = [p(X), p(Y)]$ for all $X,\ Y \in \Gamma(\mathcal{A})$.

Each Lie groupoid admits a Lie algebroid. Let us recall this construction.

Let $G \underset{r}{\overset{s}{\rightrightarrows}} G^{(0)}$ be a Lie groupoid. We denote by T^sG the subbundle of TG of s-vertical tangent vectors. In other words, T^sG is the kernel of the differential Ts of s.

For any γ in G let $R_\gamma : G_{r(\gamma)} \to G_{s(\gamma)}$ be the right multiplication by γ. A tangent vector field Z on G is *right invariant* if it satisfies:

- Z is s-vertical: $Ts(Z) = 0$.
- For all (γ_1, γ_2) in $G^{(2)}$, $Z(\gamma_1 \cdot \gamma_2) = TR_{\gamma_2}(Z(\gamma_1))$.

Note that if Z is a right invariant vector field and h^t its flow, then for any t, the local diffeomorphism h^t is a *local left translation* of G, that is, $h^t(\gamma_1 \cdot \gamma_2) = h^t(\gamma_1) \cdot \gamma_2$ when it makes sense.

The Lie algebroid $\mathcal{A}G$ of G is defined as follows:

- The fiber bundle $\mathcal{A}G \to G^{(0)}$ is the restriction of T^sG to $G^{(0)}$. In other words: $\mathcal{A}G = \bigcup_{x \in G^{(0)}} T_x G_x$ is the union of the tangent spaces to the s-fiber at the corresponding unit.
- The *anchor* $p : \mathcal{A}G \to TG^{(0)}$ is the restriction of the differential Tr of r to $\mathcal{A}G$.
- If $Y : U \to \mathcal{A}G$ is a local section of $\mathcal{A}G$, where U is an open subset of $G^{(0)}$, we define the local *right invariant vector field* Z_Y *associated* with Y by

$$Z_Y(\gamma) = TR_\gamma(Y(r(\gamma))) \quad \text{for all} \quad \gamma \in G^U.$$

The Lie bracket is then defined by

$$\begin{aligned} [\ ,\ ] : \Gamma(\mathcal{A}G) \times \Gamma(\mathcal{A}G) &\longrightarrow \Gamma(\mathcal{A}G), \\ (Y_1, Y_2) &\mapsto [Z_{Y_1}, Z_{Y_2}]_{G^{(0)}}, \end{aligned}$$

where $[Z_{Y_1}, Z_{Y_2}]$ denotes the s-vertical vector field obtained with the usual bracket, and $[Z_{Y_1}, Z_{Y_2}]_{G^{(0)}}$ is the restriction of $[Z_{Y_1}, Z_{Y_2}]$ to $G^{(0)}$.

Example If $\Pi(\mathcal{F})$ is the homotopy groupoid (or the holonomy groupoid) of a smooth foliation, its Lie algebroid is the tangent space $T\mathcal{F}$ to the foliation. The anchor is the inclusion. In particular, the Lie algebroid of the pair groupoid $M \times M$ on a smooth manifold M is TM, the anchor being the identity map.

Lie theory for groupoids is much trickier than for groups. For a long time people thought that, as for Lie algebras, every Lie algebroid integrates into a Lie groupoid [44]. In fact this assertion, named *Lie's third theorem for Lie algebroids*, is false. This was pointed out by a counterexample given by Almeida and Molino in [1]. Since then, a lot of work has been done around this problem. A few years ago Crainic and Fernandes [15] completely solved it by giving a necessary and sufficient condition for the integrability of Lie algebroids.

3.2.7 Examples of groupoids involved in index theory

Index theory is a part of noncommutative geometry where groupoids may play a crucial role. Index theory will be discussed later in this chapter, but we want to present here some of the groupoids which will arise.

Definition 3.2.6 A smooth groupoid G is called a *deformation groupoid* if

$$G = G_1 \times \{0\} \cup G_2 \times]0,1] \rightrightarrows G^{(0)} = M \times [0,1],$$

where G_1 and G_2 are smooth groupoids with unit space M. That is, G is obtained by gluing $G_2 \times]0,1] \rightrightarrows M \times]0,1]$, which is the groupoid G_2 parametrized by $]0,1]$, with the groupoid $G_1 \times \{0\} \rightrightarrows M \times \{0\}$.

Example Let G be a smooth groupoid, and let $\mathcal{A}G$ be its Lie algebroid. The *adiabatic groupoid* of G [13, 38, 39] is a deformation of G on its Lie algebroid:

$$G_{ad} = \mathcal{A}G \times \{0\} \cup G \times]0,1] \rightrightarrows G^{(0)} \times [0,1],$$

where one can put a natural smooth structure on G_{ad}. Here, the vector bundle $\pi : \mathcal{A}G \to G^{(0)}$ is considered as a groupoid in the obvious way.

3.2.7.1 The tangent groupoid

A special example of adiabatic groupoid is the *tangent groupoid* of Connes [13]. Consider the pair groupoid $M \times M$ on a smooth manifold M. We saw that its Lie algebroid is TM. In this situation, the adiabatic groupoid is called the *tangent groupoid* and is given by

$$\mathcal{G}_M^t := TM \times \{0\} \sqcup M \times M \times]0,1] \rightrightarrows M \times [0,1].$$

The Lie algebroid is the bundle $\mathcal{A}(\mathcal{G}_M^t) := TM \times [0,1] \to M \times [0,1]$ with anchor $p : (x, V, t) \in TM \times [0,1] \mapsto (x, tV, t, 0) \in TM \times T[0,1]$.

Choose a Riemannian metric on M. The smooth structure on $\mathcal{G}_M^t$ is such that the map

$$\mathcal{U} \subset TM \times [0,1] \to \mathcal{G}_M^t,$$
$$(x, V, t) \mapsto \begin{cases} (x, V, 0) & \text{if } t = 0, \\ (x, \exp_x(-tV), t) & \text{elsewhere} \end{cases}$$

is a smooth diffeomorphism on its range, where $\mathcal{U}$ is an open neighborhood of $TM \times \{0\}$.

The previous construction of the tangent groupoid of a compact manifold generalizes to the case of conical manifold. When X is a conical manifold, its tangent groupoid is a deformation of the pair groupoid over X° into the groupoid $T^{\mathsf{S}}X$. This deformation has a nice description at the level of Lie algebroids. Indeed, with the notation of Definition 3.2.5, the Lie algebroid of $\mathcal{G}_X^t$ is the (unique) Lie algebroid given by the fiber bundle $\mathcal{A}\mathcal{G}_X^t = [0,1] \times \mathcal{A}(T^{\mathsf{S}}X) = [0,1] \times TX^\circ \to [0,1] \times X^\circ$, with anchor map

$$p_{\mathcal{G}_X^t} : \mathcal{A}\mathcal{G}_X^t = [0,1] \times TX^\circ \longrightarrow T([0,1] \times X^\circ) = T[0,1] \times TX^\circ,$$
$$(\lambda, x, V) \mapsto (\lambda, 0, x, (\tilde{\tau}(x) + \lambda)V).$$

Such a Lie algebroid is almost injective; thus it is integrable [15, 17]. Moreover, it integrates into the *tangent groupoid*, which is defined by

$$\mathcal{G}_X^t = X^\circ \times X^\circ \times]0,1] \cup T^{\mathsf{S}}X \times \{0\} \rightrightarrows X^\circ \times [0,1].$$

Once again one can equip such a groupoid with a smooth structure compatible with the usual one on each piece: $X^\circ \times X^\circ \times]0,1]$ and $T^{\mathsf{S}}X \times \{0\}$ [19].

3.2.7.2 The Thom groupoid

Another important deformation groupoid for our purpose is the *Thom groupoid* [20].

Let $\pi : E \to X$ be a *conical vector* bundle. This means that X is a conical manifold (or a smooth manifold without vertices) and we have a smooth vector bundle $\pi^\circ : E^\circ \to X^\circ$ whose restriction to $X^- = L \times]0,1[$ is equal to $E_L \times]0,1[$, where $E_L \to L$ is a smooth vector bundle. If $E^+ \to X^+$ denotes the bundle E° restricted to X^+, then E is the conical manifold $E = cE_L \cup E^+$.

When X is a smooth manifold (with no conical point), this boils down to the usual notion of smooth vector bundle.

From the definition, π restricts to a smooth vector bundle map $\pi^\circ : E^\circ \to X^\circ$. We let $\pi_{[0,1]} = \pi^\circ \times \mathrm{Id} : E^\circ \times [0,1] \to X^\circ \times [0,1]$.

We consider the tangent groupoids $\mathcal{G}_X^t \rightrightarrows X^\circ \times [0,1]$ for X and $\mathcal{G}_E^t \rightrightarrows E^\circ \times [0,1]$ for E, equipped with a smooth structure constructed using the same gluing function τ (in particular $\tilde{\tau_X} \circ \pi = \tilde{\tau_E}$). We denote by ${}^*\pi^*_{[0,1]}(\mathcal{G}^t_X) \rightrightarrows E^\circ \times [0,1]$ the pullback of $\mathcal{G}^t_X$ by $\pi_{[0,1]}$.

We first associate to the conical vector bundle E a deformation groupoid $\mathcal{T}^t_E$ from ${}^*\pi^*_{[0,1]}(\mathcal{G}^t_X)$ to $\mathcal{G}^t_E$. More precisely, we define

$$\mathcal{T}^t_E := \mathcal{G}^t_E \times \{0\} \sqcup {}^*\pi^*_{[0,1]}(\mathcal{G}^t_X) \times]0,1] \rightrightarrows E^\circ \times [0,1] \times [0,1].$$

Once again, one can equip $\mathcal{T}^t_E$ with a smooth structure [20], and the restriction of $\mathcal{T}^t_E$ to $E^\circ \times \{0\} \times [0,1]$ leads to a smooth groupoid,

$$\mathcal{H}_E = T^{\mathsf{S}}E \times \{0\} \sqcup {}^*\pi^*(T^{\mathsf{S}}X) \times]0,1] \rightrightarrows E^\circ \times [0,1],$$

called a *Thom groupoid* associated to the conical vector bundle E over X.

The following example explains what these constructions become if there is no singularity.

Example Suppose that $p : E \to M$ is a smooth vector bundle over the smooth manifold M. Then we have the usual tangent groupoids $\mathcal{G}^t_E = TE \times \{0\} \sqcup E \times E \times]0,1] \rightrightarrows E \times [0,1]$ and $\mathcal{G}^t_M = TM \times \{0\} \sqcup M \times M \times]0,1] \rightrightarrows M \times [0,1]$. In this example the groupoid $\mathcal{T}^t_E$ will be given by

$$\begin{aligned}\mathcal{T}^t_E &= TE \times \{0\} \times \{0\} \sqcup {}^*p^*(TM) \times \{0\} \times]0,1] \sqcup E \times E \times]0,1] \times [0,1] \\ &\rightrightarrows E \times [0,1] \times [0,1]\end{aligned}$$

and is smooth. Similarly, the Thom groupoid will be given by $\mathcal{H}_E := TE \times \{0\} \sqcup {}^*p^*(TM) \times]0,1] \rightrightarrows E \times [0,1]$.

3.2.8 Haar systems

A locally compact groupoid $G \rightrightarrows G^{(0)}$ can be viewed as a family of locally compact spaces

$$G_x = \{\gamma \in G \mid s(\gamma) = x\}$$

parametrized by $x \in G^{(0)}$. Moreover, right translations act on these spaces. Precisely, to any $\gamma \in G$ one associates the homeomorphism

$$\begin{aligned} R_\gamma : G_y &\to G_x, \\ \eta &\mapsto \eta \cdot \gamma. \end{aligned}$$

This picture enables us to define the right analogue of Haar measure on locally compact groups to locally compact groupoids, namely *Haar systems*. The following definition is due to Renault [46].

Definition 3.2.7 A *Haar system* on G is a collection $\nu = \{\nu_x\}_{x \in G^{(0)}}$ of positive regular Borel measure on G satisfying the following conditions:

(i) *Support:* For every $x \in G^{(0)}$, the support of ν_x is contained in G_x.
(ii) *Invariance:* For any $\gamma \in G$, the right-translation operator $R_\gamma : G_y \to G_x$ is measure-preserving. That is, for all $f \in C_c(G)$,

$$\int f(\eta) d\nu_y(\eta) = \int f(\eta \cdot \gamma) d\nu_x(\eta).$$

(iii) *Continuity:* For all $f \in C_c(G)$, the map

$$G^{(0)} \to \mathbb{C},$$
$$x \mapsto \int f(\gamma) d\nu_x(\gamma)$$

is continuous.

In contrast to the case of locally compact groups, Haar systems on groupoids may not exist. Moreover, when such a Haar system exists, it may not be unique. In the special case of a smooth groupoid, a Haar system always exists [40, 45], and any two Haar systems $\{\nu_x\}$ and $\{\mu_x\}$ differ by a continuous and positive function f on $G^{(0)}$: $\nu_x = f(x)\mu_x$ for all $x \in G^{(0)}$.

Example When the source and range maps are local homeomorphisms, a possible choice for ν_x is the counting measure on G_x.

3.3 C^*-algebras of groupoids

This second part starts with the definition of a C^*-algebra together with some results. Then we construct the maximal and minimal C^*-algebras associated to a groupoid and compute explicit examples.

3.3.1 C^*-algebras – Basic definitions

In this subsection we introduce the terminology and give some examples and properties of C^*-algebras. We refer the reader to [3, 21, 41] for a more complete overview on this subject.

Definition 3.3.1 A *C^*-algebra* A is a complex Banach algebra with an involution $x \mapsto x^*$ such that:

(i) $(\lambda x + \mu y)^* = \bar{\lambda} x^* + \bar{\mu} y^*$ for $\lambda,\ \mu \in \mathbb{C}$ and $x,\ y \in A$,
(ii) $(xy)^* = y^* x^*$ for $x,\ y \in A$, and
(iii) $\|x^* x\| = \|x\|^2$ for $x \in A$.

Note that it follows from the definition that $*$ is isometric.

The element x in A is *self-adjoint* if $x^* = x$, and *normal* if $xx^* = x^*x$. When 1 belongs to A, x is *unitary* if $xx^* = x^*x = 1$.

Given two C^*-algebras A, B, a homomorphism respecting the involution is a called a $*$-homomorphism.

Examples

1. Let $\mathcal{H}$ be a Hilbert space. The algebra $\mathcal{L}(\mathcal{H})$ of all continuous linear transformations of $\mathcal{H}$ is a C^*-algebra. The involution of $\mathcal{L}(\mathcal{H})$ is given by the usual adjunction of bounded linear operators.
2. Let $\mathcal{K}(\mathcal{H})$ be the norm closure of finite-rank operators on $\mathcal{H}$. It is the C^*-algebra of compact operators on $\mathcal{H}$.
3. The algebra $M_n(\mathbb{C})$ is a C^*-algebra. It is a special example of the previous kind, when $dim(\mathcal{H}) = n$.
4. Let X be a locally compact, Hausdorff, topological space. The algebra $C_0(X)$ of continuous functions vanishing at ∞, endowed with the supremum norm and the involution $f \mapsto \bar{f}$, is a commutative C^*-algebra. When X is compact, 1 belongs to $C(X) = C_0(X)$.

Conversely, Gelfand's theorem asserts that every commutative C^*-algebra A is isomorphic to $C_0(X)$ for some locally compact space X (and it is compact precisely when A is unital). Precisely, a *character* $\mathcal{X}$ of A is a continuous homomorphism of algebras $\mathcal{X} : A \to \mathbb{C}$. The set X of characters of A, called the *spectrum* of A, can be endowed with a locally compact space topology. The *Gelfand transform* $\mathcal{F} : A \to C_0(X)$ given by $\mathcal{F}(x)(\mathcal{X}) = \mathcal{X}(x)$ is the desired $*$-isomorphism.

Let A be a C^*-algebra and $\mathcal{H}$ a Hilbert space.

Definition 3.3.2 A $*$-*representation* of A in $\mathcal{H}$ is a $*$-homomorphism $\pi : A \to \mathcal{L}(\mathcal{H})$. The representation is *faithful* if π is injective.

Theorem 3.3.3 *(Gelfand–Naimark) If A is a C^*-algebra, there exists a Hilbert space $\mathcal{H}$ and a faithful representation $\pi : A \to \mathcal{L}(\mathcal{H})$.*

In other words, any C^*-algebra is isomorphic to a norm-closed involutive subalgebra of $\mathcal{L}(\mathcal{H})$. Moreover, when A is separable, $\mathcal{H}$ can be taken to be the (unique up to isometry) separable Hilbert space of infinite dimension.

3.3.1.1 Enveloping algebra

Given a Banach $*$-algebra A, consider the family π_α of all continuous $*$-representations for A. The Hausdorff completion of A for the seminorm $\|x\| = \sup_\alpha(\|\pi_\alpha(x)\|)$ is a C^*-algebra called the *enveloping C^*-algebra* of A.

3.3.1.2 Units

A C^*-algebra may or may not have a unit, but it can always be embedded into a unital C^*-algebra $\tilde{A}$:

$$\tilde{A} := \{x + \lambda \mid x \in A,\ \lambda \in \mathbb{C}\}$$

with the obvious product and involution. The norm on $\tilde{A}$ is given for all $x \in \tilde{A}$ by $\|x\|^{\sim} = \sup\{\|xy\|,\ y \in A;\ \|y\| = 1\}$. On A we have $\|\cdot\| = \|\cdot\|^{\sim}$. The algebra A is a closed two-sided ideal in $\tilde{A}$ and $\tilde{A}/A = \mathbb{C}$.

3.3.1.3 Functional calculus

Let A be a C^*-algebra. If x belongs to A, the *spectrum* of x in A is the compact set

$$Sp(x) = \{\lambda \in \mathbb{C} \mid x - \lambda \text{ is not invertible in } \tilde{A}\}$$

The *spectral radius* of X is the number

$$\nu(x) = \sup\{|\lambda|;\ \lambda \in Sp(x)\}.$$

We have

$$Sp(x) \subset \mathbb{R} \qquad \text{when } x \text{ is self-adjoint } (x^* = x),$$

$$Sp(x) \subset \mathbb{R}_+ \qquad \text{when } x \text{ is positive } (x = y^*y \text{ with } y \in A),$$

$$Sp(x) \subset U(1) \qquad \text{when } x \text{ is unitary } (x^*x = xx^* = 1).$$

When x is *normal* ($x^*x = xx^*$), these conditions on the spectrum are equivalent.

When x is normal, $\nu(x) = \|x\|$. From these, one infers that for any polynomial $P \in \mathbb{C}[x]$ one has $\|P(x)\| = \sup\{P(t) \mid t \in Sp(x)\}$ (using that $Sp(P(x)) = P(Sp(x))$). We can then define $f(x) \in A$ for every continuous function $f : Sp(x) \to \mathbb{C}$. Precisely, according to Weierstrass's theorem, there is a sequence (P_n) of polynomials which converges uniformly to f on $Sp(x)$. We simply define $f(x) = \lim P_n(x)$.

3.3.2 The reduced and maximal C^*-algebras of a groupoid

We restrict our study to the case of Hausdorff locally compact groupoids. For the non-Hausdorff case (which is also important and not exceptional), in particular when dealing with foliations, we refer the reader to [11, 13, 32].

From now on, $G \rightrightarrows G^{(0)}$ is a locally compact Hausdorff groupoid equipped with a fixed Haar system $\nu = \{\nu_x\}_{x \in G^{(0)}}$. We let $C_c(G)$ be the space of complex-valued functions with compact support on G. It is provided with a structure of involutive algebra as follows. If f and g belong to $C_c(G)$, we define the *involution* by

$$\text{for } \gamma \in G, \qquad f^*(\gamma) = \overline{f(\gamma^{-1})},$$

and the *convolution product* by

$$\text{for } \gamma \in G, \qquad f * g(\gamma) = \int_{\eta \in G_x} f(\gamma\eta^{-1}) g(\eta) d\nu_x(\eta),$$

where $x = s(\gamma)$. The 1-norm on $C_c(G)$ is defined by

$$\|f\|_1 = \sup_{x \in G^{(0)}} \max \left(\int_{G_x} |f(\gamma)| d\nu_x(\gamma), \int_{G_x} |f(\gamma^{-1})| d\nu_x(\gamma) \right).$$

The *groupoid full* C^*-algebra $C^*(G, \nu)$ is defined to be the enveloping C^*-algebra of the Banach $*$-algebra $\overline{C_c(G)}^{\|\cdot\|_1}$ obtained by completion of $C_c(G)$ with respect to the norm $\|\cdot\|_1$.

Given x in $G^{(0)}$, f in $C_c(G)$, ξ in $L^2(G_x, \nu_x)$, and γ in G_x, we set

$$\pi_x(f)(\xi)(\gamma) = \int_{\eta \in G_x} f(\gamma\eta^{-1}) \xi(\eta) d\nu_x(\eta).$$

One can show that π_x defines a $*$-representation of $C_c(G)$ on the Hilbert space $L^2(G_x, \nu_x)$. Moreover, for any $f \in C_c(G)$, the inequality $\|\pi_x(f)\| \leq \|f\|_1$ holds. The *reduced norm* on $C_c(G)$ is

$$\|f\|_r = \sup_{x \in G^{(0)}} \{\|\pi_x(f)\|\},$$

which defines a C^*-norm. The *reduced* C^*-algebra $C_r(G, \nu)$ is defined to be the C^*-algebra obtained by completion of A with respect to $\|\cdot\|_r$.

When G is smooth, the reduced and maximal C^*-algebras of the groupoid G do not depend up to isomorphism on the choice of the Haar system ν. In the general case they do not depend on ν up to Morita equivalence [46]. When there is no ambiguity on the Haar system, we write $C^*(G)$ and $C_r^*(G)$ for the maximal and reduced C^*-algebras.

The identity map on $C_c(G)$ induces a surjective homomorphism from $C^*(G)$ to $C_r^*(G)$. Thus $C_r^*(G)$ is a quotient of $C^*(G)$.

For a quite large class of groupoids, *amenable* groupoids [2], the reduced and maximal C^*-algebras are equal. This will be the case for all the groupoids we will meet in the last part of this course devoted to index theory.

Examples

1. When $X \rightrightarrows X$ is a locally compact space, $C^*(X) = C_r^*(X) = C_0(X)$.
2. When $G \rightrightarrows e$ is a group and ν a Haar measure on G, we recover the usual notion of reduced and maximal C^*-algebras of a group.

3. Let M be a smooth manifold, and $TM \rightrightarrows M$ the tangent bundle. Let us equip the vector bundle TM with a Euclidean structure. The Fourier transformation
$$f \in C_c(TM),\ (x, w) \in T^*M, \quad \hat{f}(x, w) = \frac{1}{(2\pi)^{n/2}} \int_{X \in T_x M} e^{-iw(X)} f(X) dX$$
gives rise to an isomorphism between $C^*(TM) = C_r^*(TM)$ and $C_0(T^*M)$. Here, n denotes the dimension of M, and T^*M the cotangent bundle of M.
4. Let X be a locally compact space, with μ a measure on X, and consider the pair groupoid $X \times X \rightrightarrows X$. If $f,\ g$ belongs to $C_c(X \times X)$, the convolution product is given by
$$f * g(x, y) = \int_{z \in X} f(x, z) g(z, y) d\mu(z),$$
and a representation of $C_c(X \times X)$ by
$$\pi : C_c(X \times X) \to \mathcal{L}(L^2(X, \mu)); \qquad \pi(f)(\xi)(x) = \int_{z \in X} f(x, z)\xi(z) d\mu(z)$$
when $f \in C_c(X \times X),\ \xi \in L^2(X, \mu)$ and $x \in X$. It turns out that $C^*(X \times X) = C_r^*(X \times X) \simeq \mathcal{K}(L^2(X, \mu))$.
5. Let M be a compact smooth manifold, and $\mathcal{G}_M^t \rightrightarrows M \times [0, 1]$ its tangent groupoid. In this situation $C^*(\mathcal{G}_M^t) = C_r^*(\mathcal{G}_M^t)$ is a continuous field $(A_t)_{t \in [0,1]}$ of C^*-algebras [21] with $A_0 \simeq C_0(T^*M)$ a commutative C^*-algebra, and for any $t \in]0, 1]$, $A_t \simeq \mathcal{K}(L^2(M))$ [13].

In the sequel we will need the two following properties of C^*-algebras of groupoids.

1. Let G_1 and G_2 be two locally compact groupoids equipped with Haar systems, and suppose for instance that G_1 is amenable. Then according to [2], $C^*(G_1) = C_r^*(G_1)$ is *nuclear* – which implies that for any C^*-algebra B there is only one tensor product C^*-algebra $C^*(G_1) \otimes B$. The groupoid $G_1 \times G_2$ is locally compact, and
$$C^*(G_1 \times G_2) \simeq C^*(G_1) \otimes C^*(G_2) \text{ and } C_r^*(G_1 \times G_2) \simeq C^*(G_1) \otimes C_r^*(G_2).$$
2. Let $G \rightrightarrows G^{(0)}$ be a locally compact groupoid with a Haar system ν. An open subset $U \subset G^{(0)}$ is *saturated* if U is a union of orbits of G, in other words, if $U = s(r^{-1}(U)) = r(s^{-1}(U))$. The set $F = G^{(0)} \setminus U$ is then a closed saturated subset of $G^{(0)}$. The Haar system ν can be restricted to the restrictions $G|_U := G_U^U$ and $G|_F := G_F^F$, and we have the following exact sequence of C^*-algebras [27,45]:
$$0 \to C^*(G|_U) \xrightarrow{i} C^*(G) \xrightarrow{r} C^*(G|_F) \to 0,$$
where $i : C_c(G|_U) \to C_c(G)$ is the extension of functions by 0, and $r : C_c(G) \to C_c(G|_F)$ is the restriction of functions.

II. *KK*-theory

This part on *KK*-theory starts with a historical introduction. In order to motivate our purpose we list most of the properties of the *KK*-functor. Sections 3.5 and 3.6 are devoted to a detailed description of the ingredients involved in *KK*-theory. In order to write this review we have made intensive use of the references [26,48,49,54]. Moreover, a significant part of this chapter was written by Jorge Plazas from the lectures held in Villa de Leyva, and we would like to thank him for his great help.

3.4 Introduction to *KK*-theory

3.4.1 Historical comments

The story begins with several studies by Atiyah [4,5].

Firstly, recall that if X is a compact space, the *K-theory* of X is constructed in the following way: let $\mathcal{E}v$ be the set of isomorphism classes of continuous vector bundles over X. Thanks to the direct sum of bundles, the set $\mathcal{E}v$ is naturally endowed with the structure of an abelian semigroup. One can then symmetrize $\mathcal{E}v$ in order to get a group; this gives the *K*-theory group of X:

$$K^0(X) = \{[E] - [F];\ [E], [F] \in \mathcal{E}v\}.$$

For example, the *K*-theory of a point is $\mathbb{Z}$, for a vector bundle on a point is just a vector space, and vector spaces are classified, up to isomorphism, by their dimension.

A first step towards *KK*-theory is the discovery, made by Atiyah [4] and independently by Jänich [29], that *K*-theory of a compact space X can be described with Fredholm operators.

When $\mathcal{H}$ is an infinite-dimensional separable Hilbert space, the set $\mathcal{F}(\mathcal{H})$ of *Fredholm operators* on $\mathcal{H}$ is the open subset of $\mathcal{L}(\mathcal{H})$ made of bounded operators T on $\mathcal{H}$ such that the dimensions of the kernel and cokernel of T are finite. The set $\mathcal{F}(\mathcal{H})$ is stable under composition. We set

$$[X, \mathcal{F}(\mathcal{H})] = \{\text{homotopy classes of continuous maps: } X \to \mathcal{F}(\mathcal{H})\}.$$

The set $[X, \mathcal{F}(\mathcal{H})]$ is naturally endowed with a semigroup structure. Atiyah and Jänich showed that $[X, \mathcal{F}(\mathcal{H})]$ is actually (a group) isomorphic to $K^0(X)$ [4]. The idea of the proof is the following. If $f : X \to \mathcal{F}(\mathcal{H})$ is a continuous map, one can choose a subspace V of $\mathcal{H}$ of finite codimension such that

$$\forall x \in X,\ V \cap \ker f_x = \{0\} \quad \text{and} \quad \bigcup_{x\in X} \mathcal{H}/f_x(V) \text{ is a vector bundle.} \tag{3.2}$$

Denoting by $\mathcal{H}/f(V)$ the vector bundle arising in (3.2) and by $\mathcal{H}/V$ the product bundle $X \times \mathcal{H}/V$, the Atiyah–Janich isomorphism is then given by

$$\begin{aligned} [X, \mathcal{F}(\mathcal{H})] &\to K^0(X), \\ [f] &\mapsto [\mathcal{H}/V] - [\mathcal{H}/f(V)]. \end{aligned} \tag{3.3}$$

Note that choosing V amounts to modifying f inside its homotopy class into $\widetilde{f}$ (defined to be equal to f on V and to 0 on a supplement of V) such that

$$\operatorname{Ker}\tilde{f} := \bigcup_{x\in X} \operatorname{Ker}(\tilde{f}_x) \text{ and } \operatorname{CoKer}\tilde{f} := \bigcup_{x\in X} \mathcal{H}/\tilde{f}_x(\mathcal{H}) \tag{3.4}$$

are vector bundles over X. These constructions contain relevant information for the sequel: the map f arises as a *generalized Fredholm* operator on the *Hilbert $C(X)$-module $C(X, \mathcal{H})$*.

Later, Atiyah tried to describe the dual functor $K_0(X)$, the K-homology of X, with the help of Fredholm operators. This gave rise to Ell(X), whose cycles are triples (H, π, F) where:

- $H = H_0 \oplus H_1$ is a $\mathbb{Z}_2$ graded Hilbert space.
- $\pi : C(X) \to \mathcal{L}(H)$ is a representation by operators of degree 0, which means that

$$\pi(f) = \begin{pmatrix} \pi_0(f) & 0 \\ 0 & \pi_1(f) \end{pmatrix}.$$

- F belongs to $\mathcal{L}(H)$, is of degree 1 and thus is of the form

$$F = \begin{pmatrix} 0 & G \\ T & 0 \end{pmatrix},$$

and satisfies

$$F^2 - 1 \in \mathcal{K}(H) \quad \text{and} \quad [\pi, F] \in \mathcal{K}(H)\cdot$$

In particular, G is an inverse of T modulo compact operators.

Elliptic operators on closed manifolds produce natural examples of such cycles. Moreover, there exists a natural pairing between Ell(X) and $K^0(X)$, justifying the choice of Ell(X) as a candidate for the cycles of the K-homology of X:

$$\begin{aligned} K^0(X) \times \mathrm{Ell(X)} &\to \mathbb{Z}, \\ ([E], (H, \pi, F)) &\mapsto \mathrm{Index}(F_E), \end{aligned} \tag{3.5}$$

where $\mathrm{Index}(F_E) = \dim(\mathrm{Ker}(F_E)) - \dim(\mathrm{CoKer}(F_E))$ is the *index* of a Fredholm operator associated to a vector bundle E on X and a cycle (H, π, F), as follows. Let E' be a vector bundle on X such that $E \oplus E' \simeq \mathbb{C}^N \times X$, and let e be the projection of $\mathbb{C}^N \times X$ onto E. We can identify $C(X, \mathbb{C}^N) \underset{C(X)}{\otimes} H$ with H^N. Let $\tilde{e}$

be the image of $e \otimes 1$ under this identification. We define $F_E := \tilde{e}F^N|_{\tilde{e}(H^N)}$, where F^N is the diagonal operator with F in each diagonal entry. The operator F_E is the desired Fredholm operator on $\tilde{e}(H^N)$.

Now, we should recall that to any C^*-algebra A (actually, to any ring) is associated a group $K_0(A)$. When A is unital, it can be defined as follows:

$$K_0(A) = \{[\mathcal{E}] - [\mathcal{F}];\ [\mathcal{E}], [\mathcal{F}] \text{ are isomorphism classes of finitely generated projective } A\text{-modules}\}.$$

Recall that an A-module $\mathcal{E}$ is finitely generated and projective if there exists another A-module $\mathcal{G}$ such that $\mathcal{E} \oplus \mathcal{G} \simeq A^N$ for some integer N.

The Swan–Serre theorem asserts that for any compact space X, the category of (complex) vector bundles over X is equivalent to the category of finitely generated projective modules over $C(X)$; in particular, $K^0(X) \simeq K_0(C(X))$. This fact and the (C^*-)algebraic flavor of the preceding constructions lead to the natural attempt to generalize them for noncommutative C^*-algebras.

During 1979 and the 1980s G. Kasparov defined with great success, for any pair of C^*-algebras, a bivariant theory, the KK-theory. This theory generalizes both K-theory and K-homology and carries a product generalizing the pairing (3.5). Moreover, in many cases $KK(A, B)$ contains all the morphisms from $K_0(A)$ to $K_0(B)$. To understand this bifunctor, we will study the notions of Hilbert modules, of adjointable operators acting on them and of generalized Fredholm operators which generalize to arbitrary C^*-algebras the notions already encountered for $C(X)$. Before going to this functional-analytic part, we end this introduction by listing most of the properties of the bifunctor KK.

3.4.2 Abstract properties of $KK(A, B)$

Let A and B be two C^*-algebras. In order to simplify our presentation, we assume that A and B are separable. Here is a list of the most important properties of the KK functor:

$KK(A, B)$ is an abelian group.

Functorial properties. The functor KK is covariant in B and contravariant in A: if $f : B \to C$ and $g : A \to D$ are two homomorphisms of C^*-algebras, there exist two homomorphisms of groups,

$$f_* : KK(A, B) \to KK(A, C) \quad \text{and} \quad g^* : KK(D, B) \to KK(A, B).$$

In particular $\mathrm{Id}_* = \mathrm{Id}$ and $\mathrm{Id}^* = \mathrm{Id}$.

Each $*$-morphism $f : A \to B$ defines an element, denoted by $[f]$, in $KK(A, B)$. We set $1_A := [\mathrm{Id}_A] \in KK(A, A)$.

Homotopy invariance. $KK(A, B)$ is homotopy invariant. Recall that the C^*-algebras A and B are *homotopic* if there exist two $*$-morphisms $f : A \to B$ and $g : B \to A$ such that $f \circ g$ is homotopic to Id_B and $g \circ f$ is homotopic to Id_A. Two homomorphisms $F, G : A \to B$ are homotopic when there exists a $*$-morphism $H : A \to C([0, 1], B)$ such that $H(a)(0) = F(a)$ and $H(a)(1) = G(a)$ for any $a \in A$.

Stability. If $\mathcal{K}$ is the algebra of compact operators on a Hilbert space, there are isomorphisms

$$KK(A, B \otimes \mathcal{K}) \simeq KK(A \otimes \mathcal{K}, B) \simeq KK(A, B).$$

More generally, the bifunctor KK is invariant under *Morita equivalence.*

Suspension. If E is a C^*-algebra, there exists a homomorphism

$$\tau_E : KK(A, B) \to KK(A \otimes E, B \otimes E)$$

which satisfies $\tau_E \circ \tau_D = \tau_{E \otimes D}$ for any C^*-algebra D.

Kasparov product. There is a well-defined bilinear coupling

$$KK(A, D) \times KK(D, B) \to KK(A, B),$$
$$(x, y) \mapsto x \otimes y,$$

called the *Kasparov product.* It is associative, covariant in B and contravariant in A: if $f : C \to A$ and $g : B \to E$ are two homomorphisms of C^*-algebras, then

$$f^*(x \otimes y) = f^*(x) \otimes y \quad \text{and} \quad g_*(x \otimes y) = x \otimes g_*(y).$$

If $g : D \to C$ is another $*$-morphism, $x \in KK(A, D)$ and $z \in KK(C, B)$, then

$$h_*(x) \otimes z = x \otimes h^*(z).$$

Moreover, the following equalities hold:

$$f^*(x) = [f] \otimes x, \qquad g_*(z) = z \otimes [g] \quad \text{and} \quad [f \circ h] = [h] \otimes [f].$$

In particular

$$x \otimes 1_D = 1_A \otimes x = x.$$

The Kasparov product behaves well with respect to suspensions. If E is a C^*-algebra,

$$\tau_E(x \otimes y) = \tau_E(x) \otimes \tau_E(y).$$

This enables us to extend the Kasparov product:

$$\underset{D}{\otimes} : KK(A, B \otimes D) \times KK(D \otimes C, E) \to KK(A \otimes C, B \otimes E),$$
$$(x, y) \mapsto x \underset{D}{\otimes} y := \tau_C(x) \otimes \tau_B(y).$$

The Kasparov product $\underset{\mathbb{C}}{\otimes}$ is commutative.

Higher groups. For any $n \in \mathbb{N}$, let

$$KK_n(A, B) := KK(A, C_0(\mathbb{R}^n) \otimes B).$$

An alternative definition, leading to isomorphic groups, is

$$KK_n(A, B) := KK(A, C_n \otimes B),$$

where C_n is the Clifford algebra of $\mathbb{C}^n$. This will be explained later. The functor KK satisfies *Bott periodicity*: there is an isomorphism

$$KK_2(A, B) \simeq KK(A, B).$$

Exact sequences. Consider the following exact sequence of C^*-algebras:

$$0 \to J \xrightarrow{i} A \xrightarrow{p} Q \to 0,$$

and let B be another C^*-algebra. Under a few more assumptions (e.g., all the C^*-algebras are nuclear or K-nuclear, or the preceding exact sequence admits a completely positive norm-decreasing cross section [50]), we have the following two periodic exact sequences:

$$\begin{array}{ccccc}
KK(B, J) & \xrightarrow{i_*} & KK(B, A) & \xrightarrow{p_*} & KK(B, Q) \\
\delta \uparrow & & & & \downarrow \delta \\
KK_1(B, Q) & \xleftarrow[p_*]{} & KK_1(B, A) & \xleftarrow[i_*]{} & KK_1(B, J)
\end{array}$$

$$\begin{array}{ccccc}
KK(Q, B) & \xrightarrow{p^*} & KK(A, B) & \xrightarrow{i^*} & KK(J, B) \\
\delta \uparrow & & & & \downarrow \delta \\
KK_1(J, B) & \xleftarrow[i^*]{} & KK_1(A, B) & \xleftarrow[p^*]{} & KK_1(Q, B)
\end{array}$$

where the connecting homomorphisms δ are given by Kasparov products.

Final remarks. Let us go back to the end of the introduction in order to make it more precise.

The usual K-theory groups appears as special cases of KK-groups:

$$KK(\mathbb{C}, B) \simeq K_0(B),$$

and the *K-homology* of a C^*-algebra A is defined by

$$K^0(A) = KK(A, \mathbb{C}).$$

Any $x \in KK(A, B)$ induces a homomorphism of groups:

$$KK(\mathbb{C}, A) \simeq K_0(A) \to K_0(B) \simeq KK(\mathbb{C}, B),$$

$$\alpha \mapsto \alpha \otimes x.$$

In most situations, the induced homomorphism

$$KK(A, B) \to \mathrm{Mor}(K_0(A), K_0(B))$$

is surjective. Thus one can think of KK-elements as homomorphisms between K-groups.

When X is a compact space, one has $K^0(X) \simeq K_0(C(X)) \simeq KK(\mathbb{C}, C(X))$ and, as we will see shortly, $K^0(C(X)) = KK(C(X), \mathbb{C})$ is a quotient of the set $\mathrm{Ell}(X)$ introduced by Atiyah. Moreover the pairing $K^0(X) \times \mathrm{Ell}(X) \to \mathbb{Z}$ coincides with the Kasparov product $KK(\mathbb{C}, C(X)) \times KK(C(X), \mathbb{C}) \to KK(\mathbb{C}, \mathbb{C}) \simeq \mathbb{Z}$.

3.5 Hilbert modules

We review the main properties of Hilbert modules over C^*-algebras, necessary for a correct understanding of bivariant K-theory. We closely follow the presentation given by Skandalis [48]. Most of the proofs are taken from his lectures on the subject. We are indebted to him for allowing us to use his lecture notes. Some of the following material can also be found in [54], where the reader will find a guide to the literature and a more detailed presentation.

3.5.1 Basic definitions and examples

Let A be a C^*-algebra and E be a A-right module.

A sesquilinear form $(\cdot, \cdot) : E \times E \to A$ is *positive* if for all $x \in E$ one has $(x, x) \in A_+$. Here A_+ denotes the set of positive elements in A. It is *positive definite* if moreover $(x, x) = 0$ if and only if $x = 0$.

Let $(\cdot, \cdot) : E \times E \to A$ be a positive sesquilinear form, and set $Q(x) = (x, x)$. By the polarization identity

$$\forall x, y \in E, \quad (x, y) = \frac{1}{4}\left(Q(x + y) - i\,Q(x + iy) - Q(x - y) + i\,Q(x - iy)\right),$$

we get

$$\forall x, y \in E, \quad (x, y) = (y, x)^*.$$

Definition 3.5.1 A *pre-Hilbert A-module* is a right A-module E with a positive definite sesquilinear map $(\cdot, \cdot) : E \times E \to A$ such that $y \mapsto (x, y)$ is A-linear.

Proposition 3.5.2 *Let $(E, (\cdot, \cdot))$ be a pre-Hilbert A-module. Then*

$$\forall x \in E, \quad \|x\| = \sqrt{\|(x, x)\|} \tag{3.6}$$

defines a norm on E.

The only nontrivial fact is the triangle inequality, which results from

Lemma 3.5.3 *(Cauchy–Schwarz inequality)*

$$\forall x, y \in E, \quad (x, y)^*(x, y) \leq \|x\|^2(y, y).$$

In particular, $\|(x, y)\| \leq \|x\|\|y\|$.

Set $a = (x, y)$. We have for all $t \in \mathbb{R}$ that $(xa + ty, xa + ty) \geq 0$; thus

$$2ta^*a \leq a^*(x, x)a + t^2(y, y). \tag{3.7}$$

Because $(x, x) \geq 0$, we have $a^*(x, x)a \leq \|x\|^2 a^*a$ (use the equivalence $z^*z \leq w^*w$ if and only if $\|zx\| \leq \|wx\|$ for all $x \in A$), and choosing $t = \|x\|^2$ in (3.7) gives the result.

Definition 3.5.4 A *Hilbert A-module* is a pre-Hilbert A-module which is complete for the norm defined in (3.6).

A *Hilbert A-submodule* of a Hilbert A-module is a closed A-submodule provided with the restriction of the A-valued scalar product.

When there is no ambiguity about the base C^*-algebra A, we simply say pre-Hilbert module and Hilbert module.

Let $(E, (\cdot, \cdot))$ be a pre-Hilbert A-module. From the continuity of the sesquilinear form $(\cdot, \cdot) : E \times E \to A$ and of the right multiplication $E \to E,\ x \mapsto xa$ for any $a \in A$, we infer that the completion of E for the norm (3.6) is a Hilbert A-module.

Remark 3.5.5 In the definition of a pre-Hilbert A-module, one can remove the hypothesis that $(\cdot, \cdot)$ is *definite*. In that case, (3.6) defines a seminorm, and one checks that the Hausdorff completion of a pre-Hilbert A-module, in this extended sense, is a Hilbert A-module.

We continue this subsection with classical examples.

1. The algebra A is a Hilbert A-module with its obvious right A-module structure and

$$(a, b) := a^*b.$$

2. For any positive integer n, A^n is a Hilbert A-module with its obvious right A-module structure and

$$((a_i), (b_i)) := \sum_{i=1}^{n} a_i^* b_i.$$

Observe that $\sum_{i=1}^{n} a_i^* a_i$ is a sum of positive elements in A, which implies that

$$\|(a_i)\| = \sqrt{\left\| \sum_{i=1}^{n} a_i^* a_i \right\|} \geq \|a_k\|$$

for all k. It follows that if $(a_1^m, \dots, a_n^m)_m$ is a Cauchy sequence in A^n and that the sequences $(a_k^m)_m$ are Cauchy in A and thus convergent, and we conclude that A^n is complete.

3. Example 2 can be extended to the direct sum of n Hilbert A-modules $E_1, \dots, E_n$ with the Hilbertian product:

$$((x_i), (y_i)) := \sum_{i=1}^{n} (x_i, y_i)_{E_i}.$$

4. If F is a closed A-submodule of a Hilbert A-module E, then F is a Hilbert A-module. For instance, a closed right ideal in A is a Hilbert A-module.
5. The *standard Hilbert A-module* is defined by

$$\mathcal{H}_A = \left\{ x = (x_k)_{k \in \mathbb{N}} \in A^{\mathbb{N}} \mid \sum_{k \in \mathbb{N}} x_k^* x_k \text{ converges} \right\}. \tag{3.8}$$

The right A-module structure is given by $(x_k)a = (x_k a)$, and the Hilbertian A-valued product is

$$((x_k), (y_k)) = \sum_{k=0}^{+\infty} x_k^* y_k. \tag{3.9}$$

This sum converges for elements of $\mathcal{H}_A$; indeed, for all $q > p \in \mathbb{N}$ we have

$$\begin{aligned} \left\| \sum_{k=p}^{q} x_k^* y_k \right\| &= \left\| \left((x_k)_p^q, (y_k)_p^q \right)_{A^{q-p}} \right\| \\ &\leq \left\| (x_k)_p^q \right\|_{A^{q-p}} \left\| (y_k)_p^q \right\|_{A^{q-p}} \qquad \text{(Cauchy–Schwarz inequality in } A^{q-p}) \\ &= \sqrt{\left\| \sum_{k=p}^{q} x_k^* x_k \right\|} \sqrt{\left\| \sum_{k=p}^{q} y_k^* y_k \right\|}. \end{aligned}$$

This implies that $\sum_{k \geq 0} x_k^* y_k$ satisfies the Cauchy criterion, and therefore converges, so that (3.9) makes sense. Because for all (x_k), (y_k) in $\mathcal{H}_A$

$$\sum_{k \geq 0} (x_k + y_k)^* (x_k + y_k) = \sum_{k \geq 0} x_k^* x_k + \sum_{k \geq 0} y_k^* x_k + \sum_{k \geq 0} x_k^* y_k + \sum_{k \geq 0} y_k^* y_k$$

is the sum of four convergent series, we find that $(x_k) + (y_k) = (x_k + y_k)$ is in $\mathcal{H}_A$. We also have, as before, that for all $a \in A$ and $(x_k) \in \mathcal{H}_A$,

$$\left\| \sum_{k=0}^{+\infty} (x_k a)^* (x_k a) \right\| \leq \|a\|^2 \left\| \sum_{k=0}^{+\infty} x_k^* x_k \right\|.$$

Hence, $\mathcal{H}_A$ is a pre-Hilbert A-module, and we need to check that it is complete. Let $(u_n)_n = ((u_i^n))_n$ be a Cauchy sequence in $\mathcal{H}_A$. We get, as in example 2, that for all $i \in \mathbb{N}$ the sequence $(u_i^n)_n$ is Cauchy in A and thus converges to an element denoted v_i. Let us check that (v_i) belongs to $\mathcal{H}_A$.

Let $\varepsilon > 0$. Choose n_0 such that

$$\forall p > q \geq n_0, \qquad \|u_q - u_p\|_{\mathcal{H}_A} \leq \varepsilon/2.$$

Choose i_0 such that

$$\forall k > j \geq i_0, \qquad \left\| \sum_{i=j}^{k} u_i^{n_0 *} u_i^{n_0} \right\|^{1/2} \leq \varepsilon/2.$$

Then thanks to the triangle inequality in A^{k-j}, we get for all $p, q \geq n_0$ and $j, k \geq i_0$

$$\left\| \sum_{i=j}^{k} u_i^{p*} u_i^{p} \right\|^{1/2} \leq \left\| \sum_{i=j}^{k} (u_i^p - u_i^{n_0})^* (u_i^p - u_i^{n_0}) \right\|^{1/2} + \left\| \sum_{i=j}^{k} u_i^{n_0 *} u_i^{n_0} \right\|^{1/2} \leq \varepsilon.$$

Taking the limit $p \to +\infty$, we get $\| \sum_{i=j}^{k} v_i^* v_i \|^{1/2} \leq \varepsilon$ for all $j, k \geq i_0$, which implies that $(v_i) \in \mathcal{H}_A$. It remains to check that $(u_n)_n$ converges to $v = (v_i)$ in $\mathcal{H}_A$. With the notation just defined,

$$\forall p, q \geq n_0, \ \forall I \in \mathbb{N}, \qquad \left\| \sum_{i=0}^{I} (u_i^p - u_i^q)^* (u_i^p - u_i^q) \right\|^{1/2} \leq \varepsilon.$$

Taking the limit $p \to +\infty$, we have

$$\forall q \geq n_0, \ \forall I \in \mathbb{N}, \qquad \left\| \sum_{i=0}^{I} (v_i - u_i^q)^* (v_i - u_i^q) \right\|^{1/2} \leq \varepsilon,$$

and taking the limit $I \to +\infty$,

$$\forall q \geq n_0, \quad \|v - u_q\| \leq \varepsilon,$$

which ends the proof.

The standard Hilbert module $\mathcal{H}_A$ is maybe the most important Hilbert module. Indeed, Kasparov proved:

Theorem 3.5.6 *Let E be a countably generated Hilbert A-module. Then $\mathcal{H}_A$ and $E \oplus \mathcal{H}_A$ are isomorphic.*

The proof can be found in [54]. This means that there exists an A-linear unitary map $U : E \oplus \mathcal{H}_A \to \mathcal{H}_A$. The notion of unitary uses the notion of adjoint, which will be explained later.

Remark 3.5.7

1. The algebraic sum $\bigoplus_{\mathbb{N}} A$ is dense in $\mathcal{H}_A$.
2. In $\mathcal{H}_A$ we can replace the summand A by any sequence of Hilbert A-modules $(E_i)_{i \in \mathbb{N}}$, and the Hilbertian A-valued product by

$$((x_k), (y_k)) = \sum_{k=0}^{+\infty} (x_k, y_k)_{E_k}.$$

If $E_i = E$ for all $i \in \mathbb{N}$, the resulting Hilbert A-module is denoted by $l^2(\mathbb{N}, E)$.

3. We can generalize the construction to any family $(E_i)_{i\in I}$ using summable families instead of convergent series.

We end this subsection with two concrete examples.

a. Let X be a locally compact space and E a Hermitian vector bundle. The space $C_0(X, E)$ of continuous sections of E vanishing at infinity is a Hilbert $C_0(X)$-module with the module structure given by

$$\xi \cdot a(x) = \xi(x)a(x), \qquad \xi \in C_0(X, E), \quad a \in C_0(X),$$

and the $C_0(X)$-valued product given by

$$(\xi, \eta)(x) = (\xi(x), \eta(x))_{E_x}.$$

b. Let G be a locally compact groupoid with a Haar system, λ, and E a Hermitian vector bundle over $G^{(0)}$. Then

$$f, g \in C_c(G, r^*E), \quad (f, g)(\gamma) = \int_{G_{s(\gamma)}} (f(\eta\gamma^{-1}), g(\eta))_{E_{r(\eta)}} d\lambda^{s(\gamma)}(\eta) \tag{3.10}$$

gives a positive definite sesquilinear $C_c(G)$-valued form which has the correct behavior with respect to the right action of $C_c(G)$ on $C_c(G, r^*E)$. This leads to two norms $\|f\| = \|(f, f)\|^{1/2}_{C^*(G)}$ and $\|f\|_r = \|(f, f)\|^{1/2}_{C^*_r(G)}$ and two completions of $C_c(G, r^*E)$, denoted $C^*(G, r^*E)$ and $C^*_r(G, r^*E)$, which are Hilbert modules, respectively, over $C^*(G)$ and $C^*_r(G)$.

3.5.2 Homomorphisms of Hilbert A-modules

Let E, F be Hilbert A-modules. We will need the orthogonality in Hilbert modules:

Lemma 3.5.8 *Let S be a subset of E. The orthogonal of S,*

$$S^\perp = \{x \in E \mid \forall y \in S, \ (y, x) = 0\},$$

is a Hilbert A-submodule of E.

3.5.2.1 Adjoints

Let $T : E \to F$ be a map. T is *adjointable* if there exists a map $S : F \to E$ such that

$$\forall (x, y) \in E \times F, \quad (Tx, y) = (x, Sy). \tag{3.11}$$

Definition 3.5.9 Adjointable maps are called *homomorphisms of Hilbert A-modules.* The set of adjointable maps from E to F is denoted by $\mathrm{Mor}(E, F)$, and $\mathrm{Mor}(E) = \mathrm{Mor}(E, E)$. The space of linear continuous maps from E to F is denoted by $\mathcal{L}(E, F)$ and $\mathcal{L}(E) = \mathcal{L}(E, E)$.

The terminology will become clear after the next proposition.

Proposition 3.5.10 *Let* $T \in \mathrm{Mor}(E, F)$.

(a) *The operator satisfying (3.11) is unique. It is denoted by* T^* *and called the adjoint of* T. *One has* $T^* \in \mathrm{Mor}(F, E)$ *and* $(T^*)^* = T$.

(b) T *is linear, A-linear and continuous.*

(c) $\|T\| = \|T^*\|$, $\|T^*T\| = \|T\|^2$, *and* $\mathrm{Mor}(E, F)$ *is a closed subspace of* $\mathcal{L}(E, F)$. *In particular* $\mathrm{Mor}(E)$ *is a* C^**-algebra.*

(d) *If* $S \in \mathrm{Mor}(E, F)$ *and* $T \in \mathrm{Mor}(F, G)$, *then* $TS \in \mathrm{Mor}(E, G)$ *and* $(TS)^* = S^*T^*$.

Proof (a) Let R, S be two maps satisfying (3.11) for T. Then

$$\forall x \in E, y \in F, \quad (x, Ry - Sy) = 0,$$

and taking $x = Ry - Sy$ yields $Ry - Sy = 0$. The remaining part of the assertion is obvious.

(b) $\forall x, y \in E, z \in F, \lambda \in \mathbb{C}$,

$$(T(x + \lambda y), z) = (x + \lambda y, T^*z) = (x, T^*z) + \overline{\lambda}(y, T^*z) = (Tx, z)(\lambda Ty, z);$$

thus $T(x + \lambda y) = Tx + \lambda Ty$, and T is linear. Moreover,

$$\forall x \in E, y \in F, a \in A, \quad (T(xa), y) = (xa, T^*y) = a^*(x, T^*y) = ((Tx)a, y),$$

which gives the A-linearity. Consider the set

$$S = \{(-T^*y, y) \in E \times F \mid y \in F\}.$$

Then

$$\begin{aligned}(x_0, y_0) \in S^\perp &\Leftrightarrow \forall y \in F,\ (x_0, -T^*y) + (y_0, y) = 0\\ &\Leftrightarrow \forall y \in F,\ (y_0 - Tx_0, y) = 0.\end{aligned}$$

Thus $G(T) = \{(x, y) \in E \times F \mid y = Tx\} = S^\perp$ is closed, and the closed-graph theorem implies that T is continuous.

(c) We have

$$\|T\|^2 = \sup_{\|x\| \le 1} \|Tx\|^2 = \sup_{\|x\| \le 1} (x, T^*Tx) \le \|T^*T\| \le \|T^*\|\|T\|.$$

Thus $\|T\| \le \|T^*\|$, and switching T and T^* gives the equality. We have also proved

$$\|T\|^2 \le \|T^*T\| \le \|T^*\|\|T\| = \|T\|^2;$$

thus $\|T^*T\| = \|T\|^2$, and the norm of $\mathrm{Mor}(E)$ satisfies the C^*-algebraic equation.

Let $(T_n)_n$ be a sequence in $\mathrm{Mor}(E, F)$ which converges to $T \in \mathcal{L}(E, F)$. Because $\|T\| = \|T^*\|$ and because $T \to T^*$ is (anti)linear, the sequence $(T_n^*)_n$ is a Cauchy sequence, and therefore converges to an operator $S \in \mathcal{L}(F, E)$. It then immediately follows that S is the adjoint of T. This proves that $\mathrm{Mor}(E, F)$ is closed; in particular, $\mathrm{Mor}(E)$ is a C^*-algebra.

(d) Easy. □

Remark 3.5.11 There exist continuous linear and A-linear maps $T : E \to F$ which do not have an adjoint. For instance, take $A = C([0, 1])$, $J = C_0(]0, 1])$ and $T : J \hookrightarrow A$ the inclusion. Assuming that T is adjointable, a one-line computation proves that $T^*1 = 1$. But 1 does not belong to J. Thus $J \hookrightarrow A$ has no adjoint.

One can also take $E = C([0, 1]) \oplus C_0(]0, 1])$ and $T : E \to E, x + y \mapsto y + 0$ to produce an example of $T \in \mathcal{L}(E)$ and $T \notin \mathrm{Mor}(E)$.

One can characterize self-adjoint and positive elements in the C^*-algebra $\mathrm{Mor}(E)$ as follows.

Proposition 3.5.12 *Let* $T \in \mathrm{Mor}(E)$.

(a) $T = T^* \Leftrightarrow \forall x \in E,\ (x, Tx) = (x, Tx)^*$,

(b) $T \geq 0 \Leftrightarrow \forall x \in E,\ (x, Tx) \geq 0$.

Proof (a) The implication $\Rightarrow$ is obvious. Conversely, set $Q_T(x) = (x, Tx)$. Using the polarization identity

$$(x, Ty) = \frac{1}{4}\left(Q_T(x + y) - iQ_T(x + iy) - Q_T(x - y) + iQ_T(x - iy)\right),$$

one easily gets $(x, Ty) = (Tx, y)$ for all $x, y \in E$; thus T is self-adjoint.

(b) If T is positive, there exists $S \in \mathrm{Mor}(E)$ such that $T = S^*S$. Then $(x, Tx) = (Sx, Sx)$ is positive for all x. Conversely, if $(x, Tx) \geq 0$ for all x, then T is self-adjoint by (a), and there exist positive elements T_+, T_- such that

$$T = T_+ - T_-, \qquad T_+T_- = T_-T_+ = 0.$$

It follows that

$$\begin{aligned} &\forall x \in E,\ (x, T_+x) \geq (x, T_-x), \\ &\forall z \in E,\ (T_-z, T_+T_-z) \geq (T_-z, T_-T_-z), \\ &\forall z \in E,\ (z, (T_-)^3 z) \leq 0. \end{aligned}$$

Because T_- is positive, T_-^3 is also positive and the last inequality implies $T_-^3 = 0$. It follows that $T_- = 0$ and then $T = T_+ \geq 0$. □

3.5.2.2 Orthocompletion

Recall that for any subset S of E, $S^{\perp}$ is a Hilbert submodule of E. It is also worth noticing that any orthogonal submodules $F \perp G$ of E are direct summands.

The following properties are left to check as an exercise:

Proposition 3.5.13 *Let F, G be A-submodules of E.*

- $E^{\perp} = \{0\}$ *and* $\{0\}^{\perp} = E$.
- $F \subset G \Rightarrow G^{\perp} \subset F^{\perp}$.
- $F \subset F^{\perp\perp}$.
- *If $F \perp G$ and $F \oplus G = E$, then $F^{\perp} = G$ and $G^{\perp} = F$. In particular, F and G are Hilbert submodules.*

Definition 3.5.14 A Hilbert A-submodule F of E is said to be *orthocomplemented* if $F \oplus F^{\perp} = E$.

Remark 3.5.15 A Hilbert submodule is not necessarily orthocomplemented, even if it can be topologically complemented. For instance, consider $A = C([0, 1])$ and $J = C_0(]0, 1])$ as a Hilbert A-submodule of A. One easily checks that $J^{\perp} = \{0\}$; thus J is not orthocomplemented. On the other hand, $A = J \oplus \mathbb{C}$.

Lemma 3.5.16 *Let $T \in \mathrm{Mor}(E)$. Then*

- $\ker T^* = (\operatorname{Im} T)^{\perp}$,
- $\overline{\operatorname{Im} T} \subset (\ker T^*)^{\perp}$.

The proof is obvious. Note the difference in the second point from the case of bounded operators on Hilbert spaces (where equality always occurs). Thus, in general, $\ker T^* \oplus \overline{\operatorname{Im} T}$ is not the whole of E. Such a situation can occur when $\overline{\operatorname{Im} T}$ is not orthocomplemented.

Let us point out that we can have T^* injective without having $\operatorname{Im} T$ dense in E (for instance, $T : C[0, 1] \to C[0, 1]$, $f \mapsto tf$). Nevertheless, we have:

Theorem 3.5.17 *Let $T \in \mathrm{Mor}(E, F)$. The following assertions are equivalent:*

(i) $\operatorname{Im} T$ *is closed,*
(ii) $\operatorname{Im} T^*$ *is closed,*
(iii) 0 *is isolated in* $\mathrm{spec}(T^*T)$ *(or* $0 \notin \mathrm{spec}(T^*T)$*),*
(iv) 0 *is isolated in* $\mathrm{spec}(TT^*)$ *(or* $0 \notin \mathrm{spec}(TT^*)$*),*

and in that case $\operatorname{Im} T$, $\operatorname{Im} T^*$ *are orthocomplemented.*

Thus, under the assumption of the theorem $\ker T^* \oplus \operatorname{Im} T = F$, $\ker T \oplus \operatorname{Im} T^* = E$. Before proving the theorem, we gather some technical preliminaries into a lemma:

Lemma 3.5.18 *Let $T \in \mathrm{Mor}(E, F)$. Then:*

(i) $T^*T \geq 0$. *We set* $|T| = \sqrt{T^*T}$.
(ii) $\overline{\mathrm{Im}\, T^*} = \overline{\mathrm{Im}\, |T|} = \overline{\mathrm{Im}\, T^*T}$.
(iii) Assume that $T(E_1) \subset F_1$ *for some Hilbert submodules* E_1, F_1. *Then* $T|_{E_1} \in \mathrm{Mor}(E_1, F_1)$.
(iv) If T is onto, then TT^* is invertible (in $\mathrm{Mor}(F)$*) and* $E = \ker T \oplus \mathrm{Im}\, T^*$.

Proof Proof of the lemma: (i) is obvious.

(ii) One has $T^*T(E) \subset T^*(F)$. Conversely,

$$T^* = \lim T^*(1/n + TT^*)^{-1}TT^*.$$

This is a convergence in norm, because

$$\|T^*(1/n + TT^*)^{-1}TT^* - T^*\| = \left\| \frac{1}{n}T^* \left(\frac{1}{n} + TT^*\right)^{-1} \right\| = O(1/\sqrt{n}).$$

It follows that $T^*(F) \subset \overline{T^*T(E)}$ and thus $\overline{\mathrm{Im}\, T^*} = \overline{\mathrm{Im}\, T^*T}$. Replacing T by $|T|$ yields the other equality.

(iii) Easy.

(iv) By the open-mapping theorem, there exists a positive real number $k > 0$ such that each $y \in F$ has a preimage x_y by T with $\|y\| \geq k\|x_y\|$. Using the Cauchy–Schwarz inequality for T^*y and x_y, we get

$$\|T^*y\| \geq k\|y\| \quad \forall y \in F. \tag{$*$}$$

Recall that in a C^*-algebra, the inequality $a^*a \leq b^*b$ is equivalent to $\|ax\| \leq \|bx\|$ for all $x \in A$. This can be adapted to Hilbert modules to show that $(*)$ implies $TT^* \geq k^2$ in $\mathrm{Mor}(F)$, so that TT^* is invertible. Then $p = T^*(TT^*)^{-1}T$ is an idempotent and $E = \ker p \oplus \mathrm{Im}\, p$. Moreover, $(TT^*)^{-1}T$ is onto, from which it follows that $\mathrm{Im}\, p = \mathrm{Im}\, T^*$. On the other hand, $T^*(TT^*)^{-1}$ is injective, so that $\ker p = \ker T$. □

Proof Proof of the theorem: Let us start with the implication (i) $\Rightarrow$ (iv). By point (iii) of the lemma, $S := (T : E \to TE) \in \mathrm{Mor}(E, TE)$, and by point (iv) of the lemma SS^* is invertible. Because the spectra of SS^* and S^*S coincide outside 0 and because $S^*S = T^*T$, we get (iii).

The implication (iv) $\Rightarrow$ (i): Consider the functions $f, g : \mathbb{R} \to \mathbb{R}$ defined by $f(0) = g(0) = 0$, $f(t) = 1$, $g(t) = 1/t$ for $t \neq 0$. Thus f and g are continuous on the spectrum of TT^*. Using the equalities $f(t)t = t$ and $tg(t) = f(t)$, we get $f(TT^*)TT^* = TT^*$ and $TT^*g(TT^*) = f(TT^*)$, from which we deduce $\mathrm{Im}\, f(TT^*) = \mathrm{Im}\, TT^*$. But $f(TT^*)$ is a projector (self-adjoint idempotent); hence $\mathrm{Im}\, TT^*$ is closed and orthocomplemented. Using point (ii) of the lemma and the inclusion $\mathrm{Im}\, TT^* \subset \mathrm{Im}\, T$ yields (i) (and also the orthocomplementability of $\mathrm{Im}\, T$).

At this point we have the following equivalences: (i) $\Leftrightarrow$ (iii) $\Leftrightarrow$ (iv). Replacing T by T^*, we get (ii) $\Leftrightarrow$ (iii) $\Leftrightarrow$ (iv). □

Another result which deserves to be stated is:

Proposition 3.5.19 *Let H be a Hilbert submodule of E, and $T : E \to F$ a A-linear map.*

- *H is orthocomplemented if and only if $i : H \hookrightarrow E \in \mathrm{Mor}(H, E)$.*
- *$T \in \mathrm{Mor}(E, F)$ if and only if the graph of T,*

$$\{(x, y) \in E \times F \mid y = Tx\},$$

is orthocomplemented.

3.5.2.3 Partial isometries

The following easy result is left as an exercise:

Proposition 3.5.20 *(and definition). Let $u \in \mathrm{Mor}(E, F)$. The following assertions are equivalent:*

(i) u^*u *is an idempotent,*
(ii) uu^* *is an idempotent,*
(iii) $u^* = u^*uu^*$,
(iv) $u = uu^*u$.

u is then called a partial isometry, *with initial support $I = \mathrm{Im}\, u^*$ and final support $J = \mathrm{Im}\, u$.*

Remark 3.5.21 If u is a partial isometry, then $\ker u = \ker u^*u$, $\ker u^* = \ker uu^*$, $\mathrm{Im}\, u = \mathrm{Im}\, uu^*$ and $\mathrm{Im}\, u^* = \mathrm{Im}\, u^*u$. In particular, u has closed range, and $E = \ker u \oplus \mathrm{Im}\, u^*$, $F = \ker u^* \oplus \mathrm{Im}\, u$, where the direct sums are orthogonal.

3.5.2.4 Polar decompositions

All homomorphisms do not admit a polar decomposition. For instance, consider $T \in \mathrm{Mor}(C[-1, 1])$ defined by $Tf = t \cdot f$ (here $C[-1, 1]$ is regarded as a Hilbert $C[-1, 1]$-module). T is self-adjoint, and $|T| : f \mapsto |t| \cdot f$. The equation $T = u|T|$, $u \in \mathrm{Mor}(C[-1, 1])$, leads to the constraint $u(1)(t) = \mathrm{sign}(t)$, so $u(1) \notin C[-1, 1]$ and u does not exist.

The next result clarifies the requirements for a polar decomposition to exist:

Theorem 3.5.22 *Let $T \in \mathrm{Mor}(E, F)$ such that $\overline{\mathrm{Im}\, T}$ and $\overline{\mathrm{Im}\, T^*}$ are orthocomplemented. Then there exists a unique $u \in \mathrm{Mor}(E, F)$, vanishing on $\ker T$, such that*

$$T = u|T|.$$

Moreover, u is a partial isometry with initial support $\overline{\mathrm{Im}\, T^*}$ and final support $\overline{\mathrm{Im}\, T}$.

Proof We first assume that T and T^* have dense range. Setting $u_n = T(1/n + T^*T)^{-1/2}$, we get a bounded sequence ($\|u_n\| \leq 1$) such that for all $y \in F$ we have $u_n(T^*y) = T(1/n + T^*T)^{-1/2}T^*y \to \sqrt{TT^*}(y)$. Thus, by density of $\operatorname{Im} T^*$, $u_n(x)$ converges for all $x \in E$. Let $v(x)$ denotes the limit. Replacing T by T^*, we also have that $u_n^*(y)$ converges for all $y \in F$, which yields $v \in \operatorname{Mor}(E, F)$. A careful computation shows that $u_n|T| - T$ goes to 0 in norm. Thus $v|T| = T$. The homomorphism v is unique by density of $\operatorname{Im}|T|$, and is unitary because $u_n^*u_n(x) \to x$ for all $x \in \operatorname{Im} T^*T$; this proves $v^*v = 1$, and similarly for vv^*.

Now consider the general case, and set $E_1 = \overline{\operatorname{Im} T^*}$, $F_1 = \overline{\operatorname{Im} T}$. One applies the first step to the restriction $T_1 \in \operatorname{Mor}(E_1, F_1)$ of T, and we denote by v_1 the unitary homomorphism constructed. We set $u(x) = v_1(x)$ if $x \in E_1$, and $u(x) = 0$ if $x \in E_1^\perp = \ker T$. This definition forces the uniqueness, and it is clear that u is a partial isometry with the claimed initial and final supports. □

Remark 3.5.23 u is the strong limit of $T(1/n + T^*T)^{-1/2}$.

3.5.2.5 Compact homomorphisms

Let $x \in E,\ y \in F$, and define $\theta_{y,x} \in \operatorname{Mor}(E, F)$ by

$$\theta_{y,x}(z) = y \cdot (x, z).$$

The adjoint is given by $\theta_{y,x}^* = \theta_{x,y}$. Then

Definition 3.5.24 We define $\mathcal{K}(E, F)$ to be the closure of the linear span of $\{\theta_{y,x};\ x \in E, y \in F\}$ in $\operatorname{Mor}(E, F)$.

One easily checks that

- $\|\theta_{y,x}\| \leq \|x\|\|y\|$ and $\|\theta_{x,x}\| = \|x\|^2$,
- $T\theta_{y,x} = \theta_{Ty,x}$ and $\theta_{y,x}S = \theta_{y,S^*x}$,
- $\mathcal{K}(E) := \mathcal{K}(E, E)$ is a closed two-sided ideal of $\operatorname{Mor}(E)$ (and hence a C^*-algebra).

We also prove:

Proposition 3.5.25

$$\mathcal{M}(\mathcal{K}(E)) \simeq \operatorname{Mor}(E),$$

where $\mathcal{M}(A)$ denotes the multiplier algebra of a C^*-algebra A.

Proof One can show that for all $x \in E$ there is a unique $y \in E$ such that $x = y \cdot \langle y, y\rangle$ (a technical exercise: show that the limit $y = \lim x \cdot f_n(\sqrt{(x, x)})$ with $f_n(t) = t^{1/3}(1/n + t)^{-1}$ exists and satisfies the desired assertion). Consequently, E is a nondegenerate $\mathcal{K}(E)$-module (i.e., $\mathcal{K}(E) \cdot E = E$); indeed, $x = y \cdot \langle y, y\rangle = \theta_{y,y}(y)$. Using an approximate unit $(u_\lambda)_\Lambda$ for $\mathcal{K}(E)$, we can extend the $\mathcal{K}(E)$-module structure of E into an $\mathcal{M}(\mathcal{K}(E))$-module structure:

$$\forall T \in \mathcal{M}(\mathcal{K}(E)), x \in E, \quad T \cdot x = \lim_{\Lambda} T(u_\lambda) \cdot x.$$

The existence of the limit is a consequence of $x = \theta_{y,y}(y)$ and $T(u_\lambda) \cdot \theta_{y,y} = T(u_\lambda \theta_{y,y}) \to T(\theta_{y,y})$. The limit is $T(\theta_{y,y}) \cdot y$. By the uniqueness of y, this module structure, extending that of $\mathcal{K}(E)$, is unique.

Hence each $m \in \mathcal{M}(\mathcal{K}(E))$ gives rise to a map $M : E \to E$. For any x, z in E,

$$(z, M \cdot x) = (z, (m\theta_{y,y}) \cdot y) = ((m\theta_{y,y})^*(z), y);$$

thus M has an adjoint: $M \in \mathrm{Mor}(E)$, and M^* corresponds to m^*. The map $\rho : m \to M$ provides a $*$-homomorphism from $\mathcal{M}(\mathcal{K}(E))$ to $\mathrm{Mor}(E)$, which is the identity on $\mathcal{K}(E)$. On the other hand, let $\pi : \mathrm{Mor}(E) \to \mathcal{M}(\mathcal{K}(E)$ be the unique $*$-homomorphism, equal to the identity on $\mathcal{K}(E)$, associated to the inclusion $\mathcal{K}(E) \subset \mathrm{Mor}(E)$ as a closed ideal. We have $\pi \circ \rho = \mathrm{Id}$, and by uniqueness of the $\mathcal{M}(\mathcal{K}(E))$-module structure of E, $\rho \circ \pi = \mathrm{Id}$. □

Let us give some generic examples:

(i) Consider A as a Hilbert A-module. We know that for any $a \in A$, there exists $c \in A$ such that $a = cc^*c$. It follows that the map $\gamma_a : A \to A, b \mapsto ab$ is equal to θ_{c,c^*c} and thus is compact. We get a $*$-homomorphism $\gamma : A \to \mathcal{K}(A), a \mapsto \gamma_a$, which has dense image (the linear span of the θs is dense in $\mathcal{K}(A)$) and is clearly injective, because $yb = 0$ for all $b \in A$ implies $y = 0$. Thus γ is an isomorphism:

$$\mathcal{K}(A) \simeq A.$$

In particular, $\mathrm{Mor}(A) \simeq \mathcal{M}(A)$, and if $1 \in A$, then $A \simeq \mathrm{Mor}(A) = \mathcal{K}(A)$.

(ii) For any n, one has in a similar way $\mathcal{K}(A^n) \simeq M_n(A)$ and $\mathrm{Mor}(A^n) \simeq M_n(\mathcal{M}(A))$. If moreover $1 \in A$, then

$$\mathrm{Mor}(A^n) = \mathcal{K}(A^n) \simeq M_n(A). \tag{i}$$

For any Hilbert A-module E, we also have $\mathcal{K}(E^n) \simeq M_n(\mathcal{K}(E))$.

The relations (i) can be extended to arbitrary finitely generated Hilbert A-modules:

Proposition 3.5.26 *Let A be a unital C^*-algebra, and E a A-Hilbert module. Then the following are equivalent:*

(i) E is finitely generated.
(ii) $\mathcal{K}(E) = \mathrm{Mor}(E)$.
(iii) Id_E *is compact.*

In that case, E is also projective (i.e., it is a direct summand of A^n for some n).

For the proof we refer to [54].

3.5.3 Generalized Fredholm operators

Atkinson's theorem claims that for any bounded linear operator on a Hilbert space H, the assertion

$$\ker F \text{ and } \ker F^* \text{ are finite dimensional}$$

is equivalent to the following: There exists a linear bounded operator G such that $FG - \mathrm{Id}$, $GF - \mathrm{Id}$ are compact. This situation is a little more subtle on Hilbert A-modules, in that firstly all the kernel of homomorphisms are A-modules which are not necessarily free, and secondly, replacing the condition "finite dimensional" by "finitely generated" is not enough to recover the previous equivalence. This is why one uses the second assertion as a definition of Fredholm operator in the context of Hilbert modules, and we will see how to adapt Atkinson's classical result to this new setup.

Definition 3.5.27 The homomorphism $T \in \mathrm{Mor}(E, F)$ is a *generalized Fredholm operator* if there exists $G \in \mathrm{Mor}(F, E)$ such that

$$GF - \mathrm{Id} \in \mathcal{K}(E) \quad \text{and} \quad FG - \mathrm{Id} \in \mathcal{K}(F).$$

The following theorem is important to understand the next chapter on *KK*-theory.

Theorem 3.5.28 *Let A be a unital C^*-algebra, $\mathcal{E}$ a countably generated Hilbert A-module and F a generalized Fredholm operator on $\mathcal{E}$.*

(i) If $\mathrm{Im}\, F$ *is closed, then* $\ker F$ *and* $\ker F^*$ *are finitely generated Hilbert modules.*
(ii) There exists a compact perturbation G of F such that $\mathrm{Im}\, G$ *is closed.*

Proof (1) Because $\mathrm{Im}\, F$ is closed, so is $\mathrm{Im}\, F^*$, and both are orthocomplemented by, respectively, $\ker F^*$ and $\ker F$. Let $P \in \mathrm{Mor}(\mathcal{E})$ be the orthogonal projection on $\ker F$. Because F is a generalized Fredholm operator, there exists $G \in \mathrm{Mor}(\mathcal{E})$ such that $Q = 1 - GF$ is compact. In particular, Q is equal to Id on $\ker F$, and

$$QP : \mathcal{E} = \ker F \oplus \mathrm{Im}\, F^* \to \mathcal{E}, \qquad x \oplus y \mapsto x \oplus 0.$$

Because QP is compact, its restriction $QP|_{\ker F} : \ker F \to \ker F$ is also compact, but $QP|_{\ker F} = \mathrm{id}_{\ker F}$; hence Proposition 3.5.26 implies that $\ker F$ is finitely generated. The same argument works for $\ker F^*$.

(2) Let us denote by π the projection homomorphism

$$\pi : \mathrm{Mor}(\mathcal{E}) \to C(\mathcal{E}) := \mathrm{Mor}(\mathcal{E})/\mathcal{K}(\mathcal{E}).$$

Because $\pi(F)$ is invertible in $C(\mathcal{E})$, it has a polar decomposition: $\pi(F) = \omega \cdot |\pi(F)|$. Any unitary of $C(\mathcal{E})$ can be lifted to a partial isometry of $\mathrm{Mor}(\mathcal{E})$ [54]. Let U be such a lift of the unitary ω. Using $|\pi(F)| = \pi(|F|)$, it follows that

$$F = U|F| \mod \mathcal{K}(\mathcal{E}).$$

Because $\pi(|F|)$ is also invertible, and positive, we can form $\log(\pi(|F|))$ and choose a self-adjoint $H \in \mathrm{Mor}(\mathcal{E})$ with $\pi(H) = \log(\pi(|F|))$. Then

$$\pi(Ue^H) = \omega\pi(|F|) = \pi(F),$$

i.e., Ue^H is a compact perturbation of F (and thus is a generalized Fredholm operator). The operator U is a partial isometry and hence has a closed image; and e^H is invertible in $\mathrm{Mor}(\mathcal{E})$, whence Ue^H has a closed image, and the theorem is proved. □

3.5.4 Tensor products

3.5.4.1 Inner tensor products

Let E be a Hilbert A-module, F a Hilbert B-module, and $\pi : A \to \mathrm{Mor}(F)$ a $*$-homomorphism. We define a sesquilinear form on $E \otimes_A F$ by setting

$$\forall x, x' \in E, y, y' \in F, \quad (x \otimes y, x' \otimes y')_{E\otimes F} := (y, (x, x')_E \cdot y')_F,$$

where we have set $a \cdot y = \pi(a)(y)$ to lighten the formula. This sesquilinear form is a B-valued scalar product: only the positivity axiom needs some explanation. Set

$$b = \left(\sum_i x_i \otimes y_i, \sum_i x_i \otimes y_i \right) = \sum_{i,j} (y_i, (x_i, x_j) \cdot y_j),$$

where π has been omitted. Let us set $P = ((x_i, x_j))_{i,j} \in M_n(A)$. The matrix P provides a (self-adjoint) compact homomorphism of A^n, which is positive because

$$\forall a \in A^n, \qquad (a, Pa)_{A^n} = \sum_{i,j} a_i^*(x_i, x_j)a_j = \left(\sum_i x_i a_i, \sum_j x_j a_j \right) \geq 0.$$

This means that $P = Q^*Q$ for some $Q \in M_n(A)$. On the other hand, one can consider P as a homomorphism on F^n, and setting $y = (y_1, \ldots, y_n) \in F^n$, we have

$$b = (y, Py) = (Qy, Qy) \geq 0.$$

Thus $E \otimes_A F$ is a pre-Hilbert module in the generalized sense (i.e., we do not require the inner product to be definite), and the Hausdorff completion of $E \otimes_A F$ is a Hilbert B-module denoted in the same way.

Proposition 3.5.29 *Let $T \in \mathrm{Mor}(E)$ and $S \in \mathrm{Mor}(F)$.*

- *$T \otimes 1 : x \otimes y \mapsto Tx \otimes y$ defines a homomorphism of $E \otimes_A F$.*
- *If S commutes with π, then $1 \otimes S : x \otimes y \mapsto x \otimes Sy$ is a homomorphism which commutes with any $T \otimes 1$.*

Remark 3.5.30

1. Even if T is compact, $T \otimes 1$ is not compact in general. The same is true for $1 \otimes S$ when defined.
2. In general $1 \otimes S$ is not even defined.

3.5.4.2 Outer tensor products

Now forget the homomorphism π, and consider the tensor product over $\mathbb{C}$ of E and F. We set

$$\forall x, x' \in E, y, y' \in F, \quad (x \otimes y, x' \otimes y')_{E \otimes F} := (x, x')_E \otimes (y, y')_F \in A \otimes B.$$

This defines a pre-Hilbert $A \otimes B$-module in the generalized sense (the proof of positivity uses similar arguments), where $A \otimes B$ denotes the spatial tensor product (as it will in the following, when not otherwise specified). The Hausdorff completion will be denoted $E \otimes_{\mathbb{C}} F$.

Example 3.5.31 Let H be a separable Hilbert space. Then

$$H \otimes_{\mathbb{C}} A \simeq H_A$$

3.5.4.3 Connections

We turn back to inner tensor products. We keep the notation of the Section 3.5.4.1. Connes and Skandalis [14] introduced the notion of connection to bypass the general nonexistence of $1 \otimes S$.

Definition 3.5.32 Consider two C^*-algebras A and B. Let E be a Hilbert A-module and F be a Hilbert B-module. Assume there is a $*$-morphism

$$\pi : A \to \mathcal{L}(F),$$

and take the inner tensor product $E \otimes_A F$. Given $x \in E$, we define a homomorphism

$$T_x : E \to E \otimes_A F, \qquad y \mapsto x \otimes y,$$

whose adjoint is given by

$$T_x^* : E \otimes_A F \to F, \qquad z \otimes y \mapsto \pi((x, z))y.$$

If $S \in \mathcal{L}(F)$, an *S-connection* on $E \otimes_A F$ is given by an element

$$G \in \mathcal{L}(E \otimes_A F)$$

such that for all $x \in E$,

$$T_x S - G T_x \in \mathcal{K}(F, E \otimes_A F),$$
$$S T_x^* - T_x^* G \in \mathcal{K}(E \otimes_A F, F).$$

Proposition 3.5.33

(1) If $[\pi, S] \subset \mathcal{K}(F)$, then there are S-connections.
(2) If G_i, $i = 1, 2$, are S_i-connections, then $G_1 + G_2$ is an $S_1 + S_2$-connection and $G_1 G_2$ is an $S_1 S_2$-connection.
(3) For any S-connection G, $[G, \mathcal{K}(E) \otimes 1] \subset \mathcal{K}(E \otimes_A F)$.
(4) The space of 0-connections is exactly

$$\{G \in \mathrm{Mor}(F, E \otimes_A F) \mid (\mathcal{K}(E) \otimes 1)G \text{ and } G(\mathcal{K}(E) \otimes 1) \text{ are subsets of } \mathcal{K}(E \otimes_A F)\}.$$

All these assertions are important for the construction of the Kasparov product. For the proof, see [14].

3.6 *KK*-theory

3.6.1 Kasparov modules and homotopies

Given two C^*-algebras A and B, a *Kasparov A–B-module* (abbreviated "Kasparov module") is given by a triple

$$x = (\mathcal{E}, \pi, F),$$

where $\mathcal{E} = \mathcal{E}^0 \oplus \mathcal{E}^1$ is a $(\mathbb{Z}/2\mathbb{Z})$-graded countably generated Hilbert B-module, $\pi : A \to \mathcal{L}(\mathcal{E})$ is a $*$-morphism of degree 0 with respect to the grading, and $F \in \mathcal{L}(\mathcal{E})$ is of degree 1. These data are required to satisfy the following properties:

$$\pi(a)(F^2 - 1) \in \mathcal{K}(\mathcal{E}) \qquad \text{for all } a \in A,$$
$$[\pi(a), F] \in \mathcal{K}(\mathcal{E}) \qquad \text{for all } a \in A.$$

We denote the set of Kasparov A–B-modules by $E(A, B)$.

Let us immediately define the equivalence relation leading to *KK*-groups. We denote $B([0, 1]) := C([0, 1], B)$.

Definition 3.6.1 A *homotopy* between two Kasparov A–B-modules $x = (\mathcal{E}, \pi, F)$ and $x' = (\mathcal{E}', \pi', F')$ is a Kasparov A–$B([0, 1])$-module $\tilde{x}$ such that

$$(ev_{t=0})_*(\tilde{x}) = x,$$
$$(ev_{t=1})_*(\tilde{x}) = x'. \tag{3.12}$$

Here $ev_{t=\cdot}$ is the evaluation map at $t = \cdot$. Homotopy between Kasparov A–B-modules is an equivalence relation. If there exists a homotopy between x and x', we write $x \sim_h x'$.

The set of homotopy classes of Kasparov A–B-modules is denoted $KK(A, B)$.

There is a natural *sum* on $E(A, B)$: if $x = (\mathcal{E}, \pi, F)$ and $x' = (\mathcal{E}', \pi', F')$ belong to $E(A, B)$, their sum $x + x' \in E(A, B)$ is defined by

$$x + x' = (\mathcal{E} \oplus \mathcal{E}', \pi \oplus \pi', F \oplus F').$$

A Kasparov A–B-module $x = (\mathcal{E}, \pi, F)$ is called *degenerate* if for all $a \in A$ one has $\pi(a)(F^2 - 1) = 0$ and $[\pi(a), F] = 0$. Then:

Proposition 3.6.2 *Degenerate elements of $E(A, B)$ are homotopic to* $(0, 0, 0)$.

The sum of Kasparov A–B-modules provides $KK(A, B)$ with an abelian group structure.

Proof Let $x = (\mathcal{E}, \pi, F) \in E(A, B)$ be a degenerate element. Set $\tilde{x} = (\tilde{\mathcal{E}}, \tilde{\pi}, \tilde{F}) \in E(A, B([0, 1]))$ with

$$\begin{aligned} \tilde{\mathcal{E}} &= C_0([0, 1[\,, \mathcal{E}), \\ \tilde{\pi}(a)\xi(t) &= \pi(a)\xi(t), \\ \tilde{F}\xi(t) &= F\xi(t). \end{aligned}$$

Then $\tilde{x}$ is a homotopy between x and $(0, 0, 0)$.

One can easily show that the sum of Kasparov modules makes sense at the level of their homotopy classes. Thus $KK(A, B)$ admits a commutative semigroup structure with $(0, 0, 0)$ as a neutral element. Finally, the opposite in $KK(A, B)$ of $x = (\mathcal{E}, \pi, F) \in E(A, B)$ may be represented by

$$(\mathcal{E}^{op}, \pi, -F).$$

where $\mathcal{E}^{op}$ is $\mathcal{E}$ with the opposite graduation: $(\mathcal{E}^{op})^i = \mathcal{E}^{1-i}$. Indeed, the module $(\mathcal{E}, \pi, F) \oplus (\mathcal{E}^{op}, \pi, -F)$ is homotopically equivalent to the degenerate module

$$\left(\mathcal{E} \oplus \mathcal{E}^{op}, \pi \oplus \pi, \begin{pmatrix} 0 & \mathrm{Id} \\ \mathrm{Id} & 0 \end{pmatrix}\right).$$

This can be realized with the homotopy

$$G_t = \cos\left(\frac{\pi t}{2}\right) \begin{pmatrix} F & 0 \\ 0 & -F \end{pmatrix} + \sin\left(\frac{\pi t}{2}\right) \begin{pmatrix} 0 & \mathrm{Id} \\ \mathrm{Id} & 0 \end{pmatrix}. \qquad \square$$

3.6.2 Operations on Kasparov modules

Let us explain the functoriality of KK-groups with respect to its variables. The following two operations on Kasparov modules make sense on KK-groups:

- *Pushforward along $*$-morphisms: covariance in the second variable.* Let $x = (\mathcal{E}, \pi, F) \in E(A, B)$, and let $g : B \to C$ be a $*$-morphism. We define an element $g_*(x) \in E(A, C)$ by

$$g_*(x) = (\mathcal{E} \otimes_g C, \pi \otimes 1, F \otimes \mathrm{Id}),$$

where $\mathcal{E} \otimes_g C$ is the inner tensor product of the Hilbert B-module $\mathcal{E}$ with the Hilbert C-module C endowed with the left action of B given by g.
- *Pullback along $*$-morphisms: contravariance in the first variable.* Let $x = (\mathcal{E}, \pi, F) \in E(A, B)$, and let $f : C \to A$ be a $*$-morphism. We define an element $f^*(x) \in E(C, B)$ by

$$f^*(x) = (\mathcal{E}, \pi \circ f, F).$$

Provided with these operations, KK-theory is a bifunctor from the category (of pairs) of C^*-algebras to the category of abelian groups.

We recall another useful operation in KK-theory:

- *Suspension*: Let $x = (\mathcal{E}, \pi, F) \in E(A, B)$, and let D be a C^*-algebra. We define an element $\tau_D(x) \in E(A \otimes D, B \otimes D)$ by

$$\tau_D(x) = (\mathcal{E} \otimes_{\mathbb{C}} D, \pi \otimes 1, F \otimes \mathrm{Id}).$$

Here we take the external tensor product $\mathcal{E} \otimes_{\mathbb{C}} D$, which is a $B \otimes D$-Hilbert module.

3.6.3 Examples of Kasparov modules and of homotopies between them

3.6.3.1 Kasparov modules coming from homomorphisms between C^*-algebras

Let A, B be two C^*-algebras, and $f : A \to B$ a $*$-homomorphism. Because $\mathcal{K}(B) \simeq B$, the expression

$$[f] := (B, f, 0)$$

defines a Kasparov A–B-module. If A and B are $\mathbb{Z}_2$-graded, f has to be a homomorphism of degree 0 (i.e., respecting the grading).

3.6.3.2 *Atiyah's* Ell

Let X be a compact Hausdorff topological space. Take $A = C(X)$ to be the algebra of continuous functions on X and let $B = \mathbb{C}$. Then

$$E(A, B) = \mathrm{Ell}(X),$$

the ring of generalized elliptic operators on X as defined by Atiyah. We give two concrete examples of such Kasparov modules:

- Assume X is a compact smooth manifold, let $A = C(X)$ as before, and let $B = \mathbb{C}$. Let E and E' be two smooth vector bundles over X, and denote by π the action of $A = C(X)$ by multiplication on $L^2(X, E) \oplus L^2(X, E')$. Given a zero-order elliptic pseudodifferential operator

$$P : C^\infty(E) \to C^\infty(E')$$

with parametrix $Q : C^\infty(E') \to C^\infty(E)$, the triple

$$x_P = \left(L^2(X, E) \oplus L^2(X, E'),\ \pi,\ \begin{pmatrix} 0 & Q \\ P & 0 \end{pmatrix} \right)$$

defines an element in $E(A, B) = E(C(X), \mathbb{C})$.
- Let X be a compact spin^c manifold of dimension $2n$, let $A = C(X)$ be as before, and let $B = \mathbb{C}$. Denote by $S = S^+ \oplus S^-$ the complex spin bundle over X, and let

$$\not{D} : L^2(X, S) \to L^2(X, S)$$

be the corresponding Dirac operator. Let π be the action of $A = C(X)$ by multiplication on $L^2(X, S)$. Then the triple

$$x_{\not{D}} = \left(L^2(X, S),\ \pi,\ \frac{\not{D}}{\sqrt{1 + \not{D}^2}} \right)$$

defines an element in $E(A, B) = E(C(X), \mathbb{C})$.

3.6.3.3 Compact perturbations

Let $x = (\mathcal{E}, \pi, F) \in E(A, B)$. Let $P \in \mathrm{Mor}(\mathcal{E})$ satisfy

$$\forall a \in A, \qquad \pi(a) \cdot P \in \mathcal{K}(\mathcal{E}) \text{ and } P \cdot \pi(a) \in \mathcal{K}(\mathcal{E}). \tag{3.13}$$

Then

$$x \sim_h (\mathcal{E}, \pi, F + P).$$

The homotopy is the obvious one: $(\mathcal{E} \otimes C([0, 1]), \pi \otimes \mathrm{Id}, F + tP)$. In particular, when B is unital, we can always choose a representative $(\mathcal{E}, \pi, G)$ with $\mathrm{Im}\, G$ closed (cf. Theorem 3.5.28).

3.6.3.4 (Quasi) Self-adjoint representatives

There exists a representative $(\mathcal{E}, \pi, G)$ of $x = (\mathcal{E}, \pi, F) \in E(A, B)$ satisfying

$$\pi(a)(G - G^*) \in \mathcal{K}(\mathcal{E}). \tag{3.14}$$

Just take $(\mathcal{E} \otimes C([0,1]), \pi \otimes \mathrm{Id}, F_t)$ as a homotopy, where

$$F_t = (tF^*F+1)^{1/2}F(tF^*F+1)^{-1/2}.$$

Then $G = F_1$ satisfies (3.14). Now, $H = (G+G^*)/2$ is self-adjoint, and $P = (G-G^*)/2$ satisfies (3.13); thus $(\mathcal{E}, \pi, H)$ is another representative of x.

Note that (3.14) is often useful in practice and is added as an axiom in many definitions of *KK*-theory, like the original one of Kasparov. It was observed in [49] that it could be omitted.

3.6.3.5 *Stabilization and unitarily equivalent modules*

Any Kasparov module $(E, \pi, F) \in E(A, B)$ is homotopic to a Kasparov module $(\widehat{\mathcal{H}}_B, \rho, G)$, where $\widehat{\mathcal{H}}_B = \mathcal{H}_B \oplus \mathcal{H}_B$ is the standard graded Hilbert B-module. Indeed, add to (E, π, F) the degenerate module $(\widehat{\mathcal{H}}_B, 0, 0)$, and consider a grading-preserving isometry $u : E \oplus \widehat{\mathcal{H}}_B \to \widehat{\mathcal{H}}_B$ provided by Kasparov's stabilization theorem. Then, set $\widetilde{E} = E \oplus \widehat{\mathcal{H}}_B$, $\widetilde{F} = F \oplus 0$, $\widetilde{\pi} = \pi \oplus 0$, $\rho = u\widetilde{\pi}u^*$, $G = u\widetilde{F}u^*$, and consider the homotopy

$$\left(\widetilde{E} \oplus \widehat{\mathcal{H}}_B, \widetilde{\pi} \oplus \rho, \begin{pmatrix} \cos(\frac{t\pi}{2}) & -u^*\sin(\frac{t\pi}{2}) \\ u\sin(\frac{t\pi}{2}) & \cos(\frac{t\pi}{2}) \end{pmatrix} \begin{pmatrix} \widetilde{F} & 0 \\ 0 & J \end{pmatrix} \begin{pmatrix} \cos(\frac{t\pi}{2}) & u^*\sin(\frac{t\pi}{2}) \\ -u\sin(\frac{t\pi}{2}) & \cos(\frac{t\pi}{2}) \end{pmatrix}\right) \tag{3.15}$$

between $(E, \pi, F) \oplus (\widehat{\mathcal{H}}_B, 0, 0) = (\widetilde{E}, \widetilde{\pi}, \widetilde{F})$ and $(\widehat{\mathcal{H}}_B, \rho, G)$. Above, J denotes the operator

$$\begin{pmatrix} 0 & 1 \\ 1 & 0 \end{pmatrix}$$

defined on $\widehat{\mathcal{H}}_B$.

One says that two Kasparov modules $(E_i, \pi_i, F_i) \in E(A, B)$, $i = 1, 2$, are unitarily equivalent when there exists a grading-preserving isometry $v : E_1 \to E_2$ such that

$$vF_1v^* = F_2 \quad \text{and} \quad \forall a \in A,\ v\pi_1(a)v^* - \pi_2(a) \in \mathcal{K}(\mathcal{E}_2).$$

Unitarily equivalent Kasparov modules are homotopic. Indeed, one can replace (E_i, π_i, F_i), $i = 1, 2$, by homotopically, equivalent modules $(\widehat{\mathcal{H}}_B, \rho_i, G_i)$, $i = 1, 2$. It follows from the preceding construction that the new modules $(\widehat{\mathcal{H}}_B, \rho_i, G_i)$ remain unitarily equivalent, and one immediately adapts (3.15) to a homotopy between then.

3.6.3.6 *Relationship with ordinary K-theory*

Let B be a unital C^*-algebra. A finitely generated ($\mathbb{Z}/2\mathbb{Z}$-graded) projective B-module $\mathcal{E}$ is a submodule of some $B^N \oplus B^N$ and can then be endowed with a Hilbert

B-module structure. On the other hand, $\mathrm{Id}_{\mathcal{E}}$ is a compact morphism (Proposition 3.5.26); thus

$$(\mathcal{E}, \iota, 0) \in E(\mathbb{C}, B),$$

where ι is just multiplication by complex numbers. This provides a group homomorphism $K_0(B) \to KK(\mathbb{C}, B)$.

Conversely, let $(\mathcal{E}, 1, F) \in E(\mathbb{C}, B)$ be any Kasparov module where we have chosen F with closed range (see Theorem 3.5.28): $\ker F$ is then a finitely generated $\mathbb{Z}/2\mathbb{Z}$-graded projective B-module. Consider $\widetilde{\mathcal{E}} = \{\xi \in C([0,1], \mathcal{E}) \mid \xi(1) \in \ker F\}$ and $\widetilde{F}(\xi) : t \mapsto F(\xi(t))$. The triple $(\widetilde{\mathcal{E}}, 1, \widetilde{F})$ provides a homotopy between $(\mathcal{E}, 1, F)$ and $(\ker F, 1, 0)$. This also gives an inverse of the previous group homomorphism.

3.6.3.7 A nontrivial generator of $KK(\mathbb{C}, \mathbb{C})$

In the special case $B = \mathbb{C}$, we get $KK(\mathbb{C}, \mathbb{C}) \simeq K_0(\mathbb{C}) \simeq \mathbb{Z}$, and under this isomorphism, the triple

$$\left(L^2(\mathbb{R})^2, 1, \frac{1}{\sqrt{1+H}} \begin{pmatrix} 0 & -\partial_x + x \\ \partial_x + x & 0 \end{pmatrix} \right), \qquad \text{where} \quad H = -\partial_x^2 + x^2, \tag{3.16}$$

corresponds to $+1$. The reader can check as an exercise that $\partial_x + x$ and H are essentially self-adjoint as unbounded operators on $L^2(\mathbb{R})$, that H has a compact resolvent and that $\partial_x + x$ has a Fredholm index equal to $+1$. It follows that the Kasparov module in (3.16) is well defined and satisfies the required claim.

3.6.4 Ungraded Kasparov modules and KK_1

Triples $(\mathcal{E}, \pi, F)$ satisfying the properties (3.12) can arise with no natural grading for $\mathcal{E}$, and consequently with no diagonal–antidiagonal decompositions for π, F. We refer to those as ungraded Kasparov A–B-modules, and the corresponding set is denoted by $E^1(A, B)$. The direct sum is defined in the same way, as well as the homotopy, which this time is an element of $E^1(A, B[0,1])$. The homotopy defines an equivalence relation on $E^1(A, B)$, and the quotient inherits an abelian group structure as before.

Let C_1 be the complex Clifford algebra of the vector space $\mathbb{C}$ provided with the obvious quadratic form [33]. It is the C^*-algebra $\mathbb{C} \oplus \varepsilon\mathbb{C}$ generated by ε satisfying $\varepsilon^* = \varepsilon$ and $\varepsilon^2 = 1$. Assigning to ε the degree 1 yields a $\mathbb{Z}/2\mathbb{Z}$-grading on C_1. We have:

Proposition 3.6.3 *The map*

$$\begin{aligned} E^1(A,B) &\longrightarrow E(A, B\otimes C_1),\\ (\mathcal{E},\pi,F) &\longmapsto (\mathcal{E}\otimes C_1, \pi\otimes \mathrm{Id}, F\otimes\varepsilon) \end{aligned} \tag{3.17}$$

induces an isomorphism between the quotient of $E^1(A,B)$ *under homotopy and* $KK_1(A,B) = KK(A, B\otimes C_1)$.

Proof The grading of C_1 gives the one of $\mathcal{E}\otimes C_1$, and the map (3.17) easily gives a homomorphism c from $KK_1(A,B)$ to $KK(A,B\otimes C_1)$.

Now let $y = (\mathcal{E},\pi,F)\in E(A,B\otimes C_1)$. Multiplication by ε on the right of $\mathcal{E}$ makes sense, even if B is not unital, and one has $\mathcal{E}_1 = \mathcal{E}_0\varepsilon$. It follows that $\mathcal{E} = \mathcal{E}_0\oplus\mathcal{E}_1 \simeq \mathcal{E}_0\oplus\mathcal{E}_0$ and that any $T\in \mathrm{Mor}(\mathcal{E})$, thanks to the $B\otimes C_1$-linearity, has the following expression:

$$T = \begin{pmatrix} Q & P\\ P & Q\end{pmatrix}, \qquad P, Q\in \mathrm{Mor}_B(\mathcal{E}_0).$$

Thus $F = \begin{pmatrix} 0 & P\\ P & 0\end{pmatrix}$, $\pi = \begin{pmatrix} \pi_0 & 0\\ 0 & \pi_0\end{pmatrix}$ and $c^{-1}[y] = [\mathcal{E}_0,\pi_0,P]$. □

Remark 3.6.4 The opposite of $(\mathcal{E},\pi,F)$ in $KK_1(A,B)$ is represented by $(\mathcal{E},\pi,-F)$. One may wonder why we have to decide if a Kasparov module is graded or not. Actually, if we forget the $\mathbb{Z}/2\mathbb{Z}$ grading of a graded Kasparov A–B-module $x = (\mathcal{E},\pi,F)$ and consider it as an ungraded module, then we get the trivial class in $KK_1(A,B)$. Let us prove this claim.

The grading of x implies that $\mathcal{E}$ has a decomposition $\mathcal{E} = \mathcal{E}_0\oplus\mathcal{E}_1$ for which F has degree 1, that is,

$$F = \begin{pmatrix} 0 & Q\\ P & 0\end{pmatrix}.$$

Now

$$G_t = \cos(t\pi/2)\, F + \sin(t\pi/2)\begin{pmatrix} 1 & 0\\ 0 & -1\end{pmatrix} \tag{3.18}$$

provides an homotopy in KK_1 between x and

$$\left(\mathcal{E},\pi,\begin{pmatrix} 1 & 0\\ 0 & -1\end{pmatrix}\right).$$

Because the latter is degenerate, the claim is proved.

Example 3.6.5 Take again the example of the Dirac operator $\not{D}$ introduced in Section 3.6.3.2 on a spin^c manifold X whose dimension is odd. There is no natural

$\mathbb{Z}/2\mathbb{Z}$ grading for the spinor bundle. The previous triple $x_{\not{D}}$ provides this time an interesting class in $E^1(C(X), \mathbb{C})$.

3.6.5 The Kasparov product

In this subsection we construct the product

$$KK(A, B) \otimes KK(B, C) \to KK(A, C).$$

It satisfies the properties given in Section 3.4. Actually:

Theorem 3.6.6 *Let $x = (\mathcal{E}, \pi, F)$ in $E(A, B)$ and $x = (\mathcal{E}', \pi', F')$ in $E(B, C)$ be two Kasparov modules. Set*

$$\mathcal{E}'' = \mathcal{E} \otimes_B \mathcal{E}'$$

and

$$\pi'' = \pi \otimes 1.$$

Then there exists a unique – up to homotopy – F'-connection on $\mathcal{E}''$ denoted by F'' such that

- $(\mathcal{E}'', \pi'', F'') \in E(A, C)$,
- $\pi''(a)\big[F'', F \otimes 1\big]\pi''(a)$ *is nonnegative modulo $\mathcal{K}(\mathcal{E}'')$ for all $a \in A$.*

$(\mathcal{E}'', \pi'', F'')$ is the Kasparov product of x and x'. It enjoys all the properties described in Section 3.4.

Proof Idea of the proof: We only explain the construction of the operator F''. For a complete proof, see for instance [14, 30]. A naive idea for F'' could be $F \otimes 1 + 1 \otimes F'$, but the trouble is that the operator $1 \otimes F'$ is in general not well defined. We can overcome this first difficulty by replacing the not well defined $1 \otimes F'$ by any F'-connection G on $\mathcal{E}''$, and try $F \otimes 1 + G$. We then stumble on a second problem, namely that the properties of Kasparov module are not satisfied in general with this candidate for F'': for instance, $(F^2 - 1) \otimes 1 \in \mathcal{K}(\mathcal{E}) \otimes 1 \not\subset \mathcal{K}(\mathcal{E}'')$ as soon as $\mathcal{E}''$ is not finitely generated.

The case of tensor products of elliptic self-adjoint differential operators on a closed manifold M gives us a hint towards the right way. If D_1 and D_2 are two such operators and H_1, H_2 the natural L^2 spaces on which they act, then the bounded operator on $H_1 \otimes H_2$ given by

$$\frac{D_1}{\sqrt{1 + D_1^2}} \otimes 1 + 1 \otimes \frac{D_2}{\sqrt{1 + D_2^2}} \tag{3.19}$$

inherits the same problem as $F \otimes 1 + G$, but

$$D'' := \frac{1}{\sqrt{2 + D_1^2 \otimes 1 + 1 \otimes D_2^2}}(D_1 \otimes 1 + 1 \otimes D_2)$$

has better properties: $D''^2 - 1$ and $[C(M), D'']$ belong to $\mathcal{K}(H_1 \otimes H_2)$. Note that

$$D'' = \sqrt{M} \cdot \frac{D_1}{\sqrt{1 + D_1^2}} \otimes 1 + \sqrt{N} \cdot 1 \otimes \frac{D_2}{\sqrt{1 + D_2^2}}$$

with

$$M = \frac{1 + D_1^2 \otimes 1}{2 + D_1^2 \otimes 1 + 1 \otimes D_2^2} \quad \text{and} \quad N = \frac{1 + 1 \otimes D_2^2}{2 + D_1^2 \otimes 1 + 1 \otimes D_2^2}.$$

The operators M, N are bounded on $H_1 \otimes H_2$, are positive, and satisfy $M + N = 1$. We thus see that in that case, the naive idea (3.19) can be corrected by combining the operators involved with some adequate "partition of unity."

Turning back to our problem, this calculation leads us to look for an adequate operator F'' in the following form:

$$F'' = \sqrt{M} \cdot F \otimes 1 + \sqrt{N} G.$$

We need to have that F'' is a F'-connection, and satisfies $a \cdot (F''^2 - 1) \in \mathcal{K}(E'')$ and $[a, F''] \in \mathcal{K}(E'')$ for all $a \in A$ (by a we mean $\pi''(a)$). Using the previous form for F'', a small computation shows that these assertions become true if all the following conditions hold:

(i) M is a 0-connection (equivalently, N is a 1-connection),
(ii) $[M, F \otimes 1]$, $N \cdot [F \otimes 1, G]$, $[G, M]$, $N(G^2 - 1)$ belong to $\mathcal{K}(E'')$,
(iii) $[a, M]$, $N \cdot [G, a]$ belong to $\mathcal{K}(E'')$.

At this point there is a miracle:

Theorem 3.6.7 (Kasparov's technical theorem) *Let J be a C^*-algebra, and denote by $\mathcal{M}(J)$ its multiplier algebra. Assume there are two subalge-bras A_1, A_2 of $\mathcal{M}(J)$ and a linear subspace $\Delta \subset \mathcal{M}(J)$ such that*

$$A_1 A_2 \subset J,$$

$$[\Delta, A_1] \subset J.$$

Then there exist two nonnegative elements $M, N \in \mathcal{M}(J)$ with $M + N = 1$ such that

$$M A_1 \subset J,$$
$$N A_2 \subset J,$$
$$[M, \triangle] \subset J.$$

For a proof, see [25].

Now, to get $(\imath)$, $(\imath\imath)$, $(\imath\imath\imath)$, we apply this theorem with

$$A_1 = C^* \langle \mathcal{K}(\mathcal{E}) \otimes 1,\ \mathcal{K}(\mathcal{E}'') \rangle,$$
$$A_2 = C^* \langle G^2 - 1,\ [G, F \otimes 1],\ \left[G, \pi''\right] \rangle,$$
$$\triangle = Vect \langle \pi''(A),\ G,\ F \otimes 1 \rangle.$$

This gives us the correct F''. □

3.6.6 Equivalence and duality in KK-theory

With the Kasparov product come the following notions:

Definition 3.6.8 Let A, B be two C^*-algebras.

- One says that A and B are *KK-equivalent* if there exist $\alpha \in KK(A, B)$ and $\beta \in KK(B, A)$ such that

  $$\alpha \otimes \beta = 1_A \in KK(A, A) \quad \text{and} \quad \beta \otimes \alpha = 1_B \in KK(B, B).$$

  In that case, the pair (α, β) is called a *KK-equivalence*, and it gives rise to isomorphisms

  $$KK(A \otimes C, D) \simeq KK(B \otimes C, D) \quad \text{and} \quad KK(C, A \otimes D) \simeq KK(C, B \otimes D)$$

  given by Kasparov products for all C^*-algebras C, D.
- One says that A and B are *KK-dual (or Poincaré dual)* if there exist $\delta \in KK(A \otimes B, \mathbb{C})$ and $\lambda \in KK(\mathbb{C}, A \otimes B)$ such that

  $$\lambda \underset{B}{\otimes} \delta = 1 \in KK(A, A) \quad \text{and} \quad \lambda \underset{A}{\otimes} \delta = 1 \in KK(B, B).$$

  In that case, the pair (λ, δ) is called a KK-duality, and it gives rise to isomorphisms

  $$KK(A \otimes C, D) \simeq KK(C, B \otimes D) \quad \text{and} \quad KK(C, A \otimes D) \simeq KK(B \otimes C, B \otimes D)$$

  given by Kasparov products for all C^*-algebras C, D.

We continue this subsection with classical computations illustrating these notions.

3.6.6.1 Bott periodicity

Let $\beta \in KK(\mathbb{C}, C_0(\mathbb{R}^2))$ be represented by the Kasparov module:

$$(\mathcal{E}, \pi, C) = \left(C_0(\mathbb{R}^2) \oplus C_0(\mathbb{R}^2),\ 1,\ \frac{1}{\sqrt{1+c^2}} \begin{pmatrix} 0 & c_- \\ c_+ & 0 \end{pmatrix} \right),$$

where c_+, c_- are the operators given by pointwise multiplication by $x - \imath y$ and $x + \imath y$, respectively, and

$$c = \begin{pmatrix} 0 & c_- \\ c_+ & 0 \end{pmatrix}.$$

Let $\alpha \in KK(C_0(\mathbb{R}^2), \mathbb{C})$ be represented by the Kasparov module:

$$(\mathcal{H}, \pi, F) = \left(L^2(\mathbb{R}^2) \oplus L^2(\mathbb{R}^2),\ \pi,\ \frac{1}{\sqrt{1+D^2}} \begin{pmatrix} 0 & D_- \\ D_+ & 0 \end{pmatrix} \right),$$

where $\pi : C_0(\mathbb{R}^2) \to \mathcal{L}(L^2(\mathbb{R}^2) \oplus L^2(\mathbb{R}^2))$ is the action given by multiplication of functions, the operators D_+ and D_- are given by

$$D_+ = \partial_x + \imath \partial y,$$
$$D_- = -\partial_x + \imath \partial y,$$

and $D = \begin{pmatrix} 0 & D_- \\ D_+ & 0 \end{pmatrix}$.

Theorem 3.6.9 *α and β provide a KK-equivalence between $C_0(\mathbb{R}^2)$ and $\mathbb{C}$.*

This is the Bott periodicity theorem in the bivariant K-theory framework.

Proof Let us begin with the computation of $\beta \otimes \alpha \in KK(\mathbb{C}, \mathbb{C})$. We have an identification

$$\mathcal{E} \underset{C_0(\mathbb{R}^2)}{\otimes} \mathcal{H} \simeq \mathcal{H} \oplus \mathcal{H}, \tag{3.20}$$

where on the right, the first copy of $\mathcal{H}$ stands for

$$\mathcal{E}_0 \underset{C_0(\mathbb{R}^2)}{\otimes} \mathcal{H}_0 \oplus \mathcal{E}_1 \underset{C_0(\mathbb{R}^2)}{\otimes} \mathcal{H}_1$$

and the second for

$$\mathcal{E}_0 \underset{C_0(\mathbb{R}^2)}{\otimes} \mathcal{H}_1 \oplus \mathcal{E}_1 \underset{C_0(\mathbb{R}^2)}{\otimes} \mathcal{H}_0.$$

One checks directly that under this identification the operator

$$G = \frac{1}{\sqrt{1+D^2}} \begin{pmatrix} 0 & 0 & D_- & 0 \\ 0 & 0 & 0 & -D_+ \\ D_+ & 0 & 0 & 0 \\ 0 & -D_- & 0 & 0 \end{pmatrix} \tag{3.21}$$

is an F-connection. On the other hand, under the identification (3.20), the operator $C \otimes 1$ gives

$$\frac{1}{\sqrt{1+c^2}} \begin{pmatrix} 0 & 0 & 0 & c_- \\ 0 & 0 & c_+ & 0 \\ 0 & c_- & 0 & 0 \\ c_+ & 0 & 0 & 0 \end{pmatrix}. \tag{3.22}$$

It immediately follows that $\beta \otimes \alpha$ is represented by

$$\delta = \left(\mathcal{H} \oplus \mathcal{H}, 1, \frac{1}{\sqrt{1+c^2+D^2}} \mathbf{D} \right), \tag{3.23}$$

where

$$\mathbf{D} = \begin{pmatrix} 0 & \mathbf{D}_- \\ \mathbf{D}_+ & 0 \end{pmatrix}$$

with

$$\mathbf{D}_+ = \begin{pmatrix} D_+ & c_- \\ c_+ & -D_- \end{pmatrix} \quad \text{and} \quad \mathbf{D}_- = \mathbf{D}_+^*.$$

Observe that, denoting by ρ the rotation in $\mathbb{R}^2$ of angle $\pi/4$, we have

$$\begin{aligned} \begin{pmatrix} \rho^{-1} & 0 \\ 0 & \rho \end{pmatrix} \begin{pmatrix} 0 & \mathbf{D}_- \\ \mathbf{D}_+ & 0 \end{pmatrix} \begin{pmatrix} \rho & 0 \\ 0 & \rho^{-1} \end{pmatrix} &= \begin{pmatrix} 0 & \rho^{-1}\mathbf{D}_-\rho^{-1} \\ \rho\mathbf{D}_+\rho & 0 \end{pmatrix} \\ &= \begin{pmatrix} 0 & 0 & \imath(\partial_y - y) & -\partial_x + x \\ 0 & 0 & \partial_x + x & -\imath(\partial_y + y) \\ \imath(\partial_y + y) & -\partial_x + x & 0 & 0 \\ \partial_x + x & \imath(-\partial_y + y) & 0 & 0 \end{pmatrix} \\ &= \begin{pmatrix} 0 & x - \partial_x \\ x + \partial_x & 0 \end{pmatrix} \otimes 1 + 1 \otimes \begin{pmatrix} 0 & \imath(\partial_y - y) \\ \imath(\partial_y + y) & 0 \end{pmatrix}. \end{aligned}$$

Of course,

$$\delta \sim_h \left(\mathcal{H} \oplus \mathcal{H}, 1, \frac{1}{\sqrt{1+c^2+D^2}} \begin{pmatrix} 0 & \rho^{-1}\mathbf{D}_-\rho^{-1} \\ \rho\mathbf{D}_+\rho & 0 \end{pmatrix} \right),$$

and the preceding computation shows that δ coincides with the Kasparov product $u \otimes u$ with $u \in KK(\mathbb{C}, \mathbb{C})$ given by

$$u = \left(L^2(\mathbb{R})^2, 1, \frac{1}{\sqrt{1 + x^2 + \partial_x^2}} \begin{pmatrix} 0 & x - \partial_x \\ x + \partial_x & 0 \end{pmatrix} \right).$$

A simple exercise shows that $\partial_x + x : L^2(\mathbb{R}) \to L^2(\mathbb{R})$ is essentially self-adjoint with one-dimensional kernel and zero-dimensional cokernel; thus $1 = u = u \otimes u \in KK(\mathbb{C}, \mathbb{C})$.

Let us turn to the computation of $\alpha \otimes \beta \in KK(C_0(\mathbb{R}^2), C_0(\mathbb{R}^2))$. It is a Kasparov product over $\mathbb{C}$, and thus it commutes:

$$\alpha \otimes \beta = \tau_{C_0(\mathbb{R}^2)}(\beta) \otimes \tau_{C_0(\mathbb{R}^2)}(\alpha), \tag{3.24}$$

but we must observe that the two copies of $C_0(\mathbb{R}^2)$ in $\tau_{C_0(\mathbb{R}^2)}(\beta)$ and $\tau_{C_0(\mathbb{R}^2)}(\alpha)$ play different roles: on should think of the first copy as functions of the variable $u \in \mathbb{R}^2$, and of the second as functions of the variable $v \in \mathbb{R}^2$. It follows that one cannot directly factorize $\tau_{C_0(\mathbb{R}^2)}$ on the right-hand side of (3.24) in order to use the value of $\beta \otimes \alpha$. This is where a classical argument, known as the rotation trick of Atiyah, is necessary:

Lemma 3.6.10 *Let $\phi : C_0(\mathbb{R}^2) \otimes C_0(\mathbb{R}^2) \to C_0(\mathbb{R}^2) \otimes C_0(\mathbb{R}^2)$ be the flip automorphism: $\phi(f)(u, v) = f(v, u)$. Then*

$$[\phi] = 1 \in KK(C_0(\mathbb{R}^2) \otimes C_0(\mathbb{R}^2), C_0(\mathbb{R}^2) \otimes C_0(\mathbb{R}^2)).$$

Proof Proof of the lemma: Let us denote by I_2 the identity matrix of $M_2(\mathbb{R})$. Use a continuous path of isometries of $\mathbb{R}^4$ connecting

$$\begin{pmatrix} 0 & I_2 \\ I_2 & 0 \end{pmatrix} \quad \text{to} \quad \begin{pmatrix} I_2 & 0 \\ 0 & I_2 \end{pmatrix}.$$

This gives a homotopy $(C_0(\mathbb{R}^2) \otimes C_0(\mathbb{R}^2), \phi, 0) \sim_h (C_0(\mathbb{R}^2) \otimes C_0(\mathbb{R}^2), \mathrm{Id}, 0)$. ☐

Now

$$\begin{aligned} \alpha \otimes \beta &= \tau_{C_0(\mathbb{R}^2)}(\beta) \otimes \tau_{C_0(\mathbb{R}^2)}(\alpha) = \tau_{C_0(\mathbb{R}^2)}(\beta) \otimes [\phi] \otimes \tau_{C_0(\mathbb{R}^2)}(\alpha) \\ &= \tau_{C_0(\mathbb{R}^2)}(\beta \otimes \alpha) = \tau_{C_0(\mathbb{R}^2)}(1) = 1 \in KK(C_0(\mathbb{R}^2), C_0(\mathbb{R}^2)). \end{aligned} \tag{3.25}$$

☐

3.6.6.2 *Self-duality of $C_0(\mathbb{R})$*

With the same notation as before, we get:

Corollary 3.6.11 *The algebra $C_0(\mathbb{R})$ is Poincaré dual to itself.*

Other examples of Poincaré-dual algebras will be given later.

Proof The automorphism ψ of $C_0(\mathbb{R})^{\otimes^3}$ given by $\psi(f)(x, y, z) = f(z, x, y)$ is homotopic to the identity; thus

$$\begin{aligned}\beta \underset{C_0(\mathbb{R})}{\otimes} \alpha &= \tau_{C_0(\mathbb{R})}(\beta) \otimes \tau_{C_0(\mathbb{R})}(\alpha) = \tau_{C_0(\mathbb{R})}(\beta) \otimes [\psi] \otimes \tau_{C_0(\mathbb{R})}(\alpha) \\ &= \tau_{C_0(\mathbb{R})}(\beta \otimes \alpha) = \tau_{C_0(\mathbb{R})}(1) = 1 \in KK(C_0(\mathbb{R}), C_0(\mathbb{R})). \quad (3.26)\end{aligned}$$

□

Exercise 3.6.12 With $C_1 = \mathbb{C} \oplus \varepsilon\mathbb{C}$ the Clifford algebra of $\mathbb{C}$, consider

$$\beta_c = \left(C_0(\mathbb{R}) \otimes C_1, 1, \frac{x}{\sqrt{x^2+1}} \otimes \varepsilon \right) \in KK(\mathbb{C}, C_0(\mathbb{R}) \otimes C_1),$$

$$\alpha_c = \left(L^2(\mathbb{R}, \Lambda^*\mathbb{R}), \pi, \frac{1}{\sqrt{1+\Delta}}(d+\delta) \right) \in KK(C_0(\mathbb{R}) \otimes C_1, \mathbb{C}),$$

where $(d+\delta)(a+bdx) = -b' + a'dx$, $\Delta = (d+\delta)^2$, and $\pi(f \otimes \varepsilon)$ sends $a + bdx$ to $f(b+adx)$. Show that β_c, α_c provide a KK-equivalence between $\mathbb{C}$ and $C_0(\mathbb{R}) \otimes C_1$. (Hints: compute directly $\beta_c \otimes \alpha_c$; then use the commutativity of the Kasparov product over $\mathbb{C}$ and check that the flip of $(C_0(\mathbb{R}) \otimes C_1)^{\otimes^2}$ is 1 to conclude about the computation of $\alpha_c \otimes \beta_c$.)

3.6.6.3 A simple Morita equivalence

Let $\imath_n = (M_{1,n}(\mathbb{C}), 1, 0) \in E(\mathbb{C}, M_n(\mathbb{C}))$, where the $M_n(\mathbb{C})$-module structure is given by multiplication by matrices on the right. Note that $[\imath_n]$ is also the class of the homomorphism $\mathbb{C} \to M_n(\mathbb{C})$ given by the upper left corner inclusion. Let also $\jmath_n = (M_{n,1}(\mathbb{C}), m, 0) \in E(M_n(\mathbb{C}), \mathbb{C})$, where m is multiplication by matrices on the left. It follows in a straightforward way that

$$\imath_n \otimes \jmath_n \sim_h (\mathbb{C}, 1, 0) \quad \text{and} \quad \jmath_n \otimes \imath_n \sim_h (M_n(\mathbb{C}), 1, 0);$$

thus $\mathbb{C}$ and $M_n(\mathbb{C})$ are KK-equivalent, and this is an example of a Morita equivalence. The map in K-theory associated with $\jmath : \cdot \otimes \jmath_n : K_0(M_n(\mathbb{C})) \to \mathbb{Z}$ is just the trace homomorphism. Similarly, let us consider the Kasparov elements $\imath \in E(\mathbb{C}, \mathcal{K}(\mathcal{H}))$ associated to the homomorphism $\imath : \mathbb{C} \to \mathcal{K}(\mathcal{H})$ given by the choice of a rank one projection, and $\jmath = (\mathcal{H}, m, 0) \in E(\mathcal{K}(\mathcal{H}), \mathbb{C})$, where m is just the action of compact operators on $\mathcal{H}$: they provide a KK-equivalence between $\mathcal{K}$ and $\mathbb{C}$.

3.6.6.4 $C_0(\mathbb{R})$ and C_1.

We leave the proof of the following result as an exercise:

Proposition 3.6.13 *The algebras $C_0(\mathbb{R})$ and C_1 are KK-equivalent.*

Proof Hint: Consider

$$\widetilde{\alpha} = \left(L^2(\mathbb{R}, \Lambda^*\mathbb{R}), m, \frac{1}{\sqrt{1+\Delta}}(d+\delta)\right) \in KK(C_0(\mathbb{R}), C_1),$$

where d, δ, Δ are defined in the previous exercise, $m(f)(\xi) = f\xi$, and the C_1-right module structure of $L^2(\mathbb{R}, \Lambda^*\mathbb{R})$ is given by $(a + bdx) \cdot \varepsilon = -ib + iadx$. Consider also

$$\widetilde{\beta} = \left(C_0(\mathbb{R})^2, \varphi, \frac{x}{\sqrt{1+x^2}}\begin{pmatrix} 0 & 1 \\ 1 & 0 \end{pmatrix}\right) \in KK(C_1, C_0(\mathbb{R})),$$

where $\varphi(\varepsilon)(f, g) = (-ig, if)$. Prove that they provide the desired KK-equivalence.
□

Exercise 3.6.14

(i) Check that $\tau_{C_1} : KK(A, B) \to KK(A \otimes C_1, B \otimes C_1)$ is an isomorphism.
(ii) Check that under τ_{C_1} and the Morita equivalence $M_2(\mathbb{C}) \sim \mathbb{C}$, the elements α_c, β_c of the previous exercise coincide with $\widetilde{\alpha}, \widetilde{\beta}$ and recover the KK-equivalence between C_1 and $C_0(\mathbb{R})$.

Remark 3.6.15 At this point, one sees that $KK_1(A, B) = KK(A, B(\mathbb{R}))$ $(B(\mathbb{R}) := C_0(\mathbb{R}) \otimes B)$ can also be presented in the following different ways:

$$E_1(A, B)/\sim_h \simeq KK(A, B \otimes C_1) \simeq KK(A \otimes C_1, B) \simeq KK(A(\mathbb{R}), B).$$

3.6.7 Computing the Kasparov product without its definition

Computing the product of two Kasparov modules is in general quite hard, but we are often in one of the following situations.

3.6.7.1 Use of the functorial properties

Thanks to the functorial properties listed in Section 3.4, many products can be deduced from known, already computed ones. For instance, in the proof of Bott periodicity (the KK-equivalence between $\mathbb{C}$ and $C_0(\mathbb{R}^2)$) one had to compute two products: the first one was directly computed; the second one was deduced from the first using the properties of the Kasparov product and a simple geometric fact. There are numerous examples of this kind.

3.6.7.2 Maps between K-theory groups

Let A, B be two unital (if not, add a unit) C^*-algebras, let $x \in KK(A, B)$ be given by a Kasparov module $(\mathcal{E}, \pi, F)$ where F has a closed range, and assume that we are interested in the map $\phi_x : K_0(A) \to K_0(B)$ associated with x in the following way:

$$y \in K_0(A) \simeq KK(\mathbb{C}, A), \qquad \phi_x(y) = y \otimes x.$$

This product takes a particularly simple form when y is represented by $(\mathcal{P}, 1, 0)$ with $\mathcal{P}$ a finitely generated projective A-module (see Section 3.6.3.6):

$$y \otimes x = \left(\mathcal{P} \underset{A}{\otimes} \mathcal{E}, 1 \otimes \pi, G\right) = (\ker(G), 1, 0),$$

where G is an arbitrary F-connection.

3.6.7.3 *Kasparov elements constructed from homomorphisms*

Sometimes, Kasparov classes $y \in KK(B, C)$ can be explicitly represented as Kasparov products of classes of homomorphisms with inverses of such classes. Assume for instance that $y = [e_0]^{-1} \otimes [e_1]$, where $e_0 : \mathcal{C} \to B$, $e_1 : \mathcal{C} \to C$ are homomorphisms of C^*-algebras and e_0 produces an invertible element in KK-theory (for instance, $\ker e_0$ is K-contractible and B is nuclear or $\mathcal{C}$, B K-nuclear; see [16, 50]). Then computing a Kasparov product $x \otimes y$ where $x \in KK(A, B)$ amounts to lifting x to $KK(A, \mathcal{C})$ (that is, finding $x' \in KK(A, \mathcal{C})$ such that $(e_0)_*(x') = x$) and restricting this lift to $KK(A, C)$ (i.e., evaluating $x'' = (e_1)_*(x')$). It follows from the properties of the product that $x'' = x \otimes y$.

Example 3.6.16 Consider the tangent groupoid $\mathcal{G}_\mathbb{R}$ of $\mathbb{R}$, and let $\delta = [e_0]^{-1} \otimes [e_1] \otimes \mu$ be the associated deformation element: $e_0 : C^*(\mathcal{G}_\mathbb{R}) \to C^*(T\mathbb{R}) \simeq C_0(\mathbb{R}^2)$ is evaluation at $t = 0$, $e_1 : C^*(\mathcal{G}_\mathbb{R}) \to C^*(\mathbb{R} \times \mathbb{R}) \simeq \mathcal{K}(L^2(\mathbb{R})) \simeq \mathcal{K}$ is evaluation at $t = 1$, and $\mu = (L^2(\mathbb{R}), m, 0) \in KK(\mathcal{K}, \mathbb{C})$ gives the Morita equivalence $\mathcal{K} \sim \mathbb{C}$.

Let $\beta \in KK(\mathbb{C}, C_0(\mathbb{R}^2))$ be the element used in Section 3.6.6.1. Then $\beta \otimes \delta \in KK(\mathbb{C}, \mathbb{C})$ is easy to compute. The lift $\beta' \in KK(\mathbb{C}, C^*(\mathcal{G}_\mathbb{R}))$ is produced using the pseudodifferential calculus for groupoids (see Section 3.7) and can be presented as a family $\beta' = (\beta_t)$ with

$$\beta_0 = \beta, \qquad t > 0,$$

$$\beta_t = \left(C^*\left(\mathbb{R} \times \mathbb{R}, \frac{dx}{t}\right), 1, \frac{1}{\sqrt{1 + x^2 + t^2\partial_x^2}} \begin{pmatrix} 0 & x - t\partial_x \\ x + t\partial_x & 0 \end{pmatrix}\right).$$

After restricting at $t = 1$ and applying the Morita equivalence, only the index of the Fredholm operator appearing in β_1 remains – i.e., $+1$ – and this proves $\beta \otimes \delta = 1$.

Observe that by uniqueness of the inverse, we can conclude that $\delta = \alpha$ in $KK(C_0(\mathbb{R}^2), \mathbb{C})$.

Example 3.6.17 (Boundary homomorphisms in long exact sequences.) Let

$$0 \to I \underset{i}{\to} A \underset{p}{\to} B \to 0$$

be a short exact sequence of C^*-algebras. We assume that either it admits a completely positive, norm-decreasing linear section or I, A, B are K-nuclear [50]. Let $C_p = \{(a, \varphi) \in A \oplus C_0([0, 1[, B) \,|\, p(a) = \varphi(0)\}$ be the cone of the homomorphism $p : A \to B$, and denote by d the homomorphism $C_0(]0, 1[, B) \hookrightarrow C_p$ given by $d(\varphi) = (0, \varphi)$, and by e the homomorphism $I \to C_p$ given by $e(a) = (a, 0)$. Thanks to the hypotheses, $[e]$ is invertible in KK-theory. One can set $\delta = [d] \otimes [e]^{-1} \in KK(C_0(\mathbb{R}) \otimes B, I)$ and use the Bott periodicity $C_0(\mathbb{R}^2) \underset{\mathrm{KK}}{\sim} \mathbb{C}$ in order to identify

$$KK_2(C, D) = KK(C_0(\mathbb{R}^2) \otimes C, D) \simeq KK(C, D).$$

Then the connecting maps in the long exact sequences

$$\begin{aligned} &\cdots \to KK_1(I, D) \to KK(B, D) \\ &\overset{i^*}{\to} KK(A, D) \overset{p^*}{\to} KK(I, D) \to KK_1(B, D) \to \cdots, \\ &\cdots \to KK_1(C, B) \to KK(C, I) \\ &\overset{i_*}{\to} KK(C, A) \overset{p_*}{\to} KK(C, B) \to KK_1(C, I) \to \cdots \end{aligned}$$

are given by the appropriate Kasparov products with δ.

III. Index Theorems

3.7 Introduction to pseudodifferential operators on groupoids

The historical motivation for developing pseudodifferential calculus on groupoids comes from Connes, who implicitly introduced this notion for foliations. Later on, this calculus was axiomatized and studied on general groupoids by several authors [38,39,52].

The following example illustrates how pseudodifferential calculus on groupoids arises in our approach to index theory. If P is a partial differential operator on $\mathbb{R}^n$,

$$P(x, D) = \sum_{|\alpha| \leq d} c_\alpha(x) D_x^\alpha,$$

we may associate to it the asymptotic operator

$$P(x, tD) = \sum_{|\alpha| \leq d} c_\alpha(x)(tD_x)^\alpha$$

by introducing a parameter $t \in]0,1]$ in front of each ∂_{x_j}. Here we use the usual convention: $D_x^\alpha = (-i\partial_{x_1})^{\alpha_1} \cdots (-i\partial_{x_n})^{\alpha_n}$. We would like to give an (interesting) meaning to the limit $t \to 0$. Of course we would not be happy with $tD \to 0$.

To investigate this question, let us look at $P(x, tD)$ as a left multiplier on $C^\infty(\mathbb{R}^n \times \mathbb{R}^n \times]0,1])$ rather as a linear operator on $C^\infty(\mathbb{R}^n)$:

$$\begin{aligned}
P(x, tD_x)u(x, y, t) &= \int e^{(x-z)\cdot\xi} P(x, t\xi)u(z, y, t)dzd\xi \\
&= \int e^{\frac{x-z}{t}\cdot\xi} P(x, \xi)u(z, y, t)\frac{dzd\xi}{t^n} \\
&= \int e^{(X-Z).\xi} P(x, \xi)u(x - t(X - Z), x - tX, t)dZd\xi.
\end{aligned}$$

In the last line we introduced the notation $X = \frac{x-y}{t}$ and performed the change of variables $Z = \frac{z-y}{t}$.

At this point, assume that u has the following behavior near $t = 0$:

$$u(x, y, t) = \widetilde{u}\left(y, \frac{x-y}{t}, t\right), \qquad \text{where } \widetilde{u} \in C^\infty(\mathbb{R}^{2n} \times [0,1]).$$

It follows that

$$\begin{aligned}
P(x, tD_x)u(x, x - tX, t) &= \int e^{(X-Z)\cdot\xi} P(x, \xi)\widetilde{u}(x - tX, Z, t)dZd\xi \\
&\overset{t\to 0}{\longrightarrow} \int e^{(X-Z)\cdot\xi} P(x, \xi)\widetilde{u}(x, Z, 0)dZd\xi \\
&= P(x, D_X)\widetilde{u}(x, X, 0).
\end{aligned}$$

Some observations:

- $P(x, D_X)$ is a partial differential operator in the variable X with constant coefficients, depending smoothly on a parameter x and with symbol coinciding with that of $P(x, D_x)$, in the sense that $\sigma(P(x, D_X)(x, X, \xi) = P(x, \xi)$. In particular, $P(x, D_X)$ is invariant under the translation $X \mapsto X + X_0$. Of course, $P(x, D_X)$ is nothing else, up to a Fourier transformation in X, than the symbol $P(x, \xi)$ of $P(x, D_x)$. In other words, denoting by $S_X(T\mathbb{R}^n)$ the space of smooth functions $f(x, X)$ rapidly decreasing in X and by $\mathcal{F}_X$ the Fourier transform with respect to the variable X, we have a commutative diagram

$$\begin{array}{ccc}
S_X(T\mathbb{R}^n) & \xrightarrow{P(x,D_X)} & S_X(T\mathbb{R}^n) \\
\mathcal{F}_X \downarrow & & \mathcal{F}_X \downarrow \\
S_\xi(T^*\mathbb{R}^n) & \xrightarrow{P(x,\xi)} & S_\xi(T^*\mathbb{R}^n),
\end{array}$$

where $P(x, D_X)$ acts as a left multiplier on the convolution algebra $S_X(T\mathbb{R}^n)$, and $P(x, \xi)$ acts as a left multiplier on the function algebra $S_\xi(T^*\mathbb{R}^n)$ (equipped with the pointwise multiplication of functions).

- u and $\widetilde{u}$ are related by the bijection

$$\begin{aligned} \phi : \mathbb{R}^{2n} \times [0,1] &\longrightarrow \mathcal{G}_{\mathbb{R}^n}, \\ (x, X, t) &\longmapsto (x - tX, x, t) \qquad \text{if } t > 0, \\ (x, X, 0) &\longmapsto (x, X, 0) \end{aligned}$$

($\phi^{-1}(x, y, t) = (y, (x-y)/t, t)$, $\phi^{-1}(x, X, 0) = (x, X, 0)$). In fact, the smooth structure of the tangent groupoid $\mathcal{G}_{\mathbb{R}^n}$ of the manifold $\mathbb{R}^n$ (see Section 3.2.7) is defined by requiring that ϕ be a diffeomorphism. Thus $\widetilde{u} \in C^\infty(\mathbb{R}^{2n} \times [0,1])$ means $u \in C^\infty(\mathcal{G}_{\mathbb{R}^n})$.

Thus $P(x, D_X)$ is another way to look at, and even another way to define, the symbol of $P(x, D_x)$. What is important for us is that it arises as a limit of a family P_t constructed with P, and the pseudodifferential calculus on the tangent groupoid of $\mathbb{R}^n$ will enable us to give a rigorous meaning to this limit and perform interesting computations.

The following material is taken from [38, 39, 52]. Let G be a Lie groupoid, with unit space $G^{(0)} = V$ and with a smooth (right) Haar system $d\lambda$. We assume that V is a compact manifold and that the s-fibers G_x, $x \in V$, have no boundary. We denote by U_γ the map induced on functions by right multiplication by γ, that is,

$$U_\gamma : C^\infty(G_{s(\gamma)}) \longrightarrow C^\infty(G_{r(\gamma)});\; U_\gamma f(\gamma') = f(\gamma'\gamma).$$

Definition 3.7.1 A G-operator is a continuous linear map $P : C_c^\infty(G) \to C^\infty(G)$ such that

(i) P is given by a family $(P_x)_{x\in V}$ of linear operators $P_x : C_c^\infty(G_x) \to C^\infty(G_x)$, and

$$\forall f \in C_c^\infty(G), \quad P(f)(\gamma) = P_{s(\gamma)} f_{s(\gamma)}(\gamma),$$

where f_x stands for the restriction $f|_{G_x}$.

(ii) The following invariance property holds

$$U_\gamma P_{s(\gamma)} = P_{r(\gamma)} U_\gamma.$$

Let P be a G-operator, and denote by $k_x \in C^{-\infty}(G_x \times G_x)$ the Schwartz kernel of P_x, for each $x \in V$, as obtained from the Schwartz kernel theorem applied to the manifold G_x provided with the measure $d\lambda_x$.

Thus, using property (i) in the definition,

$$\forall \gamma \in G, f \in C^\infty(G), \quad Pf(\gamma) = \int_{G_x} k_x(\gamma, \gamma') f(\gamma') d\lambda_x(\gamma') \qquad (x = s(\gamma)).$$

Next,

$$U_\gamma Pf(\gamma') = Pf(\gamma'\gamma) = \int_{G_x} k_x(\gamma'\gamma, \gamma'')f(\gamma'')d\lambda_x(\gamma'') \qquad (x = s(\gamma)),$$

and

$$\begin{aligned} P(U_\gamma f)(\gamma') &= \int_{G_y} k_y(\gamma', \gamma'')f(\gamma''\gamma)d\lambda_y(\gamma''), \qquad (y = r(\gamma)) \\ &\overset{\eta=\gamma''\gamma}{=} \int_{G_x} k_y(\gamma', \eta\gamma^{-1})f(\eta)d\lambda_x(\eta), \qquad (x = s(\gamma)), \end{aligned}$$

where the last line uses the invariance property of Haar systems. Property (ii) is equivalent to the following equalities of distributions on $G_x \times G_x$, for all $x \in V$:

$$\forall \gamma \in G, \quad k_x(\gamma'\gamma, \gamma'') = k_y(\gamma', \gamma''\gamma^{-1}) \qquad (x = s(\gamma),\ y = r(\gamma)).$$

Setting $k_P(\gamma) := k_{s(\gamma)}(\gamma, s(\gamma))$, we get $k_x(\gamma, \gamma') = k_P(\gamma\gamma'^{-1})$, and the linear operator $P : C_c^\infty(G) \to C^\infty(G)$ is given by

$$P(f)(\gamma) = \int_{G_x} k_P(\gamma\gamma'^{-1})d\lambda_x(\gamma') \qquad (x = s(\gamma)).$$

We may consider k_P as a single distribution on G acting on smooth functions on G by convolution. With a slight abuse of terminology, we will refer to k_P as the Schwartz (or convolution) kernel of P.

We say that P is *smoothing* if k_P lies in $C^\infty(G)$, and is *compactly supported* or *uniformly supported* if k_P is compactly supported (which implies that each P_x is properly supported).

Let us develop some examples of G-operators.

Examples 3.7.2

(i) If $G = G^{(0)} = V$ is just a set, then $G_x = \{x\}$ for all $x \in V$. Then in Definition 3.7.1, property (i) is empty, and property (ii) implies that a G-operator is given by pointwise multiplication by a smooth function $P \in C^\infty(V)$: $Pf(x) = P(x) \cdot f(x)$.

(ii) Let $G = V \times V$, the pair groupoid, and let the Haar system $d\lambda$ be given in the obvious way by a single measure dy on V:

$$d\lambda_x(y) = dy \qquad \text{under the identification } G_x = V \times \{x\} \simeq V.$$

It follows that for any G-operator P,

$$Pg(z, x) = \int_{V\times\{x\}} k_P(z, y)g(y, x)d\lambda_x(y, x) = \int_V k_P(z, y)g(y, x)dy,$$

which immediately proves that $P_x = P_y$ as linear operators on $C^\infty(V)$ under the obvious identifications $V \simeq V \times \{x\} \simeq V \times \{y\}$.

(iii) Let $p : X \to Z$ be a submersion, and $G = X \underset{Z}{\times} X = \{(x, y) \in X \times X \mid p(x) = p(y)\}$ the associated subgroupoid of the pair groupoid $X \times X$. The manifold G_x can be identified with the fiber $p^{-1}(p(x))$. Property (ii) implies that for any G-operator P, we have $P_x = P_y$ as linear operators on $p^{-1}(p(x))$ as soon as $y \in p^{-1}(p(x))$. Thus, P is actually given by a family $\tilde{P}_z$, $z \in Z$, of operators on $p^{-1}(z)$, with the relation $P_x = \tilde{P}_{p(x)}$.

(iv) Let $G = E$ be the total space of a (Euclidean, Hermitian) vector bundle $p : E \to V$, with $r = s = p$. The Haar system $d_x w$, $x \in V$, is given by the metric structure on the fibers of E. We have here

$$Pf(v) = \int_{E_x} k_P(v - w) f(w) d_x w \qquad (x = p(v)).$$

Thus, for all $x \in V$, P_x is a convolution operator on the linear space E_x.

(v) Let $G = \mathcal{G}_V = TV \times \{0\} \sqcup V \times V \times]0, 1]$ be the tangent groupoid of V. It can be viewed as a family of groupoids G_t parametrized by $[0, 1]$, where $G_0 = TV$ and $G_t = V \times V$ for $t > 0$. A $\mathcal{G}_V$-operator is given by a family P_t of G_t-operators, and $(P_t)_{t>0}$ is a family of operators on $C_c^\infty(V)$ parametrized by t, whereas P_0 is a family of translation-invariant operators on $T_x V$ parametrized by $x \in V$. The $\mathcal{G}_V$-operators are thus a blend of examples (ii) and (iv).

We now turn to the definition of pseudodifferential operators on a Lie groupoid G.

Definition 3.7.3 A G-operator P is a G-pseudodifferential operator of order m if:

(i) The Schwartz kernel k_P is smooth outside $G^{(0)}$.
(ii) For every distinguished chart $\psi : U \subset G \to \Omega \times s(U) \subset \mathbb{R}^{n-p} \times \mathbb{R}^p$ of G,

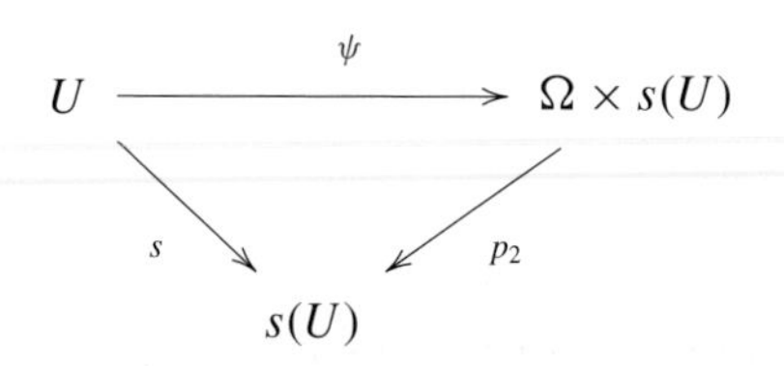

the operator $(\psi^{-1})^* P \psi^* : C_c^\infty(\Omega \times s(U)) \to C_c^\infty(\Omega \times s(U))$ is a smooth family parametrized by $s(U)$ of pseudodifferential operators of order m on Ω.

We will use few properties of this calculus and only provide some examples and a list of properties. The reader can find a complete presentation in [37–39, 51, 52].

Examples 3.7.4 In the previous five examples (Examples 3.7.3), a G-pseudodifferential operator is:

(i) an operator given by pointwise multiplication by a smooth function on V;
(ii) a single pseudodifferential operator on V;

(iii) a smooth family parametrized by Z of pseudodifferential operators in the fibers (this coincides with the notion of [7]);
(iv) a family parametrized by $x \in V$ of convolution operators in E_x such that the underlying distribution k_P identifies with the Fourier transform of a symbol on E (that is, a smooth function on E satisfying the standard decay conditions with respect to its variable in the fibers);
(v) the data provided by an asymptotic pseudodifferential operator on V together with its complete symbol, the choice of it depending on the gluing in $\mathcal{G}_V$ (this is quite close to the notions studied in [8, 22, 23]).

It turns out that the space $\Psi_c^*(G)$ of compactly supported G-pseudodifferential operators is an involutive algebra.

The principal symbol of a G-pseudodifferential operator P of order m is defined as a function $\sigma_m(P)$ on $A^*(G) \setminus G^{(0)}$ by

$$\sigma_m(P)(x, \xi) = \sigma_{pr}(P_x)(x, \xi),$$

where $\sigma_{pr}(P_x)$ is the principal symbol of the pseudodifferential operator P_x on the manifold G_x. Conversely, suppose we are given a symbol f of order m on $A^*(G)$ together with the following data:

(i) a smooth embedding $\theta : \mathcal{U} \to AG$, where $\mathcal{U}$ is a open set in G containing $G^{(0)}$, such that $\theta(G^{(0)}) = G^{(0)}$, $(d\theta)|_{G^0} = \mathrm{Id}$, and $\theta(\gamma) \in A_{s(\gamma)}G$ for all $\gamma \in \mathcal{U}$;
(ii) a smooth, compactly supported map $\phi : G \to \mathbb{R}_+$ such that $\phi^{-1}(1) = G^{(0)}$.

Then we get a G-pseudodifferential operator $P_{f,\theta,\phi}$ with the formula ($u \in C_c^\infty(G)$)

$$P_{f,\theta,\phi}u(\gamma) = \int_{\substack{\gamma' \in G_{s(\gamma)},\\ \xi \in A^*_{r(\gamma)}(G)}} e^{-i\theta(\gamma'\gamma^{-1})\cdot\xi} f(r(\gamma), \xi)\phi(\gamma'\gamma^{-1})u(\gamma')d\lambda_{s(\gamma)}(\gamma').$$

The principal symbol of $P_{f,\theta,\phi}$ is just the leading part of f.

The principal symbol map respects pointwise product, whereas the product law for total symbols is much more involved. An operator is *elliptic* when its principal symbol never vanishes, and in that case, as in the classical situation, it has a parametrix inverting it modulo $\Psi_c^{-\infty}(G) = C_c^\infty(G)$.

Operators of negative order in $\Psi_c^*(G)$ are actually in $C^*(G)$, whereas zero-order operators are in the multiplier algebra $\mathcal{M}(C^*(G))$.

All these definitions and properties immediately extend to the case of operators acting between sections of bundles on $G^{(0)}$ pulled back to G with the range map r. The space of compactly supported pseudodifferential operators on G acting on sections of r^*E and taking values in sections of r^*F will be denoted by $\Psi_c^*(G, E, F)$. If $F = E$, we get an algebra denoted by $\Psi_c^*(G, E)$.

Examples 3.7.5

(i) The family given by $P_t = P(x, tD_x)$ for $t > 0$ and $P_0 = P(x, D_X)$ described in the introduction of this section is a G-pseudodifferential operator with G the tangent groupoid of $\mathbb{R}^n$.

(ii) More generally, let V be a closed manifold endowed with a Riemannian metric. We denote by exp the exponential map associated with the metric. Let f be a symbol on V. We get a $\mathcal{G}_V$-pseudodifferential operator P by setting

$$(t > 0) \quad P_t u(x, y, t) = \int_{z \in V, \xi \in T_x^* V} e^{\frac{\exp_x^{-1}(z)}{t} \cdot \xi} f(x, \xi) u(z, y) \frac{dz d\xi}{t^n},$$

$$P_0 u(x, X, 0) = \int_{Z \in T_x V, \xi \in T_x^* V} e^{(X-Z) \cdot \xi} f(x, \xi) u(x, Z) dZ d\xi.$$

Moreover, P_1 is a pseudodifferential operator on the manifold V, which admits f as a complete symbol.

3.8 Index theorem for smooth manifolds

The purpose of this section is to present a proof of the Atiyah–Singer index theorem using deformation groupoids and show how it generalizes to conical pseudomanifolds. The results presented here come from recent works of the authors together with a joint work with V. Nistor [18–20]; we refer to [19,20] for the proofs.

3.8.1 The KK-element associated to a deformation groupoid

Before going to the description of the index maps, let us describe a useful and classical construction [13, 27].

Let G be a smooth deformation groupoid (Definition 3.2.6):

$$G = G_1 \times \{0\} \cup G_2 \times]0, 1] \rightrightarrows G^{(0)} = M \times [0, 1].$$

One can consider the saturated open subset $M \times]0, 1]$ of $G^{(0)}$. Using the isomorphisms $C^*(G|_{M \times]0,1]}) \simeq C^*(G_2) \otimes C_0(]0, 1])$ and $C^*(G|_{M \times \{0\}}) \simeq C^*(G_1)$, we obtain the following exact sequence of C^*-algebras:

$$0 \longrightarrow C^*(G_2) \otimes C_0(]0, 1]) \xrightarrow{i_{M \times]0,1]}} C^*(G) \xrightarrow{ev_0} C^*(G_1) \longrightarrow 0,$$

where $i_{M \times]0,1]}$ is the inclusion map and ev_0 is the *evaluation map* at 0, that is, ev_0 is the map coming from the restriction of functions to $G|_{M \times \{0\}}$. We assume now that $C^*(G_1)$ is nuclear. Because the C^*-algebra $C^*(G_2) \otimes C_0(]0, 1])$ is contractible, the long exact sequence in KK-theory shows that the group homomorphism

$$(ev_0)_* = \cdot \otimes [ev_0] : KK(A, C^*(G)) \to KK(A, C^*(G_1))$$

is an isomorphism for each C^*-algebra A. In particular, with $A = C^*(G)$ we get that $[ev_0]$ is invertible in KK-theory: there is an element $[ev_0]^{-1}$ in $KK(C^*(G_1), C^*(G))$ such that $[ev_0]\otimes[ev_0]^{-1} = 1_{C^*(G)}$ and $[ev_0]^{-1}\otimes[ev_0] = 1_{C^*(G_1)}$.

Let $ev_1 : C^*(G) \to C^*(G_2)$ be the evaluation map at 1, and $[ev_1]$ the corresponding element of $KK(C^*(G), C^*(G_2))$. The *KK-element associated to the deformation groupoid G* is defined by

$$\delta = [ev_0]^{-1}\otimes[ev_1] \in KK(C^*(G_1), C^*(G_2)).$$

We will meet several examples of this construction in the sequel.

3.8.2 The analytical index

Let M be a closed manifold, and consider its tangent groupoid:

$$\mathcal{G}_M^t := TM \times \{0\} \cup M \times M \times [0, 1] \rightrightarrows M \times [0, 1].$$

It is a deformation groupoid, and the preceding construction provides us a KK-element:

$$\partial_M = (e_1^M)_* \circ (e_0^M)_*^{-1} \in KK(C_0(T^*M), \mathcal{K}) \simeq KK(C_0(T^*M), \mathbb{C}),$$

where $e_i^M : C^*(\mathcal{G}_M^t) \to C^*(\mathcal{G}_M^t)|_{t=i}$ are evaluation homomorphisms.

The analytical index is then [13]

$$\begin{aligned}\mathrm{Ind}_a M := (e_1^M)_* \circ (e_0^M)_*^{-1} \; : \; & KK(\mathbb{C}, C_0(T^*M)) \to KK(\mathbb{C}, \mathcal{K}(L^2(M)) \\ & \simeq K_0(C_0(T^*M)) \simeq \mathbb{Z}\end{aligned}$$

or, in terms of the Kasparov product,

$$\mathrm{Ind}_a M = \cdot \otimes \partial_M.$$

Using the notion of pseudodifferential calculus for $\mathcal{G}_M^t$, it is easy to conclude that this map is the usual analytical index map. Indeed, let $f(x, \xi)$ be an elliptic zero-order symbol, and consider the $\mathcal{G}_M^t$-pseudodifferential operator, $P_f = (P_t)_{0\leq t\leq 1}$, defined as in Examples 3.7.5. Then f provides a K-theory class $[f] \in K_0(C^*(TM)) \simeq K_0(C_0(T^*M))$, whereas P provides a K-theory class $[P] \in K_0(C^*(\mathcal{G}_M^t))$, and

$$(e_0^M)_*([P]) = [f] \in K_0(C^*(TM)).$$

Thus

$$[f] \otimes [e_0^M]^{-1} \otimes [e_1^M] = [P_1] \in K_0(\mathcal{K}),$$

and $[P_1]$ coincides with $\mathrm{Ind}(P_1)$ under $K_0(\mathcal{K}) \simeq \mathbb{Z}$.

Because P_1 has principal symbol equal to the leading part of f, and because every class in $K_0(C_0(T^*M))$ can be obtained from a zero-order elliptic symbol, the claim is justified.

To be complete, let us explain that the analytical index map is the Poincaré dual of the homomorphism in K-homology associated with the obvious map: $M \to \{\cdot\}$. Indeed, thanks to the obvious homomorphism $\Psi : C^*(TM) \otimes C(M) \to C^*(TM)$ given by multiplication, ∂_M can be lifted into an element $D_M = \Psi_*(\partial_M) \in KK(C^*(TM) \otimes C(M), \mathbb{C}) = K^0(C^*(TM) \otimes C(M))$, called the *Dirac element*. This Dirac element yields the well-known Poincaré duality between $C_0(T^*M)$ and $C(M)$ [14, 19, 31], and in particular it gives an isomorphism

$$\cdot \underset{C^*(TM)}{\otimes} D_M : K_0(C^*(TM)) \xrightarrow{\simeq} K^0(C(M)),$$

whose inverse is induced by the principal symbol map.

One can then easily show the following proposition:

Proposition 3.8.1 *Let q: $M \to \cdot$ be the projection onto a point. The following diagram commutes:*

$$\begin{array}{ccc} K^0(T^*M) & \xrightarrow{\text{PD}} & K_0(M) \\ \text{Ind}_a \downarrow & & \downarrow q_* \\ \mathbb{Z} & \xrightarrow{=} & \mathbb{Z} \end{array}$$

3.8.3 The topological index

Take an embedding $M \to \mathbb{R}^n$, and let $p : N \to M$ be the normal bundle of this embedding. The vector bundle $TN \to TM$ admits a complex structure; thus we have a Thom isomorphism

$$T : K_0(C^*(TM)) \xrightarrow{\simeq} K_0(C^*(TN))$$

given by a KK-equivalence

$$T \in KK(C^*(TM), C^*(TN)).$$

T is called the *Thom element* [30].

The bundle N identifies with an open neighborhood of M into $\mathbb{R}^n$, so we have the excision map

$$j : C^*(TN) \to C^*(T\mathbb{R}^n).$$

Consider also $B : K_0(C^*(T\mathbb{R}^n)) \to \mathbb{Z}$ given by the isomorphism $C^*(T\mathbb{R}^n) \simeq C_0(\mathbb{R}^{2n})$ together with Bott periodicity.

The *topological index map* Ind_t is the composition

$$K(C^*(TM)) \xrightarrow{T} K(C^*(TN)) \xrightarrow{j_*} K(C^*(T\mathbb{R}^n)) \underset{\simeq}{\xrightarrow{B}} \mathbb{Z}.$$

This classical construction can be reformulated with groupoids.

First, let us give a description of T, or rather of its inverse, in terms of groupoids. Recall the construction of the Thom groupoid. We begin by pulling back TM over N in the groupoid sense. Let

$${}^*p^*(TM) = N \underset{M}{\times} TM \underset{M}{\times} N \rightrightarrows N$$

and

$$\mathcal{T}_N = TN \times \{0\} \sqcup {}^*p^*(TM) \times]0,1] \rightrightarrows N \times [0,1].$$

This *Thom groupoid* and the Morita equivalence between ${}^*p^*(TM)$ and TM provide the KK-element

$$\tau_N \in KK(C^*(TN), C^*(TM)).$$

This element is defined exactly as ∂_M is. Precisely, the evaluation map at 0, namely $\tilde{e}_0 : C^*(\mathcal{T}_N) \to C^*(TN)$, defines an invertible KK-element. We let $\tilde{e}_1 : C^*(\mathcal{T}_N) \to C^*({}^*p^*(TM))$ be the evaluation map at 1. The Morita equivalence between the groupoids TM and ${}^*p^*(TM)$ leads to a Morita equivalence between the corresponding C^*-algebra and thus to a KK-equivalence $\mathcal{M} \in KK(C^*({}^*p^*(TM)), C^*(TM))$. Then

$$\tau_N := [\tilde{e}_0]^{-1} \otimes [\tilde{e}_1] \otimes \mathcal{M}.$$

We have the following:

Proposition 3.8.2 [20] *If T is the KK-equivalence giving the Thom isomorphism, then*

$$\tau_N = T^{-1}.$$

This proposition also applies to interpret the isomorphism $B : K_0(C^*(T\mathbb{R}^n)) \to \mathbb{Z}$. Indeed, consider the embedding $\cdot \hookrightarrow \mathbb{R}^n$. The normal bundle is just $\mathbb{R}^n \to \cdot$, and we get as before

$$\tau_{\mathbb{R}^n} \in KK(C^*(T\mathbb{R}^n), \mathbb{C}).$$

Using the previous proposition, we get $B = \cdot \otimes \tau_{\mathbb{R}^n}$.

Remark also that $\mathcal{T}_{\mathbb{R}^n} = \mathcal{G}_{\mathbb{R}^n}$, so that $\tau_{\mathbb{R}^n} = [e_0^{\mathbb{R}^n}]^{-1} \otimes [e_1^{\mathbb{R}^n}]$.

Finally, the topological index

$$\mathrm{Ind}_t = \tau_{\mathbb{R}^n} \circ j_* \circ \tau_N^{-1}$$

is entirely described using (deformation) groupoids.

3.8.4 The equality of the indices

A last groupoid is necessary in order to prove the equality of index maps. Namely, this groupoid is obtained by recasting the construction of the Thom groupoid at the level of tangent groupoids:

$$\widetilde{\mathcal{T}}_N = \mathcal{G}_N \times \{0\} \sqcup {}^*(p \otimes \mathrm{Id}_{[0,1]})^*(\mathcal{G}_M) \times]0,1]. \tag{3.27}$$

As before, this yields a class

$$\widetilde{\tau}_N \in KK(C^*(\mathcal{G}_N), C^*(\mathcal{G}_M)).$$

All maps in the diagram

$$\begin{array}{ccccc}
\mathbb{Z} & = & \mathbb{Z} & = & \mathbb{Z} \\
{\scriptstyle e_1^M}\uparrow & & {\scriptstyle e_1^N}\uparrow & & {\scriptstyle e_1^{\mathbb{R}^n}}\uparrow \\
K_0(C^*(\mathcal{G}_M)) & \xleftarrow{\otimes\widetilde{\tau}_N} & K_0(C^*(\mathcal{G}_N)) & \xrightarrow{\widetilde{j}_*} & K_0(C^*(\mathcal{G}_{\mathbb{R}^n})) \\
{\scriptstyle e_0^M}\downarrow\simeq & & {\scriptstyle e_0^N}\downarrow\simeq & & {\scriptstyle e_0^{\mathbb{R}^n}}\downarrow\simeq \\
K_0(C^*(TM)) & \underset{\simeq}{\xleftarrow{\otimes\tau_N}} & K_0(C^*(TN)) & \xrightarrow{j_*} & K_0(C^*(T\mathbb{R}^n))
\end{array} \tag{3.28}$$

are given by Kasparov products with

(i) classes of homomorphisms coming from restrictions or inclusions between groupoids,
(ii) inverses of such classes,
(iii) explicit Morita equivalences.

This easily yields the commutativity of the diagram (3.28). Having in mind the previous description of index maps using groupoids, this commutativity property just implies

$$\mathrm{Ind}_a = \mathrm{Ind}_t.$$

3.9 The case of pseudomanifolds with isolated singularities

As we explained earlier, the proof of the K-theoretical form of the Atiyah–Singer index theorem presented in this chapter easily extends to the case of pseudomanifolds with isolated singularities. This is achieved provided one uses the correct notion of *tangent space* of the pseudomanifold; for a pseudomanifold X with one conical point (the case of several isolated singularities is similar), this is the noncommutative tangent space defined in Section 3.2.5:

$$T^{\mathsf{S}}X = X^- \times X^- \cup T\overline{X^+} \rightrightarrows X^\circ.$$

In the sequel, it will replace the ordinary tangent space of a smooth manifold. Moreover, it gives rise to another deformation groupoid, which will replace the ordinary tangent groupoid of a smooth manifold:

$$\mathcal{G}_X^t = T^{\mathsf{S}}X \times \{0\} \cup X^\circ \times X^\circ \times]0,1] \rightrightarrows X^\circ \times [0,1].$$

We call $\mathcal{G}_X^t$ the *tangent groupoid* of X. It can be provided with a smooth structure such that $T^{\mathsf{S}}X$ is a smooth subgroupoid. Moreover, both are amenable, so their reduced and maximal C^*-algebras coincide and are nuclear.

With these choices of $T^{\mathsf{S}}X$ as a tangent space for X and of $\mathcal{G}_X^t$ as a tangent groupoid, one can follow step by step all the constructions made in the previous section.

3.9.1 The analytical index

Using the partition $X^\circ \times [0,1] = X^\circ \times \{0\} \cup X^\circ \times]0,1]$ into saturated open and closed subsets of the units space of the tangent groupoid, we define the *KK*-element associated to the tangent groupoid of X:

$$\partial_X := [e_0]^{-1} \otimes [e_1] \in KK(C^*(T^{\mathsf{S}}X), \mathcal{K}) \simeq KK(C^*(T^{\mathsf{S}}X), \mathbb{C}),$$

where $e_0 : C^*(\mathcal{G}_X^t) \to C^*(\mathcal{G}_X^t|_{X^\circ \times \{0\}}) \simeq C^*(T^{\mathsf{S}}X)$ is the evaluation at 0, and $e_1 : C^*(\mathcal{G}_X^t) \to C^*(\mathcal{G}_X^t|_{X^\circ \times \{1\}}) \simeq \mathcal{K}(L^2(X))$ is the evaluation at 1.

Now we can define the analytical index exactly as we did for closed smooth manifolds. Precisely, the *analytical index* for X is set to be the map

$$\mathrm{Ind}_a^X = \cdot \otimes \partial_X : KK(\mathbb{C}, C^*(T^{\mathsf{S}}X)) \to KK(\mathbb{C}, \mathcal{K}(L^2(X^\circ))) \simeq \mathbb{Z}.$$

The interpretation of this map as the Fredholm index of an appropriate class of elliptic operators is possible and carried out in [34].

3.9.2 The Poincaré duality

Pursuing the analogy with smooth manifolds, we explain in this subsection that the analytical index map for X is Poincaré dual to the index map in *K*-homology associated to the obvious map : $X \to \{.\}$.

The algebras $C(X)$ and $C^\bullet(X) := \{f \in C(X) \mid f \text{ is constant on } cL\}$ are homotopic. If g belongs to $C^\bullet(X)$ and f to $C_c(T^{\mathsf{S}}X)$, let $g \cdot f$ be the element of $C_c(T^{\mathsf{S}}X)$ defined by $g \cdot f(\gamma) = g(r(\gamma))f(\gamma)$. This induces a $*$-morphism

$$\Psi : C(X) \otimes C^*(T^{\mathsf{S}}X) \to C^*(T^{\mathsf{S}}X).$$

The *Dirac element* is defined to be

$$D_X := [\Psi] \otimes \partial_X \in KK(C(X) \otimes C^*(T^{\mathsf{S}}X), \mathbb{C}).$$

We recall

Theorem 3.9.1 [19] *There exists a (dual-Dirac) element* $\lambda_X \in KK(\mathbb{C}, C(X) \otimes C^*(T^{\mathsf{S}}X))$ *such that*

$$\lambda_X \underset{C(X)}{\otimes} D_X = 1_{C^*(T^{\mathsf{S}}X)} \in KK(C^*(T^{\mathsf{S}}X), C^*(T^{\mathsf{S}}X)),$$

$$\lambda_X \underset{C^*(T^{\mathsf{S}}X)}{\otimes} D_X = 1_{C(X)} \in KK(C(X), C(X)).$$

This means that $C(X)$ *and* $C^*(T^{\mathsf{S}}X)$ *are Poincaré dual.*

Remark 3.9.2 The explicit construction of λ_X, which is heavy going and technical, can be avoided. In fact, the definitions of $T^{\mathsf{S}}X, \mathcal{G}_X^t$ and thus of D_X can be extended in a natural way to the case of an arbitrary pseudomanifold, and the proof of Poincaré duality can be done using a recursive argument on the depth of the stratification, starting with the case depth $= 0$, that is, with the case of smooth closed manifolds. This is the subject of [18].

The theorem implies that

$$KK(\mathbb{C}, C^*(T^{\mathsf{S}}X)) \simeq K_0(C^*(T^{\mathsf{S}}X)) \to K(C(X), \mathbb{C}) \simeq K^0(C(X)),$$
$$x \mapsto x \underset{C^*(T^{\mathsf{S}}X)}{\otimes} D_X$$

is an isomorphism. In [34], it is explained how to interpret its inverse as a principal symbol map, and one also get the analogue of Proposition 3.8.1:

Proposition 3.9.3 *Let* $q : X \to \cdot$ *be the projection onto a point. The following diagram commutes:*

$$\begin{array}{ccc} K_0(C^*(T^{\mathsf{S}}X)) & \xrightarrow{\text{PD}} & K_0(X) \\ \text{Ind}_a^X \downarrow & & \downarrow q_* \\ \mathbb{Z} & \xrightarrow{=} & \mathbb{Z} \end{array}$$

3.9.3 The topological index

3.9.3.1 Thom isomorphism

Take an *embedding* $X \hookrightarrow c\mathbb{R}^n = \mathbb{R}^n \times [0, +\infty[/\mathbb{R}^n \times \{0\}$. This means that we have a map which restricts to an embedding $X^\circ \to \mathbb{R}^n \times]0, +\infty[$ in the usual sense and which sends c to the image of $\mathbb{R}^n \times \{0\}$ in $c\mathbb{R}^n$. Moreover, we require the embedding on $X^- = L \times]0, 1[$ to be of the form $j \times \text{Id}$ where j is an embedding of L in $\mathbb{R}^n$.

Such an embedding provides a *conical normal bundle*. Precisely, let $p : N^\circ \to X^\circ$ be the normal bundle associated with $X^\circ \hookrightarrow \mathbb{R}^n \times]0, +\infty[$. We can identify

$N^\circ|_{X^-} \simeq N^\circ|_L \times]0,1[$, and set

$$N = \bar{c}N^\circ|_L \cup N^\circ|_{X^+}.$$

Thus N is the pseudomanifold with an isolated singularity obtained by gluing the closed cone $\bar{c}N^\circ|_L := N^\circ|_L \times [0,1]/N^\circ|_L \times \{0\}$ with $N^\circ|_{X^+}$ along their common boundary $N^\circ|_L \times \{1\} = N^\circ|_{\partial X^+}$. Moreover, $p : N \to X$ is a conical vector bundle.

The *Thom groupoid* is then

$$\mathcal{T}_N = T^{\mathsf{S}}N \times \{0\} \sqcup {}^*p^*(T^{\mathsf{S}}X) \times]0,1].$$

It is a deformation groupoid. The corresponding KK-element gives the *inverse Thom element*:

$$\tau_N \in KK(C^*(T^{\mathsf{S}}N), C^*(T^{\mathsf{S}}X)).$$

Proposition 3.9.4 [20] *The following map is an isomorphism:*

$$K(C^*(T^{\mathsf{S}}N)) \overset{\cdot \otimes \tau_N}{\longrightarrow} K(C^*(T^{\mathsf{S}}X)).$$

Roughly speaking, the inverse of $\cdot \otimes \tau_N$ is the *Thom isomorphism* for the "vector bundle" $T^{\mathsf{S}}N$ "over" $T^{\mathsf{S}}X$. One can show that it really restricts to usual Thom homomorphism on regular parts.

3.9.3.2 Excision

The groupoid $T^{\mathsf{S}}N$ is identified with an open subgroupoid of $T^{\mathsf{S}}c\mathbb{R}^n$, and we have an excision map

$$j : C^*(T^{\mathsf{S}}N) \to C^*(T^{\mathsf{S}}\mathbb{R}^n).$$

3.9.3.3 Bott element

Consider $c \hookrightarrow c\mathbb{R}^n$. The (conical) normal bundle is $c\mathbb{R}^n$ itself. Remark that $\mathcal{G}^t_{c\mathbb{R}^n} = \mathcal{T}_{c\mathbb{R}^n}$. Then

$$\tau_{c\mathbb{R}^n} \in KK(C^*(T^{\mathsf{S}}c\mathbb{R}^n), \mathbb{C})$$

gives an isomorphism

$$B = (\cdot \otimes \tau_{c\mathbb{R}^n}) : K_0(C^*(T^{\mathsf{S}}c\mathbb{R}^n)) \to \mathbb{Z}.$$

Definition 3.9.5 The *topological index* is the morphism

$$\mathrm{Ind}_t^X = B \circ j_* \circ \tau_N^{-1} : \; K_0(C^*(T^{\mathsf{S}}X)) \to \mathbb{Z}.$$

The following index theorem can be proved along the same lines as in the smooth case:

Theorem 3.9.6 *The following equality holds:*

$$\mathrm{Ind}_a^X = \mathrm{Ind}_t^X.$$

Bibliography

[1] R. Almeida and P. Molino. Suites d'Atiyah et feuilletages transversalement complets. *C.R.A.S. série I*, 300(1):13–15, 1985.

[2] C. Anantharaman-Delaroche and J. Renault. *Amenable groupoids*, Monographies de L'Enseignement Mathématique, vol. 36. L'Enseignement Mathématique, Geneva, 2000. With a foreword by Georges Skandalis and Appendix B by E. Germain.

[3] W. Arveson. *An invitation to C^*-algebras*. Graduate Texts in Mathematics, No. 39. Springer-Verlag, New York, 1976.

[4] M. Atiyah. *K-theory*. Benjamin, New York and Amsterdam, 1967.

[5] M. Atiyah. *Collected Works, vols. 2, 4*. Oxford Sciences Publications, 1988.

[6] M. Atiyah and I. Singer. The index of elliptic operators I, III. *Annals of Math.*, 87:484–530, 546–604, 1968.

[7] M. Atiyah and I. Singer. The index of elliptic operators IV. *Annals of Math.*, 93:119–138, 1971.

[8] J. Block and J. Fox. Asymptotic pseudodifferential operators and index theory. *Contemp. Math.*, 105:1–31, 1990.

[9] J. P. Brasselet, G. Hector, and M. Saralegi. Théorème de de Rham pour les variétés stratifiées. *Ann. Global Anal. Geom.*, 9(3):211–243, 1991.

[10] A. Connes. Sur la théorie non commutative de l'intégration. In *Algèbres d'opérateurs*, Lecture Notes in Math., vol. 725, pages 19–143. Springer-Verlag, 1979.

[11] A. Connes. A survey of foliations and operators algebras. In *Operator algebras and applications, Part 1*, Proc. Sympos. Pure Math., vol. 38, pages 521–628. Amer. Math. Soc., Providence, 1982.

[12] A. Cannas da Silva and A. Weinstein. *Geometric models for noncommutative algebras*. Berkeley Math. Lecture Notes series, 1999.

[13] A. Connes. *Noncommutative geometry*. Academic Press, 1994.

[14] A. Connes and G. Skandalis. The longitudinal index theorem for foliations. *Publ. R.I.M.S. Kyoto Univ.*, 20:1139–1183, 1984.

[15] M. Crainic and R. L. Fernandes. Integrability of Lie brackets. *Ann. of Math.*, 157:575–620, 2003.

[16] J. Cuntz and G. Skandalis. Mapping cones and exact sequences in *KK*-theory. *J. Operator Theory*, 15(1):163–180, 1986.

[17] C. Debord. Holonomy groupoids for singular foliations. *J. Diff. Geom.*, 58:467–500, 2001.

[18] C. Debord and J.-M. Lescure. *K*-duality for stratified pseudo-manifolds. *Geometry and Topology*, 13(1):49–96, 2009.

[19] C. Debord and J.-M. Lescure. *K*-duality for pseudomanifolds with isolated singularities. *J. Functional Analysis*, 219(1):109–133, 2005.

[20] C. Debord, J.-M. Lescure, and V. Nistor. Index theorem for stratified pseudomanifolds. *J. Reine Angew. Math.*, 628:1–36, 2009.

[21] J. Dixmier. *Les C^*-algèbres et leurs représentations*. Gauthier-Villars, 1969.

[22] G. Elliott, T. Natsume, and R. Nest. The Atiyah–Singer index theorem as passage to the classical limit in quantum mechanics. *Comm. Math. Phys.*, 182(3):505–533, 1996.

[23] E. Getzler. Pseudodifferential operators on supermanifolds and the Atiyah–Singer index theorem. *Commun. Math. Phys.*, 92:163–178, 1983.

[24] M. Goresky and R. MacPherson. Intersection homology theory. *Topology*, 19:135–162, 1980.

[25] N. Higson. On the technical theorem of Kasparov. *J. Functional Analysis*, 73:107–112, 1987.

[26] N. Higson. *A primer on KK-theory*. Proc. Symp. Pure Math., Amer. Math. Soc., Providence, 1990.

[27] M. Hilsum and G. Skandalis. Stabilité des C^*-algèbres de feuilletages. *Ann. Inst. Fourier*, 33:201–208, 1983.

[28] B. Hughes and S. Weinberger. Surgery and stratified spaces. In *Surveys on surgery theory, vol. 2*, Ann. of Math. Stud., vol. 149, pages 319–352. Princeton Univ. Press, Princeton, NJ, 2001.

[29] K. Jänich. Vektorraumbündel und der Raum der Fredholm-Operatoren. *Math. Ann.*, 161:129–142, 1965.

[30] G. G. Kasparov. The operator K-functor and extensions of C^*-algebras. *Izv. Akad. Nauk SSSR, Ser. Math.*, 44:571–636, 1980.

[31] G. G. Kasparov. Equivariant *KK*-theory and the Novikov conjecture. *Invent. Math.*, 91:147–201, 1988.

[32] M. Khoshkam and G. Skandalis. Crossed products of C^*-algebras by groupoids and inverse semigroups. *J. Operator Theory*, 51(2):255–279, 2004.

[33] H. B. Lawson, Jr., and M.-L. Michelsohn. *Spin geometry*, Princeton Mathematical Series, vol. 38. Princeton Univ. Press, Princeton, NJ, 1989.

[34] J. M. Lescure. Elliptic symbols, elliptic operators and Poincaré duality on conical pseudomanifolds. *J. K-Theory*, 4(02):263–297, 2009.

[35] K. C. H. Mackenzie. *General theory of Lie groupoids and Lie algebroids, London Mathematical Society Lecture Note Series*, vol. 213. Cambridge Univ. Press, Cambridge, 2005.

[36] J. N. Mather. Stratifications and mappings. In *Dynamical systems (Proc. Sympos., Univ. Bahia, Salvador, 1971)*, pages 195–232. Academic Press, New York, 1973.

[37] B. Monthubert. Groupoids and pseudodifferential calculus on manifolds with corners. *J. Funct. Anal.*, 199(1):243–286, 2003.

[38] B. Monthubert and F. Pierrot. Indice analytique et groupoïde de Lie. *C.R.A.S Série 1*, 325:193–198, 1997.

[39] V. Nistor, A. Weinstein, and P. Xu. Pseudodifferential operators on differential groupoids. *Pacific J. Math.*, 181(1):117–152, 1999.

[40] A. L. T. Paterson. *Groupoids, inverse semigroups, and their operator algebras*, Progress in Mathematics, vol. 170. Birkhäuser Boston, 1999.

[41] G. K. Pedersen. *C^*-algebras and their automorphism groups*. Academic Press, 1979.

[42] J. Pradines. Théorie de Lie pour les groupoïdes différentiables. *C.R.A.S.*, 264:245–248, 1967.

[43] J. Pradines. Géométrie différentielle au-dessus d'un groupoïde. *C.R.A.S.*, 266:1194–1196, 1968.

[44] J. Pradines. Troisième théorème de Lie pour les groupoïdes différentiables. *C.R.A.S.*, 267:21–23, 1968.

[45] B. Ramazan. *Deformation, quantization of Lie–Poisson manifolds*. PhD thesis, Université d'Orléans, 1998.

[46] J. Renault. *A groupoid approach to C^*-algebras*, Lecture Notes in Math., vol. 793. Springer-Verlag, 1980.

[47] J. Renault. C^*-algebras of groupoids and foliations. *Proc. Sympos. Pure Math.*, 38:339–350, 1982.

[48] G. Skandalis. Hilbert modules. Cours de DEA, Paris 7, 1996.

[49] G. Skandalis. Kasparov's bivariant K-theory and applications. *Expositiones mathematicae*, 9:193–250, 1991.

[50] Georges Skandalis. Une notion de nucléarité en K-théorie (d'après J. Cuntz). *K-theory*, 1(6):549–573, 1988.

[51] S. Vassout. *Feuilletages et résidu non commutatif longitudinal*. PhD thesis, Université Paris VI, 2001.

[52] S. Vassout. Unbounded pseudodifferential calculus on Lie groupoids. *J. Funct. Anal.*, 236(1):161–200, 2006.

[53] A. Verona. *Stratified mappings – structure and triangulability*, Lecture Notes in Mathematics, vol. 1102. Springer-Verlag, Berlin, 1984.

[54] N. E. Wegge-Olsen. *K-theory and C^*-algebras: A friendly approach*. Oxford Science Publications. The Clarendon Press, Oxford University Press, New York, 1993.

4

Renormalization Hopf algebras and combinatorial groups

ALESSANDRA FRABETTI*

Abstract

These are the contents of five lectures given at the Summer School "Geometric and Topological Methods for Quantum Field Theory," held in Villa de Leyva (Colombia), July 2–20, 2007. The lectures are meant for graduate or advanced undergraduate students in physics or mathematics. They include references, many examples and some exercises.

Part I is a short introduction to algebraic and proalgebraic groups, based on some examples of groups of matrices and groups of formal series, and their Hopf algebras of coordinate functions.

Part II presents a greatly condensed review of classical and quantum field theory, from the Lagrangian formalism to the Euler–Lagrange equation and the Dyson–Schwinger equation for Green functions. It poses the main problem of solving some nonlinear differential equations for interacting fields.

In Part III we study the perturbative solution of the previous equations, expanded in Feynman graphs, in the simplest case of the scalar ϕ^3 theory.

Part IV introduces the problem of divergent integrals appearing in quantum field theory, the renormalization procedure for the graphs, and how the renormalization affects the Lagrangian and the Green functions given as perturbative series.

Part V presents the Connes–Kreimer Hopf algebra of renormalization for the scalar theory and its associated proalgebraic group of formal series.

I. Groups and Hopf algebras

In this part we review the classical duality between groups and Hopf algebras of certain types. Details can be found, for instance, in [16].

* These lectures are based on a course for Ph.D. students in mathematics, held at Université Lyon 1 in spring 2006, by Alessandra Frabetti and Denis Perrot. Thanks, Denis!

During the summer school "Geometric and Topological Methods for Quantum Field Theory," many students provided interesting questions and comments, which greatly helped the writing of these notes. Thanks to all of them!

4.1 Algebras of representative functions

Let G be a group, for instance, a group of real or complex matrices, a topological group or a Lie group. Let

$$F(G) = \{f : G \longrightarrow \mathbb{C} \text{ (or } \mathbb{R})\}$$

denote the set of functions on G, possibly continuous or differentiable. Then $F(G)$ has a lot of algebraic structures, which I describe in detail.

4.1.1 Product

The natural vector space $F(G)$ is a unital associative and commutative algebra over $\mathbb{C}$, with *product* $(fg)(x) = f(x)g(x)$, where $f, g \in F(G)$ and $x \in G$, and *unit* given by the constant function $1(x) = 1$.

4.1.2 Coproduct

For any $f \in F(G)$, the group law $G \times G \overset{\cdot}{\longrightarrow} G$ induces an element $\Delta f \in F(G \times G)$ defined by $\Delta f(x, y) = f(x \cdot y)$. Can we characterize the algebra $F(G \times G) = \{f : G \times G \longrightarrow \mathbb{C}\}$ starting from $F(G)$?

Of course, we can consider the tensor product

$$F(G) \otimes F(G) = \left\{ \sum_{\text{finite}} f_i \otimes g_i, \ f_i, g_i \in F(G) \right\},$$

with componentwise product $(f_1 \otimes g_1)(f_2 \otimes g_2) = f_1 g_1 \otimes f_2 g_2$, but in general this algebra is a strict subalgebra of $F(G \times G) = \{\sum_{\text{infinite}} f_i \otimes g_i\}$ (it is equal to $F(G \times G)$ for finite groups). For example, $f(x, y) = \exp(x + y) \in F(G) \otimes F(G)$, but $f(x, y) = \exp(xy) \notin F(G) \otimes F(G)$. Similarly, if $\delta(x, y)$ is the function equal to 1 when $x = y$ and equal to 0 when $x \neq y$, then $\delta \notin F(G) \otimes F(G)$. To avoid this problem we could use the *completed* or *topological* tensor product $\hat{\otimes}$ such that $F(G) \hat{\otimes} F(G) = F(G \times G)$. However, this tensor product is difficult to handle, and for our purpose we want to avoid it. As an alternative, we can consider the subalgebras $R(G)$ of $F(G)$ such that $R(G) \otimes R(G) = R(G \times G)$. Such algebras are of course much easier to describe then a completed tensor product. For our purpose, we are interested in the case when one of these subalgebras is big enough to describe the group completely. That is, it does not lose too much information about the group with respect to $F(G)$. This condition will be specified later on.

Let us therefore assume that there exists a subalgebra $R(G) \subset F(G)$ such that $R(G) \otimes R(G) = R(G \times G)$. Then, the group law $G \times G \overset{\cdot}{\longrightarrow} G$ induces a *coproduct* $\Delta : R(G) \longrightarrow R(G) \otimes R(G)$ defined by $\Delta f(x, y) = f(x \cdot y)$. We denote it by $\Delta f = \sum_{\text{finite}} f_{(1)} \otimes f_{(2)}$. The coproduct has two main properties:

(i) Δ is a homomorphism of algebras; in fact,

$$\Delta(fg)(x, y) = (fg)(x \cdot y) = f(x \cdot y)g(x \cdot y) = \Delta f(x, y)\Delta g(x, y),$$

that is, $\Delta(fg) = \Delta(f)\Delta(g)$. This can also be expressed as

$$\sum (fg)_1 \otimes (fg)_2 = \sum f_1 g_1 \otimes f_2 g_2.$$

(ii) Δ is coassociative, that is, $(\Delta \otimes \mathrm{Id})\Delta = (\mathrm{Id} \otimes \Delta)\Delta$, because of the associativity $(x \cdot y) \cdot z = x \cdot (y \cdot z)$ of the group law in G.

4.1.3 Counit

The neutral element e of the group G induces a *counit* $\varepsilon : R(G) \longrightarrow \mathbb{C}$ defined by $\varepsilon(f) = f(e)$. The counit has two main properties:

(i) ε is a homomorphism of algebras; in fact,

$$\varepsilon(fg) = (fg)(e) = f(e)g(e) = \varepsilon(f)\varepsilon(g).$$

(ii) ε satisfies the equality $\sum f_{(1)}\varepsilon(f_{(2)}) = \sum \varepsilon(f_{(1)})f_{(2)}$, induced by the equality $x \cdot e = x = e \cdot x$ in G.

4.1.4 Antipode

The inversion operation in G, that is, $x \to x^{-1}$, induces the *antipode* $S : R(G) \longrightarrow R(G)$ defined by $S(f)(x) = f(x^{-1})$. The counit has four main properties:

(i) S is a homomorphism of algebras; in fact,

$$S(fg)(x) = (fg)(x^{-1}) = f(x^{-1})g(x^{-1}) = S(f)(x)S(g)(x).$$

(ii) S satisfies the five-term equality

$$m(S \otimes \mathrm{Id})\Delta = u\varepsilon = m(\mathrm{Id} \otimes S)\Delta,$$

where $m : R(G) \otimes R(G) \longrightarrow R(G)$ denotes the product and $u : \mathbb{C} \longrightarrow R(G)$ denotes the unit. This is induced by the equality $x \cdot x^{-1} = e = x^{-1} \cdot x$ in G.

(iii) S is anti-comultiplicative, that is, $\Delta \circ S = (S \otimes S) \circ \tau \circ \Delta$, where $\tau(f \otimes g) = g \otimes f$ is the twist operator. This property is induced by the equality $(x \cdot y)^{-1} = y^{-1} \cdot x^{-1}$ in G.

(iv) S is nilpotent, that is, $S \circ S = \mathrm{Id}$, because of the identity $(x^{-1})^{-1} = x$ in G.

4.1.5 Abelian groups

Finally, G is abelian (that is, $x \cdot y = y \cdot x$ for all $x, y \in G$) if and only if the coproduct is cocommutative, that is, $\Delta = \Delta \circ \tau$, that is, $\sum f_{(1)} \otimes f_{(2)} = \sum f_{(2)} \otimes f_{(1)}$.

4.1.6 Hopf algebras

A unital, associative and commutative algebra $\mathbb{H}$ endowed with a coproduct Δ, a counit ε and an antipode S, satisfying all the properties listed in the preceding subsections, is called a *commutative Hopf algebra.*

In conclusion, we have shown that if G is a (topological) group, and $R(G)$ is a subalgebra of (continuous) functions on G such that $R(G) \otimes R(G) = R(G \times G)$, and sufficiently big to contain the image of Δ and of S, then $R(G)$ is a commutative Hopf algebra. Moreover, $R(G)$ is cocommutative if and only if G is abelian.

4.1.7 Representative functions

We now turn to the existence of such a Hopf algebra $R(G)$. If G is a finite group, then the largest such algebra is simply the linear dual $R(G) = F(G) = (\mathbb{C}G)^*$ of the group algebra.

If G is a topological group, then the condition $R(G) \otimes R(G) = R(G \times G)$, roughly speaking, forces $R(G)$ to be a polynomial algebra, or a quotient of it. The generators are the *coordinate functions* on the group, but we do not always know how to find them.

For compact Lie groups, $R(G)$ always exists, and we can be more precise. We say that a function $f : G \longrightarrow \mathbb{C}$ is *representative* if there exist a finite number of functions $f_1, \ldots, f_k$ such that any translation of f is a linear combination of them. If we denote by $(L_x f)(y) = f(x \cdot y)$ the left translation of f by $x \in G$, this means that $L_x f = \sum l_i(x) f_i$. Denote by $R(G)$ the set of all representative functions on G. Then, using representation theory, and, in particular, the *Peter–Weyl theorem*, one can show the following facts:

(i) $R(G) \otimes R(G) = R(G \times G)$;
(ii) $R(G)$ is dense in the set of continuous functions;
(iii) as an algebra, $R(G)$ is generated by the matrix elements of all the representations of G of finite dimension;
(iv) $R(G)$ is also generated by the matrix elements of one faithful representation of G, and therefore it is finitely generated.

Moreover, for compact Lie groups, the algebra $R(G)$ has two additional structures:

(i) because the group G is a real manifold and the functions have complex values, $R(G)$ has an *involution*, that is, a map $* : R(G) \longrightarrow R(G)$ such that $(f^*)^* = f$ and $(fg)^* = g^* f^*$;

(ii) because G is compact, $R(G)$ has a *Haar measure*, that is, a linear map $\mu : R(G) \longrightarrow \mathbb{R}$ such that $\mu(aa^*) > 0$ for all $a \neq 0$.

Similar results hold in general for groups of matrices, even if they are complex manifolds, and even if they are not compact. In particular, the algebra generated by the matrix elements of one faithful representation of G satisfies the required properties.

For other groups than those of matrices, a suitable algebra $R(G)$ can exist, but there is no general procedure to find it. The best hint is to look for a faithful representation, possibly with infinite dimension. This may work also for groups which are not locally compact, as shown in the examples in Sections 4.2.8 and 4.2.9, but in general not for groups of diffeomorphisms on a manifold.

4.2 Examples

4.2.1 Complex affine plane

Let $G = (\mathbb{C}^n, +)$ be the additive group of the complex affine plane. A complex group is assumed to be a holomorphic manifold. The functions are also assumed to be holomorphic, that is, they do not depend on the complex conjugate of the variables. The map

$$\rho : (\mathbb{C}^n, +) \longrightarrow GL_{n+1}(\mathbb{C}) = \mathrm{Aut}(\mathbb{C}^{n+1}),$$

$$(t_1, \dots, t_n) \mapsto \begin{pmatrix} 1 & t_1 & \cdots & t_n \\ 0 & 1 & \cdots & 0 \\ \vdots & \vdots & & \vdots \\ 0 & 0 & \dots & 1 \end{pmatrix}$$

is a faithful representation, in fact

$$\begin{aligned} \rho\big((t_1, \dots, t_n) + (s_1, \dots, s_n)\big) &= \begin{pmatrix} 1 & t_1 + s_1 & \cdots & t_n + s_n \\ 0 & 1 & \cdots & 0 \\ \vdots & \vdots & & \vdots \\ 0 & 0 & \cdots & 1 \end{pmatrix} \\ &= \begin{pmatrix} 1 & t_1 & \cdots & t_n \\ 0 & 1 & \cdots & 0 \\ \vdots & \vdots & & \vdots \\ 0 & 0 & \cdots & 1 \end{pmatrix} \begin{pmatrix} 1 & s_1 & \cdots & s_n \\ 0 & 1 & \cdots & 0 \\ \vdots & \vdots & & \vdots \\ 0 & 0 & \cdots & 1 \end{pmatrix} \\ &= \rho(t_1, \dots, t_n)\rho(s_1, \dots, s_n). \end{aligned}$$

Therefore, there are n local coordinates $x_i(t_1, \ldots, t_n) = t_i$, for $i = 1, \ldots, n$, which are free of mutual relations. Hence the algebra of local coordinates on the affine line is the polynomial ring $R(\mathbb{C}^n, +) = \mathbb{C}[x_1, \ldots, x_n]$. The Hopf-structure is the following:

- Coproduct: $\Delta x_i = x_i \otimes 1 + 1 \otimes x_i$ and $\Delta 1 = 1 \otimes 1$. The group is abelian, and the coproduct is indeed cocommutative.
- Counit: $\varepsilon(x_i) = x(0) = 0$ and $\varepsilon(1) = 1$.
- Antipode: $Sx_i = -x_i$ and $S1 = 1$.

This Hopf algebra is usually called the *unshuffle Hopf algebra*, because the coproduct on a generic monomial

$$\Delta(x_{i_1} \cdots x_{i_l}) = \sum_{p+q=l} \sum_{\sigma \in \Sigma_{p,q}} x_{\sigma(i_1)} \cdots x_{\sigma(i_p)} \otimes x_{\sigma(i_{p+1})} \cdots x_{\sigma(i_{p+q})}$$

makes use of the shuffle permutations $\sigma \in \Sigma_{p,q}$, that is, the permutations of Σ_{p+q} such that $\sigma(i_1) < \cdots < \sigma(i_p)$ and $\sigma(i_{p+1}) < \cdots < \sigma(i_{p+q})$.

4.2.2 Real affine plane

Let $G = (\mathbb{R}^n, +)$ be the additive group of the real affine plane. A real group is assumed to be a differentiable manifold. The functions with values in $\mathbb{C}$ are the complexification of the functions with values in $\mathbb{R}$, that is, $R_{\mathbb{C}}(G) = R_{\mathbb{R}}(G) \otimes \mathbb{C}$. In principle, then, the functions depend also on the complex conjugates, but the generators must be real: we expect that the algebra $R_{\mathbb{C}}(G)$ has an involution $*$. In fact, we have the following results:

- *Real functions*: The map

$$\rho : (\mathbb{R}^n, +) \longrightarrow GL_{n+1}(\mathbb{R}) = \mathrm{Aut}(\mathbb{R}^{n+1}),$$

$$(t_1, \ldots, t_n) \mapsto \begin{pmatrix} 1 & t_1 & \cdots & t_n \\ 0 & 1 & \cdots & 0 \\ \vdots & \vdots & & \vdots \\ 0 & 0 & \ldots & 1 \end{pmatrix}$$

  is a faithful representation. The local coordinates are $x_i(t_1, \ldots, t_n) = t_i$, for $i = 1, \ldots, n$, and the algebra of real local coordinates is the polynomial ring $R_{\mathbb{R}}(\mathbb{R}^n, +) = \mathbb{R}[x_1, \ldots, x_n]$. The Hopf structure is exactly as in the previous example.
- *Complex functions:* Complex faithful representation as before, but local coordinates $x_i(t_1, \ldots, t_n) = t_i$ subject to an involution defined by $x_i^*(t_1, \ldots, t_n) = \overline{t_i}$ and such that $x_i^* = x_i$. Then the algebra of complex local coordinates is the quotient

$$R_{\mathbb{C}}(\mathbb{R}^n, +) = \frac{\mathbb{C}[x_1, x_1^*, \ldots, x_n, x_n^*]}{\langle x_i^* - x_i, i = 1, \ldots, n \rangle},$$

which is isomorphic to $\mathbb{C}[x_1, \dots, x_n]$ as an algebra, but not as an algebra with involution. Of course, the Hopf structure is always the same.

4.2.3 Complex simple linear group

The group

$$SL(2, \mathbb{C}) = \left\{ M = \begin{pmatrix} m_{11} & m_{12} \\ m_{21} & m_{22} \end{pmatrix} \in M_2(\mathbb{C}),\ \det M = 1 \right\}$$

has a lot of finite-dimensional representations, and the smallest faithful one is the identity:

$$\rho = \mathrm{Id} : \quad SL(2, \mathbb{C}) \longrightarrow GL_2(\mathbb{C}),$$
$$M \mapsto \begin{pmatrix} m_{11} = a(M) & m_{12} = b(M) \\ m_{21} = c(M) & m_{22} = d(M) \end{pmatrix}.$$

Therefore there are four local coordinates $a, b, c, d : SL(2, \mathbb{C}) \longrightarrow \mathbb{C}$, given by $a(M) = m_{11}$, etc., related by $\det M = 1$. Hence the algebra of local coordinates of $SL(2, \mathbb{C})$ is the quotient

$$R(SL(2, \mathbb{C})) = \frac{\mathbb{C}[a, b, c, d]}{\langle ad - bc - 1 \rangle}.$$

The Hopf structure is the following:

- Coproduct: $\Delta f(M, N) = f(MN)$; therefore

$$\Delta a = a \otimes a + b \otimes c, \qquad \Delta b = a \otimes b + b \otimes d,$$
$$\Delta c = c \otimes a + d \otimes c, \qquad \Delta d = c \otimes b + d \otimes d.$$

  To shorten the notation, we can write

$$\Delta \begin{pmatrix} a & b \\ c & d \end{pmatrix} = \begin{pmatrix} a & b \\ c & d \end{pmatrix} \otimes \begin{pmatrix} a & b \\ c & d \end{pmatrix}.$$

- Counit: $\varepsilon(f) = f(1)$; hence $\varepsilon \begin{pmatrix} a & b \\ c & d \end{pmatrix} = \begin{pmatrix} 1 & 0 \\ 0 & 1 \end{pmatrix}$.
- Antipode: $Sf(M) = f(M^{-1})$; therefore $S \begin{pmatrix} a & b \\ c & d \end{pmatrix} = \begin{pmatrix} d & -b \\ -c & a \end{pmatrix}$.

4.2.4 Complex general linear group

For the group

$$GL(2, \mathbb{C}) = \{ M \in M_2(\mathbb{C}),\ \det M \neq 0 \},$$

the identity $GL(2,\mathbb{C}) \longrightarrow GL(2,\mathbb{C}) \equiv \mathrm{Aut}(\mathbb{C}^2)$ is of course a faithful representation. We then have four local coordinates, as for $SL(2,\mathbb{C})$. However, this time they satisfy the condition $\det M \neq 0$, which is not closed. To express this relation we use a trick: Because $\det M \neq 0$ if and only if there exists the inverse of $\det M$, we add a variable $t(M) = (\det M)^{-1}$. Therefore the algebra of local coordinates of $GL(2,\mathbb{C})$ is the quotient

$$R(GL(2,\mathbb{C})) = \frac{\mathbb{C}[a,b,c,d,t]}{\langle (ad-bc)t - 1\rangle}.$$

The Hopf structure is the same as that of $SL(2,\mathbb{C})$ on the local coordinates a, b, c, d, and on the new variable t is as follows:

- Coproduct: $\Delta t = t \otimes t$; in fact,

$$\begin{aligned}\Delta t(M,N) = t(MN) &= (\det(MN))^{-1} = (\det M)^{-1}(\det N)^{-1}\\ &= t(M)t(N).\end{aligned}$$

- Counit: $\varepsilon(t) = t(1) = 1$.
- Antipode: $St = ad - bc$, because

$$St(M) = t(M^{-1}) = (\det(M^{-1}))^{-1} = \det M.$$

4.2.5 Simple unitary group

The group

$$SU(2) = \left\{ M \in M_2(\mathbb{C}),\ \det M = 1,\ M^{-1} = \overline{M}^t \right\}$$

is a real group; in fact, it is one *real form* of $SL(2,\mathbb{C})$, the other one being $SL(2,\mathbb{R})$, and it is also the maximal compact subgroup of $SL(2,\mathbb{C})$. As a real manifold, $SU(2)$ is isomorphic to the three-dimensional sphere S^3; in fact,

$$\begin{aligned} M = \begin{pmatrix} a & b \\ c & d \end{pmatrix} \in SU(2) \quad &\Longleftrightarrow \quad ad - bc = 1, \quad \overline{a} = d\,,\ \overline{b} = c \\ &\Longleftrightarrow \quad M = \begin{pmatrix} a & b \\ -\overline{b} & \overline{a} \end{pmatrix} \quad \text{with } a\overline{a} + b\overline{b} = 1. \end{aligned}$$

If we set $a = x + iy$ and $b = u + iv$, with $x, y, u, v \in \mathbb{R}$, we then have

$$\begin{aligned} a\overline{a} + b\overline{b} = 1 \quad &\Longleftrightarrow \quad x^2 + y^2 + u^2 + v^2 = 1 \quad \text{in } \mathbb{R}^4 \\ &\Longleftrightarrow \quad (x, y, u, v) \in S^3. \end{aligned}$$

We then expect that the algebra of complex functions on $SU(2)$ has an involution:

$$R(SU(2)) = \frac{\mathbb{C}[a,b,c,d,a^*,b^*,c^*,d^*]}{\langle a^* - d, b^* + c, ad - bc - 1\rangle} \cong \frac{\mathbb{C}[a,b,a^*,b^*]}{\langle aa^* + bb^* - 1\rangle}.$$

The Hopf structure is the same as that of $SL(2,\mathbb{C})$, but expressed in terms of the proper coordinate functions of $SU(2)$, that is:

- Coproduct: $\Delta\begin{pmatrix} a & b \\ -b^* & a^* \end{pmatrix} = \begin{pmatrix} a & b \\ -b^* & a^* \end{pmatrix} \otimes \begin{pmatrix} a & b \\ -b^* & a^* \end{pmatrix}$.
- Counit: $\varepsilon\begin{pmatrix} a & b \\ -b^* & a^* \end{pmatrix} = \begin{pmatrix} 1 & 0 \\ 0 & 1 \end{pmatrix}$.
- Antipode: $S\begin{pmatrix} a & b \\ -b^* & a^* \end{pmatrix} = \begin{pmatrix} a^* & -b \\ b^* & a \end{pmatrix}$.

4.2.6 Exercise: Heisenberg group

The Heisenberg group H_3 is the group of complex 3×3 (upper) triangular matrices with all the diagonal elements equal to 1, that is,

$$H_3 = \left\{ \begin{pmatrix} 1 & a & b \\ 0 & 1 & c \\ 0 & 0 & 1 \end{pmatrix} \in GL(3,\mathbb{C}) \right\}.$$

Describe the Hopf algebra of complex representative (algebraic) functions on H_3.

4.2.7 Exercise: Euclidean group

The group of rotations on the plane $\mathbb{R}^2$ is the special orthogonal group

$$SO(2,\mathbb{R}) = \left\{ A \in GL(2,\mathbb{R}),\ \det A = 1,\ A^{-1} = A^t \right\}.$$

The group of rotations acts on the group of translations $T_2 = (\mathbb{R}^2, +)$ as a product Av of a matrix $A \in SO(2,\mathbb{R})$ by a vector $v \in \mathbb{R}^2$.

The Euclidean group is the semidirect product $E_2 = T_2 \rtimes SO(2,\mathbb{R})$. That is, E_2 is the set of all the couples $(v, A) \in T_2 \times SO(2,\mathbb{R})$, with the group law

$$(v, A) \cdot (u, B) := (v + Au, AB).$$

(i) Describe the Hopf algebra of real representative functions on $SO(2,\mathbb{R})$.
(ii) Find a real faithful representation of T_2 of dimension 3.
(iii) Describe the Hopf algebra of real representative functions on E_2.

4.2.8 Group of invertible formal series

The set

$$G^{\text{inv}}(\mathbb{C}) = \left\{ f(z) = \sum_{n=0}^{\infty} f_n\, z^n,\ f_n \in \mathbb{C},\ f_0 = 1 \right\}$$

of formal series in one variable, with constant term equal to 1, is an Abelian group with

- product: $(fg)(z) = f(z)g(z) = \sum_{n=0}^{\infty} \left(\sum_{p+q=n} f_p\, g_q \right) z^n;$
- unit: $1(z) = 1;$
- inverse: by recursion; in fact,

$$(ff^{-1})(z) = \sum_{n=0}^{\infty} \left(\sum_{p+q=n} f_p\, (f^{-1})_q \right) z^n = 1$$

if and only if for $n = 0$ we have

$$f_0 (f^{-1})_0 = 1 \quad \Leftrightarrow \quad (f^{-1})_0 = 1 \quad \Leftrightarrow \quad f^{-1} \in G^{\text{inv}}(\mathbb{C}),$$

and for $n \geq 1$ we have

$$\sum_{p=0}^{n} f_p\, (f^{-1})_{n-p} = f_0 (f^{-1})_n + f_1 (f^{-1})_{n-p} + \cdots + f_n (f^{-1})_0 \ = \ 0,$$

that is, $(f^{-1})_1 = -f_1$, $(f^{-1})_2 = f_1^2 - f_2$, etc.

This group has many finite-dimensional representations, of the form

$$\rho : G^{\text{inv}}(\mathbb{C}) \longrightarrow GL_n(\mathbb{C}),$$

$$f(z) = \sum_{n=0}^{\infty} f_n\, z^n \mapsto \begin{pmatrix} 1 & f_1 & f_2 & f_3 & \cdots & f_{n-1} \\ 0 & 1 & f_1 & f_2 & \cdots & f_{n-2} \\ 0 & 0 & 1 & f_1 & \cdots & f_{n-3} \\ \vdots & \vdots & & \ddots & & \vdots \\ 0 & 0 & & \cdots & & 1 \end{pmatrix},$$

but they are never faithful! To have a faithful representation, we need to consider the map

$$\rho : G^{\text{inv}}(\mathbb{C}) \longrightarrow GL_\infty(\mathbb{C}) = \lim_{\leftarrow} GL_n(\mathbb{C}),$$

$$f(z) \mapsto \begin{pmatrix} 1 & f_1 & f_2 & f_3 & \cdots \\ 0 & 1 & f_1 & f_2 & \cdots \\ 0 & 0 & 1 & f_1 & \cdots \\ \vdots & \vdots & & \ddots & \\ 0 & 0 & \cdots & & \end{pmatrix},$$

where $\lim\limits_{\leftarrow} GL_n(\mathbb{C})$ is the projective limit of the groups $(GL_n(\mathbb{C}))_n$, that is, the limit of the groups such that each $GL_n(\mathbb{C})$ is identified with the quotient of $GL_{n+1}(\mathbb{C})$ by its last column and row. Because $\lim\limits_{\leftarrow} GL_n(\mathbb{C})$ is not a group, it is necessary to restrict the image of the map ρ to the triangular matrices $T_n(\mathbb{C})$, whose projective limit $\lim\limits_{\leftarrow} T_n(\mathbb{C})$ indeed forms a group.[1]

Therefore there are infinitely many local coordinates $x_n : G^{\text{inv}}(\mathbb{C}) \longrightarrow \mathbb{C}$, given by $x_n(f) = f_n$, which are free from each other. Hence the algebra of local coordinates of $G^{\text{inv}}(\mathbb{C})$ is the polynomial ring

$$R(G^{\text{inv}}(\mathbb{C})) = \mathbb{C}[x_1, x_2, \ldots, x_n, \ldots].$$

The Hopf structure is the following (with $x_0 = 1$):

- Coproduct: $\Delta x_n = \sum_{k=0}^{n} x_k \otimes x_{n-k}$.
- Counit: $\varepsilon(x_n) = \delta(n, 0)$.
- Antipode: recursively, from the five-term identity. In fact, for any $n > 0$ we have

$$\varepsilon(x_n)1 = 0 = \sum_{k=0}^{n} S(x_k)x_{n-k}$$
$$= S(1)x_n + S(x_1)x_{n-1} + S(x_2)x_{n-2} + \cdots + S(x_n)1,$$

and because $S(1) = 1$, we obtain $S(x_n) = -x_n - \sum_{k=1}^{n-1} S(x_k)x_{n-k}$.

This Hopf algebra is isomorphic to the so-called *algebra of symmetric functions*; cf. [19].

4.2.9 Group of formal diffeomorphisms

The set

$$G^{\text{dif}}(\mathbb{C}) = \left\{ f(z) = \sum_{n=0}^{\infty} f_n\, z^{n+1},\ f_n \in \mathbb{C},\ f_0 = 1 \right\}$$

[1] Thanks to B. Richter and R. Holtkamp for pointing this cut to me.

of formal series in one variable, with zero constant term and linear term equal to 1, is a (non-abelian) group with

- product given by the composition (or substitution)

$$
\begin{aligned}
(f \circ g)(z) = f(g(z)) &= \sum_{n=0}^{\infty} f_n \, g(z)^n \\
&= z + (f_1 + g_1)\, z^2 + (f_2 + 2 f_1 g_1 + g_2)\, z^3 \\
&\quad + (f_3 + 3 f_2 g_1 + 2 f_1 g_2 + f_1 g_1^2 + g_3)\, z^4 + \mathrm{O}(z^5);
\end{aligned}
$$

- unit $\mathrm{id}(z) = z$;
- inverse given by the reciprocal series f^{-1} such that $f \circ f^{-1} = \mathrm{id} = f^{-1} \circ f$, which can be found recursively, using for instance the *Lagrange formula*; cf. [22].

This group also has many finite-dimensional representations, which are not faithful, and a faithful representation of infinite dimension:

$$
\rho : G^{\mathrm{dif}}(\mathbb{C}) \longrightarrow T_\infty(\mathbb{C}) = \lim_{\leftarrow} T_n(\mathbb{C}) \subset GL_\infty(\mathbb{C}),
$$

$$
f(z) \mapsto \begin{pmatrix}
1 & f_1 & f_2 & f_3 & f_4 & \cdots \\
0 & 1 & 2f_1 & 2f_2 + f_1^2 & 2f_3 + 2f_1 f_2 & \cdots \\
0 & 0 & 1 & 3f_1 & 3f_2 + 3f_1^2 & \cdots \\
0 & 0 & 0 & 1 & 4f_1 & \cdots \\
\vdots & \vdots & \vdots & & \ddots & \\
0 & 0 & 0 & \cdots & &
\end{pmatrix}.
$$

Therefore there are infinitely many local coordinates $x_n : G^{\mathrm{dif}}(\mathbb{C}) \longrightarrow \mathbb{C}$, given by $x_n(f) = f_n$, which are free one from another. As in the previous example, the algebra of local coordinates of $G^{\mathrm{dif}}(\mathbb{C})$ is then the polynomial ring

$$
R(G^{\mathrm{dif}}(\mathbb{C})) = \mathbb{C}[x_1, x_2, \ldots].
$$

The Hopf structure is the following (with $x_0 = 1$):

- Coproduct: $\Delta x_n(f, g) = x_n(f \circ g)$; hence

$$
\Delta x_n = x_n \otimes 1 + 1 \otimes x_n + \sum_{m=1}^{n-1} x_m \otimes \sum_{\substack{p_0 + p_1 + \cdots + p_m = n-m \\ p_0, \ldots, p_m \geq 0}} x_{p_0} x_{p_1} \cdots x_{p_m}.
$$

- Counit: $\varepsilon(x_n) = \delta(n, 0)$.

- Antipode: recursively, using

$$S(x_n) = -x_n - \sum_{m=1}^{n-1} S(x_m) \sum_{\substack{p_0+p_1+\cdots+p_m=n-m \\ p_0,\ldots,p_m \geq 0}} x_{p_0} x_{p_1} \cdots x_{p_m}.$$

This Hopf algebra is the so-called *Faà di Bruno* Hopf algebra, because the computation of the coefficients of the Taylor expansion of the composition of two functions was first carried out by F. Faà di Bruno [12] (in 1855!).

4.3 Groups of characters and duality

Let $\mathbb{H}$ be a commutative Hopf algebra over $\mathbb{C}$, with product m, unit u, coproduct Δ, counit ε, antipode S and possibly an involution $*$.

4.3.1 Group of characters

We define a *character* of the Hopf algebra $\mathbb{H}$ as a linear map $\alpha : \mathbb{H} \longrightarrow \mathbb{C}$ such that

(i) α is a homomorphism of algebras, that is, $\alpha(ab) = \alpha(a)\alpha(b)$;
(ii) α is unital, that is, $\alpha(1) = 1$.

Denote by $G_{\mathbb{H}} = \mathrm{Hom}_{Alg}(\mathbb{H}, \mathbb{C})$ the set of characters of $\mathbb{H}$. Given two characters $\alpha, \beta \in G_{\mathbb{H}}$, we define the *convolution* of α and β as the linear map $\alpha \star \beta : \mathbb{H} \longrightarrow \mathbb{C}$ defined by $\alpha \star \beta = m_{\mathbb{C}} \circ (\alpha \otimes \beta) \circ \Delta$, that is, $\alpha \star \beta(a) = \sum \alpha(a_{(1)})\beta(a_{(2)})$ for any $a \in \mathbb{H}$. Applying the definitions, it is easy to prove the following properties:

(i) For any $\alpha, \beta \in G_{\mathbb{H}}$, the convolution $\alpha \star \beta$ is a unital algebra homomorphism, that is $\alpha \star \beta \in G_{\mathbb{H}}$.
(ii) The convolution product $G_{\mathbb{H}} \otimes G_{\mathbb{H}} \longrightarrow G_{\mathbb{H}}$ is associative.
(iii) The counit $\varepsilon : \mathbb{H} \longrightarrow \mathbb{C}$ is the unit of the convolution.
(iv) For any $\alpha \in G_{\mathbb{H}}$, the homomorphism $\alpha^{-1} = \alpha \circ S$ is the inverse of α.
(v) The convolution product is commutative if and only if the coproduct is cocommutative.

In other words, the set of characters $G_{\mathbb{H}}$ forms a group with the convolution product.

4.3.2 Real subgroups

If $\mathbb{H}$ is a commutative Hopf algebra endowed with an involution $* : \mathbb{H} \longrightarrow \mathbb{H}$ compatible with the Hopf structure, in the sense that

$$(ab)^* = b^* a^*, \qquad 1^* = 1,$$

$$\Delta(a^*) = (\Delta a)^*, \qquad \varepsilon(a^*) = \varepsilon(a), \qquad S(a^*) = (Sa)^*,$$

then the subset

$$G_{\mathbb{H}}^{*} = \mathrm{Hom}_{*Alg}(\mathbb{H}, \mathbb{C}) = \left\{ \alpha \in G_{\mathbb{H}}, \alpha(a^{*}) = \overline{\alpha(a)} \right\}$$

is a (real) subgroup of $G_{\mathbb{H}}$.

4.3.3 Compact subgroups

If, furthermore, $\mathbb{H}$ is a commutative *Hopf algebra, finitely generated and endowed with a Haar measure compatible with the Hopf structure, that is, a linear map $\mu : \mathbb{H} \longrightarrow \mathbb{R}$ such that

$$(\mu \otimes \mathrm{Id})\Delta = (\mathrm{Id} \otimes \mu)\Delta = u \circ \mu,$$

$$\mu(aa^{*}) > 0 \qquad \text{for all } a \neq 0,$$

then $G_{\mathbb{H}}^{*}$ is a compact Lie group.

4.3.4 Comparison of $SL(2,\mathbb{C})$, $SL(2,\mathbb{R})$ and $SU(2)$

Consider the commutative algebra

$$\mathbb{H} = \frac{\mathbb{C}[a, b, c, d]}{\langle ad - bc - 1 \rangle}.$$

If on $\mathbb{H}$ we consider the Hopf structure

$$\Delta \begin{pmatrix} a & b \\ c & d \end{pmatrix} = \begin{pmatrix} a & b \\ c & d \end{pmatrix} \otimes \begin{pmatrix} a & b \\ c & d \end{pmatrix},$$

$$\varepsilon \begin{pmatrix} a & b \\ c & d \end{pmatrix} = \begin{pmatrix} 1 & 0 \\ 0 & 1 \end{pmatrix},$$

$$S \begin{pmatrix} a & b \\ c & d \end{pmatrix} = \begin{pmatrix} d & -b \\ -c & a \end{pmatrix},$$

then $G_{\mathbb{H}} = SL(2,\mathbb{C})$. If in addition we consider the involution

$$\begin{pmatrix} a & b \\ c & d \end{pmatrix}^{*} = \begin{pmatrix} a & b \\ c & d \end{pmatrix},$$

then $G_{\mathbb{H}}^{*} = SL(2,\mathbb{R})$. If, instead, we consider the involution

$$\begin{pmatrix} a & b \\ c & d \end{pmatrix}^{*} = \begin{pmatrix} d & -c \\ -b & a \end{pmatrix},$$

then $G_{\mathbb{H}}^{*} = SU(2)$.

4.3.5 Duality

We have seen first how to associate a Hopf algebra to a group, through a functor R, and then how to associate a group to a Hopf algebra, through a functor G. In general, these two functors are *adjoint* to each other, that is,

$$\mathrm{Hom}_{Groups}(G, G_{\mathbb{H}}) \cong \mathrm{Hom}_{Alg}(\mathbb{H}, R(G)).$$

Sometimes, these two functors are dual to each other. In particular, we have the following results:

- Given a complex group G and its Hopf algebra $R(G)$ of representative functions, the map

$$\Phi : G \longrightarrow G_{R(G)} = \mathrm{Hom}_{Alg}(R(G), \mathbb{C}),$$
$$x \mapsto \Phi_x : \mathbb{R}(G) \to \mathbb{C},\ \Phi_x(f) = f(x)$$

  defines an isomorphism of groups to the characters group of $R(G)$. This result must be refined to the group $G^*_{R(G)}$ if G is real. It is known as *Tannaka duality* for compact Lie groups.
- Vice versa, given a commutative Hopf algebra over $\mathbb{C}$, the complex group G can be defined as the group of characters of $\mathbb{H}$, that is, by stating that its coordinate functions are given by $\mathbb{H}$. If the Hopf algebra $\mathbb{H}$ has an involution and a Haar measure, and it is finitely generated, then the map

$$\Psi : \mathbb{H} \longrightarrow R(G^*(\mathbb{H})),$$
$$a \mapsto \Psi_a : \mathrm{Hom}_{*Alg}(\mathbb{H}, \mathbb{C}) \to \mathbb{C},\ \Psi_a(\alpha) = \alpha(a)$$

  defines an isomorphism of Hopf algebras. The underlying group is compact, and this result is known as the *Krein duality*.

4.3.6 Algebraic and proalgebraic groups

As we saw in most of the examples, the group structure of many groups does not depend on the field where the coefficients take value. This is the case for matrix groups, but also for the groups of formal series. Apart from the coefficients, such groups have in common the form of their coordinate ring, that is, the Hopf algebra $\mathbb{H}$. They are better described as follows.

Given a commutative Hopf algebra $\mathbb{H}$ which is finitely generated, we define the *algebraic group* associated to $\mathbb{H}$ as the functor

$$G_{\mathbb{H}} : \quad \{\text{Commutative, associative algebras}\} \longrightarrow \{\text{Groups}\},$$
$$A \mapsto G_{\mathbb{H}}(A) = \mathrm{Hom}_{Alg}(\mathbb{H}, A),$$

where $G_{\mathbb{H}}(A)$ is a group with the convolution product. If $\mathbb{H}$ is not finitely generated, we call the same functor the *proalgebraic group*.

In particular, all the matrix groups SL_n, GL_n, etc., can have matrix coefficients in any commutative algebra A, not only $\mathbb{C}$ or $\mathbb{R}$, and therefore are algebraic groups. Similarly, the groups of formal series G^{inv}, G^{dif}, with coefficients in any commutative algebra A, are proalgebraic groups.

II. Review of field theory

4.4 Review of classical field theory

In this section we briefly review the standard Lagrangian tools applied to fields, and the main examples of solutions of the Euler–Lagrange equations.

4.4.1 *Space-time*

The space-time coordinates are points in the Minkowski space $\mathbb{R}^{1,3}$, that is, the space endowed with the flat diagonal metric $g = (1, -1, -1, -1)$. A transformation, called *Wick's rotation*, allows us to reformulate the problems on the Euclidean space $\mathbb{R}^4$. For more generality, we then consider an Euclidean space $\mathbb{R}^D$ of dimension D, and we denote the space-time coordinates by $x = (x^\mu)$, with $\mu = 0, 1, \ldots, D-1$.

4.4.2 *Classical fields*

A *field* is a section of a bundle on a base space. If the base space is flat, as in the case we consider here, a field is just a vector-valued function. By *classical field*, we mean a real function $\phi : \mathbb{R}^D \longrightarrow \mathbb{R}$ of class C^∞, with compact support and rapidly decreasing. To be precise, we can take the function ϕ in the Schwartz space $S(\mathbb{R}^D)$, that is, ϕ is a C^∞ function such that all its derivatives $\partial^n_\mu \phi$ converge rapidly to zero for $|x| \to \infty$.

The *observables* of the system described by a field ϕ, that is, the observable quantities, are real functionals F of the field ϕ, and what can be measured of these observables are the values $F(\phi) \in \mathbb{R}$. To determine all the observables it is enough to know the field itself.

When the field $\phi : \mathbb{R}^D \longrightarrow \mathbb{C}$ has complex (nonreal) values, it is called a *wave function*. In this case, what can be measured is not the value $\phi(x)$ itself, for any $x \in \mathbb{R}^D$, but rather the real value $|\phi(x)|^2$, which describes the probability of finding the particle at the position x.

4.4.3 *Euler–Lagrange equation*

A classical field is determined as the solution of a partial differential equation, called the *field equation*, which encodes its evolution. To any system is associated a

Lagrangian density, that is, a real function $\mathcal{L}: \mathbb{R}^D \longrightarrow \mathbb{R}, x \mapsto \mathcal{L}(x, \phi(x), \partial\phi(x))$, where $\partial\phi$ denotes the gradient of ϕ. By *Noether's theorem*, the dynamics of the field ϕ is such that the *symmetries* of the field (i.e., the transformations that leave the Lagrangian invariant) are conserved. The conservation conditions are turned into a field equation by means of the *action* of the field ϕ: it is the functional S of ϕ given by

$$\phi \mapsto S[\phi] = \int_{\mathbb{R}^D} d^D x \; \mathcal{L}(x, \phi(x), \partial\phi(x)).$$

The action S is *stationary* in $\phi \in \mathrm{S}(\mathbb{R}^D)$ if for any other function $\delta\phi \in \mathrm{S}(\mathbb{R}^D)$ we have $\frac{d}{dt} S[\phi + t\delta\phi]_{|t=0} = 0$. Then, *Hamilton's principle of least* (or *stationary*) *action* states that a field ϕ satisfies the field equation if and only if the action S is stationary in ϕ. In terms of the Lagrangian, the field equation results in the so-called *Euler–Lagrange equation*

$$\left[\frac{\partial \mathcal{L}}{\partial \phi} - \sum_\mu \partial_\mu \left(\frac{\partial \mathcal{L}}{\partial(\partial_\mu \phi)} \right) \right] (x, \phi(x)) = 0. \tag{4.1}$$

This is the equation that we need to solve to find the classical field ϕ. In general, it is a nonhomogeneous and nonlinear partial differential equation, where the nonhomogeneous terms appear if the system is not isolated, and the nonlinear terms appear if the field is self-interacting.

For example, a field with Lagrangian density

$$\mathcal{L}(x, \phi, \partial\phi) = \frac{1}{2} \left(|\partial_\mu \phi(x)|^2 + m^2 \phi(x)^2 \right) - J(x)\phi(x) - \frac{\lambda}{3!}\phi(x)^3 - \frac{\mu}{4!}\phi(x)^4 \tag{4.2}$$

is subject to the Euler–Lagrange equation

$$(-\Delta + m^2)\phi(x) = J(x) + \frac{\lambda}{2}\phi(x)^2 + \frac{\mu}{3!}\phi(x)^3, \tag{4.3}$$

where we denote $\Delta\phi(x) = \sum_\mu \partial_\mu(\partial_\mu \phi(x))$. This equation is called the *Klein–Gordon equation*, because the operator $-\Delta + m^2$ is called the *Klein–Gordon operator*.

4.4.4 Free and interacting Lagrangian

A generic relativistic particle with mass m, described by a field ϕ, can have a Lagrangian density of the form

$$\mathcal{L}(x, \phi, \partial\phi) = \frac{1}{2}\, \phi^t(x) A \phi(x) - J(x)\phi(x) - \frac{\lambda}{3!}\phi(x)^3 - \frac{\mu}{4!}\phi(x)^4, \tag{4.4}$$

where A is a differential operator such as the Dirac operator or the Laplacian, typically added to the operator of multiplication by the mass or its square. The term $\frac{1}{2}\phi^t A\phi$ (quadratic in ϕ) is the kinetic term. It is also called the *free Lagrangian density* and denoted by $\mathcal{L}_0$.

The field J is an external field, which may represent a source for the field ϕ. If $J = 0$, the system described by ϕ is *isolated*, that is, it is placed in the *vacuum*. The term of the Lagrangian containing J (linear in ϕ) is the same for any field theory.

The parameters λ, μ are called *coupling constants*, because they express the self-interactions of the field. They are usually measurable parameters such as the electric charge or the flavor, but can also be unphysical parameters added for convenience. The sum of the terms which are nonquadratic in ϕ (and nonlinear) is called *interacting Lagrangian density* and denoted by $\mathcal{L}_{int}$.

4.4.5 Free fields

A free field, which we shall denote by ϕ_0, has the dynamics of a free Lagrangian $\mathcal{L}(\phi_0) = \frac{1}{2}\phi_0^t A\phi_0 - J\phi_0$. The Euler–Lagrange equation is easily written in the form

$$A\phi_0(x) = J(x). \tag{4.5}$$

The general solution of this equation is well known to be the sum $\phi_0^g + \phi_0^p$ of the general solution of the homogeneous equation $A\phi_0^g(x) = 0$ and a particular solution $\phi_0^p(x)$ of the nonhomogeneous one. In the Minkowski space-time the function ϕ_0^g is a wave (superposition of plane waves); in the Euclidean space-time the formal solution ϕ_0^g is not a Schwartz function, and we do not consider it. Therefore the function ϕ_0 is the convolution

$$\phi_0(x) = \int d^D y\; G_0(x-y)\; J(y),$$

where $G_0(x)$ is the *Green function* of the operator A, that is, the distribution such that $AG_0(x) = \delta(x)$. The physical interpretation of the convolution is that from each point y of its support, the source J affects the field ϕ at the position x through the action of $G_0(x-y)$, which is then regarded as the *field propagator*.

For instance, if $A = -\Delta + m^2$ is the Klein–Gordon operator, the Green function G_0 is the distribution defined by the Fourier transformation

$$G_0(x-y) = \int_{\mathbb{R}^D} \frac{d^D p}{(2\pi)^D}\, \frac{1}{p^2+m^2}\, e^{-ip\cdot(x-y)}. \tag{4.6}$$

4.4.6 Self-interacting fields

A field ϕ with Lagrangian density of the form (4.4) satisfies the Euler–Lagrange equation

$$A\phi(x) = J(x) + \frac{\lambda}{2}\,\phi(x)^2 + \frac{\mu}{3!}\,\phi(x)^3.$$

This differential equation is nonlinear, and in general cannot be solved exactly. If the coupling constants λ and μ are suitably small, we solve it *perturbatively*, that is, we regard the interacting terms as perturbations of the free ones. In fact, the Euler–Lagrange equation can be expressed as a recursive equation

$$\phi(x) = \int_{\mathbb{R}^D} d^D y\; G_0(x-y)\,\left[J(y) + \frac{\lambda}{2}\,\phi(y)^2 + \frac{\mu}{3!}\,\phi(y)^3\right],$$

where G_0 is the Green function of A. This equation can then be solved as a formal series in the powers of λ and μ.

For instance, let us consider the simplest Lagrangian (4.4) with $\mu = 0$, whose Euler–Lagrange equation is

$$\phi(x) = \int_{\mathbb{R}^D} d^D y\; G_0(x-y)\,\left[J(y) + \frac{\lambda}{2}\phi(y)^2\right]. \tag{4.7}$$

If on the right-hand side of Equation (4.7) we replace $\phi(y)$ by its value, and we repeat the substitutions recursively, we obtain the following perturbative solution:

$$\begin{aligned}
\phi(x) &= \int d^D y\; G_0(x-y)\; J(y) \\
&+ \frac{\lambda}{2}\int d^D y\, d^D z\, d^D u\; G_0(x-y)\; G_0(y-z)\; G_0(y-u)\; J(z)\; J(u) \\
&+ \frac{2\lambda^2}{4}\int d^D y\, d^D z\, d^D u\, d^D v\, d^D w\; G_0(x-y)\; G_0(y-z) \\
&\qquad \times G_0(y-u)\; G_0(z-v)\; G_0(z-w)\; J(z)\; J(u)\; J(v)\; J(w) \\
&+ \frac{\lambda^3}{8}\int d^D y\, d^D z\, d^D u\, d^D v\, d^D w\, d^D s\, d^D t\; G_0(x-y)\; G_0(y-z) \\
&\qquad \times G_0(y-u)\; G_0(z-v)\; G_0(z-w)\; G_0(u-s)\; G_0(u-t) \\
&\qquad \times J(z)\; J(u)\; J(v)\; J(w)\; J(s)\; J(t)\; + \mathrm{O}(\lambda^4),
\end{aligned} \tag{4.8}$$

which describes the self-interacting field in presence of an external field J.

4.4.7 Conclusion

To summarize, a typical classical field ϕ with Lagrangian density of the form

$$\mathcal{L}(\phi) = \frac{1}{2}\,\phi^t A\phi - J(x)\,\phi(x) - \frac{\lambda}{3!}\,\phi(x)^3$$

can be described perturbatively as a formal series

$$\phi(x) = \sum_{n=0}^{\infty} \lambda^n\ \phi_n(x)$$

in the powers of the coupling constant λ. Each coefficient $\phi_n(x)$ is a finite sum of integrals involving only the field propagator and the source. I describe these coefficients in Part III, using Feynman graphs.

4.5 Review of quantum field theory

In this section we briefly review the standard tools to describe quantum fields.

4.5.1 Minkowski versus Euclidean approach

In the Minkowski space-time coordinates, the quantization procedure is the so-called *canonical quantization*, based on the principle that the observables of a quantum system are self-adjoint operators acting on a Hilbert space whose elements are the *states* in which the system can be found. The probability that the measurement of an observable F yields the value carried by a state v is given by the expectation value $\langle v|F|v\rangle \in \mathbb{R}$. In this procedure, the quantum fields are *field operators*, which must be defined together with the Hilbert space of states on which they act.

A standard way to deal with quantum fields is to Wick-rotate the time, through the transformation $t \mapsto -it$, and thereby transform the Minkowski space-time into a Euclidean space. The quantum fields are then treated as *statistical fields*, that is, classical fields or wave functions ϕ which fluctuate around their expectation values. The result is equivalent to that of the Minkowski approach, and this quantization procedure is the so-called *path integral quantization*.

4.5.2 Green functions through path integrals

The first interesting expectation value is the mean value $\langle\phi(x)\rangle$ of the field ϕ at the point x. More generally, we wish to compute the Green functions $\langle\phi(x_1)\cdots\phi(x_k)\rangle$, which represent the probability that the quantum field ϕ moves from the point x_k to x_{k-1}, and so on, and reaches x_1.

A quantum field does not properly satisfy the principle of stationary action, but can be interpreted as a fluctuation around the classical solution of the Euler–Lagrange equation. On the Euclidean space, the probability of observing the quantum field at the value ϕ is proportional to $\exp\left(-\frac{S[\phi]}{\hbar}\right)$,[2] where $\hbar = \frac{h}{2\pi}$ is the reduced Planck's constant. When $\hbar \to 0$ (classical limit), we recover a maximal probability to find the field ϕ at the minimum of the action, that is, to recover the classical solution of the Euler–Lagrange equation. The Green functions can then be computed as the *path integrals*

$$\langle \phi(x_1)\cdots\phi(x_k)\rangle = \frac{\int d\phi\ \phi(x_1)\cdots\phi(x_k)\ e^{-S[\phi]/\hbar}|_{J=0}}{\int d\phi\ e^{-S[\phi]/\hbar}|_{J=0}}.$$

This approach presents a major problem: on the infinite-dimensional set of classical fields, which we may fix as the Schwartz space $S(\mathbb{R}^D)$, for $D > 1$ there is no measure $d\phi$ suitable for giving a meaning to such an integral. (For $D = 1$ the problem is solved on continuous functions by the Wiener measure.) However, assuming that we can give a meaning to the path integrals, this formulation allows us to recover the classical values, for instance $\langle\phi(x)\rangle \sim \phi(x)$, when $\hbar \to 0$.

4.5.3 Free fields

The quantization of a classical free field ϕ_0 is easy. In fact, the action $S_0[\phi_0] = \frac{1}{2}\int d^Dx\ \phi_0(x)A\phi_0(x)$ is quadratic in ϕ_0 and gives rise to a Gaussian measure, $\exp(-S[\phi_0]/\hbar)d\phi_0$. If the field is isolated, the Green functions are then easily computed:

- the mean value $\langle\phi_0(x)\rangle$ is zero;
- the two-point Green function $\langle\phi_0(x)\phi_0(y)\rangle$ coincides with the Green function $G_0(x - y)$;
- all the Green functions on an odd number of points are zero;
- the Green functions on an even number of points are products of Green functions exhausting all the points.

If the field is not isolated, or if it is self-interacting, the computation of the Green functions is more involved.

4.5.4 Dyson–Schwinger equation

In general, the Green functions satisfy an integrodifferential equation which generalizes the Euler–Lagrange equation, written in the form $\frac{\partial S[\phi]}{\partial\phi(x)} = 0$. To obtain this equation, in analogy with the analysis that one would perform on a finite-dimensional set of paths, one can proceed by introducing a generating functional

[2] On the Minkowski space this value is $\exp\left(i\,\frac{S[\phi]}{\hbar}\right)$.

for Green functions. The consistency of the results is considered sufficient for accepting the meaningless intermediate steps.

To do it, let us regard the action as a function also of the classical source field J, that is, $S[\phi] = S[\phi, J]$. Then we define the *partition function*

$$Z[J] = \int d\phi \; e^{-S[\phi]/\hbar},$$

and impose the normalization condition $Z[J]|_{J=0} = 1$. It is then easy to verify that the Green functions can be derived from the partition function, as

$$\langle \phi(x_1) \cdots \phi(x_k) \rangle = \frac{\hbar^k}{Z[J]} \left. \frac{\delta^k Z[J]}{\delta J(x_1) \cdots \delta J(x_k)} \right|_{J=0},$$

where $\frac{\delta}{\delta J(x)}$ is the functional derivative. The Dyson–Schwinger equation for Green functions, then, can be deduced from a functional equation which constrains the partition function:

$$\frac{\delta S}{\delta \phi(x)} \left[\hbar \frac{\delta}{\delta J} \right] Z[J] = 0.$$

The notation used on the left-hand side means that in the functional $\frac{\delta S}{\delta \phi(x)}$ of ϕ, we replace the variable ϕ with the operator $\hbar \frac{\delta}{\delta J}$. Because $S[\phi]$ is a polynomial, we obtain an operator which contains higher derivatives with respect to J, and which can then act on $Z[J]$.

4.5.5 Connected Green functions

If, starting from the partition function, we define the *free energy*

$$W[J] = \hbar \log Z[J], \qquad \text{that is,} \qquad Z[J] = e^{W[J]/\hbar},$$

with normalization condition $W[J]|_{J=0} = 0$, we see that the Green functions are sums of recursive terms (products of Green functions on a smaller number of points) and additional terms which involve the derivatives of the free energy:

$$\langle \phi(x) \rangle = \frac{\hbar}{Z[J]} \left. \frac{\delta Z[J]}{\delta J(x)} \right|_{J=0} = \left. \frac{\delta W[J]}{\delta J(x)} \right|_{J=0},$$

$$\langle \phi(x)\phi(y) \rangle = \langle \phi(x) \rangle \, \langle \phi(y) \rangle + \hbar \left. \frac{\delta^2 W[J]}{\delta J(x) \delta J(y)} \right|_{J=0}, \ldots.$$

These additional terms

$$G(x_1, \ldots, x_k) = \left. \frac{\delta^k W[J]}{\delta J(x_1) \cdots \delta J(x_k)} \right|_{J=0}$$

are called *connected Green functions*, for reasons which will be clear after we introduce the Feynman diagrams. Of course, knowing the connected Green functions $G(x_1, \ldots, x_k)$ is enough to recover the *full* Green functions $\langle \phi(x_1) \cdots \phi(x_k) \rangle$, through the relations

$$
\begin{aligned}
\langle \phi(x) \rangle &= G(x), \\
\langle \phi(x)\phi(y) \rangle &= G(x)\, G(y) + \hbar\, G(x, y), \\
\langle \phi(x)\phi(y)\phi(z) \rangle &= G(x)\, G(y)\, G(z) \\
&\quad + \hbar\, [G(x)\, G(y, z) + G(y)\, G(x, z) + G(z)\, G(x, y)] \\
&\quad + \hbar^2\, G(x, y, z), \qquad (4.9) \\
\langle \phi(x)\phi(y)\phi(z)\phi(u) \rangle &= G(x)\, G(y)\, G(z)\, G(u) \\
&\quad + \hbar\, [G(x)\, G(y)\, G(z, u) + \text{terms}] \\
&\quad + \hbar^2\, [G(x, y)\, G(z, u) + G(x)\, G(y, z, u) + \text{terms}] \\
&\quad + \hbar^3\, G(x, y, z, u),
\end{aligned}
$$

and so on, where by "terms" we mean the same products evaluated on suitable permutations of the points (x, y, z, u).

4.5.6 Self-interacting fields

The Dyson–Schwinger equation can be expressed in terms of the connected Green functions. To be precise, we consider the typical quantum field with classical action

$$S[\phi] = \frac{1}{2}\phi^t A\phi - J^t\phi - \frac{\lambda}{3!}\int d^D x\; \phi(x)^3,$$

and we denote by $G_0 = A^{-1}$ the resolvent of the operator A. Then, the Dyson–Schwinger equation for the one-point Green function of a field in an external field J is

$$\langle \phi(x) \rangle_J = \frac{\delta W[J]}{\delta J(x)} = \int d^D u\; G_0(x-u) \left[J(u) + \frac{\lambda}{2}\left[\left(\frac{\delta W[J]}{\delta J(u)} \right)^2 + \hbar \frac{\delta^2 W[J]}{\delta J(u)^2} \right] \right]. \tag{4.10}$$

If we evaluate Equation (4.10) at $J = 0$, we obtain the Dyson–Schwinger equation for the one-point Green function of an isolated field:

$$\langle \phi(x) \rangle = G(x) = \frac{\lambda}{2} \int d^D u\; G_0(x-u) \big[G(u)^2 + \hbar\, G(u, u) \big]. \tag{4.11}$$

If we differentiate Equation (4.10) by $\frac{\delta}{\delta J(y)}$, and evaluate the result at $J = 0$, we obtain the Dyson–Schwinger equation for the two-point connected Green function:

$$G(x, y) = G_0(x - y) + \frac{\lambda}{2} \int d^D u \; G_0(x - u) \, [2 \, G(u) \, G(u, y) + \hbar \; G(u, u, y)] \, , \tag{4.12}$$

which involves the three-point Green function. Repeating the differentiation, we get the Dyson–Schwinger equation for the n-point connected Green function.

As for classical interacting fields, these equations can be solved perturbatively. For instance, the solution of Equation (4.10), which is the mean value of a field ϕ in an external field J, is

$$\begin{aligned}
\langle \phi(x) \rangle_J = & \int d^D u \; G_0(x - u) J(u) \\
& + \frac{\lambda}{2} \int d^D y \, d^D z \, d^D u \; G_0(x - y) \, G_0(y - z) \, G_0(y - u) \; J(z) \; J(u) \\
& + \frac{2\lambda^2}{4} \int d^D y \, d^D z \, d^D u \, d^D v \, d^D w \; G_0(x - y) \, G_0(y - z) \\
& \qquad \times G_0(y - u) \, G_0(z - v) \, G_0(z - w) \; J(z) \; J(u) \; J(v) \; J(w) \\
& + \hbar \frac{\lambda}{2} \int d^D y \; G_0(x - y) \, G_0(y - y) \\
& + \hbar \frac{\lambda^2}{2} \int d^D y \, d^D z \, d^D u \; G_0(x - y) \, G_0(y - z)^2 \; G_0(z - u) \; J(u)) \\
& + O(\lambda^3).
\end{aligned} \tag{4.13}$$

Of course, the mean value of the isolated field, which is the solution of Equation (4.11), is then obtained by setting $J = 0$:

$$G(x) = \hbar \frac{\lambda}{2} \int d^D y \; G_0(x - y) \, G_0(y - y) + O(\lambda^3). \tag{4.14}$$

4.5.7 Exercise: two-point connected Green function

Compute the first perturbative terms of the solution of Equation (4.12), which represents the Green function $G(x, y)$ for an isolated field ($J = 0$).

4.5.8 Conclusion

For a typical quantum field ϕ with classical Lagrangian density of the form

$$\mathcal{L}(\phi) = \frac{1}{2} \, \phi^t A \phi - \frac{\lambda}{3!} \, \phi(x)^3,$$

- the full k-point Green function $\langle \phi(x_1) \cdots \phi(x_k) \rangle$ is the sum of the products of the connected Green functions exhausting the k external points;
- the connected k-point Green function can be described perturbatively as a power series

$$G(x_1, \ldots, x_k) = \sum_{n=0}^{\infty} \lambda^n \, G_n(x_1, \ldots, x_k)$$

in the powers of the coupling constant λ;
- the constant coefficient $G_0(x_1, \ldots, x_k)$ is the Green function of the free field;
- each higher-order coefficient $G_n(x_1, \ldots, x_k)$ is a finite sum of integrals involving only the free propagator.

I describe the sums appearing in $G_n(x_1, \ldots, x_k)$ in Part III using Feynman graphs.

III. Formal series expanded over Feynman graphs

In this part we consider a quantum field ϕ with classical Lagrangian density of the form

$$\mathcal{L}(\phi) = \frac{1}{2} \, \phi^t A \phi - J(x) \, \phi(x) - \frac{\lambda}{3!} \, \phi(x)^3,$$

where A is a differential operator, typically the Klein–Gordon operator. We denote by G_0 the Green function of A. We saw in Section 4.5 that the Green functions of this field are completely determined by the connected Green functions, and that these can only be described as formal series in the powers of the coupling constant,

$$G(x_1, \ldots, x_k) = \sum_{n=0}^{\infty} \lambda^n \, G_n(x_1, \ldots, x_k).$$

In the next section I describe the coefficients $G_n(x_1, \ldots, x_k)$ using Feynman diagrams. I begin by describing the coefficients of the perturbative solution $\phi(x) = \sum \lambda^n \, \phi_n(x)$ for the classical field.

4.6 Interacting classical fields

4.6.1 *Feynman notation*

We adopt the following Feynman notation for the field ϕ:

- field $\phi(x) = x$ •—⊘;
- source $J(y) =$ —× y;
- propagator $G_0(x - y) = x$ •—• y.

For each graphical object resulting this notation, we call its analytical value its *amplitude*.

4.6.2 Euler–Lagrange equation

The Euler–Lagrange equation (4.7) is represented by the following diagrammatic equation:

$$x \;\text{—}\!\oslash = x \;\text{—}\!\times\, y + \frac{\lambda}{2}\, x \;\text{—}\!\!<^{\oslash}_{\oslash} . \tag{4.15}$$

4.6.3 Perturbative expansion on trees

Inserting the value of y —⊘ on the right hand-side of Equation (4.15), and repeating the insertion until all the black boxes there have disappeared, we obtain a perturbative solution given by a formal series expanded on *trees*, which are graphs without loops in the space:

$$x \;\text{—}\!\oslash = x \;\text{—}\!\times + \frac{\lambda}{2}\, x \;\text{—}\!\!<^{\times}_{\times} + \frac{\lambda^2}{2}\, x \;\text{—}\!\!<^{<^{\times}_{\times}}_{\times} + \frac{\lambda^3}{8}\, x \;\text{—}\!\!<^{<^{\times}_{\times}}_{<^{\times}_{\times}} + \cdots . \tag{4.16}$$

The coefficient of each tree t contains a factor $\lambda^{V(t)}$, where $V(t)$ is the number of internal vertices of the tree, and in its denominator the symmetry factor $\mathrm{Sym}(t)$ of the tree, that is, the number of permutations of the external crosses (the sources) which leave the tree invariant.

If we compare the diagrammatic solution (4.16) with the explicit solution (4.8), we can write explicitly the value $\phi_t(x)$ of each tree t, for instance,

$$t = \;\text{—}\!\times \;\Longrightarrow\; \phi_t(x) = \int d^D y\, G_0(x-y)\, J(y),$$

$$t = \;\text{—}\!\!<^{\times}_{\times} \;\Longrightarrow\; \phi_t(x) = \int d^D y\, d^D z\, d^D u\, G_0(x-y)\, G_0(y-z) \times G_0(y-u)\, J(z)\, J(u).$$

Finally note that the valence of the internal vertices of the trees depends directly on the interacting term of the Lagrangian. In the preceding example this term was $-\frac{\lambda}{3!}\phi^3$. If the Lagrangian contains the interacting term $-\frac{\mu}{4!}\phi^4$, the internal vertices of the trees turn out to have valence 4, that is, the trees are of the form

$$\frac{\mu}{3!}\ x \ .$$

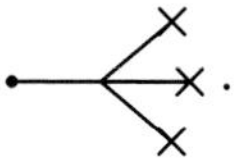

4.6.4 Feynman rules

We can therefore conclude that the field $\phi(x) = \sum_n \lambda^n\ \phi_n(x)$ has perturbative coefficients $\phi_n(x)$ given by the finite sum of the amplitudes $\phi_t(x)$ of all the trees t with n internal vertices, constructed according to the following *Feynman rules*:

- consider all the trees with internal vertices of valence 3, and external vertices of valence 1;
- fix one external vertex called the *root* (therefore the trees are called *rooted*), and call the other external vertices the *leaves*;
- label the root by x;
- label the internal vertices and the leaves by free variables $y, z, u, v, \ldots$;
- assign a weight $G_0(y-z)$ to each edge joining the vertices y and z;
- assign a weight λ to each internal vertex 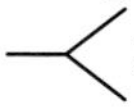 ;
- assign a weight $J(y)$ to each leaf;
- to obtain $\phi_t(x)$ for a given tree t, multiply all the weights and integrate over the free variables;
- divide by the symmetry factor $\mathrm{Sym}(t)$ of the tree.

4.6.5 Conclusion

A typical classical field ϕ with Lagrangian density of the form

$$\mathcal{L}(\phi) = \frac{1}{2}\ \phi^t A\phi - J(x)\,\phi(x) - \frac{\lambda}{3!}\ \phi(x)^3$$

can be described as a formal series in the coupling constant λ,

$$\phi(x) = \sum_{n=0}^{\infty} \lambda^n\ \phi_n(x),$$

where each coefficient $\phi_n(x)$ is a finite sum

$$\phi_n(x) = \sum_{V(t)=n} \frac{1}{\mathrm{Sym}(t)}\ \phi_t(x)$$

of amplitudes $\frac{1}{\mathrm{Sym}(t)}\ \phi_t(x)$ associated to each tree t with n internal vertices of valence 3. Note that, in this chapter, the amplitude of a tree is considered up to the factor $\frac{1}{\mathrm{Sym}(t)}$.

4.7 Interacting quantum fields

4.7.1 Feynman notation

We adopt the following Feynman notation:

- k-point full Green's function $\langle\phi(x_1)\cdots\phi(x_k)\rangle =$ [diagram: black vertex with legs $x_1, x_2, \ldots, x_k$] ;
- k-point connected Green function $G(x_1,\ldots,x_k) =$ [diagram: hatched vertex with legs $x_1, x_2, \ldots, x_k$] ;
- source $J(y) = \longrightarrow\!\times\, y$;
- propagator $G_0(x-y) = x \bullet\!\!-\!\!\bullet\; y$.

4.7.2 Exercise: Diagrammatic expression of the full Green functions

Using the equations (4.9), draw the diagrammatic expression of the full Green functions in terms of the connected ones.

4.7.3 Dyson–Schwinger equations

The Dyson–Schwinger equation for the one-point connected Green function of a field in the presence of an external field J (cf. Equation (4.10)) is the following:

$$x \bullet\!\!-\!\!\oslash = x \bullet\!\!\longrightarrow\!\!\times + \frac{\lambda}{2}\, x \bullet\!\!-\!\!<^{\oslash}_{\oslash} + \hbar\,\frac{\lambda}{2}\, x \bullet\!\!-\!\!\bigcirc\!\!\!\oslash\,. \tag{4.17}$$

Note that in the limit $\hbar \to 0$, we recover the Euler–Lagrange equation (4.15) for the field.

The Dyson–Schwinger equation for the one-point connected Green function of an isolated field (cf. Eq. (4.11)) is the following:

$$x \bullet\!\!-\!\!\oslash = \frac{\lambda}{2}\, x \bullet\!\!-\!\!<^{\oslash}_{\oslash} + \hbar\,\frac{\lambda}{2}\, x \bullet\!\!-\!\!\bigcirc\!\!\!\oslash\,. \tag{4.18}$$

For the two-point connected Green function, the Dyson–Schwinger equation is (cf. Equation (4.12))

$$x \bullet\!\!-\!\!\oslash\!\!-\!\!\bullet\; y = x \bullet\!\!-\!\!\bullet\; y + \lambda\; x \bullet\!\!-\!\!\overset{\oslash}{\top}\!\!-\!\!\oslash\!\!-\!\!\bullet\; y + \hbar\,\frac{\lambda}{2}\; x \bullet\!\!-\!\!\bigcirc\!\!\!\oslash\!\!-\!\!\bullet\; y\,. \tag{4.19}$$

For the three-point Green function,

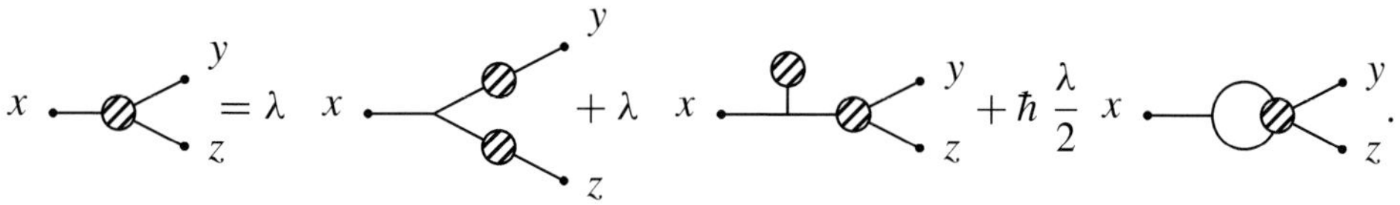

$$x \quad {}^{y}_{z} = \lambda \quad x \quad {}^{y}_{z} + \lambda \quad x \quad {}^{y}_{z} + \hbar\,\frac{\lambda}{2} \quad x \quad {}^{y}_{z}. \tag{4.20}$$

4.7.4 Perturbative expansion on graphs

Then the perturbative solution of the Dyson–Schwinger equation is given by a formal series expanded in Feynman diagrams, which are graphs in the space. For the one-point Green function, the solution of (4.17) is ($J \neq 0$)

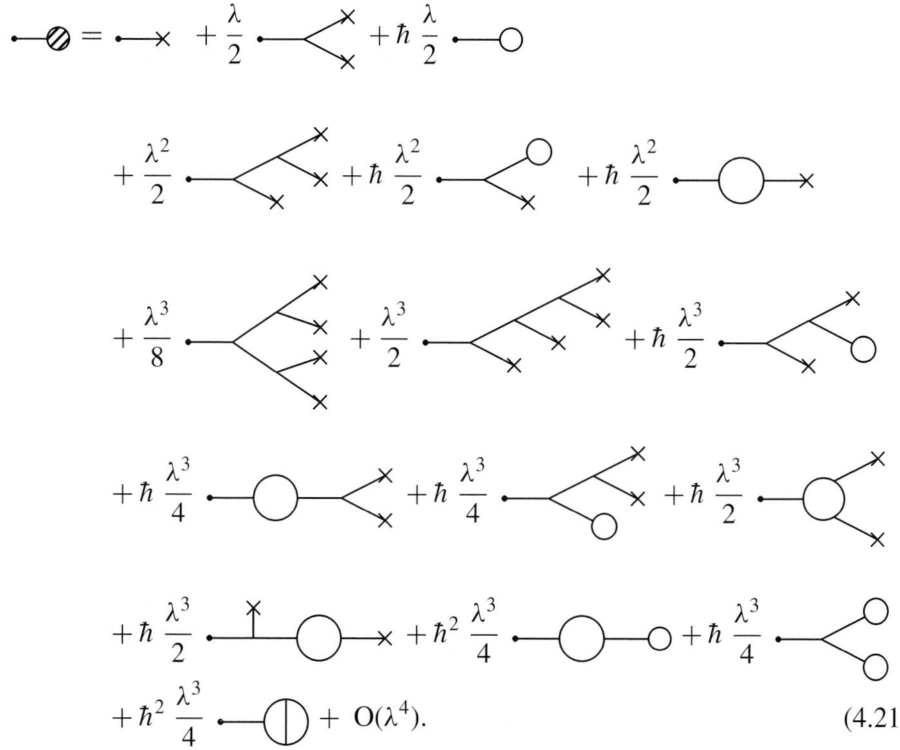

$$= \quad + \frac{\lambda}{2} \quad + \hbar\,\frac{\lambda}{2}$$

$$+ \frac{\lambda^2}{2} \quad + \hbar\,\frac{\lambda^2}{2} \quad + \hbar\,\frac{\lambda^2}{2}$$

$$+ \frac{\lambda^3}{8} \quad + \frac{\lambda^3}{2} \quad + \hbar\,\frac{\lambda^3}{2}$$

$$+ \hbar\,\frac{\lambda^3}{4} \quad + \hbar\,\frac{\lambda^3}{4} \quad + \hbar\,\frac{\lambda^3}{2}$$

$$+ \hbar\,\frac{\lambda^3}{2} \quad + \hbar^2\,\frac{\lambda^3}{4} \quad + \hbar\,\frac{\lambda^3}{4}$$

$$+ \hbar^2\,\frac{\lambda^3}{4} \quad + \mathrm{O}(\lambda^4). \tag{4.21}$$

The coefficient of each graph Γ contains a factor $\lambda^{V(\Gamma)}$, where $V(\Gamma)$ is the number of internal vertices of the graph, and in the denominator the symmetry factor $\mathrm{Sym}(\Gamma)$ of the graph, that is, the number of permutations of the external crosses (the sources) and of the internal edges (joined to the same internal vertices) which leave the graph invariant, multiplied by a factor 2 for each bubble (an internal edge connected to a single vertex).

Of course, the solution of Equation (4.18) is ($J = 0$)

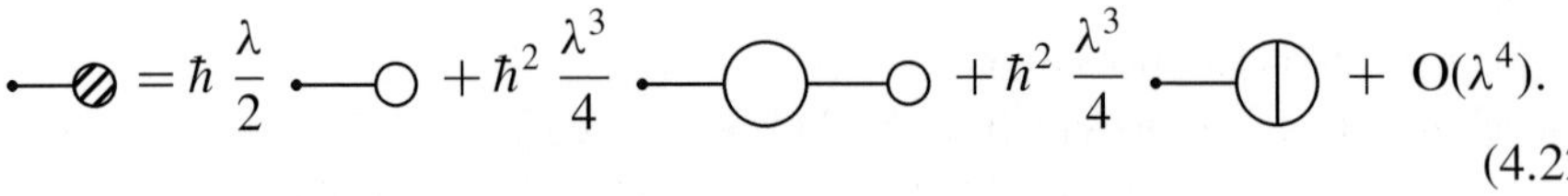

$$= \hbar \frac{\lambda}{2} \quad + \hbar^2 \frac{\lambda^3}{4} \quad + \hbar^2 \frac{\lambda^3}{4} \quad + O(\lambda^4). \tag{4.22}$$

For the two-point Green function, the solution of Equation (4.19) is

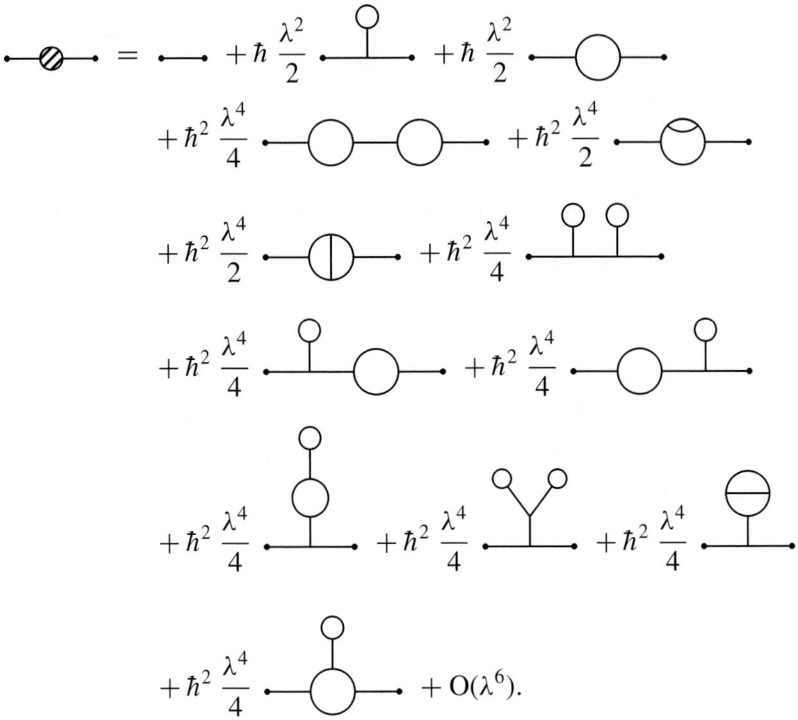

$$= \quad + \hbar \frac{\lambda^2}{2} \quad + \hbar \frac{\lambda^2}{2}$$
$$+ \hbar^2 \frac{\lambda^4}{4} \quad + \hbar^2 \frac{\lambda^4}{2}$$
$$+ \hbar^2 \frac{\lambda^4}{2} \quad + \hbar^2 \frac{\lambda^4}{4}$$
$$+ \hbar^2 \frac{\lambda^4}{4} \quad + \hbar^2 \frac{\lambda^4}{4}$$
$$+ \hbar^2 \frac{\lambda^4}{4} \quad + \hbar^2 \frac{\lambda^4}{4} \quad + \hbar^2 \frac{\lambda^4}{4}$$
$$+ \hbar^2 \frac{\lambda^4}{4} \quad + O(\lambda^6). \tag{4.23}$$

Note that the graphs appearing in Equations (4.21), (4.22), and (4.23) are connected. This motivates the name *connected Green functions*.

4.7.5 Exercise: Three-point connected Green function

Write the diagrammatic expansion of the three-point connected Green function, that is, the solution of Equation (4.20).

4.7.6 Feynman rules

We can therefore conclude that each connected Green's function $G(x_1, \ldots, x_k) = \sum_n \lambda^n \, G_n(x_1, \ldots, x_k)$ has perturbative coefficients $G_n(x_1, \ldots, x_k)$ given by the finite sum of the amplitudes $A(\Gamma; x_1, \ldots, x_k)$ of all the Feynman graphs with n internal vertices, constructed according to the following Feynman rules (valid for $J = 0$):

- consider all the graphs with internal vertices of valence 3, and k external vertices of valence 1;
- label the external vertices by $x_1, \ldots, x_k$;
- label the internal vertices by free variables $y, z, u, v, \ldots$;
- assign a weight $G_0(y - z)$ to each edge joining the vertices y and z;
- assign a weight λ to each internal vertex —< ;
- assign a weight $\hbar$ to each loop ◯ ;
- to obtain $A(\Gamma; x_1, \ldots x_k)$ for a given graph Γ, multiply all the weights and integrate over the free variables;
- divide by the symmetry factor $\mathrm{Sym}(\Gamma)$ of the graph.

4.7.7 Exercise: Feynman rules in the presence of an external source

Modify the preceding Feynman rules so that they are valid when $J \neq 0$.

4.7.8 Exercise: Compute some amplitudes

Compute the amplitudes of the first Feynman graphs appearing in the expansions of the two-point Green function (4.23), using the Feynman rules, and compare them with the results of the exercise in Section 4.5.7.

4.7.9 Conclusion

For a typical quantum field ϕ with Lagrangian density of the form

$$\mathcal{L}(\phi) = \frac{1}{2}\,\phi^t A \phi - J(x)\,\phi(x) - \frac{\lambda}{3!}\,\phi(x)^3,$$

the connected k-point Green function can be described as a formal series

$$G(x_1, \ldots, x_k) = \sum_{n=0}^{\infty} \lambda^n G_n(x_1, \ldots, x_k),$$

where each coefficient $G_n(x_1, \ldots, x_k)$ is a finite sum

$$G_n(x_1, \ldots, x_k) = \sum_{V(\Gamma)=n} \frac{\hbar^{L(\Gamma)}}{\mathrm{Sym}(\Gamma)}\, A(\Gamma; x_1, \ldots, x_k)$$

of amplitudes $A(\Gamma; x_1, \ldots, x_k)$ associated to each connected Feynman diagram Γ with n internal vertices of valence 3. Note that, in this chapter, the amplitude of a graph is considered up to the factor $\hbar^{L(\Gamma)}/\mathrm{Sym}(\Gamma)$.

4.8 Field theory on the momentum space

4.8.1 Momentum coordinates

In relativistic quantum mechanics, the *four-momentum* p, which we call simply the *momentum* here, is the conjugate variable of the four-position x, seen as an operator of multiplication on the wave function. Therefore the momentum is the Fourier transform of the operator of derivation by the position, and belongs to the Fourier space $\widehat{\mathbb{R}^D}$.

To express the field theory in momentum variables, we Fourier transform all the components of the equation of motion:

$$\widehat{\phi}(p) = \int_{\mathbb{R}^D} d^D x\, \phi(x)\, e^{ip\cdot x},$$

$$\widehat{J}(p) = \int_{\mathbb{R}^D} d^D x\, J(x)\, e^{ip\cdot x},$$

$$\widehat{G_0}(p) = \int_{\mathbb{R}^D} d^D x\, G_0(x-y)\, e^{ip\cdot(x-y)};$$

for instance, for the Klein–Gordon field, $\widehat{G_0}(p) = 1/(p^2+m^2)$ is the Fourier transform of the free propagator (4.6). The classical Euler–Lagrange equation (4.7) is then transformed into

$$\widehat{\phi}(p) = \widehat{G_0}(p)\,\widehat{J}(p) + \frac{\lambda}{2}\,\widehat{G_0}(p) \int \frac{d^D q}{(2\pi)^D}\,\widehat{\phi}(q)\,\widehat{\phi}(p-q).$$

The Fourier transform of the Green functions is

$$\widehat{G}^{(k)}(p_1,\ldots,p_k) = \int_{(\mathbb{R}^D)^k} d^D x_1 \ldots d^D x_k\, G(x_1,\ldots,x_k)\, e^{ip_1\cdot(x_1-x_k)} \cdots e^{ip_k\cdot(x_{k-1}-x_k)},$$

where the translation invariance of $G(x_1,\ldots,x_k)$ implies that $\sum_{i=1,\ldots,k} p_i = 0$, and the Dyson–Schwinger equations (4.11), (4.12), etc., can easily be expressed in terms of external momenta:

$$\widehat{G}^{(1)}(0) = \frac{\lambda}{2}\widehat{G_0}(0) \int \frac{d^D q}{(2\pi)^D} \widehat{G}^{(1)}(q)\widehat{G}^{(1)}(-q) + \hbar\frac{\lambda}{2}\widehat{G_0}(0) \int \frac{d^D q}{(2\pi)^D} \widehat{G}^{(2)}(q),$$

$$\widehat{G}^{(2)}(p) = \widehat{G_0}(p) + \lambda\,\widehat{G_0}(p)\,\widehat{G}^{(1)}(0)\,\widehat{G}^{(2)}(p) + \hbar\,\frac{\lambda}{2}\,\widehat{G_0}(p) \int \frac{d^D q}{(2\pi)^D}\,\widehat{G}^{(3)}(q, p-q, -p),$$

and so on.

4.8.2 Feynman graphs on the momentum space

The Feynman graphs in momentum variables look exactly like those in space-time coordinates, except that the external legs are not fixed in the dotted positions $x_1, \ldots, x_k$, but have oriented edges, and in particular oriented external legs labeled by momenta $p_1, \ldots, p_k$, where the arrows give the direction of the propagation. The Feynman notation is as follows:

- field $\widehat{\phi}(p) = $ [diagram: arrow labeled p into hatched blob],

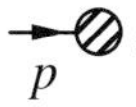

  or k-point connected Green's function $\widehat{G}^{(k)}(p_1, \ldots, p_k) = $ [diagram: hatched blob with legs $p_1, p_2, \ldots, p_k$];
- propagator $\widehat{G_0}(p) = $ [diagram: arrow labeled p];
- source $\widehat{J}(p) = $ [diagram: short leg labeled p ending in ×] (short leg labeled by p), such that $\widehat{G_0}(p)\widehat{J}(p)$ has the same dimension as $\widehat{\phi}(p)$.

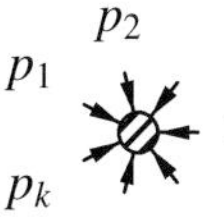

The Feynman graphs with short external legs are sometimes called *truncated* or *amputated*. Except for these few differences, the Euler–Lagrange equation, the Dyson–Schwinger equations, and their perturbative solutions are the same as those already given on the space-time coordinates.

To simplify the notation, from now on we denote by $G_0(p)$ the free propagator also in the momentum space, instead of $\widehat{G_0}(p)$, and in general we omit the hat symbol. Similarly, we omit the orientation of the propagators unless it is necessary.

4.8.3 One-particle irreducible graphs

The Feynman rules, which allow us to write the amplitude of a Feynman graph, implicitly state that the amplitude of a nonconnected graph is the product of the amplitudes of all its connected components (cf. the equations (4.9) and Section 4.7.2). If we work in the momentum space, then from the Feynman rules it also follows that if a graph Γ is the junction of two subgraphs Γ_1 and Γ_2 through a simple edge, that is,

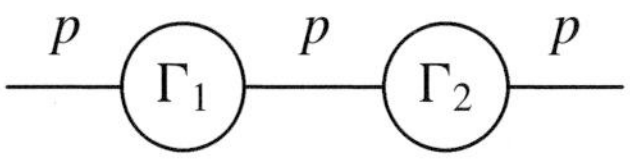

$$\Gamma = \overset{p}{—}(\Gamma_1)\overset{p}{—}(\Gamma_2)\overset{p}{—},$$

then the amplitude of Γ is the product of the amplitudes of the single graphs, that is,

$$A(\Gamma; p) = G_0(p)\, A(\Gamma_1; p)\, G_0(p)\, A(\Gamma_2; p)\, G_0(p),$$

where Γ_1 and Γ_2 are truncated on both sides. (Note that the internal edge must have momentum p because of the conservation of total momentum at each vertex.)

We say that a connected Feynman graph Γ is *one-particle irreducible*, for short 1PI, if it remains connected when we cut one of its edges. In particular, the free propagator —— is not 1PI; therefore the 1PI graphs in the momentum space are truncated. For instance, the graphs

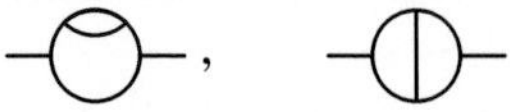

are 1PI, whereas the graphs

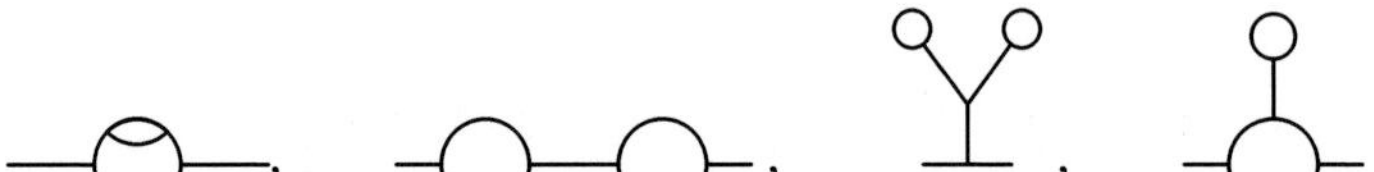

are not 1PI. If we denote the junction of graphs through one of their external legs by concatenation – for instance,

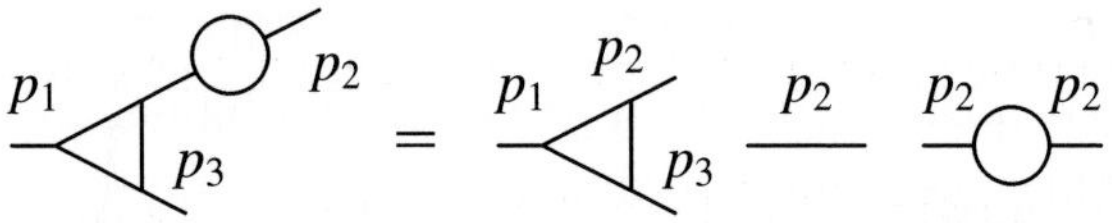

– then any connected graph can then be seen as the concatenation of its 1PI components and the free propagators necessary to join them. To avoid these free propagators popping out at any cut, we can consider graphs which are truncated only on some of their external legs, and allow joining truncated legs with full ones, for instance,

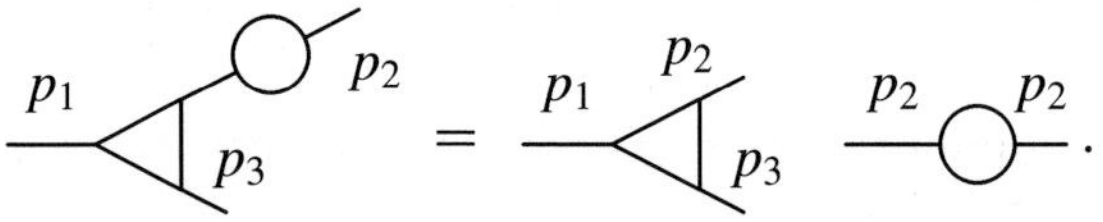

With this trick, any connected graph Γ can be seen as the junction $\Gamma = \Gamma_1 \cdots \Gamma_s$ of its 1PI components (modulo some free propagators).

4.8.4 Proper or 1PI Green functions

The fact that any connected Feynman graph can be reconstructed from its 1PI components implies that the connected Green function

$$G^{(k)}(p_1, \ldots, p_k) = \sum_{E(\Gamma)=k} \lambda^{V(\Gamma)} \frac{\hbar^{L(\Gamma)}}{\mathrm{Sym}(\Gamma)} A(\Gamma; p_1, \ldots, p_k),$$

where the sum is over all connected graphs with k external legs, can be reconstructed from the set of *proper*, or *1PI, Green functions*

$$G^{(k)}_{1\mathrm{PI}}(p_1, \ldots, p_k) = \sum_{\substack{E(\Gamma)=k \\ 1\mathrm{PI}\ \Gamma}} \lambda^{V(\Gamma)} \frac{\hbar^{L(\Gamma)}}{\mathrm{Sym}(\Gamma)} A(\Gamma; p_1, \ldots, p_k),$$

where the sum is now over 1PI graphs suitably truncated. The precise relation between connected and proper Green functions can be given easily only for the two-point Green functions: in this case we have

$$G^{(2)}(p) = G_0(p) \left[1 - G^{(2)}_{1\mathrm{PI}}(p)\, G_0(p)\right]^{-1}.$$

The general case is much more involved, and was treated recently using algebraic tools by Mestre and Oeckl in [20].

4.8.5 Conclusion

In summary, for a typical quantum field ϕ with Lagrangian density of the form

$$\mathcal{L}(\phi) = \frac{1}{2}\, \phi^t A \phi - J(x)\, \phi(x) - \frac{\lambda}{3!}\, \phi(x)^3,$$

the connected k-point Green function on the momentum space can be described as a formal series

$$G(p_1, \ldots, p_k) = \sum_{n=0}^{\infty} \lambda^n G_n(p_1, \ldots, p_k),$$

where each coefficient $G_n(p_1, \ldots, p_k)$ is a finite sum of amplitudes associated to each (partially amputated) connected Feynman diagram with n internal vertices of valence 3, and the amplitude of each graph Γ is the product of the amplitudes of its 1PI components Γ_i, that is,

$$\begin{aligned} G_n(p_1, \ldots, p_k) &= \sum_{V(\Gamma)=n} \frac{\hbar^{L(\Gamma)}}{\mathrm{Sym}(\Gamma)} A(\Gamma; p_1, \ldots, p_k) \\ &= \sum_{V(\Gamma)=n} \prod_{\Gamma=\Gamma_1\cdots\Gamma_s} \frac{\hbar^{L(\Gamma_i)}}{\mathrm{Sym}(\Gamma_i)} A(\Gamma_i; p^{(i)}_1, \ldots, p^{(i)}_{k_i}). \end{aligned}$$

IV. Renormalization

In Part II we computed the first terms of the perturbative solution of the classical and the quantum interacting fields. As we saw in Part III, these terms can be regarded as the amplitudes of some useful combinatorial objects: the rooted trees and the Feynman graphs. These analytic expressions, the amplitudes, are constructed as repeated integrals of products of the field propagator G_0 and possibly an external field J. The field propagator $G_0(x)$ is a distribution of the point x, and it is singular in $x = 0$ if $n > 1$. Then, the square $G_0(x)^2$ is a continuous function for $x \neq 0$, but it is not defined at $x = 0$. On the momentum space, this problem is translated into the divergence of the integral containing powers of the free propagator.

The powers of a free propagator never occur in the amplitude of the trees labeling the perturbative expansion of classical fields; see Equation (4.8). Similarly, they do not occur in the classical part of the perturbative expansion of Green functions for a quantum field (that is, in those terms which do not contain factors of $\hbar$). Instead, such terms occur in the quantum corrections, that is, the terms which contain factors of $\hbar$. For instance, the last two terms in Equation (4.13) contain $G_0(y - y) = G_0(0)$ and the square $G_0(y - z)^2$, which is meaningless for $y = z$.

In this part I explain some tools developed to give a meaning to the ill-defined terms appearing in the perturbative expansions of the Green functions. This technique is known as the theory of renormalization.

4.9 Renormalization of Feynman amplitudes

The renormalization of the ill-defined amplitudes can be done for graphs on the momentum variables as well as on the space-time variables. On the space-time variables, the renormalization program has been described by Epstein and Glaser in [11], in the context of the *causal perturbation theory*. However, to describe renormalization it is convenient to work on the momentum space and to consider 1PI graphs.

4.9.1 Problem of divergent integrals: ultraviolet and infrared divergences

In dimension $D = 1$, all the integrals appearing in the perturbative expansion of the Green functions are convergent. For example, if we consider the Klein–Gordon field ϕ, the free propagator

$$G_0(x - y) = \int_{\mathbb{R}} \frac{dp}{2\pi} \frac{e^{-ip(x-y)}}{p^2 + m^2}$$

is a continuous function. Therefore all the products of propagators are also continuous functions, and the integrals are well defined.

In dimension $D > 1$, the free propagator $G_0(x - y)$ is a singular distribution on the diagonal $x = y$, and the product with other distributions which are singular at the same points, such as its powers $G_0(x-y)^m$, makes no sense. For the Klein–Gordon field, for example, this already happens in the simple loop case

$$\Gamma = x \;\bullet\!\!-\!\!\bigcirc\!\!-\!\!\bullet\; y,$$

whose amplitude

$$A(\Gamma; x, y) = \int d^D u \, d^D v \; G_0(x-u)\, G_0(u-v)^2\, G_0(v-z)$$

contains the square $G_0(u-v)^2$. To understand how the integral is affected by the singularity, we better write the simple loop on the momentum space. The Fourier transform of Γ gives the (truncated) simple loop

$$\overset{p}{-}\!\!\bigcirc\!\!\overset{p}{-}\;.$$

To compute its amplitude, we write the integrated momentum q in spherical coordinates, with $|q|$ denoting the module. Then we see that for $|q| \to \infty$ the integral

$$\int \frac{d^D q}{(2\pi)^D} \frac{1}{q^2+m^2} \frac{1}{(p-q)^2+m^2}$$

behaves roughly like

$$\int_{|q|_{min}}^{\infty} d|q|^D \frac{1}{|q|^4} \simeq \int_{|q|_{min}}^{\infty} d|q| \, \frac{1}{|q|^{4-(D-1)}}.$$

This integral converges if and only if $4-(D-1) > 1$, that is, $D < 4$. Therefore $A(\Gamma; x, y)$ diverges when the dimension of the base space is $D \geq 4$.

The divergence of an amplitude $A(\Gamma; p)$ which occurs when an integrated variable q has modulus $|q| \to \infty$ is called *ultraviolet*. The divergence which occurs when $|q| \to |q|_{min}$ is called *infrared*. The infrared divergences typically appear when the mass m is zero and $|q|_{min} = 0$ (for instance, for photons). In this lecture we only deal with ultraviolet divergences.

To simplify notation, if Γ is a graph with k external legs, we denote its amplitudes $A(\Gamma; x_1, \ldots, x_k)$ or $A(\Gamma; p_1, \ldots, p_k)$ simply by $A(\Gamma)$ when the dependence on the external parameters $x_1, \ldots, x_k$ or $p_1, \ldots, p_k$ is not relevant.

4.9.2 Renormalized amplitudes, normalization conditions and renormalizable theories

There is a general procedure to estimate which integrals are divergent, and then to extract from each infinite value a finite contribution which has a physical meaning. This program is called the *renormalization* of the amplitude of Feynman graphs.

Given a graph Γ with divergent amplitude $A(\Gamma)$, the aim of the renormalization program is to find a finite contribution $A^{ren}(\Gamma)$, called the *renormalized amplitude*, which satisfies some physical requirements. In contrast to the renormalized amplitude, the original divergent amplitude is often called *bare* or *nude*.

The physical conditions required, called *normalization conditions*, are those which guarantee that the connected Green function and its derivatives have a precise value at a given point. The theory is called *renormalizable* if the number of conditions that we have to impose to determine the amplitude of all Feynman graphs is finite. For instance, the ϕ^3 theory is renormalizable in dimension $D \leq 6$.

4.9.3 Power counting: classification of one loop divergences

The *superficial degree of divergence* of a 1PI graph Γ measures the degree of singularity $\omega(\Gamma)$ of the integral in $A(\Gamma)$ with respect to the integrated variables $q_1, q_2, \ldots$. By definition, $\omega(\Gamma)$ is the integer such that, under the transformation of momentum $q_i \to tq_i$, with $t \in \mathbb{R}$, the amplitude is transformed as

$$A(\Gamma) \longrightarrow t^{\omega(\Gamma)}\, A(\Gamma).$$

The superficial degree of divergence detects the *real* divergence only for diagrams with one single loop: in this case $A(\Gamma)$ converges if and only if $\omega(\Gamma)$ is negative. The divergences for single-loop graphs are then classified according to $\omega(\Gamma)$:

- A graph Γ has a *logarithmic* divergence if $\omega(\Gamma) = 0$.
- It has a *polynomial* divergence of degree $\omega(\Gamma)$ if $\omega(\Gamma) > 0$. In particular, the divergence is *linear* if $\omega(\Gamma) = 1$, it is *quadratic* if $\omega(\Gamma) = 2$, and so on.

If instead the graph contains many loops, it can have a negative value of $\omega(\Gamma)$ and at the same time contain some divergent subgraphs. Therefore $\omega(\Gamma)$ cannot be used to estimate the real (not superficial) divergence of a graph Γ with many loops. In this case, we first need to compute $\omega(\gamma)$ for each single 1PI subgraph γ of Γ, starting from the subgraphs with a simple loop and proceeding by enlarging the subgraphs until we reach Γ itself. This recursive procedure on the subgraphs will be discussed in detail for the renormalization of a graph with many loops.

The superficial degree of divergence can easily be computed knowing only the combinatorial data of each graph. If we denote by

- I the number of internal edges of a given graph,
- E the number of external edges,
- V the number of vertices, and
- L the number of loops ($L = I - V + 1$ because of conservation of momentum at each vertex),

then for the Klein–Gordon field we have

$$\omega(\Gamma) = D\,L - 2\,I = D + (D-2)\,I - D\,V, \tag{4.24}$$

where D is the dimension of the base space. In fact, the transformation $q \to tq$ gives

$$\frac{d^D q}{(2\pi)^D} \longrightarrow t^D \frac{d^D q}{(2\pi)^D},$$
$$\frac{1}{q^2 + m^2} \longrightarrow t^{-2} \frac{1}{q^2 + m^2};$$

therefore, to compute $\omega(\Gamma)$ we have to add a term D for each loop, and a term -2 for each internal edge.

In particular, for the ϕ^3 theory (the field ϕ with interacting Lagrangian proportional to ϕ^3), we have an additional relation $3V = E + 2I$, and therefore

$$\omega(\Gamma) = D + \frac{D-6}{2}\,V - \frac{D-2}{2}\,E.$$

4.9.4 Regularization: yes or no?

Let Γ be a divergent graph, that is, assume that the amplitude $A(\Gamma)$ presents an ultraviolet divergence. In order to extract the renormalized amplitude $A^{ren}(\Gamma)$, we cannot work directly with $A(\Gamma)$, which is infinite. Instead, there are the following two main possibilities.

4.9.4.1 Regularization

We can modify $A(\Gamma)$ to a new integral $A_\rho(\Gamma)$, called the *regularized amplitude*, by introducing a *regularization parameter* ρ such that

- $A_\rho(\Gamma)$ converges,
- $A_\rho(\Gamma)$ reproduces the divergence of $A(\Gamma)$ in a certain limit $\rho \to \rho_0$.

The regularized amplitude $A_\rho(\Gamma)$ is then a well-defined function of the external momenta with values which depend on the parameter ρ. Let us denote by R_ρ the ring of such values. Then we can modify the function $A_\rho(\Gamma)$ to a new function $A^{ren}_\rho(\Gamma)$ such that the limit

$$A^{ren}(\Gamma) = \lim_{\rho \to \rho_0} A^{ren}_\rho(\Gamma)$$

is finite and compatible with the normalization conditions.

Because we are dealing here with ultraviolet divergences, it suffices to choose as regularization parameter a *cutoff* $\Lambda \in \mathbb{R}^+$ which bounds the integrated variables by above. If we denote by $I(\Gamma; q_1, \ldots, q_\ell)$ the integrand of $A(\Gamma)$, that is

$$A(\Gamma) = \int \frac{d^D q_1}{(2\pi)^D} \cdots \frac{d^D q_\ell}{(2\pi)^D} \, I(\Gamma; q_1, \ldots, q_\ell),$$

the regularized amplitude can be chosen as

$$A_\Lambda(\Gamma) = \int_{|q_i| \leq \Lambda} \frac{d^D q_1}{(2\pi)^D} \cdots \frac{d^D q_\ell}{(2\pi)^D} \, I(\Gamma; q_1, \ldots, q_\ell),$$

which reproduces the divergence of $A(\Gamma)$ for $\Lambda \to \infty$. Alternatively, the regularized amplitude $A_\Lambda(\Gamma)$ can also be described as

$$A_\Lambda(\Gamma) = \int \frac{d^D q_1}{(2\pi)^D} \cdots \frac{d^D q_\ell}{(2\pi)^D} \, \chi_\Lambda(|q_1|, \ldots, |q_\ell|) \, I(\Gamma; q_1, \ldots, q_\ell),$$

where $\chi_\Lambda(|q_1|, \ldots, |q_\ell|)$ is the step function with value 1 for $|q_1|, \ldots, |q_\ell| \leq \Lambda$ and value 0 for $|q_1|, \ldots, |q_\ell| > \Lambda$.

Besides the cutoff, there exist other possible regularizations. One of the most frequently used is the *dimensional regularization*, which modifies the real dimension D by a complex parameter ε such that $A_\varepsilon(\Gamma)$ reproduces the divergence of $A(\Gamma)$ for $\varepsilon \to 0$. Because this regularization demands many explanations, and we are not going to use it here, I omit the details, which can be found in [23] or [18].

4.9.4.2 Integrand functions

The integrand $I(\Gamma; q_1, \ldots, q_\ell)$ of $A(\Gamma)$ is a well-defined (rational) function of the variables $q_1, \ldots, q_\ell$. Therefore we can work directly with the integrand in order to modify it into a new function $I^{ren}(\Gamma; q_1, \ldots, q_\ell)$, called the *renormalized integrand*, such that

$$A^{ren}(\Gamma) = \int \frac{d^D q_1}{(2\pi)^D} \cdots \frac{d^D q_\ell}{(2\pi)^D} \, I^{ren}(\Gamma; q_1, \ldots, q_\ell)$$

is finite. This method was used by Bogoliubov in his first formulation of the renormalization, and by Zimmermann in the final proof of the so-called *BPHZ formula* (cf. Section 4.9.9). Its main advantage is that it is independent of the choice of a regularization. For these reasons we adopt it here.

4.9.5 Renormalization of a simple loop: Bogoliubov's subtraction scheme

Let Γ be a 1PI graph with one loop and superficial degree of divergence $\omega(\Gamma) \geq 0$. We assume that Γ has k external legs with external momentum $\mathbf{p} = (p_1, \ldots, p_k)$;

then the bare amplitude of the graph is

$$A(\Gamma;\mathbf{p}) = \int \frac{d^D q}{(2\pi)^D}\, I(\Gamma;\mathbf{p};q).$$

Let $T^{\omega(\Gamma)}$ denote the operator which computes the Taylor expansion in the external momentum variables $\mathbf{p}$ around the point $\mathbf{p}=0$, up to the degree $\omega(\Gamma)$. Then Bogoliubov and Parasiuk proved in [1,21] (see also [2]) that the integral

$$A^{ren}(\Gamma;\mathbf{p}) = \int \frac{d^D q}{(2\pi)^D}\Big(I(\Gamma;\mathbf{p};q) - T^{\omega(\Gamma)}[I(\Gamma;\mathbf{p};q)]\Big)$$

is finite. Changing the value $\mathbf{p}=0$ to another value $\mathbf{p}=\mathbf{p}_0$ amounts to changing $A^{ren}(\Gamma)$ by a finite amount. Possibly, the parameter $\mathbf{p}_0$ can then be chosen according to the normalization conditions.

4.9.6 Local counterterms

If we fix some regularization ρ, the renormalized (finite) amplitude can be expressed as a sum

$$A^{ren}_\rho(\Gamma;\mathbf{p}) = A_\rho(\Gamma;\mathbf{p}) - T^{\omega(\Gamma)}\big[A_\rho(\Gamma;\mathbf{p})\big], \tag{4.25}$$

where the removed divergence is contained in a polynomial of the external momenta $\mathbf{p}$,

$$-T^{\omega(\Gamma)}\big[A_\rho(\Gamma;\mathbf{p})\big] = -\int \frac{d^D q}{(2\pi)^D} I_\rho(\Gamma)\Big|_{\mathbf{p}=0} - \sum_{i,\mu} p_i^\mu \int \frac{d^D q}{(2\pi)^D}\frac{\partial I_\rho(\Gamma)}{\partial p_i^\mu}\Big|_{\mathbf{p}=0}$$
$$- \frac{1}{2}\sum_{\substack{i,j\\ \mu,\nu}} p_i^\mu p_j^\nu \int \frac{d^D q}{(2\pi)^D}\frac{\partial^2 I_\rho(\Gamma)}{\partial p_i^\mu \partial p_j^\nu}\Big|_{\mathbf{p}=0} - \cdots.$$

In matrix notation, with $\mathbf{p}=(p_1,\ldots,p_k)$, we can write

$$-T^{\omega(\Gamma)}\big[A_\rho(\Gamma;\mathbf{p})\big] = C_0^\rho(\Gamma) + C_1^\rho(\Gamma)\,\mathbf{p} + \cdots + C^\rho_{\omega(\Gamma)}(\Gamma)\,\mathbf{p}^{\omega(\Gamma)},$$

where the coefficients

$$C_r^\rho(\Gamma) = -\frac{1}{r!}\int \frac{d^D q}{(2\pi)^D}\,\partial^r_{\mathbf{p}} I_\rho(\Gamma)\Big|_{\mathbf{p}=0} \tag{4.26}$$

are called the *counterterms* of the graph Γ. If $\omega(\Gamma)=0$, we denote by $C(\Gamma)$ the unique counterterm in degree 0.

The counterterms are usually directly related to the normalization conditions; therefore, having a finite number of counterterms is equivalent to the renormalizability of the theory.

From now on, any time we mention the counterterms, we assume that a regularization has been fixed a priori, and we omit the regularization parameter ρ in the notation.

4.9.7 Examples: renormalization of a simple loop

4.9.7.1

Let us consider the graph $\Gamma =$ (one-loop graph with external momenta p, p), in dimension $D = 4$. Its amplitude (which we assume regularized) is

$$A(\Gamma; p) = \int \frac{d^4 q}{(2\pi)^4} \frac{1}{q^2 + m^2} \frac{1}{(p-q)^2 + m^2}.$$

Because $E = 2$ and $V = 2$, we have $\omega(\Gamma) = 0$; therefore the graph Γ has a logarithmic divergence. According to the subtraction scheme, its renormalized amplitude is $A^{ren}(\Gamma; p) = A(\Gamma; p) + C(\Gamma)$, where the counterterm is

$$C(\Gamma) = -\int \frac{d^4 q}{(2\pi)^4} I(\Gamma)\Big|_{p=0} = -\int \frac{d^4 q}{(2\pi)^4} \frac{1}{(q^2 + m^2)^2}.$$

The integral $A^{ren}(\Gamma; p)$ is indeed finite, because

$$I(\Gamma) - I(\Gamma)\Big|_{p=0} = \frac{1}{(q^2 + m^2)^2} \frac{2pq - p^2}{(p-q)^2 + m^2}$$

behaves like $1/|q|^5$ for $|q| \to \infty$, and therefore $A^{ren}(\Gamma; p) = \int (d^4 q/(2\pi)^4)$ $(I(\Gamma) - I(\Gamma)|_{p=0})$ behaves like

$$\int_{|q|_{min}}^{\infty} \frac{d^4|q|}{|q|^5} \simeq \int_{|q|_{min}}^{\infty} \frac{d|q|}{|q|^{5-3}} = \left[-\frac{1}{|q|}\right]_{|q|_{min}}^{\infty} = \frac{1}{|q|_{min}}.$$

4.9.7.2

Let us consider the same graph $\Gamma =$ (one-loop graph with external momenta p, p), but in dimension $D = 6$. Its amplitude is

$$A(\Gamma; p) = \int \frac{d^6 q}{(2\pi)^6} \frac{1}{q^2 + m^2} \frac{1}{(p-q)^2 + m^2}.$$

Because $E = 2$ and $V = 2$, we have $\omega(\Gamma) = 2$; therefore the graph Γ has a quadratic divergence. Then $A^{ren}(\Gamma; p) = A(\Gamma; p) - T^2\big[A(\Gamma; p)\big]$ with

$$-T^2\big[A(\Gamma; p)\big] = C_0(\Gamma) + p\, C_1(\Gamma) + p^2\, C_2(\Gamma),$$

and the local counterterms of Γ are

$$C_0(\Gamma) = -\int \frac{d^6q}{(2\pi)^6} \frac{1}{(q^2+m^2)^2},$$

$$C_1(\Gamma) = -\int \frac{d^6q}{(2\pi)^6} \frac{2q}{(q^2+m^2)^3} = 0 \quad \text{(because the integrand is odd)},$$

$$C_2(\Gamma) = -\int \frac{d^6q}{(2\pi)^6} \frac{3q^2-m^2}{(q^2+m^2)^4}.$$

Because the function

$$I(\Gamma) - T^2[I(\Gamma)] = \frac{4p^3q^3 - 3p^4q^2 - 4m^2p^3q + m^2p^4}{(q^2+m^2)^4[(p-q)^2+m^2]}$$

has leading term of order $|q|^3/|q|^{10} = 1/|q|^7$ for $|q| \to \infty$, its integral $A^{ren}(\Gamma; p)$ behaves like

$$\int_{|q|_{min}}^{\infty} \frac{d^6|q|}{|q|^7} \simeq \int_{|q|_{min}}^{\infty} \frac{d|q|}{|q|^{7-5}} = \left[-\frac{1}{|q|}\right]_{|q|_{min}}^{\infty} = \frac{1}{|q|_{min}},$$

and therefore it converges.

Exercise Check that the counterterm $C_0(\Gamma)$ alone is not sufficient to make the amplitude converge.

4.9.7.3

Let us consider the graph $\Gamma = p_1$ [graph with external legs p_2 and $p_1 - p_2$] in dimension $D = 6$. Its amplitude is

$$A(\Gamma; p_1, p_2) = \int \frac{d^6q}{(2\pi)^6} \frac{1}{q^2+m^2} \frac{1}{(q+p_2)^2+m^2} \frac{1}{(q-p_1)^2+m^2}.$$

Because $E = 3$ and $V = 3$, we have $\omega(\Gamma) = 0$; therefore, Γ has a logarithmic divergence. Then the renormalized amplitude is $A^{ren}(\Gamma; p_1, p_2) = A(\Gamma; p_1, p_2) + C(\Gamma)$, where the counterterm is

$$C(\Gamma) = -\int \frac{d^6q}{(2\pi)^6} I(\Gamma; p_1, p_2; q)\Big|_{p_i=0} = -\int \frac{d^6q}{(2\pi)^6} \frac{1}{(q^2+m^2)^3}.$$

In fact, the function

$$I(\Gamma) - I(\Gamma)\Big|_{p_i=0} = \frac{1}{q^2+m^2}\left(\frac{1}{[(q-p_1)^2+m^2]\,[(q+p_2)^2+m^2]} - \frac{1}{(q^2+m^2)^2}\right)$$
$$= \frac{2(p_1-p_2)q^3+\cdots}{(q^2+m^2)^3\,[(q-p_1)^2+m^2]\,[(q+p_2)^2+m^2]}$$

has leading term $|q|^3/|q|^{10} = 1/|q|^7$, and therefore its integral in dimension 6 converges, as in the preceding example.

4.9.8 Divergent subgraphs

The subtraction scheme employed for graphs with one loop does not work for graphs with many loops, because of the possible presence of divergent subgraphs.

For instance, consider the graph

$$\Gamma = \text{[graph]}$$

in dimension $D = 4$. Because $E = 2$ and $V = 6$, the graph has negative superficial degree of divergence, $\omega(\Gamma) = -4$. According to the subtraction scheme, it should therefore have a zero counterterm $C(\Gamma)$. However, the graph Γ contains the 1PI subgraph $\gamma = $ [graph], which has $\omega(\gamma) = 0$ in dimension $D = 4$ (as we computed in the first example of Section 4.9.7). Because γ diverges, the graph Γ diverges too, even if $\omega(\Gamma)$ is strictly negative.

4.9.9 Renormalization of many loops: BPHZ algorithm

Let Γ be a 1PI graph with many loops and superficial degree of divergence $\omega(\Gamma) \geq 0$, and/or containing some divergent subgraphs. Let $A(\Gamma)$ be its amplitude (we omit the external momenta $\mathbf{p}$), and $I(\Gamma)$ or $I(\Gamma;\mathbf{q})$ its integrand, where $\mathbf{q} = (q_1,\ldots,q_\ell)$ are the integrated momenta and ℓ is the number of loops of Γ.

Then, the *BPHZ formula* states that the renormalized (that is, finite) amplitude of Γ is given by

$$A^{ren}(\Gamma) = \int \frac{d^D q_1}{(2\pi)^D}\cdots\frac{d^D q_\ell}{(2\pi)^D}\left(I^{prep}(\Gamma;\mathbf{q}) - T^{\omega(\Gamma)}\Big[I^{prep}(\Gamma;\mathbf{q})\Big]\right), \qquad (4.27)$$

where $I^{prep}(\Gamma)$ denotes a *prepared term* where all the divergent subgraphs have been renormalized. The prepared term is defined recursively on the 1PI divergent subgraphs of Γ by the formula

$$I^{prep}(\Gamma;\mathbf{q}) = I(\Gamma;\mathbf{q}) + \sum_{\gamma_i}\prod_i \left(-T^{\omega(\gamma_i)}\left[I^{prep}(\gamma_i;\mathbf{q}_i)\right]\right) \frac{I(\Gamma;\mathbf{q})}{\prod_i I(\gamma_i;\mathbf{q}_i)},$$

where the sum is over all 1PI divergent proper subgraphs γ_i of Γ (that is, the subgraphs different from Γ itself) such that $\gamma_i \cap \gamma_j = \emptyset$ (that is, they are disjoint). The proof was first partially given by Bogoliubov and Parasiuk in 1957 [1], then improved by Hepp in 1966 [15], and finally established in 1969 by Zimmermann [26], who gave a nonrecursive formulation in terms of *forests* of divergent subgraphs.

The formula (4.27) is usually expressed in a more uniform way. Assume that in the quotient $I(\Gamma;\mathbf{q})/\prod_i I(\gamma_i;\mathbf{q}_i)$ there remain the first ℓ' momenta $\mathbf{q}' = (q_1, \ldots, q_{\ell'})$ appearing explicitly. If we set

$$I(\Gamma/\{\gamma_i\};\mathbf{q}') := I(\Gamma;\mathbf{q})/\prod_i I(\gamma_i;\mathbf{q}_i)$$

and we integrate over the momenta $\mathbf{q}'$, we define a new graph $\Gamma/\{\gamma_i\}$ through its amplitude

$$A(\Gamma/\{\gamma_i\}) = \int \frac{d^D q_1}{(2\pi)^D} \cdots \frac{d^D q_{\ell'}}{(2\pi)^D}\, I(\Gamma/\{\gamma_i\};\mathbf{q}').$$

This graph can be defined graphically by squeezing each vertex subgraph γ_i of Γ to the corresponding usual vertex point, and each propagator subgraph γ_j (with two external legs) to a new kind of vertex point

which separates two distinguished free propagators (and therefore it is not considered to be 1PI). Then the prepared term can be written

$$I^{prep}(\Gamma;\mathbf{q}) = I(\Gamma;\mathbf{q}) + \sum_{\gamma_i}\prod_i \left(-T^{\omega(\gamma_i)}\left[I^{prep}(\gamma_i;\mathbf{q}_i)\right]\right) I(\Gamma/\{\gamma_i\};\mathbf{q}'), \tag{4.28}$$

and the integrand of the renormalized amplitude can be given in a recursive manner,

$$I^{ren}(\Gamma) = I(\Gamma) + \sum_{\gamma_i}\left\{\prod_i \left(-T^{\omega(\gamma_i)}\left[I^{prep}(\gamma_i)\right]\right) I(\Gamma/\{\gamma_i\})\right\} - T^{\omega(\Gamma)}\left[I^{prep}(\Gamma)\right]. \tag{4.29}$$

4.9.10 Recursive definition of the counterterms

The definition of the counterterms given for one-loop graphs by (4.26) can then be naturally extended to graphs with many loops by applying the Taylor expansion to the prepared integrand $I^{prep}(\Gamma)$ instead of the bare integrand $I(\Gamma)$, that is, by considering

$$-T^{\omega(\Gamma)}\left[\int \frac{d^D\,\mathbf{q}}{(2\pi)^D}\, I^{prep}(\Gamma;\mathbf{q})\right] = C_0(\Gamma) + C_1(\Gamma)\,\mathbf{p} + \cdots + C_{\omega(\Gamma)}(\Gamma)\,\mathbf{p}^{\omega(\Gamma)},$$

where we symbolically denote by $d^D\,\mathbf{q}/(2\pi)^D$ the full expression

$$\frac{d^D\,q_1}{(2\pi)^D}\cdots\frac{d^D\mathbf{q}_\ell}{(2\pi)^D}.$$

To express the counterterms $C_r(\Gamma)$ in a recursive way, we must separate the integrals of each component appearing in the prepared term $I^{prep}(\Gamma)$. Consider the complete integral

$$\int \frac{d^D\,\mathbf{q}}{(2\pi)^D}\, I^{prep}(\Gamma;\mathbf{q}) = \int \frac{d^D\,\mathbf{q}}{(2\pi)^D}\, I(\Gamma;\mathbf{q}) + \sum_{\gamma_i}\prod_i \int \frac{d^D\,\mathbf{q}'}{(2\pi)^D}\int \frac{d^D\,\mathbf{q}_i}{(2\pi)^D}\left(-T^{\omega(\gamma_i)}\Big[I^{prep}(\gamma_i;\mathbf{q}_i)\Big]\right) I(\Gamma/\{\gamma_i\};\mathbf{q}').$$

If we denote by $\mathbf{p}_i$ the external momenta of the subgraph γ_i, we have

$$\int \frac{d^D\,\mathbf{q}_i}{(2\pi)^D}\left(-T^{\omega(\gamma_i)}\Big[I^{prep}(\gamma_i;\mathbf{q}_i)\Big]\right) = C_0(\gamma_i) + C_1(\gamma_i)\,\mathbf{p}_i + \cdots + C_{\omega(\gamma_i)}(\gamma_i)\,\mathbf{p}_i^{\omega(\gamma_i)}.$$

Of course the momenta $\mathbf{p}_i$ are integrated over $\mathbf{q}'$, because they are internal in Γ. To separate the integrals, it suffices to modify the amplitude of the graph $\Gamma/\{\gamma_i\}$ by multiplying it by each remaining momentum $\mathbf{p}_i^r$. In practice, it suffices to label each new crossed vertex obtained by squeezing γ_i with a label (r), with $r = 0, 1, \ldots, \omega(\gamma_i)$, and to define its amplitude by

$$A(\Gamma/\{\gamma_{i\,(r)}\}) = \int \frac{d^D\,\mathbf{q}'}{(2\pi)^D}\prod_i \mathbf{p}_i^r\; I(\Gamma/\{\gamma_i\};\mathbf{q}'). \tag{4.30}$$

Finally, if we label each scratched subgraph γ_i with the same label (r) used in its associated crossed vertex, and we define its counterterm by

$$C(\gamma_{i\,(r)}) = C_r(\gamma_i), \tag{4.31}$$

we can describe the counterterms in a recursive way as

$$C(\Gamma_{(r)}) = -\frac{1}{r!}\,\partial^r_{\mathbf{p}}\Big|_{\mathbf{p}=0}\left[A(\Gamma;\mathbf{p}) + \sum_{\gamma_i}\prod_i\sum_{r_i=0}^{\omega(\gamma_i)} C(\gamma_{i\,(r_i)})\;A(\Gamma/\{\gamma_{i\,(r_i)}\};\mathbf{p})\right]. \tag{4.32}$$

The labels (r) are useful only for graphs with positive superficial degree of divergence. If $\omega(\Gamma) = 0$, the subscript (0) is systematically omitted.

As a consequence, the extension of (4.25) to graphs with many loops is given by

$$A^{ren}(\Gamma;\mathbf{p}) = A(\Gamma;\mathbf{p}) + \sum_{\gamma_i}\prod_i\sum_{r_i=0}^{\omega(\gamma_i)} C(\gamma_{i\,(r_i)})\;A(\Gamma/\{\gamma_{i\,(r_i)}\};\mathbf{p}) + C(\Gamma_{(0)}) + \cdots + \mathbf{p}^{\omega(\Gamma)}\;C(\Gamma_{(\omega(\Gamma))}). \tag{4.33}$$

4.9.11 Examples: renormalization of many loops

4.9.11.1

Let us consider the graph $\Gamma = $ [graph with external momenta p_1, p_2, $p_1 - p_2$ and internal momenta q_1, q_2] in dimension $D = 6$. Its amplitude is

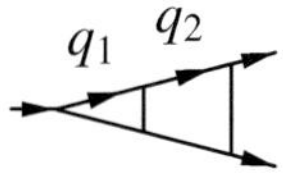

$$A(\Gamma;p_1,p_2) = \int \frac{d^6q_1}{(2\pi)^6}\frac{d^6q_2}{(2\pi)^6}\,\frac{1}{q_1^2+m^2}\,\frac{1}{(p_1-q_1)^2+m^2}\,\frac{1}{(q_1-q_2)^2+m^2} \times \frac{1}{q_2^2+m^2}\,\frac{1}{(p_1-q_2)^2+m^2}\,\frac{1}{(q_2-p_2)^2+m^2}.$$

Because $E = 3$ and $V = 5$, we have $\omega(\Gamma) = 0$; therefore the graph Γ has a logarithmic superficial divergence. Besides this, the graph Γ has two 1PI subgraphs:

- the graph $\gamma = $ [graph with external momenta p_1, q_2] has a logarithmic divergence;
- the graph $\gamma' = $ [graph with external momenta q_1, p_2] has $E = 4$ and $V = 4$, and therefore $\omega(\gamma') = -2$: it converges.

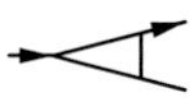

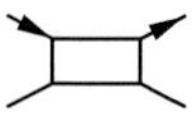

In conclusion, Γ has one divergent 1PI subgraph, γ. According to the BPHZ formula (4.28), the prepared amplitude of Γ is

$$I^{prep}(\Gamma;p_1,p_2;q_1,q_2) = I(\Gamma) - T^0[I(\gamma)]\,I(\Gamma/\gamma),$$

where for the graph γ we have

$$-T^0[I(\gamma)] = -I(\gamma; p_1, q_2; q_1)\Big|_{p_1, q_2=0} = -\frac{1}{(q_1^2+m^2)^3},$$

and for the graph $\Gamma/\gamma = p_1$ 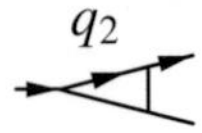 p_2 we have

$$I(\Gamma/\gamma; p_1, p_2; q_2) = \frac{1}{q_2^2+m^2}\,\frac{1}{(p_1-q_2)^2+m^2}\,\frac{1}{(q_2-p_2)^2+m^2}.$$

Therefore

$$A^{prep}(\Gamma; p_1, p_2) = \int \frac{d^6 q_1}{(2\pi)^6}\frac{d^6 q_2}{(2\pi)^6}\left(I(\Gamma) - I(\gamma)\Big|_{p_1, q_2=0} I(\Gamma/\gamma)\right),$$

and the overall counterterm $C(\Gamma) = -T^0\Big[A^{prep}(\Gamma; p_1, p_2)\Big]$ of Γ is then

$$C(\Gamma) = -\int \frac{d^6 q_1}{(2\pi)^6}\frac{d^6 q_2}{(2\pi)^6}\left(\frac{1}{(q_1^2+m^2)^2}\,\frac{1}{(q_1-q_2)^2+m^2}\,\frac{1}{(q_2^2+m^2)^3} - \frac{1}{(q_1^2+m^2)^3}\,\frac{1}{(q_2^2+m^2)^3}\right).$$

4.9.11.2

Let us consider the graph $\Gamma = p$ 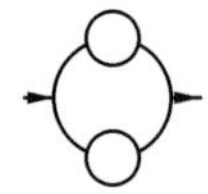 p in dimension $D = 6$. The integrand of its amplitude is

$$I(\Gamma; p; q_1, q_2, q_3) = \frac{1}{(q_1^2+m^2)^2}\,\frac{1}{q_2^2+m^2}\,\frac{1}{(q_1-q_2)^2+m^2} \times \frac{1}{\big((p-q_1)^2+m^2\big)^2}\,\frac{1}{q_3^2+m^2}\,\frac{1}{(p-q_1-q_3)^2+m^2}.$$

Because $E = 2$ and $V = 6$, we have $\omega(\Gamma) = 2$; therefore the graph Γ has a quadratic superficial divergence. Moreover, the graph Γ has two 1PI subgraphs, $\gamma_1 = \gamma_2 =$ 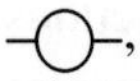, which have a quadratic divergence. Let us compute the counterterms of Γ using the BPHZ formula. We have

$$I^{prep}(\Gamma) = I(\Gamma) - T^2\Big[I(\gamma_1)I(\Gamma/\gamma_1) + I(\gamma_2)I(\Gamma/\gamma_2) + I(\gamma_1)I(\gamma_2)I(\Gamma/\gamma_1\gamma_2)\Big],$$

where

$\Gamma/\gamma_1 =$ 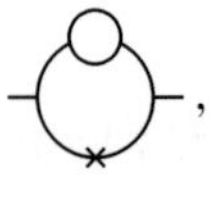, $\quad\Gamma/\gamma_2 =$ 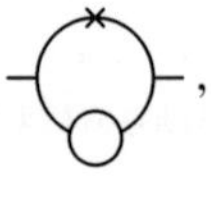, and $\Gamma/\gamma_1\gamma_2 =$ 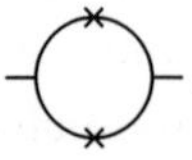.

Because the graphs γ_1 and γ_2 give the same contribution when integrated, we can call them both γ and sum them. The counterterms $C(\Gamma_{(r)}) = -\frac{1}{r!}\partial^r_{\mathbf{p}}\big|_{\mathbf{p}=0}\left[A^{prep}(\Gamma)\right]$, for $r = 0, 2$, are then given explicitly as follows:

$$C(\Gamma_{(r)}) = -\frac{1}{r!}\partial^r_{\mathbf{p}}\Big|_{\mathbf{p}=0} \Big[A(\Gamma) + 2C(\gamma_{(0)})\, A(\Gamma/\gamma_{(0)}) + 2C(\gamma_{(2)})\, A(\Gamma/\gamma_{(2)})$$
$$+ C(\gamma_{(0)})^2\, A(\Gamma/(\gamma_{(0)})^2) + 2C(\gamma_{(0)})\, C(\gamma_{(2)})\, A(\Gamma/\gamma_{(0)}\gamma_{(2)})$$
$$+ C(\gamma_{(2)})^2\, A(\Gamma/(\gamma_{(2)})^2)\Big].$$

4.10 Dyson's renormalization formulas for Green functions

As I mentioned in Section 4.5, the aim of quantum field theory is to compute the full Green functions $\langle\phi(x_1)\cdots\phi(x_k)\rangle$. To do this, we need to compute the connected Green functions $G(x_1,\ldots,x_k)$, which can only be found perturbatively, as formal series in the powers of the coupling constant λ. In Part III, I showed that the coefficients of these series can be labeled by suitable Feynman graphs. Therefore the connected Green functions can be written as

$$G(x_1,\ldots,x_k) = \sum_{n=0}^{\infty} \lambda^n \sum_{V(\Gamma)=n} \frac{\hbar^{L(\Gamma)}}{\mathrm{Sym}(\Gamma)}\, A(\Gamma; x_1,\ldots,x_k),$$

where the sum is over all the connected Feynman graphs with k external legs. In Section 4.9, then, I pointed out the problem of divergences, which affects some graphs with loops, and showed how to extract a finite contribution for each graph, the renormalized amplitude. Summing all the renormalized amplitudes, we obtain the renormalized connected Green functions

$$G^{ren}(x_1,\ldots,x_k) = \sum_{n=0}^{\infty} \lambda^n \sum_{V(\Gamma)=n} \frac{\hbar^{L(\Gamma)}}{\mathrm{Sym}(\Gamma)}\, A^{ren}(\Gamma; x_1,\ldots,x_k),$$

and finally the searched renormalized full Green functions $\langle\phi(x_1)\cdots\phi(x_k)\rangle^{ren}$.

In this section, we discuss the direct way from the bare Green functions $G(x_1,\ldots,x_k)$ to the renormalized ones, $G^{ren}(x_1,\ldots,x_k)$, without making use of Feynman graphs.

4.10.1 *Bare and renormalized Lagrangian*

From the BPHZ formula (4.33), it is clear that the passage from the bare to the renormalized amplitudes amounts to adding many terms which contain the counterterms of the divergent subgraphs,

$$A^{ren}(\Gamma) = A(\Gamma) + \text{terms}.$$

Inserting these terms in the connected Green functions, then, amounts to adding a series in λ,

$$G^{ren}(x_1, \ldots, x_k; \lambda) = G(x_1, \ldots, x_k; \lambda) + \text{series}(\lambda).$$

Because the connected Green functions are completely determined from the Lagrangian $\mathcal{L}(\phi)$ as we saw in Section 4.5, the new terms added to $G(x_1, \ldots, x_k; \lambda)$ must correspond to new terms added to $\mathcal{L}(\phi)$:

$$\mathcal{L}^{ren}(\phi, \lambda) = \mathcal{L}(\phi, \lambda) + \Delta\mathcal{L}(\phi, \lambda).$$

This Lagrangian is called *renormalized*, in contrast with the original Lagrangian $\mathcal{L}(\phi, \lambda)$, called *bare*.

Let us stress that, besides its name, the renormalized Lagrangian has no particular physical meaning: it is only a formal Lagrangian which gives rise to the renormalized (hence physically meaningful) Green functions, through the standard procedure described in Section 4.5.

The number of terms appearing in $\Delta\mathcal{L}(\phi)$ tells us if the theory is renormalizable or not: the theory is not renormalizable if the number of terms to be added is infinite.

4.10.2 Renormalization factors

If the theory is renormalizable, then $\Delta\mathcal{L}(\phi)$ contains exactly one term proportional to each term of $\mathcal{L}(\phi)$. The factors appearing in each term of the renormalized Lagrangian are called *renormalization factors*.

To be precise, let us consider again the interacting Klein–Gordon Lagrangian

$$\mathcal{L}(\phi, m, \lambda) = \frac{1}{2}|\partial_\mu \phi(x)|^2 + \frac{m^2}{2}\phi(x)^2 - \frac{\lambda}{3!}\phi(x)^3,$$

as a function of the field ϕ and of the physical parameters m (the mass) and λ (the coupling constant). Then the terms added by the renormalization can be organized as follows:

$$\Delta\mathcal{L}(\phi, m, \lambda) = \frac{1}{2}\Delta_k(\lambda)|\partial_\mu \phi(x)|^2 + \frac{m^2}{2}\Delta_m(\lambda)\phi(x)^2 - \frac{\lambda}{3!}\Delta_\lambda(\lambda)\phi(x)^3,$$

where Δ_k, Δ_m and Δ_λ are series in λ containing the counterterms of all Feynman graphs. Hence the renormalized Lagrangian is of the form

$$\begin{aligned}\mathcal{L}^{ren} = \mathcal{L} + \Delta\mathcal{L} &= \frac{1}{2}|\partial_\mu\phi(x)|^2 + \frac{m^2}{2}\phi(x)^2 - \frac{\lambda}{3!}\phi(x)^3 \\ &\quad + \Delta_k(\lambda)\frac{1}{2}|\partial_\mu\phi(x)|^2 + \Delta_m(\lambda)\frac{m^2}{2}\phi(x)^2 - \Delta_\lambda(\lambda)\frac{\lambda}{3!}\phi(x)^3 \\ &= \frac{1}{2}\, Z_3(\lambda)\, |\partial_\mu\phi(x)|^2 + \frac{m^2}{2}\, Z_m(\lambda)\, \phi(x)^2 - \frac{\lambda}{3!}\, Z_1(\lambda)\, \phi(x)^3,\end{aligned}$$

where $Z_3(\lambda) = 1 + \Delta_k(\lambda)$, $Z_m(\lambda) = 1 + \Delta_m(\lambda)$ and $Z_1(\lambda) = 1 + \Delta_\lambda(\lambda)$ are the renormalization factors.

The renormalization factors are completely determined by the counterterms of the divergent graphs. For the ϕ^3 theory in dimension $D = 6$, for instance, a graph Γ with $E = 2$ is quadratically divergent (as we saw in the example of Section 4.9.7.2), and its counterterms are of the form $C_0(\Gamma) + p^2\ C_2(\Gamma)$. According to our previous notation, and up to the scalar factor m^2, this can also be written as $m^2\, C(\Gamma_{(0)}) + p^2\, C(\Gamma_{(2)})$. Instead, a graph Γ with $E = 3$ is logarithmically divergent (as we saw in the example of Section 4.9.7.3) and has a single counterterm $C(\Gamma)$. It turns out that in this case the renormalization factors are organized as follows:

$$\begin{aligned}Z_3(\lambda) &= 1 - \sum_{E(\Gamma)=2} \frac{C(\Gamma_{(2)})}{\mathrm{Sym}(\Gamma)}\, \lambda^{V(\Gamma)}, \\ Z_m(\lambda) &= 1 - \sum_{E(\Gamma)=2} \frac{C(\Gamma_{(0)})}{\mathrm{Sym}(\Gamma)}\, \lambda^{V(\Gamma)}, \\ \lambda\, Z_1(\lambda) &= \lambda + \sum_{E(\Gamma)=3} \frac{C(\Gamma)}{\mathrm{Sym}(\Gamma)}\, \lambda^{V(\Gamma)}.\end{aligned} \tag{4.34}$$

4.10.3 Bare and effective parameters

If we define $\phi_b = Z_3(\lambda)^{\frac{1}{2}}\phi$, then we have

$$\begin{aligned}\mathcal{L}^{ren}(\phi, m, \lambda) = \frac{1}{2}\, |\partial_\mu\phi_b(x)|^2 + \frac{m^2}{2}\, Z_m(\lambda)\, Z_3(\lambda)^{-1}\phi_b(x)^2 \\ - \frac{\lambda}{3!}\, Z_1(\lambda)\, Z(\lambda)^{-\frac{3}{2}}\phi_b(x)^3,\end{aligned}$$

and if we set also

$$m_b = m\ Z_m(\lambda)^{\frac{1}{2}} Z_3(\lambda)^{-\frac{1}{2}}, \tag{4.35}$$

$$\lambda_b = \lambda\ Z_1(\lambda)\ Z_3(\lambda)^{-\frac{3}{2}}, \tag{4.36}$$

we finally obtain

$$\mathcal{L}^{ren}(\phi, m, \lambda) = \frac{1}{2}\ |\partial_\mu \phi_b(x)|^2 + \frac{1}{2}\ m_b^2\ \phi_b(x)^2 - \frac{\lambda_b}{3!}\ \phi_b(x)^3 = \mathcal{L}(\phi_b, m_b, \lambda_b). \tag{4.37}$$

In other words, the *formal* Lagrangian in ϕ, m, λ which produces the "real" (renormalized) Green functions, is exactly the original Lagrangian, but on "unreal" values of the field (ϕ_b), of the mass (m_b), and of the coupling constant (λ_b). By definition, the parameters ϕ_b, m_b, λ_b are formal series in λ with coefficients given by the counterterms of the graphs. The are called *bare*, in contrast with the physical ones, ϕ, m, λ, which are called *effective* because they are the measured ones.

4.10.4 Dyson's formulas

According to Equation (4.37), the renormalized Lagrangian in the effective parameters, $\mathcal{L}^{ren}(\phi, m, \lambda)$, is equal to the bare Lagrangian in the bare parameters, $\mathcal{L}(\phi_b, m_b, \lambda_b)$. Therefore, the renormalized Green functions in the effective parameters, $G^{ren}(x_1, \ldots, x_k; m, \lambda)$, must be related to the bare Green functions in the bare parameters, $G(x_1, \ldots, x_k; m_b, \lambda_b)$. This relation is given by the following formula:

$$G^{ren}(p_1, \ldots, p_k; m, \lambda) = Z_3^{-\frac{k}{2}}(\lambda)\ G(p_1, \ldots, p_k; m_b, \lambda_b), \tag{4.38}$$

where $m_b = m_b(m, \lambda)$ and $\lambda_b = \lambda_b(\lambda)$ are the formal series in the powers of λ given by Equations (4.35) and (4.36).

In this part of the chapter, this equality is called *Dyson's formula*, because it was firstly introduced by F. Dyson for quantum electrodynamics in 1949 [9].

4.10.5 Renormalization and semidirect product of series

Dyson's formula (4.38), together with the formulas (4.35) and (4.36), answers the question posed at the beginning of this section. Combining all of them, in fact, we get the explicit expression of the renormalized Green functions from the bare ones, by means of a product and a substitution of suitable formal series in λ. The transformation from bare to renormalized Green functions is a *semidirect product* law.

To show this, let us rewrite Dyson's formula by expressing only the dependence of the formal series on the parameters m and λ:

$$G^{ren}(m,\lambda) = Z_3^{-\frac{k}{2}}(\lambda)\, G(m_b(m,\lambda), \lambda_b(\lambda)). \tag{4.39}$$

In this formula, the quantities

$$G^{ren}(m,\lambda) = G_0 + O(\lambda),$$
$$G(m_b, \lambda_b) = G_0 + O(\lambda_b),$$
$$Z_3^{-\frac{k}{2}}(\lambda) = (1 + O(\lambda))^{-\frac{k}{2}} = 1 + O(\lambda)$$

are invertible series in λ (with respect to the multiplication; cf. Section 4.2.8), and the two bare parameters

$$m_b = m + O(\lambda),$$
$$\lambda_b = \lambda + O(\lambda^2)$$

are formal diffeomorphisms in λ (with respect to the substitution or composition; cf. Section 4.2.9). Therefore Equation (4.39) tells us that the renormalized Green function can be found as a semidirect product of suitable series in λ.

The relationship between the renormalization of the Green functions and the renormalization of each single graph appearing in the perturbative expansions is the main topic of these lectures. It is described in detail in the next section.

V. Hopf algebra of Feynman graphs and combinatorial groups of renormalization

In Part I, I described the Hopf algebra canonically associated to an algebraic or to a proalgebraic group, and gave some examples, for the most common groups. In this part, we start from a Hopf algebra on graphs related to the renormalization, and discuss the physical meaning of its associated proalgebraic group.

4.11 Connes–Kreimer Hopf algebra of Feynman graphs and diffeographisms

In the context of renormalization, a Hopf algebra is suitable to describe the combinatorics of the BPHZ formula, and can be given for any quantum field theory which is renormalizable by local counterterms. Its aim is precisely to describe the recursive definition of the counterterms.

Following the works [7, 8] of Connes and Kreimer, we choose as a toy model the ϕ^3 theory in dimension $D = 6$, for a scalar field ϕ. In this theory the superficial divergent graphs are those with a number of exterior legs $E \leq 3$. Among these, the tadpole graphs, which have $E = 1$, are not considered, because we assume that the one-point Green function $\langle\phi(x)\rangle$ vanishes.

4.11.1 Graded algebra of Feynman graphs

Let $\mathcal{H}^{\mathrm{CK}}$ be the polynomial algebra over $\mathbb{C}$ generated by the Feynman graphs which describe the local counterterms of the ϕ^3 theory. These are the 1PI graphs with two or three external legs, constructed on three types of vertices:

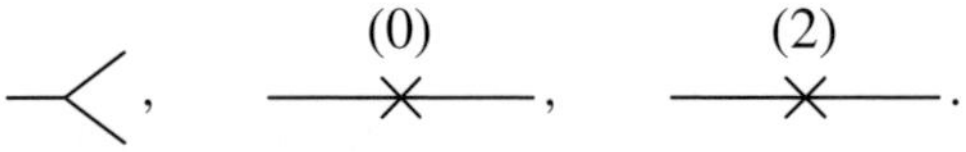

$$-\!\!<\,, \quad \overset{(0)}{-\!\!\times\!\!-}\,, \quad \overset{(2)}{-\!\!\times\!\!-}\,.$$

The free commutative multiplication between graphs is denoted by the concatenation, and the formal unit is denoted by 1.

On the algebra $\mathcal{H}^{\mathrm{CK}}$ we consider the grading induced by the number L of loops of the Feynman graphs: the degree of a monomial $\Gamma_1 \cdots \Gamma_s$ in $\mathcal{H}^{\mathrm{CK}}$ is given by $L(\Gamma_1) + \cdots + L(\Gamma_s)$. Then in degree 0 we have only the scalars (multiples of the unit 1), and therefore $\mathcal{H}^{\mathrm{CK}}$ is a connected graded algebra. In degree 1 we have only linear combinations of the one-loop graphs $-\!\bigcirc\!-$ and $-\!\!\lhd$, possibly containing some crossed vertices. In degree 2 we have linear combinations of products of two one-loop graphs and graphs with two loops, and so on for all higher degrees.

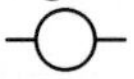

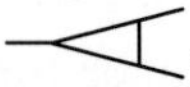

The number of non-crossed vertices V of Feynman graphs can be used as an alternative grading of $\mathcal{H}^{\mathrm{CK}}$. Note, however, that it is not equivalent to the grading by L. In fact, according to Section 4.9.3, if E is the number of external legs of a ϕ^3 graph in $D = 6$, then the number of its vertices is $V = 2L + E - 2$. Then, at a given degree L by loops, the degree by vertices is $V = 2L$ for graphs with two external legs, and $V = 2L + 1$ for graphs with three external legs. Therefore the grading induced by V is finer then that induced by L.

4.11.2 Hopf algebra of Feynman graphs

On the graded algebra $\mathcal{H}^{\mathrm{CK}}$ we consider the coproduct $\Delta : \mathcal{H}^{\mathrm{CK}} \longrightarrow \mathcal{H}^{\mathrm{CK}} \otimes \mathcal{H}^{\mathrm{CK}}$ defined as the multiplicative and unital map given on a generator Γ by

$$\Delta(\Gamma) = \Gamma \otimes 1 + 1 \otimes \Gamma + \sum_{\gamma_i, r_i} \Gamma/\{\gamma_{i\,(r_i)}\} \otimes \prod_i \gamma_{i\,(r_i)} \tag{4.40}$$

where the sum is over every possible choice of 1PI proper and disjoint divergent subgraphs γ_i of Γ, and $r_i = 0, \ldots, \omega(\gamma_i)$. The notation used here was fixed in Section 4.9:

- The term $\Gamma/\{\gamma_{i\,(r_i)}\}$ is the graph obtained from Γ by replacing each subgraph γ_i having two external legs with a labeled crossed vertex $\overset{(r_i)}{\longrightarrow\!\!\!\times\!\!\!\longrightarrow}$, and each subgraph γ_j having three external legs with a vertex graphs $-\!\!<$.
- Each graph $\gamma_{i\,(r_i)}$ means in fact the graph γ_i with a prescribed counterterm map given as the partial derivative of order r_i (evaluated at zero external momenta).
- The term $\prod_i \gamma_{i\,(r_i)}$ is a monomial in $\mathcal{H}^{\mathrm{CK}}$, that is, a free product of graphs.

On $\mathcal{H}^{\mathrm{CK}}$ we also consider the counit $\varepsilon : \mathcal{H}^{\mathrm{CK}} \longrightarrow \mathbb{C}$ defined as the multiplicative and unital map which annihilates the generators, that is, such that $\varepsilon(1) = 1$ and $\varepsilon(\Gamma) = 0$.

The coproduct and the counit so defined are graded algebra maps. Because the algebra $\mathcal{H}^{\mathrm{CK}}$ is connected, we can use the five-term equality of Section 4.1.4 to recursively define the antipode $S : \mathcal{H}^{\mathrm{CK}} \longrightarrow \mathcal{H}^{\mathrm{CK}}$. Explicitly, it is the multiplicative and unital map defined on the generators as

$$S(\Gamma) = -\Gamma - \sum_{\gamma_i, r_i} \Gamma/\{\gamma_{i\,(r_i)}\} \prod_i S(\gamma_{i\,(r_i)}).$$

In [7], Connes and Kreimer showed that $\mathcal{H}^{\mathrm{CK}}$ is a commutative and connected graded Hopf algebra, that is, the coproduct, the counit and the antipode satisfy all the compatibility properties listed in Section 4.1.

4.11.3 Group of diffeographisms and renormalization

The Hopf algebra $\mathcal{H}^{\mathrm{CK}}$ is commutative, but of course it is not finitely generated. Then, according to Section 4.3.6, $\mathcal{H}^{\mathrm{CK}}$ defines a proalgebraic group: for any associative and commutative algebra A, the set $G^{\mathrm{CK}}(A)$ of A-valued characters on $\mathcal{H}^{\mathrm{CK}}$ is a group with the convolution product $\alpha \star \beta = m_A \circ (\alpha \otimes \beta) \circ \Delta$.

Connes and Kreimer showed in [7] that if $\mathcal{A}_\rho$ is the algebra of regularized amplitudes for the ϕ^3 theory in dimension $D = 6$, then the BPHZ renormalization recursion takes place in the so-called *diffeographisms group*

$$G^{\mathrm{CK}}(\mathcal{A}_\rho) = \mathrm{Hom}_{Alg}(\mathcal{H}^{\mathrm{CK}}, \mathcal{A}_\rho). \tag{4.41}$$

More precisely, this means that the bare amplitude map A, the regularized amplitude map A^{ren} and the counterterm map C are characters $\mathcal{H}^{\mathrm{CK}} \longrightarrow \mathcal{A}_\rho$, and moreover that the BPHZ renormalization formula (4.33) is equivalent to

$$A^{ren} = A \star C. \tag{4.42}$$

In fact, for a given 1PI ϕ^3 graph Γ, Equation (4.42) means that

$$(A \star C)(\Gamma) = A(\Gamma)C(1) + A(1)C(\Gamma) + \sum_{\gamma_i, r_i} A(\Gamma/\{\gamma_{i(r_i)}\}) \prod_i C(\gamma_{i(r_i)})$$
$$= A^{ren}(\Gamma).$$

Then, comparing the BPHZ formula (4.33) with the expression (4.40) for the coproduct in $\mathcal{H}^{\mathrm{CK}}$, we see that Equation (4.42) is trivially verified provided that the counterterm map C is indeed an algebra homomorphism, and therefore $C(\gamma_1 \cdots \gamma_s) = C(\gamma_1) \cdots C(\gamma_s)$. This fact is due to a peculiar property of the truncated Taylor operator $T^{\omega(\Gamma)}$ which appears in the counterterm of any graph Γ. Namely, if we denote by T the truncated Taylor expansion, then for any functions f and g of the external momenta we have

$$T[fg] + T[f]T[g] = T\big[T[f]g + fT[g]\big].$$

An operator having this property is called a *Rota–Baxter operator*. The relationship between Rota–Baxter operators and renormalization has been extensively investigated by K. Ebrahimi-Fard and L. Guo; see for instance [10].

4.11.4 Diffeographisms and diffeomorphisms

In [8], Connes and Kreimer showed that the renormalization of the coupling constant – that is, the formula (4.36),

$$\lambda_b(\lambda) = \lambda\ Z_1(\lambda)\ Z_3(\lambda)^{-\frac{3}{2}}$$

– defines an inclusion of the coordinate ring of the group of formal diffeomorphisms into the Hopf algebra $\mathcal{H}^{\mathrm{CK}}$.

Let us denote by $\mathcal{H}^{\mathrm{dif}}$ the complex coordinate ring of the proalgebraic group G^{dif} of formal diffeomorphisms in one variable, as illustrated in Section 4.2.9. Recall that $\mathcal{H}^{\mathrm{dif}} = \mathbb{C}[x_1, x_2, \ldots]$ is an infinitely generated commutative Hopf algebra with coproduct

$$\Delta x_n = x_n \otimes 1 + 1 \otimes x_n + \sum_{m=1}^{n-1} x_m \otimes \sum_{\substack{p_0+p_1+\cdots+p_m=n-m \\ p_0,\ldots,p_m \geq 0}} x_{p_0} x_{p_1} \cdots x_{p_m}$$

and counit $\varepsilon(x_n) = 0$. Then, the inclusion $\mathcal{H}^{\mathrm{dif}} \hookrightarrow \mathcal{H}^{\mathrm{CK}}$ is defined as follows: consider the expansion (4.34) of the renormalization factors in terms of the counterterms of the divergent graphs, namely

$$Z_1(\lambda) = 1 + \sum_{E(\Gamma)=3} \frac{C(\Gamma_{(0)})}{\mathrm{Sym}(\Gamma)} \lambda^{V(\Gamma)},$$

$$Z_3(\lambda) = 1 - \sum_{E(\Gamma)=2} \frac{C(\Gamma_{(2)})}{\mathrm{Sym}(\Gamma)} \lambda^{V(\Gamma)},$$

and assign to a generator x_n of $\mathcal{H}^{\mathrm{dif}}$ the combination of Feynman graphs appearing in the coefficient of λ^{n+1} in the series $\lambda_b = \lambda\ Z_1(\lambda)\ Z_3(\lambda)^{-\frac{3}{2}}$. In [8], Connes and Kreimer proved that this map preserves the coproduct, and therefore it is a morphism of Hopf algebras.

4.11.5 Diffeographisms as generalized series

Connes and Kreimer's result summarized in the preceding subsection means in particular that the group of diffeographisms $G^{\mathrm{CK}}(\mathcal{A}_\rho)$ is projected onto the group of formal diffeomorphisms $G^{\mathrm{dif}}(\mathcal{A}_\rho)$ in one variable, with coefficients in the algebra of regularized amplitudes. In this context, formal diffeomorphisms are formal series in the powers of the coupling constant λ, that is, series of the form

$$f(\lambda) = \sum_{n=0}^{\infty} f_n\ \lambda^{n+1},$$

endowed with the composition law.

A useful way to understand the map $G^{\mathrm{CK}}(\mathcal{A}_\rho) \longrightarrow G^{\mathrm{dif}}(\mathcal{A}_\rho)$ is to represent the diffeographisms as a generalization of usual series of the form

$$f(\lambda) = \sum_{\Gamma} f_\Gamma\ \lambda^\Gamma, \tag{4.43}$$

where the sum is over suitable Feynman diagrams Γ, the coefficients f_Γ are taken in the algebra $\mathcal{A}_\rho$, and the powers λ^Γ are not monomials in a possibly complex variable λ, but just formal symbols. The projection $\pi : G^{\mathrm{CK}}(\mathcal{A}_\rho) \longrightarrow G^{\mathrm{dif}}(\mathcal{A}_\rho)$ is simply the dual map of the inclusion $\mathcal{H}^{\mathrm{dif}} \longrightarrow \mathcal{H}^{\mathrm{CK}}$, and sends a diffeographism of the form (4.43) into the formal diffeomorphism

$$\pi(f)(\lambda) = \sum_{n=0}^{\infty} \left(\sum_{V(\Gamma)=n+1} f_\Gamma \right) \lambda^{n+1}. \tag{4.44}$$

In other words, the projection is induced on the series by the map which sends a graph Γ to the number $V(\Gamma)$ of its internal vertices.

Series of the form (4.43) are unreal, and of course have no physical meaning. Instead, their images (4.44) are usual series, and have a physical meaning in the context of perturbative quantum field theory: the coupling constants are exactly series of this form, summed over suitable sets of Feynman diagrams. Moreover, the Green functions and the renormalization factors are series of this form up to a constant term which makes them invertible series instead of formal diffeomorphisms. In conclusion, the meaning of Connes and Kreimer's results is that the renormalization procedure takes place in the group $G^{\mathrm{CK}}(\mathcal{A}_\rho)$, even if the physical results are read in the group $G^{\mathrm{dif}}(\mathcal{A}_\rho)$.

4.11.6 Diffeographisms and Dyson's formulas

According to Section 4.10, the result of renormalization is described by Dyson's formulas (4.38) directly on usual series in the powers of the coupling constant λ. As we noted, this happens in the semidirect product $G^{\mathrm{dif}}(\mathcal{A}_\rho) \ltimes G^{\mathrm{inv}}(\mathcal{A}_\rho)$ of the groups of formal diffeomorphisms with that of invertible series.

However, these formulas require knowledge of the renormalization factors. According to (4.34), these are known through the computations of the counterterms of all Feynman graphs. In other words, the physical results given by Dyson's formulas seem to be the projection of computations which take place in the semidirect product $G^{\mathrm{CK}}(\mathcal{A}_\rho) \ltimes G^{\mathrm{inv}}_{graphs}(\mathcal{A}_\rho)$, where $G^{\mathrm{CK}}(\mathcal{A}_\rho)$ is the diffeographism group dual to the Connes–Kreimer Hopf algebra, and $G^{\mathrm{inv}}_{graphs}(\mathcal{A}_\rho)$ is a suitable lifting of the group of invertible series whose coordinate ring is spanned by Feynman graphs.

This conjecture has been proved for quantum electrodynamics in the series of works [4], [5] and [6]. In those works, the Green functions are expanded over *planar binary trees*, that is, planar trees with internal vertices of valence 3, which were used by Brouder in [3] as intermediate summation terms between integer numbers and Feynman graphs. It has also been proved by van Suijlekom in [25] for any gauge theory. For the ϕ^3 theory the work is in progress.

4.11.7 Groups of combinatorial series

If the diffeographisms are represented as generalized series of the form (4.43), the group law dual to the coproduct in $\mathcal{H}^{\mathrm{CK}}$ should be represented as a *composition*. This operation was abstractly defined by van der Laan in [24], using operads. An *operad* is the set of all possible operations of a given type that one can do on any algebra of that type. A particular algebra is then a representation of the corresponding operad. For instance, there exist the operad of associative algebras, that of Lie algebras, and many other examples of operads giving rise to corresponding types of algebras. By assumption, operads are endowed with an intrinsic *operadic composition* which allows one to perform the operations one after another in the corresponding algebras, and still get the result of an operation. The group G^{dif} of formal diffeomorphisms is deeply related to the operad $\mathcal{A}s$ of associative algebras, and in particular the composition of formal series in one variable can be directly related to the operadic composition in $\mathcal{A}s$. Based on this observation, Van der Laan had the idea to realize the composition among diffeographisms as the operadic composition of a suitable operad constructed on Feynman graphs. In [24], he indeed defined an *operad of all Feynman graphs*, but did not explicitly describe how to restrict the general construction to the particular case of Feynman graphs for a given theory. In particular, the explicit form of the group $G^{\mathrm{CK}}(\mathcal{A}_\rho)$ related to the renormalization of the ϕ^3 theory is not achieved.

A complete description of the generalized series and their composition law is given in [13] for the renormalization of quantum electrodynamics, on the intermediate coordinate rings spanned by planar binary trees. However, trees are combinatorial objects much simpler to handle then Feynman graphs, and the generalization of this construction to diffeographisms is still incomplete.

Groups of series expanded over other combinatorial objects, such as rooted (nonplanar) trees, also appear in the context of renormalization. Such trees, in fact, can be used to describe the perturbative expansion of Green functions, and were used by Kreimer in [17] to describe the first Hopf algebra of renormalization appearing in the literature. The dual group of tree-expanded series was then used by Girelli, Krajewski and Martinetti in [14], in their study of Wilson's continuous renormalization group.

Furthermore, the series expanded over various combinatorial objects make sense not only in the context of the renormalization of a quantum field theory, but already for classical interacting fields. In fact, as I pointed out in Section 4.6, these fields are described perturbatively as series expanded over trees. Then, any result on usual series which has a physical meaning should be the projection of computations which take place in the corresponding set of combinatorial series.

Finally, all the Hopf algebras constructed on combinatorial objects which appear in physics share some properties which are investigated in various branches of mathematics. On one side, as I already mentioned, these Hopf algebras seem to be deeply related to operads or to some generalization of them; see for instance the works by Loday and Ronco [27–29]. On the other side they turn out to be related to the various generalizations of the algebras of symmetric functions (see for instance the several works by J. Y. Thibon and colleagues [30, 31], or those by M. Aguiar and F. Sottile [32], and seem related to the so-called *combinatorial Hopf algebras*.

Bibliography

[1] N. N. Bogoliubov and O. S. Parasiuk, Acta Math. **97** (1957), 227.
[2] N. N. Bogoliubov and D. V. Shirkov, *Introduction to the theory of quantized fields*, Interscience, New York, 1959.
[3] C. Brouder, *On the trees of quantum fields*, Eur. Phys. J. C **12** (2000), 535–549.
[4] C. Brouder and A. Frabetti, *Renormalization of QED with planar binary trees*, Eur. Phys. J. C **19** (2001), 715–741.
[5] C. Brouder and A. Frabetti, *QED Hopf algebras on planar binary trees*, J. Alg. **267** (2003), 298–322.
[6] C. Brouder, A. Frabetti and C. Krattenthaler, *Non-commutative Hopf algebra of formal diffeomorphisms*, Adv. Math. **200** (2006), 479–524.
[7] A. Connes and D. Kreimer, *Renormalization in quantum field theory and the Riemann–Hilbert problem. I. The Hopf algebra structure of graphs and the main theorem*, Comm. Math. Phys. **210** (2000), 249–273.
[8] A. Connes and D. Kreimer, *Renormalization in quantum field theory and the Riemann–Hilbert problem. II. The β-function, diffeomorphisms and the renormalization group*, Comm. Math. Phys. **216** (2001), 215–241.
[9] F. J. Dyson, *The S matrix in quantum electrodynamics*, Phys. Rev. **76** (1949), 1736–1755.
[10] K. Ebrahimi-Fard and L. Guo, *Rota–Baxter algebras in renormalization of perturbative quantum field theory*, Fields Inst. Comm. 50 (2007), 47–105.
[11] H. Epstein and V. J. Glaser, *The role of locality in perturbation theory*, Ann. Poincaré Phys. Theor. A **19** (1973), 211.
[12] F. Faà di Bruno, *Sullo sviluppo delle funzioni*, Ann. Sci. Mat. Fis., Roma **6** (1855), 479–480.
[13] A. Frabetti, *Groups of tree-expanded formal series*, J. Algebra 319 (2008), 377–413.
[14] F. Girelli, T. Krajewski and P. Martinetti, *An algebraic Birkhoff decomposition for the continuous renormalization group*, J. Math. Phys. **45** (2004), 4679–4697.
[15] K. Hepp, *Proof of the Bogoliubov–Parasiuk theorem on renormalization*, Comm. Math. Phys. **2** (1966), 301–326.
[16] G. Hochschild, *La structure des groupes de Lie*, Dunod, 1968.
[17] D. Kreimer, *On the Hopf algebra structure of perturbative quantum field theories*, Adv. Theor. Math. Phys. **2** (1998), 303–334.
[18] C. Itzykson and J.-B. Zuber, *Quantum Field Theory*, McGraw-Hill, 1980.
[19] I. G. Macdonald, *Symmetric Functions and Hall Polynomials*, Oxford University Press, 1979.
[20] Â. Mestre and R. Oeckl, *Combinatorics of n-point functions via Hopf algebra in quantum field theory*, J. Math. Phys. **47** (2006), 052301.
[21] O. S. Parasiuk, Ukrainskii Math. J. **12** (1960), 287.
[22] R. P. Stanley, *Enumerative Combinatorics*, Cambridge University Press, 1997.
[23] G. 't Hooft and M. Veltman, *Regularization and renormalization of gauge fields*, Nucl. Phys. B **44** (1972), 189–213.
[24] P. van der Laan, *Operads and the Hopf algebras of renormalization*, preprint (2003), http://www.arxiv.org/abs/math-ph/0311013.
[25] W. van Suijlekom, *Multiplicative renormalization and Hopf algebras*, preprint (2007), arXiv:0707.0555.
[26] W. Zimmermann, *Convergence of Bogoliubov's method of renormalization in momentum space*, Comm. Math. Phys. **15** (1969), 208–234.

[27] J.-L. Loday and M. O. Ronco, *On the structure of cofree Hopf algebras*, J. Reine Angew. Math. **592** (2006), 123–155.
[28] J.-L. Loday, *Generalized bialgebras and triples of operads*, Astérisque **320** (2008), vi + 114 pp.
[29] J.-L. Loday and M. O. Ronco, *Combinatorial Hopf algebras*, preprint arXiv: 0810.0435.
[30] F. Hivert, J.-C. Novelli and J.-Y. Thibon, *Commutative combinatorial Hopf algebras*, J. Alg. Combin. **28** (2008), no. 1, 65–95.
[31] F. Hivert, J.-C. Novelli and J.-Y. Thibon, *Trees, functional equations, and combinatorial Hopf algebras*, Eur. J. Combin. **29** (2008), no. 7, 1682–1695.
[32] M. Aguiar and F. Sottile, *Structure of the Loday–Ronco Hopf algebra on trees*, J. Algebra **295** (2006), no. 2, 473–511.

5

BRS invariance for massive boson fields

JOSÉ M. GRACIA-BONDÍA*

Abstract

This chapter corresponds to lectures given at the Villa de Leyva Summer School in Colombia (July 2007). The main purpose in this short treatment of BRS invariance of gauge theories is to illuminate corners of the theory left in the shade by standard treatments. The plan is as follows. First, a review is given of Utiyama's general gauge theory. Immediately we find a counterexample to it in the shape of the massive spin-1 Stückelberg gauge field. This is not fancy, as the massive case is the most natural one to introduce BRS invariance in the context of free quantum fields. Mathematically speaking, the first part of the chapter uses Utiyama's notation, and thus has the flavour and nonintrinsic notation of standard physics textbooks. Next we deal with boson fields on Fock space and BRS invariance in connection with the existence of Krein operators; the attending rigour points are then addressed.

5.1 Utiyama's method in classical gauge theory

5.1.1 A historical note

Ryoyu Utiyama developed non-abelian gauge theory early in 1954 in Japan, almost at the same time that Yang and Mills [1] did at the Institute for Advanced Study (IAS) in Princeton, which Utiyama was to visit later in the year. Unfortunately, Utiyama chose not to publish immediately, and upon his arrival at IAS in September of that year, he was greatly discouraged to find he had apparently just been 'scooped'.

* I acknowledge partial support from CICyT, Spain, through the grant FIS2005-02309. This work was mostly done at the Departamento de Física Teórica I of the Universidad Complutense, to which I remain gratefully indebted.

In fact, he had not, or not entirely. He writes [2, p. 209]:

> [In March 1955], I decided to return to the general gauge theory, and took a closer look at Yang's paper, which had been published in 1954. At this moment I realized for the first time that there was a significant difference between Yang's theory and mine. The difference was that Yang had merely found an example of non-abelian gauge theory whereas I had developed a general idea of gauge theory that would contain gravity as well as electromagnetic theory. Then I decided to publish my work by translating it into English, and adding an extra section where Yang's theory is discussed as an example of my general theory.

Utiyama's article appeared on the March 1, 1956 issue of the *Physical Review* [3], and is also is reprinted in the book by the late Lochlainn O'Raifeartaigh [2], where the foregoing (and other) interesting historical remarks are made.

As Utiyama himself did in the statement quoted, most people who read his paper focused on the kinship there shown between gravity and gauge theory. This is in some sense a pity, because in contrast with textbook treatments of Yang–Mills theories – see [4] for just one example – which manage to leave, despite disguises of sophisticated language, a strong impression of arbitrariness, Utiyama strenuously tried to *derive* gauge theory from first principles. The most important trait of [3] is that he asks the right questions from the outset, as to what happens when a Lagrangian invariant with respect to a global Lie group G is required to become invariant with respect to the local group $G(x)$. What kind of new (gauge) fields need be introduced to maintain the symmetry? What is the form of the new Lagrangian, including the interaction? His answer is that the gauge field *must* be a spacetime vector field on which $G(x)$ acts by the adjoint representation, transforming in such a way that a covariant derivative exists. To our knowledge, Utiyama's argument is reproduced only in a couple of modern texts; such are [5] and [6]. I have profited from the excellent notes [7] as well.

One can speculate that, if the sequence of events had been slightly different, more attention would have been devoted to the theoretical underpinnings of the accepted dogma. It is revealing, and another pity, that Utiyama's later book in Japanese on the general gauge theory has never been translated.

5.1.2 Utiyama's analysis, first part

The starting point for Utiyama's analysis is a Lagrangian

$$\mathcal{L}(\varphi_k, \partial_\mu \varphi_k),$$

depending on a multiplet of fields φ_k and their first derivatives, *globally* invariant under a group G (of *gauge transformations of the first class*) with n independent parameters θ^a. The group is supposed to be compact. We denote by f^{abc} the structure constants of its Lie algebra $\mathfrak{g}$; that is, $\mathfrak{g}$ possesses generators T^a with commutation relations

$$[T^a, T^b] = f^{abc}T^c, \quad \text{with} \quad f^{abc} = -f^{bca},$$

and the Jacobi identity

$$f^{abd}f^{dce} + f^{bcd}f^{dae} + f^{cad}f^{dbe} = 0 \tag{5.1}$$

holds. We assume that the T^a can be chosen in such a way that f^{abc} is antisymmetric in all the three indices. This means that the adjoint representation of $\mathfrak{g}$ is semisimple, that is, $\mathfrak{g}$ is reductive [8, Chapter 15]. Close to the identity, an element $g \in G$ is of the form $\exp(T^a\theta^a)$.

The invariance is to be extended to a group $G(x)$ – of *gauge transformations of the second class* – depending on local parameters $\theta^a(x)$, in such a way that a new Lagrangian $\mathcal{L}(\varphi_k, \partial_\mu\varphi_k, A)$ invariant under the wider class of transformations is uniquely determined. Utiyama's questions are

- What new field $A(x)$ needs to be introduced?
- How does $A(x)$ transform under $G(x)$?
- What are the form of the interaction and the new Lagrangian?
- What are the allowed field equations for $A(x)$?

The global invariance is given to us under the form $\delta\varphi_k(x) = T^a_{kl}\varphi_l(x)\theta^a$; now we want to consider

$$\delta\varphi_k(x) = T^a_{kl}\varphi_l(x)\theta^a(x), \tag{5.2}$$

for $1 \le a \le n$. This last transformation in general does not leave $\mathcal{L}$ invariant. Let us first learn about the constraints imposed on the Lagrangian density by the assumed global invariance. One has

$$0 = \delta\mathcal{L} = \frac{\partial\mathcal{L}}{\partial\varphi_k}\,\delta\varphi_k + \frac{\partial\mathcal{L}}{\partial(\partial_\mu\varphi_k)}\,\delta\,\partial_\mu\varphi_k, \tag{5.3}$$

where now

$$\delta\,\partial_\mu\varphi_k = \partial_\mu\,\delta\varphi_k = T^a_{kl}\partial_\mu\varphi_l(x)\theta^a(x) + T^a_{kl}\varphi_l(x)\partial_\mu\theta^a(x). \tag{5.4}$$

With a glance at (5.3) and (5.4), we see that

$$\delta\mathcal{L} = \frac{\partial\mathcal{L}}{\partial(\partial_\mu\varphi_k)}T^a_{kl}\varphi_l(x)\partial_\mu\theta^a(x) \neq 0. \tag{5.5}$$

Then it is necessary to add new fields A'_p, $p = 1, \dots, M$, in the Lagrangian, a process which we write as

$$\mathcal{L}(\varphi_k, \partial_\mu\varphi_k) \longrightarrow \mathcal{L}'(\varphi_k, \partial_\mu\varphi_k, A'_p).$$

The question is, how do the new fields transform? We assume not only a term of the form (5.4), but also a derivative term in $\theta^a(x)$ – indeed, the latter will be needed to compensate the right-hand side of (5.5):

$$\delta A'_p = U^a_{pq} A'_q \theta^a + C^{a\mu}_p \partial_\mu \theta^a. \tag{5.6}$$

Here $C^{a\mu}_p$ and the U^a_{pq} are constant matrices, for the moment unknown. The requirement is

$$0 = \delta \mathcal{L}' = \frac{\partial \mathcal{L}'}{\partial \varphi_k} \delta\varphi_k + \frac{\partial \mathcal{L}'}{\partial(\partial_\mu \varphi_k)} \partial_\mu \delta\varphi_k + \frac{\partial \mathcal{L}'}{\partial A'_p} \delta A'_p,$$

boiling down to

$$\begin{aligned}\delta \mathcal{L}' = &\left[\frac{\partial \mathcal{L}'}{\partial \varphi_k} T^a_{kl}\varphi_l + \frac{\partial \mathcal{L}'}{\partial(\partial_\mu \varphi_k)} T^a_{kl}\partial_\mu\varphi_l + \frac{\partial \mathcal{L}'}{\partial A'_p} U^a_{pq} A'_q\right]\theta^a \\ &+ \left[\frac{\partial \mathcal{L}'}{\partial(\partial_\mu \varphi_k)} T^a_{kl}\varphi_l + \frac{\partial \mathcal{L}'}{\partial A'_p} C^{a\mu}_p\right]\partial_\mu\theta^a = 0.\end{aligned} \tag{5.7}$$

The coefficients must vanish separately, as the θ^a and their derivatives are arbitrary. The coefficient of $\partial_\mu\theta^a$ gives $4n$ equations involving A'_p, and hence to determine the A' dependence uniquely one needs $M = 4n$ components. Furthermore, the matrix $C^{a\mu}_p$ must be nonsingular. We have then an inverse:

$$C^{a\mu}_p C^{-1a}{}_{\mu q} = \delta_{pq}, \qquad C^{-1a}{}_{\mu p} C^{b\nu}_p = \delta^\nu_\mu \delta^{ab}.$$

Define the gauge (potential) field

$$A^a_\mu = \frac{1}{g} C^{-1a}{}_{\mu p} A'_p, \quad \text{with inverse} \quad A'_p = g C^{a\mu}_p A^a_\mu. \tag{5.8}$$

Before proceeding, note that (5.6) and (5.8) together imply

$$\delta A^a_\mu = \left(C^{-1a}{}_{\mu p} U^c_{pq} C^{b\nu}_q\right) A^b_\nu \theta^c + \frac{\partial_\mu \theta^a}{g} =: (S^a_\mu)^{cb\nu} A^b_\nu \theta^c + \frac{\partial_\mu \theta^a}{g}.$$

Clearly, from (5.7) we have

$$\frac{\partial \mathcal{L}'}{\partial(\partial_\mu \varphi_k)} T^a_{kl}\varphi_l + \frac{1}{g}\frac{\partial \mathcal{L}'}{\partial A^a_\mu} = 0.$$

Hence only the combination (called the ‘covariant derivative’)

$$D_\mu \varphi_k := \partial_\mu \varphi_k - g T^a_{kl} \varphi_l A^a_\mu$$

occurs in $\mathcal{L}'(\varphi_k, \partial_\mu\varphi_k, A'_p)$, and we rewrite:

$$\mathcal{L}'(\varphi_k, \partial_\mu\varphi_k, A'_p) \longrightarrow \mathcal{L}''(\varphi_k, D_\mu\varphi_k).$$

Moreover, it follows that

$$\frac{\partial\mathcal{L}'}{\partial\varphi_k} = \frac{\partial\mathcal{L}''}{\partial\varphi_k} - g\frac{\partial\mathcal{L}''}{\partial(D_\mu\varphi_l)}T^a_{lk}A^a_\mu,$$
$$\frac{\partial\mathcal{L}'}{\partial(\partial_\mu\varphi_k)} = \frac{\partial\mathcal{L}''}{\partial(D_\mu\varphi_k)},$$
$$\frac{\partial\mathcal{L}'}{\partial A'_p} = -\frac{\partial\mathcal{L}''}{\partial(D_\mu\varphi_k)}T^a_{kl}\varphi_l C^{-1}{}^a_{\mu p}.$$

Now we look at the vanishing coefficient of θ^a occurring in $\delta\mathcal{L}'$ in (5.7). By use of the last set of equations, we have

$$\begin{aligned}
0 &= \frac{\partial\mathcal{L}''}{\partial\varphi_k}T^a_{kl}\varphi_l - g\frac{\partial\mathcal{L}''}{\partial(D_\mu\varphi_m)}T^b_{mk}T^a_{kl}A^b_\mu\varphi_l \\
&\quad + \frac{\partial\mathcal{L}''}{\partial D_\mu\varphi_k}T^a_{kl}\partial_\mu\varphi_l - g\frac{\partial\mathcal{L}''}{\partial(D_\mu\varphi_m)}T^c_{ml}\varphi_l C^{-1}{}^c_{\mu p}U^a_{pq}C^{bv}_q A^b_v \\
&= \frac{\partial\mathcal{L}''}{\partial\varphi_k}T^a_{kl}\varphi_l + \frac{\partial\mathcal{L}''}{\partial D_\mu\varphi_k}T^a_{kl}D_\mu\varphi_l \\
&\quad - g\frac{\partial\mathcal{L}''}{\partial(D_\mu\varphi_m)}\big[T^b_{mk}T^a_{kl}A^b_\mu\varphi_l - T^a_{mk}T^b_{kl}A^b_\mu\varphi_l + T^c_{ml}(S^c_\mu)^{abv}A^b_v\varphi_l\big]. \qquad (5.9)
\end{aligned}$$

We have come thus to the crucial (and delicate) point. Remarkably, it seems that the two first terms in (5.9) cancel each other by global invariance if we identify

$$\mathcal{L}''(\varphi_k, D_\mu\varphi_k) = \mathcal{L}(\varphi_k, D_\mu\varphi_k).$$

Utiyama [3] writes here: 'This particular choice of $\mathcal{L}''$ is due to the requirement that when the field A is assumed to vanish, we must have the original Lagrangian'. It seems to me, however, that covariance of $D_\mu\varphi_k$ is implicitly required. The whole procedure is at least consistent: The vanishing of the last term in (5.9) allows us to identify

$$(S^c_\mu)^{abv} = f^{abc}\delta^v_\mu.$$

This implies in the end

$$\delta A^a_\mu = f^{cba}A^b_\mu\theta^c + \frac{\partial_\mu\theta^a}{g}. \qquad (5.10)$$

As a consequence we obtain that $D_\mu \varphi_k$ indeed is a covariant quantity, in the sense of (5.4):

$$\delta(D_\mu \varphi_k) = \delta(\partial_\mu \varphi_k - g T^a_{kl} A^a_\mu \varphi_l) = \partial_\mu (T^a_{kl} \theta^a \varphi_l) - g f^{cba} T^a_{km} A^b_\mu \theta^c \varphi_m$$
$$- T^a_{kl} \partial_\mu \theta^a \varphi_l - g T^b_{kl} T^c_{lm} A^b_\mu \theta^c \varphi_m = T^a_{kl} \theta^a \partial_\mu \varphi_l - g T^c_{kl} T^b_{lm} A^b_\mu \theta^c \varphi_m$$
$$= T^a_{kl} \theta^a (D_\mu \varphi_l).$$

(In summary, Utiyama's argument here looks a bit circular; but all is well in the end.)

5.1.3 Final touches to the Lagrangian

The local Lagrangian of the matter fields contains in the bargain the interaction Lagrangian between matter and gauge fields. The missing piece is the Lagrangian for the 'free' A-field. Next we investigate its possible type. Call the sought for Lagrangian $\mathcal{L}_0(A^a_\nu, \partial_\mu A^a_\nu)$. The postulate of invariance (under the local group of internal symmetry), together with (5.10), says in detail

$$0 = \left[\frac{\partial \mathcal{L}_0}{\partial A^a_\nu} f^{cba} A^b_\nu + \frac{\partial \mathcal{L}_0}{\partial(\partial_\mu A^a_\nu)} f^{cba} \partial_\mu A^b_\nu \right] \theta^c$$
$$+ \left[\frac{\partial \mathcal{L}_0}{\partial(\partial_\mu A^a_\nu)} f^{cba} A^b_\nu + \frac{1}{g} \frac{\partial \mathcal{L}_0}{\partial A^c_\mu} \right] \partial_\mu \theta^c$$
$$+ \frac{1}{g} \frac{\partial \mathcal{L}_0}{\partial(\partial_\mu A^c_\nu)} \partial_{\mu\nu} \theta^c.$$

As the θ^c are arbitrary again, one concludes that

$$\frac{\partial \mathcal{L}_0}{\partial A^a_\nu} f^{cba} A^b_\nu + \frac{\partial \mathcal{L}_0}{\partial(\partial_\mu A^a_\nu)} f^{cba} \partial_\mu A^b_\nu = 0, \tag{5.11}$$

$$\frac{\partial \mathcal{L}_0}{\partial(\partial_\mu A^a_\nu)} f^{cba} A^b_\nu + \frac{1}{g} \frac{\partial \mathcal{L}_0}{\partial A^c_\mu} = 0, \tag{5.12}$$

$$\frac{\partial \mathcal{L}_0}{\partial(\partial_\mu A^a_\nu)} + \frac{\partial \mathcal{L}_0}{\partial(\partial_\nu A^a_\mu)} = 0. \tag{5.13}$$

Introduce provisionally

$$\mathcal{A}^a_{\mu\nu} := \partial_\mu A^a_\nu - \partial_\nu A^a_\mu.$$

Then (5.12) is rewritten

$$\frac{\partial \mathcal{L}_0}{\partial A^c_\mu} + 2g \frac{\partial \mathcal{L}}{\partial(\mathcal{A}^a_{\mu\nu})} f^{cba} A^b_\nu = 0.$$

It ensues that the only combination occurring in the Lagrangian is

$$F^c_{\mu\nu} := \mathcal{A}^c_{\mu\nu} - \tfrac{1}{2} g f^{abc}(A^a_\mu A^b_\nu - A^a_\nu A^b_\mu). \tag{5.14}$$

One may write then

$$\mathcal{L}_0(A^a_\nu, \partial_\mu A^a_\nu) = \mathcal{L}'_0(F^a_{\mu\nu}).$$

Parenthetically we note

$$F^a_{\mu\nu} + F^a_{\nu\mu} = 0.$$

Now,

$$\frac{\partial \mathcal{L}_0}{\partial(\partial_\mu A^a_\nu)} = 2\frac{\partial \mathcal{L}'_0}{\partial F^a_{\mu\nu}}, \qquad \frac{\partial \mathcal{L}_0}{\partial A^b_\mu} = 2\frac{\partial \mathcal{L}'_0}{\partial F^c_{\mu\nu}} f^{abc} A^a_\nu.$$

Thus, by use of (5.1), Equation (5.11) yields

$$\frac{\partial \mathcal{L}'_0}{\partial F^c_{\mu\nu}} f^{abc} F^a_{\mu\nu} = 0, \tag{5.15}$$

for $1 \leq b \leq n$. This is left as an exercise. Also, by use of the identity of Jacobi again, one obtains

$$\delta F^c_{\mu\nu} = f^{abc} F^b_{\mu\nu} \theta^a. \tag{5.16}$$

This is a covariance equation similar to (5.4); its proof is an exercise as well.

Equation (5.15) is as far as we can go with the general argument. The simplest Lagrangian satisfying this condition is the one quadratic in $F^a_{\mu\nu}$:

$$\mathcal{L}_{\mathrm{YM}} := -\tfrac{1}{4} F^a_{\mu\nu} F^{a\,\mu\nu}, \qquad \text{implying} \quad F^a_{\mu\nu} = -\frac{\partial \mathcal{L}_{\mathrm{YM}}}{\partial(\partial_\mu A^a_\nu)}. \tag{5.17}$$

The last equation is consistent with (5.13). Note that $\delta\mathcal{L}_{\mathrm{YM}} = 0$ is obvious from (5.16).

If now we define

$$J^{c\mu} = g f^{abc} \frac{\partial \mathcal{L}_{\mathrm{YM}}}{\partial(\partial_\mu A^a_\nu)} A^b_\nu, \tag{5.18}$$

then from (5.11) again,

$$\partial_\mu J^{a\mu} = 0, \tag{5.19}$$

and from (5.12),

$$\partial^\nu F^a_{\mu\nu} = J^a_\mu, \tag{5.20}$$

by use of the equations of motion in both cases.

Let us take stock of what we have obtained.

- Equation (5.18) tells us that (in this non-abelian case) a self-interaction current J_μ exists, and gives us an explicit expression for it.
- Equation (5.19) furthermore shows that the current is conserved. Such a conservation equation, involving ordinary derivatives instead of covariant ones, does not look natural, perhaps, and is not so easy to prove directly – see the discussion in [9, Section 12.1.2]. This is the content of Noether's second theorem as applied in the present context.
- We observe that (5.20) is the field equation in the absence of matter fields.

The full Lagrangian is $\mathcal{L}(\varphi_k, D_\mu\varphi_k) + \mathcal{L}'_{\mathrm{YM}}$. One can proceed now to verify its invariance under the local transformation group and study the corresponding conserved currents. It should be clear that the conserved currents arising from local gauge invariance are exactly those following from global gauge invariance. The proof is left as an exercise.

5.1.4 The electromagnetic field

We illustrate only with the simplest example, as our main purpose is to produce a 'counterexample' pretty soon. Let a Dirac spinor field of mass M be given:

$$\mathcal{L} = \tfrac{i}{2}[\overline{\psi}\gamma^\mu\partial_\mu\psi - \partial_\mu\overline{\psi}\,\gamma^\mu\psi] - \overline{\psi}M\psi.$$

(Borrowing the frequent notation $A\overleftrightarrow{\partial^\alpha}B = A\partial^\alpha B - (\partial^\alpha A)B$, one can write this as

$$\tfrac{i}{2}\overline{\psi}\,\overleftrightarrow{\partial_\mu}\,\gamma^\mu\psi - \overline{\psi}M\psi$$

as well.) This is invariant under the global abelian group of phase transformations

$$\overline{\psi}(x) \mapsto e^{i\theta}\,\psi(x), \qquad \psi(x) \mapsto e^{-i\theta}\,\psi(x),$$

or, infinitesimally,

$$\delta\overline{\psi} = i\overline{\psi}\theta, \qquad \delta\psi = -i\psi\theta.$$

This leads to the covariant derivatives

$$D_\mu\overline{\psi} = \partial_\mu\overline{\psi} - igA_\mu\overline{\psi}, \qquad D_\mu\psi = \partial_\mu\psi + igA_\mu\psi.$$

In conclusion, the original Lagrangian receives an interaction piece $-g\overline{\psi}\gamma^\mu A_\mu\psi$, with invariance of the new Lagrangian thanks to $\delta A_\mu = \partial_\mu\theta/g$. The full locally invariant Lagrangian is

$$\tfrac{i}{2}[\overline{\psi}\gamma^\mu\partial_\mu\psi - \partial_\mu\overline{\psi}\gamma^\mu\psi] - g\overline{\psi}\gamma^\mu A_\mu\psi - \overline{\psi}M\psi - \tfrac{1}{4}F_{\mu\nu}F^{\mu\nu}.$$

One can find now the associated electromagnetic current. This is the last exercise of this section.

5.1.5 The original Yang–Mills field

Consider an isospin doublet of spinor fields:

$$\psi = (\psi_k) = \begin{pmatrix} \psi_1 \\ \psi_2 \end{pmatrix},$$

with free Lagrangian

$$\tfrac{i}{2}[\overline{\psi}_k \gamma^\mu \partial_\mu \psi_k - \partial_\mu \overline{\psi}_k \gamma^\mu \psi_k] - \overline{\psi}_k M \psi_k.$$

This is invariant under the global $SU(2)$ group, with σ^a denoting as usual the Pauli matrices:

$$\psi_k \mapsto e^{-ig\theta^a\sigma^a/2}\big|_{kl}\psi_l, \qquad \overline{\psi}_k \mapsto \overline{\psi}_l\, e^{ig\theta^a\sigma^a/2}\big|_{lk}.$$

Infinitesimally,

$$\delta\psi_k = T^a_{kl}\psi_l\theta^a, \quad \text{with} \quad T^a_{kl} = -\frac{ig}{2}\sigma^a_{kl}.$$

We have $f^{abc} = g\epsilon^{abc}$ for this group. The Lagrangian becomes gauge invariant through the replacement

$$\partial_\mu\psi_k \mapsto D_\mu\psi_k = \partial_\mu\psi_k + \frac{ig}{2}\sigma^a_{kl}\psi_l A^a_\mu;$$

that is, the triplet of vector fields is the gauge (potential) field, the number of gauge field components being equal to the number of symmetry generators. Note the slight difference in the introduction of the coupling constant of the gauge field with the spinor field and with itself.

The full locally invariant Lagrangian is

$$\tfrac{i}{2}[\overline{\psi}_k \gamma^\mu \partial_\mu \psi_k - \partial_\mu \overline{\psi}_k \gamma^\mu \psi_k] - \overline{\psi}_k M \psi_k - \tfrac{1}{4}F^a_{\mu\nu}F^{a\,\mu\nu} - \frac{g}{2}\overline{\psi}_k\gamma^\mu\sigma^a_{kl}\psi_l A^a_\mu,$$

with $F^a_{\mu\nu}$ given by (5.14). The current

$$\begin{aligned} J^a_\mu &= -\frac{g}{2}\overline{\psi}_k\gamma^\mu\sigma^a_{kl}\psi_l - g\epsilon^{abc}A^c_\nu\big[\partial_\mu A^b_\nu - \partial_\nu A^b_\mu - \frac{g}{2}\epsilon^{bde}(A^d_\mu A^e_\nu - A^d_\nu A^e_\mu)\big] \\ &= -\frac{g}{2}\overline{\psi}_k\gamma^\mu\sigma^a_{kl}\psi_l - g\epsilon^{abc}A^c_\nu\big(\partial_\mu A^b_\nu - \partial_\nu A^b_\mu\big) + g^2(A^a_\mu(AA) + A^c_\mu A^c_\nu A^a_\nu), \end{aligned}$$

with $AA := A^c_\nu A^{c\nu}$, is conserved.

5.2 Massive vector fields

5.2.1 What is wrong with the Proca field?

The starting point in relativistic quantum physics is Wigner's theory of particles as positive-energy irreps of the Poincaré group with finite spin (helicity) [10]. The transition to local free fields is made through intertwiners between the Wigner representation matrices and the matrices of covariant Lorentz group representations. Therefore, following standard notation [11], the general form of a quantum field is

$$\varphi_l(x) = \varphi_l^{(-)}(x) + \varphi_l^{(+)}(x)$$

with

$$\varphi_l^{(-)}(x) = (2\pi)^{-3/2} \sum_{\sigma,n} \int d\mu_m(k)\, u_l(k,\sigma,n) e^{-ikx} a(k,\sigma,n),$$

$$\varphi_l^{(+)}(x) = (2\pi)^{-3/2} \sum_{\sigma,n} \int d\mu_m(k)\, v_l(k,\sigma,n) e^{ikx} a^\dagger(k,\sigma,n),$$

where $d\mu_m(k)$ is the usual Lorentz-invariant measure on the mass m hyperboloid in momentum space, and n stands for particle species. Leaving the latter aside, the other labels are of representation-theoretic nature. Operator solutions to the wave equations carry the following labels, in all: the Poincaré representation (m, s) gives the the mass shell condition and the spin s; the (k, σ), with the range of σ determined by s, label the momentum basis states; the (u, v) are Lorentz representation labels, usually appearing as a superscript indicating the tensorial or spinorial character of the solution. The c-number functions u_l, v_l in the plane-wave expansion formulae are the coefficient functions or intertwiners, connecting the set of creation or absorption operators $a^\#(k,\sigma)$, transforming as the irreducible representation (m, s) of the Poincaré group, to the set of field operators $\varphi_l(x)$, transforming as a certain finite-dimensional – and thus nonunitary – irrep of the Lorentz group. We have thus in the vector field case

$$\varphi^{(-)\mu}(x) = (2\pi)^{-3/2} \sum_{\sigma} \int d\mu_m(k)\, u^\mu(k,\sigma) e^{-ikx} a(k,\sigma),$$

$$\varphi^{(+)\mu}(x) = (2\pi)^{-3/2} \sum_{\sigma} \int d\mu_m(k)\, v^\mu(k,\sigma) e^{ikx} a^\dagger(k,\sigma).$$

For the time being we ignore in the notation any colour quantum number.

For the spin of the particle described by the vector field, both values $j = 0$ and $j = 1$ are possible. In the first case, at $\vec{k} = 0$ only u^0, v^0 are nonzero, and, dropping the label σ, we have by Lorentz invariance

$$u^\mu(k) \propto ik^\mu, \qquad v^\mu(k) \propto -ik^\mu,$$

and therefore $\varphi^\mu(x) = \partial^\mu \varphi(x)$ for some scalar field φ. In the second case, only the space components u^j, v^j are nonvanishing at $\vec{k} = 0$, and we are led to

$$\varphi^{(-)\mu}(x) = \varphi^{(+)\mu\dagger}(x) = (2\pi)^{-3/2} \sum_{\sigma=1}^{3} \int d\mu_m(k)\, \epsilon^\mu(k,\sigma) e^{-ikx} a(k,\sigma), \quad (5.21)$$

with ϵ^μ suitable (spacelike, normalized, orthogonal to k_μ, also real) polarization vectors, so that

$$\sum_{\sigma=1}^{3} \epsilon_\mu(k,\sigma)\epsilon_\nu(k,\sigma) = -g_{\mu\nu} + \frac{k_\mu k_\nu}{m^2}. \quad (5.22)$$

On the right-hand side we have the projection matrix on the space orthogonal to the four-vector k^μ. This may be rewritten

$$\sum_{\sigma=0}^{\sigma=3} g_{\sigma\sigma} \epsilon_\mu(k,\sigma)\epsilon_\nu(k,\sigma) = g_{\mu\nu},$$

with the definition $\epsilon_\mu(k,0) = k_\mu/m$. With this treatment, we have the equations

$$(\Box + m^2)\varphi^\mu(x) = 0, \qquad \partial_\mu \varphi^\mu(x) = 0.$$

The last one ensures that one of the four degrees of freedom in φ^μ is eliminated. However, eventually (5.22) leads to the commutation relations for the Proca field of the form

$$[\varphi^\mu(x), \varphi^\nu(y)] = i\left(g^{\mu\nu} + \frac{\partial^\mu \partial^\nu}{m^2}\right) D(x-y).$$

In momentum space this is constant as $|k| \uparrow \infty$, which bodes badly for renormalizability. The Feynman propagator is proportional to

$$\frac{g_{\mu\nu} - k_\mu k_\nu/m^2}{k^2 - m^2};$$

there is moreover a troublesome extra term, which we leave aside.

The argument for nonrenormalizability is as follows. Suppose that, as in the examples of the previous section, the vector field is coupled with a conserved current made out of spinor fields. Consider an arbitrary Feynman graph with E_F external fermion lines, I_F internal ones, and respective boson lines E_B, I_B. The assumption says two fermion lines and one boson line meet at each vertex. The number of vertices is thus

$$V = 2I_B + E_B = \tfrac{1}{2}(2I_F + E_F).$$

There is a delta function for each vertex, one of them corresponding to overall momentum conservation, and each internal line has an integration over its moment.

Thus, by eliminating I_F, I_B the superficial degree of divergence is found to be

$$D = -4(V-1) + 3I_F + 4I_B = 4 + V - 3E_F/2 - 2E_B.$$

This shows that, no matter how many external lines there are, the degree of divergence can be made arbitrarily large.

The difficulty is with the intertwiners, whose dimension does not allow the usual renormalizability condition. The idea is then to cure this by a cohomological extension of the Wigner representation space for massive spin-1 particles. This involves both the Stückelberg field and the ghost fields, already at the level of the description of free fields. The nilpotency condition $s^2 = 0$ for the BRS operator s will yield a cohomological representation for the physical Hilbert space $\ker s / \operatorname{ran} s$, which, as we shall see later, is the (closure of) the space of transverse vector wave functions. On that extended Hilbert space the renormalizability problem fades away. This goes hand in hand with a philosophy of primacy of a quantum character for the gauge principle, which should be read backwards into classical field theory; fibre bundle theory is no doubt elegant, but not intrinsic from this viewpoint. (For massless particles, the situation is worse in that problematic aspects of the use of vector potentials in the local description of spin-1 particles show up already in the covariance properties of photons and gluons.)

5.2.2 What escaped through the net

Another unsung hero of quantum field theory is the Swiss physicist Ernst Carl Gerlach Stückelberg, baron von Breidenbach. He found himself among the pioneers of the 'new' quantum mechanics; at the end of the 1920s, while working in Princeton with Morse, he was the first to explain the continuous spectrum of molecular hydrogen. On his return to Europe in 1933, he met Wentzel and Pauli for the first time. Stückelberg stayed in Zurich for two years before accepting a position at Geneva. He turned to particle physics, where he would among other things contribute, according to his obituary [12], the meson hypothesis (unpublished at the time because of Pauli's criticism, and usually associated with Yukawa), the causal propagator (better known as the Feynman propagator) and the renormalization group [13, 14]. Also due to Stückelberg (not emphasized in [12]) are the first formulation of baryon number conservation; the first sketch of what is called nowadays 'Epstein–Glaser renormalization' [15] (towards which, according to the account in [16], Pauli was better disposed) and the *Stückelberg field* [17], which concerns us here.

We have seen the extreme care that Utiyama exercised in deriving the precise form of gauge theory as a theorem. However, already at the moment that he published it, his result was known to be false. The thing that escapes through Utiyama's net is Stückelberg's gauge theory for massive spin-1 particles.

In the old paper [18], Pauli had given rather dismissively a short account of that fact before plunging into the Proca field – although anyone who has tried to work with the latter rapidly realizes it is good for nothing. There are several natural ways to discover the Stückelberg gauge field, even after one has been miseducated by textbooks – like [11] – into learning exclusively about the Proca field. A principled quantum approach is contained in embryo in the paper [19], where the starting point is Wigner's picture of the unitary irreps of the Poincaré group. In the book by Itzykson and Zuber, the Stückelberg method is used time and again [9, pp. 136, 172, 610] to smooth the $m \downarrow 0$ limit and exorcise infrared troubles. A useful reference for the Stückelberg field is the review [20]. I have been inspired also by [21].

5.2.3 The Stückelberg field and Utiyama's test

Actually, there is no logical fault in the Lagrangian approach by Utiyama. Where he goes astray is only in the 'initial condition' (5.2). We next try to find the Stückelberg field by the Utiyama path; that is, to see whether we actually could have derived the existence of the field B using the arguments of Section 5.1.2. We do this for an abelian theory. Assume that a globally $G \equiv U(1)$-invariant model of a Dirac fermion of mass M and a real vector field of mass m are given:

$$\begin{aligned}\mathcal{L}_0 &= \frac{i}{2}(\overline{\psi}\gamma^\mu\partial_\mu\psi - \partial_\mu\overline{\psi}\gamma^\mu\psi) - \overline{\psi}M\psi + \tfrac{1}{2}m^2A_\mu A^\mu + \mathcal{L}_{\mathrm{kin}}(\partial_\nu A_\mu)\\ &=: \mathcal{L}_{0,\mathrm{f}} + \mathcal{L}_{0,\mathrm{phmass}} + \mathcal{L}_{\mathrm{kin}},\end{aligned}$$

with obvious notation. This is obviously a model for (noninteracting) massive photon electrodynamics. Here $\mathcal{L}_{\mathrm{kin}}$ is the kinetic energy term for the photon, of the form (5.17). This Lagrangian is invariant under the global gauge transformations:

$$A_\mu(x) \mapsto A_\mu(x), \qquad \overline{\psi}(x) \mapsto e^{i\theta}\psi(x), \qquad \psi(x) \mapsto e^{-i\theta}\psi(x),$$

or, infinitesimally,

$$\delta A_\mu = 0, \qquad \delta\overline{\psi} = i\overline{\psi}\theta, \qquad \delta\psi = -i\psi\theta.$$

Now the Utiyama questions come in: what new (gauge) fields need be introduced? How do they transform under $G(x)$? What is the form of the interaction, and what is the new Lagrangian? To save space and time, we restart from

$$\begin{aligned}&\tfrac{i}{2}[\overline{\psi}\gamma^\mu\partial_\mu\psi - \partial_\mu\overline{\psi}\gamma^\mu\psi] - \overline{\psi}\gamma^\mu A_\mu\psi - \overline{\psi}M\psi + \tfrac{1}{2}m^2A_\mu A^\mu\\ &\quad - \tfrac{1}{4}(\partial_\mu A_\nu - \partial_\nu A_\mu)(\partial^\mu A^\nu - \partial^\nu A^\mu) =: \mathcal{L}_{\mathrm{f}} + \mathcal{L}_{0,\mathrm{phmass}} + \mathcal{L}_{\mathrm{kin}}.\end{aligned}$$

The multiplet of fields includes now

$$\varphi = \begin{pmatrix} \overline{\psi} \\ \psi \\ A^\mu \end{pmatrix} \quad \text{transforming as} \quad \delta\varphi = \begin{pmatrix} i\overline{\psi}\theta(x) \\ -i\psi\theta(x) \\ \partial^\mu\theta(x) \end{pmatrix}, \tag{5.23}$$

where of course we required a variation of the QED type for the A_μ. For simplicity we have put $g = 1$. However, still

$$\delta\mathcal{L}_0 = \frac{\partial\mathcal{L}_{0,\text{phmass}}}{\partial A_\mu}\,\delta A_\mu = m\partial_\mu\theta \neq 0.$$

It seems that, when vector fields are conjured *ab initio*, further infinitesimal gauge transformations of the form

$$\delta\varphi_k = \mathcal{A}_{kc}\theta_c + \mathcal{B}^\nu_{kc}\partial_\nu\theta_c \tag{5.24}$$

need to be considered. Here we have a particular case, with a trivial colour index c, with $\varphi_k \to A_\mu$, $\mathcal{A}_\mu$ vanishing, and $\mathcal{B}^\nu_\mu = \delta^\nu_\mu$.

There is no need to involve other parts of the Lagrangian than $\mathcal{L}_{0,\text{phmass}}$ in the remaining calculation. We need an extra vector field. It is natural to propose that it be fabricated from the derivatives of a scalar B, and we write

$$\mathcal{L}_{0,\text{phmass}}(A_\mu) \longrightarrow \mathcal{L}'(A_\mu, \partial_\mu B).$$

It is immediate to note that if we assume the new field transforms like $\delta B = m\theta$, then the requirement of local gauge invariance is

$$\delta\mathcal{L}' = \left[\frac{\partial\mathcal{L}'}{\partial A_\mu} + m\frac{\partial\mathcal{L}'}{\partial(\partial_\mu B)}\right]\partial_\mu\theta = 0.$$

It follows that

$$m\frac{\partial\mathcal{L}'}{\partial(\partial_\mu B)} = -\frac{\partial\mathcal{L}'}{\partial A_\mu}.$$

Consequently, only the combination

$$A_\mu - \partial_\mu B/m$$

occurs in $\mathcal{L}'(A_\mu, \partial_\mu B)$. Thus we rewrite:

$$\mathcal{L}'(A_\mu, \partial_\mu B) \longrightarrow \mathcal{L}_{0,\text{phmass}}(A_\mu - \partial_\mu B/m).$$

The bosonic part of the Lagrangian is *in fine*

$$\mathcal{L}_\text{b} = \mathcal{L}_\text{kin} + \frac{m^2}{2}\left(A_\mu - \frac{\partial_\mu B}{m}\right)^2;$$

note that, with $V_\mu = (A_\mu - \partial_\mu B/m)$, one has $\mathcal{L}_{\rm kin}(A_\mu) = \mathcal{L}_{\rm kin}(V_\mu)$. The total Lagrangian $\mathcal{L} = \mathcal{L}_{\rm f} + \mathcal{L}_{\rm b}$ has what we want. With the multiplet of fields

$$\varphi = \begin{pmatrix} \overline{\psi} \\ \psi \\ A^\mu \\ B \end{pmatrix} \quad \text{transforming as} \quad \delta\varphi = \begin{pmatrix} i\overline{\psi}\theta(x) \\ -i\psi\theta(x) \\ \partial^\mu\theta(x) \\ m\theta(x) \end{pmatrix},$$

we plainly obtain local gauge invariance of $\mathcal{L}_{\rm f}$, $\mathcal{L}_{\rm b}$ and $\mathcal{L}$. Note the Euler–Lagrange equation

$$\partial_\mu \frac{\partial\mathcal{L}}{\partial_\mu B} = \frac{\partial\mathcal{L}}{\partial B} \quad \text{yielding} \quad \Box B = m\,\partial A.$$

Note as well that one can fix the gauge so B vanishes; this does not mean the gauge symmetry is trivial.

Maybe Utiyama missed this because [22] he only takes into account, for the original variables, infinitesimal gauge transformations typical of *matter* fields, of the form (5.2); he did not consider the possibility (5.23) – that is, (5.10) – for the vector fields acting as sources of gauge fields.

We finish this subsection by noting that $\mathcal{L}_{\rm b}$ may also be written

$$\mathcal{L}_{\rm b} = (\partial_\mu - igA_\mu)\Phi\,(\partial_\mu + igA_\mu)\Phi^*, \quad \text{with} \quad \Phi = \frac{m}{\sqrt{2}g}\exp(igB/m);$$

that is, an abelian Higgs model without self-interaction. Verification is straightforward.

5.2.4 The Stückelberg formalism for non-abelian Yang–Mills fields

The sophisticated method for this was established by Kunimasa and Goto [23]; we follow in the main [24]. For apparent simplicity, consider an isovector field A^a_μ interacting with an isospinor spinor field ψ, as in Section 5.1.5. Let us choose the notation

$$\mathbb{A}_\mu = \tfrac{1}{2}\sigma^a A^a_\mu, \qquad \mathbb{F}_{\mu\nu} = \partial_\mu\mathbb{A}_\nu - \partial_\nu\mathbb{A}_\mu + ig(\mathbb{A}_\mu\mathbb{A}_\nu - \mathbb{A}_\nu\mathbb{A}_\mu).$$

Indeed, $\frac{i}{4}\sigma^a\sigma^b = -\frac{1}{2}\epsilon^{abc}\sigma^c$, in consonance with (5.14). The Lagrangian density is written

$$-\tfrac{1}{2}\,\mathrm{tr}(\mathbb{F}_{\mu\nu}\mathbb{F}^{\mu\nu}) + \tfrac{i}{2}\overline{\psi}\,\overleftrightarrow{\partial_\mu}\,\gamma^\mu\psi - \overline{\psi}M\psi - g\overline{\psi}\gamma^\mu A_\mu\psi.$$

This is invariant under

$$\psi \to \mathbb{W}^{-1}\psi, \qquad \mathbb{A}_\mu \to \mathbb{W}^{-1}\mathbb{A}_\mu\mathbb{W} - \frac{i}{g}\mathbb{W}^{-1}\partial\mu\mathbb{W}$$

for $\mathbb{W} \in SU(2)$, which is nothing but (5.10) with

$$\mathbb{W} = \exp(T^a \theta^a(x)).$$

To make the mass term

$$m^2 \operatorname{tr}(\mathbb{A}_\mu \mathbb{A}^\mu) = \tfrac{1}{2} m^2 A^a_\mu A^{a\mu}$$

gauge invariant, it is enough to introduce a 2×2 matrix ω_μ of auxiliary vector fields, so that

$$m^2 \operatorname{tr}(\mathbb{A}_\mu - \omega_\mu / g)$$

is invariant under gauge transformations if

$$\omega_\mu \to \mathbb{W}^{-1} \omega_\mu \mathbb{W} - i \mathbb{W}^{-1} \partial_\mu \mathbb{W}. \tag{5.25}$$

Indeed, let $C \in SU(2)$ transform as $C \to C\mathbb{W}$. Then

$$\omega_\mu := -i C^{-1} \partial_\mu C$$

satisfies (5.25):

$$-i \mathbb{W}^{-1} C^{-1} \partial_\mu C \mathbb{W} = \mathbb{W}^{-1} \omega_\mu \mathbb{W} - i \mathbb{W}^{-1} \partial_\mu \mathbb{W}.$$

With $C = \exp(B^a T^a / m)$, we can think of the B^a as the auxiliary fields.

We may note, however, that the introduction of scalar Stückelberg partners for the A^a_μ by the substitution $\mathbb{A}_\mu \to \mathbb{A}_\mu - \partial_\mu B$, with $B = B^a T^a$, seems to work as well. In gauge theory, the elegant non-infinitesimal notation is a bit dangerous, in that it tends to obscure the fact that the transformation of the gauge fields (5.10) is *independent* of the considered representation of the gauge group.

5.2.5 Gauge-fixing and the Stückelberg Lagrangian

We begin to face quantization now. For that, we need to fix a gauge. Otherwise, we cannot even derive a propagator from the Lagrangian. Let us briefly recall the standard argument:

$$\mathcal{L}_{\text{QED}} = -\tfrac{1}{4} F_{\mu\nu} F^{\mu\nu} = \tfrac{1}{2} A^\mu \mathcal{D}_{\mu\nu} A^\nu,$$

with

$$\mathcal{D}_{\mu\nu}(x) = -g_{\mu\nu} \overleftarrow{\partial_\sigma} \overrightarrow{\partial^\sigma} + \overleftarrow{\partial_\nu} \overrightarrow{\partial_\mu} \quad \text{or} \quad \mathcal{D}_{\mu\nu}(k) = -g_{\mu\nu} k^2 + k_\mu k_\nu,$$

in momentum space. The matrix $\mathcal{D}_{\mu\nu}$ has null determinant and thus is not invertible; so one cannot define a Feynman propagator. This is precisely due to gauge invariance. The problem for QED was cured by Fermi long ago [25] by introduction of the piece $\frac{-1}{2\alpha}(\partial^\nu A_\nu)^2$. Here we proceed similarly, and the gauge-fixing term we take is of the 't Hooft type:

$$\mathcal{L}_{\text{gf}} = \frac{-1}{2\alpha}(\partial^\nu A_\nu + \alpha m B)^2. \tag{5.26}$$

We denote

$$\mathcal{L}_S = \mathcal{L} + \mathcal{L}_{\text{gf}},$$

the Stückelberg Lagrangian. The gauge-fixing amounts to saying that now the gauge variation θ must satisfy the Klein–Gordon (KG) equation with mass $m\sqrt{\alpha}$:

$$(\Box + \alpha m^2)\theta = 0,$$

just as in the old trick by Fermi in electrodynamics, where the new Lagrangian is still gauge-invariant provided we assume $\Box\theta = 0$ for the gauge variations. Now, instead the Euler–Lagrange equation,

$$\partial_\mu \frac{\partial \mathcal{L}_S}{\partial(\partial_\mu B)} = \frac{\partial \mathcal{L}_S}{\partial B} \quad \text{yields} \quad (\Box + \alpha m^2)B = 0.$$

Hence the gauge-fixing implies B itself now is a free field with mass $m\sqrt{\alpha}$.

Another good reason for the gauge-fixing is to keep A_ν as an honest spin-1 field in the interaction. Recall that in a quantum vector field spins 0 and 1 are possible. The scalar B extracts the spin-0 part, so the remaining part is transverse. In fact $\partial_\mu(A^\mu - \partial^\mu B/m) = \partial A + \alpha m B$ if the equation of motion is taken into account; and this gauge-fixing term is destined to vanish in an appropriate sense on the physical state space.

A word is needed on the Noether theorem now. There is now an extra term in $\partial_\mu \frac{\partial \mathcal{L}}{\partial(\partial_\mu A_\nu)}$, of the form $-\frac{g^{\mu\nu}}{\alpha}(\partial A + \alpha m B)$. This gives rise to the Euler–Lagrange equation:

$$\Box A_\mu + \left(\frac{1}{\alpha} - 1\right)\partial_\mu(\partial A) + m^2 A_\mu = g\overline{\psi}\gamma_\mu\psi, \tag{5.27}$$

where we have reestablished temporarily the coupling constant. As a consequence of (5.27) we have

$$\Box\,\partial A + \alpha m^2 \partial A = 0.$$

The simplest option now is to take $\alpha = 1$ (so the masses of A_ν and B coincide), as then the A_ν obey the KG equation at zeroth order in g. This could be termed the 'Feynman gauge'. But in some contexts it is important to keep the freedom of different mass values for the vector and the scalar bosons. (We have for the fermion the Dirac equation

$$i\gamma^\mu \partial_\mu \psi = (g\gamma^\mu A_\mu + M)\psi$$

and its conjugate. Nothing new here.)

A comment on renormalizability is in order at this point. The choice $\alpha \downarrow 0$ is the Landau gauge, in which renormalizability is almost explicit. On the other hand, it is clear that $B = 0$ (the original Proca model), where the theory is non-renormalizable by power counting, can be recovered as a sort of 'unitary gauge'. If we can prove gauge covariance of the theory, all these versions will be physically equivalent. An extra advantage of the Stückelberg field in renormalization is that, because it cures the limit $m \downarrow 0$, it allows the use of masses as infrared regulators.

To finish, I again call attention to the similarities of the model with the abelian Higgs model. Upon renormalization, a Higgs-potential–like term pops up in the Lagrangian. However, the vacuum expected value of the Stückelberg field is still zero. For non-abelian theories, the situation remains murky even now.

5.2.6 *The ghosts we summoned up?*

For completeness, I provide next a conventional discussion of BRS invariance for the Lagrangian obtained in the previous subsection. (This is not intended to be discussed in the exercises, and both the cognoscenti and the noncognoscenti may skip it in first reading.)

Nowadays BRS invariance of the (final) Lagrangian is an integral part of the quantization process. Among other things, it helps to establish gauge *covariance*, that is, independence of the chosen gauge for physical quantities; in turn this helps with renormalizability proofs. We approach the quantum context by introducing two fermionic ghosts $\omega, \tilde{\omega}$ plus an auxiliary (Nakanishi–Lautrup) field h that we add to the collection φ. From the infinitesimal gauge transformations we read off the BRS transformation:

$$s\varphi = s\begin{pmatrix} \overline{\psi} \\ \psi \\ A^\mu \\ B \\ \omega \\ \tilde{\omega} \\ h \end{pmatrix} = \begin{pmatrix} i\omega\overline{\psi} \\ -i\omega\psi \\ \partial^\mu\omega \\ m\omega \\ 0 \\ h \\ 0 \end{pmatrix}.$$

It is clear that s increases the ghost number by one. Extend s as an antiderivation; from the fact that $\omega, \tilde{\omega}$ are anticommuting we obtain (even off-shell) nilpotency of order two for the BRS transformation: $s^2 = 0$ (we will always understand 'nilpotent of order two' for 'nilpotent' in this work). Now, in the BRS approach, one takes the action to be a local action functional of matter, gauge, ghost and h-fields with ghost number zero and invariant under s. This is provided by the new form

$$\mathcal{L}_{\rm gf} = s[\mathcal{F}(\overline{\psi}, \psi, A_\mu, B)\tilde{\omega} + \tfrac{1}{2}\alpha h\tilde{\omega}]$$

for the gauge-fixing term of the Lagrangian. Here $\mathcal{F}$ is the gauge-fixing functional, like $(\partial^\mu A_\mu + \alpha m B)$ in the preceding subsection. Invariance comes from $s\mathcal{L}_{\rm gf} = 0$ on account of nilpotency, of course. We can rewrite

$$\mathcal{L}_{\rm gf} = -\tilde{\omega} s\mathcal{F} + h\mathcal{F} + \tfrac{1}{2}\alpha h^2 = -\tilde{\omega} s\mathcal{F} + \tfrac{1}{2}\Big(\frac{\mathcal{F}}{\sqrt{\alpha}} + h\sqrt{\alpha}\Big)^2 - \frac{\mathcal{F}^2}{2\alpha}.$$

One can eliminate h using its equation of motion

$$0 = \frac{\partial\mathcal{L}_{\rm gf}}{\partial h} = \mathcal{F} + h\alpha, \qquad \text{so that} \quad \mathcal{L}_{\rm gf} = -\tilde{\omega} s\mathcal{F} - \frac{\mathcal{F}^2}{2\alpha},$$

and also $s\tilde{\omega} = -\mathcal{F}/\alpha$: the BRS transformation then maps the antighosts or dual ghosts into the gauge-fixing terms (the price to pay is that s will be nilpotent off-shell only when acting on functionals independent of $\tilde{\omega}$). In our case (5.26), we have

$$s\mathcal{F} = s(\partial^\mu A_\mu + \alpha m B) = (\Box + \alpha m^2)\omega.$$

Thus the contribution of the fermionic ghosts in this abelian model to $\mathcal{L}_{\rm gf}$ is

$$-\tilde{\omega} s\mathcal{F} = -\tilde{\omega}(\Box + \alpha m^2)\omega;$$

also $\partial_\mu\tilde{\omega}\partial^\mu\omega - \tilde{\omega}\alpha m^2\omega$ would do; the ghosts turn out to be free fields with the same mass as Stückelberg's B-field. Notice that the ghost term decouples in the final effective Lagrangian. (According to [26], adding to the action a term invariant under the BRS transformation amounts to a redefinition of the fields coupled to the source in the generating functional; this has no influence on the $\mathbb{S}$-matrix.)

I have followed [8] and mainly [27] in this subsection.

At the end of the day, the Lagrangian for massive electrodynamics is of the form

$$\begin{aligned}\mathcal{L}_{\rm f} + \mathcal{L}_{\rm kin} + \mathcal{L}_{\rm b} + \mathcal{L}_{\rm gf} &= \tfrac{i}{2}[\overline{\psi}\gamma^\mu\partial_\mu\psi - \partial_\mu\overline{\psi}\gamma^\mu\psi] - \overline{\psi}\gamma^\mu A_\mu\psi - \overline{\psi}M\psi \\ &\quad - \tfrac{1}{4}(FF) + \frac{m^2}{2}(A - \partial B/m)^2 \\ &\quad - \frac{1}{2\alpha}(\partial A + \alpha m B)^2 - \tilde{\omega}(\Box + \alpha m^2)\omega \\ &= \mathcal{L}_{\rm f} + \mathcal{L}_{\rm kin} + \frac{m^2 A^2}{2} - \frac{1}{2\alpha}(\partial A)^2 + \tfrac{1}{2}(\partial B)^2 \\ &\quad - \frac{\alpha m^2}{2}B^2 - m\partial_\mu(BA^\mu) - \tilde{\omega}(\Box + \alpha m^2)\omega.\end{aligned}$$

Highlights:

- The gauge-fixing has been chosen independently of the matter field.
- The gauge sector contains first a massive vector field, with three physical components of mass m (one longitudinal and two transverse) and an unphysical spin-zero piece of mass $\sqrt{\alpha}m$.
- The cross term between A_μ and B has disappeared.
- The gauge sector also contains a (commuting) Stückelberg B-field with mass $\sqrt{\alpha}m$ and a pair of (anticommuting) ghost–antighost scalars, with mass $\sqrt{\alpha}m$ as well.
- For computing $\mathbb{S}$-matrix elements, the ghosts can be integrated out, for they are decoupled and do not appear in asymptotic states. But we cannot integrate out the B-field, because, as discussed in Section 5.3, it plays a role in the definition of the physical states – and moreover it undergoes a nontrivial renormalization.
- The only interacting piece is the $\overline{\psi} A \psi$ term in the fermionic part of the Lagrangian.
- The model is *renormalizable*.

5.3 Quantization of massive spin-1 fields

5.3.1 On the need for BRS invariance

It is impossible for us, within the narrow limits of this short chapter, to follow in any meaningful detail the tortuous chronological path to the discovery of BRS invariance in relation with gauge invariance. The story in outline is well known. By fixing the gauge, Feynman was able to generate Feynman diagrams [28] for non-abelian gauge theories; but unitarity of the $\mathbb{S}$-matrix was lost unless additional 'probability-eating' quantum fields were introduced. The auxiliary ghost fields appeared clearly in the work by Faddeev and Popov, which uses the functional integral. In the seventies it was discovered that the resulting effective Lagrangian still supports a global invariance of a new kind, the nilpotent BRS transformation, which allows one to recover unitarity, ensures gauge independence of the quantum observables and powerfully contributes to the proofs of renormalizability.

We approached quantization in Section 5.2.1 through the canonical method. So we motivate the introduction of the ghosts and the BRS symmetry and operator in our previous considerations. Now that hopefully we have broken the mental association between 'gauge principle' and 'masslessness', we can proceed to a simple and general version of gauge theory with BRS invariance. The quantization of massive vector fields is interesting in that it is conceptually simpler, although analytically more complicated, than that of massless ones. (It is true that in theories with massive gauge bosons, the masses are generated by the 'Higgs mechanism'; but this is just a poetic description that cannot be verified or falsified at present.) In the context, concretely we need the ghosts as 'renormalization catalysers'. In fact, it has been shown in [19] that for interacting massive vector field models the renormalizability condition fixes the theory completely, including the cohomological extension of the Wigner representation theory by the ghosts, and the Stückelberg field in the abelian case – even if you had never heard of that in a semiclassical study of Lagrangians, like the one performed in Section 5.2 – as well as a Higgs-like field for flavourdynamics; we shall touch upon this in the last section.

The crucial problem, illustrated by our discussion in Section 5.2.1, is to eliminate the unphysical degrees of freedom in the quantization of free vector fields in a subtler way than Proca's, particularly without giving up commutators of the form

$$[A_\mu(x), A_\nu(y)] = ig_{\mu\nu}D(x-y), \quad A_\mu^+ = A_\mu. \tag{5.28}$$

Also we ask for the KG equations $(\Box + m^2)A^\mu = 0$ to hold (in the Feynman gauge). It is impossible to realize (5.28) on Hilbert space. Let us sketch the solution in this subsection. It goes through the introduction of a distinguished symmetry η (that is, an operator both self-adjoint and unitary), called the 'Krein operator', on the Hilbert–Fock space H. Whenever such a Krein operator is considered, the η-adjoint O^+ of an operator O is defined:

$$O^+ = \eta O^\dagger \eta.$$

Let $(\cdot, \cdot)$ denote the positive definite scalar product in H. Then

$$\langle \cdot, \cdot \rangle := (\cdot, \eta \cdot)$$

gives an *indefinite scalar product*, and the definition of O^+ is just that of the adjoint with respect to $\langle \cdot, \cdot \rangle$. The algebraic properties are like those in the usual adjunction $\dagger$, but O^+O is not positive in general.

The pair (H, η), where H is the original Hilbert–Fock space, including ghosts, is called a 'Krein space'. The undesired contributions from the A-space will be cancelled by the unphysical statistics of the ghosts. The BRS operator is an (unbounded) nilpotent η-self-adjoint operator Q on H. That is, $Q^2 = 0$, $Q = Q^+$. By means of Q one shows that H (or a suitable dense domain of it) splits into the direct sum of three pairwise orthogonal subspaces (quite analogous to the Hodge–de Rham decomposition in the differential geometry of manifolds):

$$H = \operatorname{ran} Q \oplus \operatorname{ran} Q^\dagger \oplus (\ker Q \cap \ker Q^\dagger).$$

In addition we assume

$$\eta\big|_{\ker Q \cap \ker Q^\dagger} = 1.$$

That is, $\langle \cdot, \cdot \rangle$ is positive definite on

$$H_{\text{phys}} := \ker Q \cap \ker Q^\dagger,$$

which is called the physical subspace. An alternative definition for H_{phys} is the cohomological one:

$$H_{\text{phys}} = \ker Q / \operatorname{ran} Q.$$

Nilpotency of Q is the reason to introduce the anticommuting pair of ghost fields. In interaction, the $\mathbb{S}$-matrix must be physically consistent:

$$[Q, \mathbb{S}]_+ = 0, \quad \text{or at least} \quad [Q, \mathbb{S}]_+\big|_{\ker Q} = 0.$$

In the following subsections we flesh out the details of all this.

5.3.2 *Ghosts as free quantum fields*

A first step in a rigorous construction of ghosts is their understanding as quantum fields, together with the issue of the 'failure' of the spin–statistics theorem for them. We look for two operator-valued distributions u, $\tilde{u}$, acting on a Hilbert–Fock space H_{gh} and satisfying KG equations:

$$(\Box + m^2)u = (\Box + m^2)\tilde{u} = 0, \tag{5.29}$$

and the following commutation relations, in the sense of tempered distributions:

$$[u_a(x), \tilde{u}_b(y)]_+ = -i\delta_{ab} D(x-y), \qquad [u_a(x) u_b(y)]_+ = [\tilde{u}_a(x), \tilde{u}_b(y)]_+ = 0.$$

Here $D = D^+ + D^-$ is the Jordan–Pauli function. The fields live in the adjoint representation of a gauge group G (like the gauge fields themselves); the colour indices a, b most often can be omitted. The components of H_{gh} of degree n are *skew-symmetric* square-summable functions (with the Lorentz-invariant measure $d\mu_m(p)$) of n momenta on the mass hyperboloid $\mathcal{H}_m$, with their colour indices and ghost indices, where the first, say a, can run from 1 to $\dim G$, and we let the second, say i, take the values ± 1. (The reader is warned that the notation for the ghost fields in this section, and a few other notational conventions, are different from the ones we found convenient in the sections dealing with the semiclassical aspects.)

We proceed to the construction. Consider the dense domain $\mathcal{D} \subset H_{\text{gh}}$ of vectors with finitely many nonvanishing components which are Schwartz functions of their arguments. Then there exist annihilation (unbounded) operator functions $c_{a,i}(p)$ of $\mathcal{D}$ into itself, given by

$$[c_{a,i}(p)\Phi]^{(n)}_{a_1,\dots,a_n;i_1,\dots,i_n}(p_1, \dots, p_n) = \sqrt{n+1}\, \Phi^{(n+1)}_{a,a_1,\dots,a_n;i,i_1,\dots,i_n}(p, p_1, \dots, p_n).$$

Integrating this with a Schwartz function on the mass hyperboloid gives a bounded operator. The adjoint of $c_{a,i}(p)$ is defined as a sesquilinear form on $\mathcal{D} \otimes \mathcal{D}$, and we have the usual 'commutation relations' among them:

$$[c_{a,i}(p), c^\dagger_{b,j}(p')]_+ = \delta_{ab}\delta_{ij}\delta(p-p'),$$

the other anticommutators being zero. Notice that $\delta(p-p')$ is shorthand for the Lorentz-invariant Dirac distribution $2E\delta(\vec{p}-\vec{p}\,')$ corresponding to $d\mu_m(p)$.

We are set now to define the distributional ghost field operators in coordinate space in terms of the $c_{a,i}$, $c^\dagger_{b,j}$. The construction is diagonal in the G-index, so it will be omitted. The general ansatz is

$$u_i(x) = \int d\mu_m(p)\big[A_{ij}c_j(p)e^{-ipx} + B_{ij}c^\dagger_j(p)e^{+ipx}\big].$$

Here

$$A = \begin{pmatrix} A_{11} & A_{1-1} \\ A_{-11} & A_{-1-1} \end{pmatrix}, \qquad B = \begin{pmatrix} B_{11} & B_{1-1} \\ B_{-11} & B_{-1-1} \end{pmatrix}.$$

Because p is on the mass hyperboloid, the KG equations (5.29) hold. The anticommutators are

$$[u_i(x), u_j(y)]_+ = -i\big[A_{ik}B_{jk}D^+(x-y) - B_{ik}A_{jk}D^-(x-y)\big].$$

The only combinations with causal support are multiples of $D^+ + D^-$. As we want to keep causality, it must be that $AB^t + BA^t = 0$, so we obtain

$$[u_i(x), u_j(y)]_+ = -iC_{ij}D(x-y),$$

with $C := AB^t$ skew-symmetric. There are of course many possible choices of A, B with this constraint. We pick

$$C = \begin{pmatrix} & 1 \\ -1 & \end{pmatrix}.$$

This finally gives

$$u(x) = u_1(x) = \int d\mu_m(p)\big(c_1(p)e^{-ipx} + c^\dagger_{-1}(p)e^{ipx}\big),$$

$$\tilde{u}(x) = u_{-1}(x) = \int d\mu_m(p)\big(c_{-1}(p)e^{-ipx} - c^\dagger_1(p)e^{ipx}\big).$$

We remark that $[\tilde{u}(x), u(y)]_+ = iD(x-y) = -iD(y-x) = [u(y), \tilde{u}(x)]_+$.

The representation of the Poincaré group is the same as for 2dim$G$ independent scalar fields; we do not bother to write it. As we have chosen A, B invertible, the creation and annihilation operators can be expressed in terms of the ghost fields and their adjoints. Then the vacuum is cyclic with respect to these.

Defining the adjoint fields, one sees that the anticommutators of the ghost fields with their adjoints are not causal. This, according to [29, 30], allows one to escape the spin–statistics theorem. Indeed, a version of the last says that no nonvanishing scalar fields can exist satisfying

$$[u_a(x), u_b(y)]_+ = 0, \qquad [u_a(x), u_b^\dagger(y)]_+ = 0$$

for spacelike separations. Because the second anticommutator is not causal, the last condition is not violated. (There are other explanations in the literature for the same conundrum, though.)

5.3.3 Mathematical structure of BRS theories

There are several questions regarding the scheme proposed in Section 5.3.1 that we address systematically now:

(i) What is the algebraic framework?
(ii) In which mathematical sense is BRS invariance a symmetry?
(iii) When is there a BRS charge associated to a BRS symmetry?
(iv) What are the continuity properties of the generator Q?
(v) How does the Hodge–de Rham decomposition of the Hilbert space take place?
(vi) How are the physical states characterized?

The famous paper on the quark confinement problem by Kugo and Ojima [31] was the first to tackle these questions, although their answers were not quite correct. A good treatment, which we follow for the most part, was given by Horuzhy and Voronin [32].

(i) Consider a ‘general BRS theory’ on a Krein space (H, η). On a suitable common invariant dense domain $\mathcal{D} \subset H$ there is defined a system of physical quantum fields and ghost fields (the physical fields could be matter fields, Yang–Mills fields or, say, the coordinates of a first-quantized string), forming a polynomial algebra $\mathcal{A}$; the operator id $\in \mathcal{A}$ on H we denote by 1. A Krein operator has the eigenvalues ± 1, so $\eta = P_+^\eta - P_-^\eta$ with an obvious notation. We assume moreover $\dim P_\pm^\eta H = \infty$. By O° we shall mean the restriction of O^+ to $\mathcal{D}$. We say O is η-self-adjoint when $O = O^\circ$; η-unitary when $O^{-1} = O^\circ$. The field algebra has a cyclic vector, or vacuum, $|0\rangle$, that is, $\mathcal{A}|0\rangle$ is dense in $\mathcal{D}$.
(ii) Mathematically speaking, a BRS (infinitesimal) transformation is a skew-adjoint, nilpotent superderivation s acting on the field algebra of H. Let $\epsilon_O := (-)^{\mathbb{N}_{\rm gh}(O)}$, with $\mathbb{N}_{\rm gh}(O)$ the number of ghost fields in the monomial O. Typically s changes the ghost number by one. Then s is a linear map of $\mathcal{A}$ into $\mathcal{A}$ such that

$$s(OB) = s(O)B + \epsilon_O O s(B), \quad s^2 = 0, \quad \epsilon_{s(O)} = -\epsilon_O, \quad s(O)^\circ = -\epsilon_O s(O^\circ).$$

The key point for BRS invariance is obviously the nilpotency equation $s^2 = 0$.

(iii) An important question is whether the BRS transformation s possesses a generator or BRS charge Q, that is, takes the form

$$s(O) = [Q, O]_\pm, \qquad \text{where} \quad [Q, O]_\pm := QO - \epsilon_O OQ. \tag{5.30}$$

Indeed, we may try to equivalently write (5.30) as

$$QO|0\rangle = s(O)|0\rangle.$$

This equation will serve as definition of Q, at least on a dense subset of $\mathcal{D}$, provided

$$O|0\rangle = 0 \quad \text{implies} \quad s(O)|0\rangle = 0.$$

Note that $Q|0\rangle = 0$ because $s(1) = 0$. Thus (5.30) is consistent. Nilpotency of Q follows:

$$Q^2 O|0\rangle = Qs(Q)|0\rangle = 0.$$

One expects Q as defined to be η-self-adjoint. But this is not completely automatic. We have

$$\begin{aligned}\big\langle QO|0\rangle, B|0\rangle\big\rangle &= \big\langle s(O)|0\rangle, B|0\rangle\big\rangle = \big\langle B^\circ s(O)|0\rangle, |0\rangle\big\rangle \\ &= \epsilon_{B^\circ}\big\langle\big(s(B^\circ O) - s(B^\circ)O\big)|0\rangle, |0\rangle\big\rangle \\ &= \epsilon_O\langle s(O^\circ B)\rangle + \big\langle O|0\rangle, s(B)|0\rangle\big\rangle.\end{aligned} \tag{5.31}$$

This will be equal to $\langle O|0\rangle, QB|0\rangle\rangle$ if in general we have

$$\langle s(O)\rangle = 0 \quad \text{for all} \quad O \in \mathcal{A}.$$

In this case, we have η-symmetry. For passing to η-self-adjointness, consult [33].

Reciprocally, if Q is η-self-adjoint with $Q|0\rangle = 0$, is nilpotent, and generates s by (5.30), then, rather trivially,

$$\langle s(O)\rangle = \langle [Q, O]_\pm\rangle = \big\langle Q|0\rangle, O|0\rangle\big\rangle = 0.$$

Moreover, for s so defined,

$$\begin{aligned}s(O)^\circ &= (QO - \epsilon_O OQ)^\circ = O^\circ Q - \epsilon_O QO^\circ = -\epsilon_O(QO^\circ - \epsilon_O O^\circ Q) \\ &= -\epsilon_O s(O^\circ).\end{aligned}$$

We finally verify nilpotency of s:

$$s^2(O) := [Q, [Q, O]_\pm]_\pm = Q(QO - \epsilon_O OQ) + \epsilon_O(QO - \epsilon_O OQ)Q = 0.$$

(iv) In physics Q is often treated as a bounded operator. But there are large classes of nilpotent, η-self-adjoint unbounded operators. Let for instance $H = H_1 \oplus H_2$ and

$$\eta = \begin{pmatrix} 1 & 0 \\ 0 & -1 \end{pmatrix} \quad \text{with} \quad Q = \begin{pmatrix} 0 & A \\ 0 & 0 \end{pmatrix},$$

with A unbounded and skew-adjoint. Then Q is nilpotent, η-self-adjoint and unbounded. For another example, take $H = H_1 \otimes H_2$, where H_1 is an infinite-dimensional Hilbert space, H_2 is a Krein space, and $Q = O \otimes B$, with $O = O^\dagger$ unbounded and B nilpotent and η-self-adjoint. Typically BRS operators are sums of such operators.

Given an arbitrary nilpotent operator Q, such that dom Q^2 is dense, the following holds: *either Q is bounded, with* 0 *as unique point in its spectrum, or Q is unbounded and its spectrum is all of the complex plane.*

Proof Assume spec $Q \neq \mathbb{C}$. Let λ belong to the resolvent of Q. Then Q is closed, as $Q - \lambda$ is. (We recall that a Hilbert space operator is by definition closed when its graph is closed. Also by definition, $Q - \lambda$ is a one-to-one map from dom Q onto H with bounded inverse, so it is closed.) Now $(Q - \lambda)^{-1} H \subset \operatorname{dom} Q$. Therefore

$$(Q - \lambda)\big(Q + \lambda Q(Q - \lambda)^{-1}\big)$$

makes sense and is equal to Q^2. Now Q is closed and $\lambda Q(Q - \lambda)^{-1}$ is bounded; therefore $Q + \lambda Q(Q - \lambda)^{-1}$ is closed; then Q^2 is closed. Therefore its domain is all of H, so Q is bounded (by the closed-graph theorem). Then it is well known that spec $Q = \{0\}$. □

(v) Consider the subspaces ker Q, η ker Q, ran Q, η ran Q. Due to $Q^2 = 0$, we can assume ran $Q \subset$ dom Q; otherwise we extend Q to the whole ran Q by zero. Because of η-self-adjointness, ker Q is closed; also, $\eta \operatorname{ran} Q = \eta \operatorname{ran} \eta Q^\dagger \eta = \operatorname{ran} Q^\dagger$ and $\eta \ker Q = \ker \eta Q \eta = \ker Q^\dagger$. In view of nilpotency, it is immediate that

$$\operatorname{ran} Q \perp \operatorname{ran} Q^\dagger,$$

where $\perp$ indicates perpendicularity in the Hilbert space sense. We have

$$(\operatorname{ran} Q \oplus \operatorname{ran} Q^\dagger)^\perp = \ker Q^\dagger \cap \ker Q.$$

Indeed, the domain of $Q^\dagger$ is dense in H, and thus $(x, Q^\dagger y) = 0$ for all $y \in \operatorname{dom} Q^\dagger$ implies $Qx = 0$. Similarly for $(\operatorname{ran} Q)^\perp = \ker Q^\dagger$. Denoting by $[\perp]$ perpendicularity in the Krein space sense, it also clear that

$$(\ker Q^\dagger \cap \ker Q)^\perp = (\ker Q^\dagger \cap \ker Q)^{[\perp]}$$

In summary,

$$\begin{aligned} H = \overline{\operatorname{ran} Q^\dagger} \oplus \ker Q &= \overline{\operatorname{ran} Q} \oplus \ker Q^\dagger = \overline{\operatorname{ran} Q} \oplus \overline{\operatorname{ran} Q^\dagger} \\ &\oplus (\ker Q^\dagger \cap \ker Q) = \overline{\operatorname{ran} Q} \oplus \overline{\operatorname{ran} Q^\dagger} \,[+]\, (\ker Q^\dagger \cap \ker Q), \end{aligned} \tag{5.32}$$

where the last symbol means the η-orthogonal sum. This is the Hodge–de Rham decomposition of H.

(vi) Assume moreover

$$\eta\big|_{\ker Q \cap \ker Q^\dagger} = 1.$$

Then we baptize

$$H_{\text{phys}} := \ker Q \cap \ker Q^\dagger,$$

the physical subspace, on which $\langle\cdot,\cdot\rangle$ is positive. Alternative characterizations are

$$H_{\text{phys}} = \ker Q/\overline{\operatorname{ran} Q},$$

in view of (5.32), and

$$H_{\text{phys}} = \ker[Q, Q^\dagger]_+.$$

Indeed $[Q, Q^\dagger]_+ x = 0$ iff $Qx = Q^\dagger x = 0$.

5.3.4 BRS theory for massive spin-one fields

We finally turn to our physical case. When dealing with the massive vector field, instead of eliminating *ab initio* the longitudinal component as in (5.21), we keep the $a(k,0)$ and their adjoints, and proceed as follows. We recognize Krein spaces as appropriate tools to study (quantum) gauge theories. In our present case $\eta := (-)^{\mathbb{N}_l}$, where $\mathbb{N}_l$ is the particle number operator for the longitudinal modes. Now

$$A^\mu(x) = (2\pi)^{-3/2}\sum_{\sigma=0}^{3}\int d\mu_m(k)\,\big(\epsilon^\mu(k,\sigma)e^{-ikx}a(k,\sigma) + \epsilon^\mu(k,\sigma)e^{ikx}a^+(k,\sigma)\big).$$

Clearly

$$a^+(k,0) = -\eta^2 a^\dagger(k,0) = -a^\dagger(k,0);$$

however, by definition $A^\mu(x)$ is η-self-conjugate.

I hasten to indicate the main difference from the massless case. Note that a unitary representation of the Poincaré group on the original space is given by

$$U(a,\Lambda)A^\mu(x)U^{-1}(a,\Lambda) = \Lambda^\mu_\nu A^\nu(\Lambda x + a) = U^{-1+}(a,\Lambda)A^\mu(x)U^+(a,\Lambda).$$

This implies

$$[U^+(a,\Lambda)U(a,\Lambda), A^\mu(x)] = 0;$$

therefore U is η-unitary. As $\mathbb{N}_l$, and thus η, commutes with U – basically because the longitudinal polarization transforms into itself under a Lorentz transformation

$$\Lambda^\nu_\mu \epsilon^\mu(k,0) = \frac{(\Lambda k)^\nu}{m} = \epsilon^\nu(\Lambda k, 0)$$

– the representation U is also unitary. This cannot be obtained in the massless case.

The commutation relations for A-field are of the form

$$[A^\mu(x), A^\nu(y)] = i g^{\mu\nu} D(x-y),$$

as we wished for. We now employ a nilpotent gauge charge Q to characterize the physical state subspace and eliminate the unphysical longitudinal mode. For photons, the definition of Q is known to be

$$Q = \int_{x^0=\text{const}} d^3x\, (\partial \cdot A) \overleftrightarrow{\partial_0}\, u. \tag{5.33}$$

Let us accept that this is a conserved quantity, associated to the current

$$j_\mu = (\partial \cdot A) \overleftrightarrow{\partial_\mu}\, u.$$

Obviously $[Q, u] = 0$. By use of the algebraic identity

$$[AB, C]_+ = A[B, C]_+ - [A, C]B,$$

nilpotency then is checked as follows:

$$2Q^2 = [Q, Q]_+ = -\int_{x^0=\text{const}} d^3x\, [(\partial \cdot A), Q] \overleftrightarrow{\partial_0}\, u = i \int_{x^0=\text{const}} d^3x\, \Box u \overleftrightarrow{\partial_0}\, u = 0,$$

because the ghost is a free massless quantum field, that is, satisfies the wave equation.

The form (5.33) will not do for the massive case, as now, with ghost fields of the same mass as A^μ, after a relatively long calculation involving the solution of the Cauchy problem for u, we would obtain

$$2Q^2 = i \int_{x^0=\text{const}} d^3x\, \Box u \overleftrightarrow{\partial_0}\, u = -im^2 \int_{x^0=\text{const}} d^3x\, u \overleftrightarrow{\partial_0}\, u \neq 0,$$

A suitable form of Q is reached by introducing a (Bose) scalar field with the same mass, satisfying

$$(\Box + m^2)B = 0, \quad [B(x), B(y)] = -iD(x-y),$$

and then

$$Q = \int_{x^0=\text{const}} d^3x\, (\partial \cdot A + mB) \overleftrightarrow{\partial_0}\, u. \tag{5.34}$$

I leave it to the reader to check this is a conserved quantity. Now we obtain

$$2Q^2 = i \int_{x^0=\text{const}} d^3x\, \Box u \overleftrightarrow{\partial_0}\, u + im^2 \int_{x^0=\text{const}} d^3x\, u \overleftrightarrow{\partial_0}\, u = 0.$$

In this way we have recovered the Stückelberg field!

In summary, the gauge variations are

$$\begin{aligned} sA^\mu(x) &= [Q, A^\mu(x)]_\pm = i\partial^\mu u(x), \\ sB(x) &= [Q, B(x)]_\pm = imu(x), \\ su(x) &= [Q, u(x)]_\pm = 0, \\ s\tilde{u}(x) &= [Q, \tilde{u}(x)]_\pm = -i\big(\partial^\mu A_\mu(x) + mB(x)\big). \end{aligned} \tag{5.35}$$

With respect to the semiclassical analysis in Section 5.2 there is a slight change of notation; the present one is more advantageous when dealing with quantum fields. As expected, the BRS variation of the gauge field corresponds to substituting the ghost field for the infinitesimal parameter of the gauge transformation.

I finish with a little collection of remarks.

- The ghost number of Q is precisely 1.
- In view of nilpotency of Q, finite gauge variations are easily computed. We have
$$A'_\mu(x) = e^{-i\lambda Q} A_\mu(x) e^{i\lambda Q} = A_\mu(x) - i\lambda[Q, A_\mu(x)] - \tfrac{1}{2}\lambda^2[Q, [Q, A_\mu]].$$
Note that the last term is *not* zero. But certainly there are no higher-order terms.
- Only unphysical fields appear in the formula (5.34) for Q.
- A stronger BRS theory includes the anti-BRS symmetry $\bar{s}$, with the *complete nilpotency* conditions $s^2 = \bar{s}^2 = s\bar{s} + \bar{s}s = 0$ [34]. The main role of $\bar{s}$ is to ensure the closure of the classical algebra, at the level of Lagrangians. This is more or less unnecessary in Yang–Mils theories, but useful for instance in supersymmetric theories.
- It would seem that the foregoing analysis applies only to abelian fields. The cognoscenti would in general expect in (5.35) extra terms in the first equality (covariant derivative rather than the ordinary one) and in the third one (a ghost term involving the structure constants). That is,
$$\begin{aligned} sA^a_\mu(x) &= [Q, A^a_\mu(x)] = iD_\mu u^a(x), \\ su^a(x) &= [Q, u^a(x)]_+ = -\tfrac{i}{2} g f^{abc} u^b(x) u^c(x). \end{aligned} \tag{5.36}$$
However, it ain't necessarily so. By just adding the colour index, one can think of (5.35) as a first step, one in which self-interaction is neglected, for a non-abelian theory. In the causal approach to QFT [30], one approaches interacting fields by means of free fields, and then the two methods differ.

5.3.5 The ghostly Krein operator

For completeness, I include here a discussion on the *charge algebra* for ghosts. Let f_r denote an orthonormal basis of $L^2(\mathcal{H}_m, d\mu_m(p))$. Consider the charge operators

$$Q(A) := \sum_{r,b,i} c^\dagger_{b,i}(f_r) a_{ij} c_{b,j}(f_r) = \sum_{b,i} \int d\mu_m(p)\, c^\dagger_{b,i}(p) a_{ij} c_{b,j}(p),$$

for $A = (a_{ij})$ a 2×2 matrix. This is defined on a common dense domain of H_{gh}, bigger than $\mathcal{D}$, which is mapped by the charge operators into itself. This map represents $\mathfrak{gl}(2, \mathbb{C})$, as

$$Q(AB - BA) = Q(A)Q(B) - Q(B)Q(A); \quad \text{also} \quad Q(A^\dagger) = Q^\dagger(A).$$

(By the way, by $Q^\dagger(A)$ we mean its restriction to $\mathcal{D}$.) Taking for A the unit matrix and the Pauli matrix σ_3, we respectively obtain the ghost number $\mathbb{N}_{\text{gh}}$ and ghost charge Q_{gh} operators. The other two Pauli matrices yield ghost–antighost exchanging operators, respectively called here Γ, Ω. Their commutators with the local fields $u, \tilde{u}$ are

$$\begin{aligned}
[\mathbb{N}_{\text{gh}}, u] &= -\tilde{u}^\dagger, & [\mathbb{N}_{\text{gh}}, \tilde{u}] &= u^\dagger; \\
[Q_{\text{gh}}, u] &= -u, & [Q_{\text{gh}}, \tilde{u}] &= \tilde{u}; \\
[\Gamma, u] &= \tilde{u}, & [\Gamma, \tilde{u}] &= u; \\
[\Omega, u] &= -i\tilde{u}, & [\Omega, \tilde{u}] &= iu.
\end{aligned}$$

The verification of this is an exercise. The generator of $\mathbb{N}_{\text{gh}}$, which constitutes the centre of the charge algebra, gives by commutation with $u, \tilde{u}$ fields that are not relatively local. We write down the following currents:

$$\begin{aligned}
j_{\mathbb{N}_{\text{gh}}}(x) &:= i{:}u^\dagger(x)\, \overleftrightarrow{\partial^\mu}\, u(x){:}, & j_{\text{gh}}(x) &:= i{:}\tilde{u}(x)\, \overleftrightarrow{\partial^\mu}\, u(x){:}, \\
j_u(x) &:= i{:}u(x)\, \overleftrightarrow{\partial^\mu}\, u(x){:}, & j_{\tilde{u}}(x) &:= i{:}\tilde{u}(x)\, \overleftrightarrow{\partial^\mu}\, \tilde{u}(x){:}.
\end{aligned}$$

Again, $j_{\mathbb{N}_{\text{gh}}}$ is not a relatively local quantum field. They are related to the corresponding charges in the usual way; one has, moreover,

$$\Gamma = \tfrac{1}{2}(Q_u - Q_{\tilde{u}}), \quad \Omega = \tfrac{i}{2}(Q_u + Q_{\tilde{u}}).$$

We can consider as well operators $T\big(e^{iA}\big) := \exp(i\,Q(A))$. They give a representation of the general linear group. It is $T(B^\dagger) = T(B)^\dagger$. Also,

$$T(B)Q(A)T^{-1}(B) = Q(BAB^{-1}).$$

The theory with ghosts has to be constructed by using only the fields $u, \tilde{u}$, while their adjoints do not appear at all; in this way the troubles with locality are avoided. In massless Yang–Mills theories, say, one considers the interaction

$$T_1(x) = \tfrac{i}{2} f^{abc}\big({:}A^a_\mu A^b_\nu F^{c\mu\nu}{:}(x) + {:}A^a_\mu u^b \partial^\mu \tilde{u}^c{:}\big)(x). \tag{5.37}$$

This is invariant under gauge transformations generated by the differential operator (5.33). The $u^\dagger$, $\tilde{u}^\dagger$ do not appear here. But then it is right to worry about unitarity. The solution in gauge theories is as follows: η-unitarity of $\mathbb{S}$ together with gauge invariance will imply unitarity of the $\mathbb{S}$-matrix on the physical subspace.

For the theory defined by (5.37), we have

$$\eta = \eta_A \otimes \eta_{\rm gh} \qquad \text{on} \quad H = H_A \otimes H_{\rm gh}.$$

We recall η_A is given by

$$\eta_A = \prod_{a=1}^{\dim G} (-)^{\mathbb{N}_{0a}},$$

where $\mathbb{N}_{0a}$ is the number operator for gauge particles of G-colour a. The gauge potentials A^a_μ are η-Hermitian. *Grosso modo*: we expect the η-adjoint fields $u^+, \tilde{u}^+$ to enter T_1, in order to have $\eta_{\rm gh}$-Hermitian quantities. The key is causality: the latter Krein operator must be defined in such a way that $u^+, \tilde{u}^+$ are relatively local to $u, \tilde{u}$; we know $u^\dagger, \tilde{u}^\dagger$ do not have this property. With all this in mind, we search for the good $\eta_{\rm gh}$. Clearly, it cannot be relatively local itself, which is tantamount to involving $\mathbb{N}_{\rm gh}$. A natural guess would be to take the (already much used) operator

$$E := \exp(i\pi \mathbb{N}_{\rm gh}).$$

However, consider the ghost and antighost number operators:

$$N_j := \tfrac{1}{2}(\mathbb{N}_{\rm gh} + j\,Q_{\rm gh})$$

for $j = 1, -1$. They also have integer spectrum. Moreover,

$$E = (-)^{N_1 + N_{-1}} = (-)^{N_1 - N_{-1}} = (-)^{Q_{\rm gh}},$$

so E cannot be the right choice. We consider instead

$$I := (-)^{N_{-1}} = e^{\frac{i}{2}\pi(N - Q_{\rm gh})} = T(\sigma_3).$$

This is indeed a symmetry. We do have $Ic_j(p)I = jc_j(p)$, and it is then quickly seen that

$$Iu^\dagger I = \tilde{u}, \quad I\tilde{u}^\dagger I = u;$$

so we have locality.

Though this is a perfectly sensible solution to the problem, T_1 and Q are not I-Hermitian. One could write different, equivalent expressions for the terms involving ghosts in the Lagrangian (see the discussion in the next paragraph); but first we submit to convention. Consider then

$$S = T(U) := T\big(i(\sigma_1 + \sigma_3)/\sqrt{2}\big) = T\Big(e^{i\pi(\sigma_1+\sigma_3)/2\sqrt{2}}\Big)$$

and

$$\eta_{\rm gh} := SIS^{-1} = T(\sigma_1) = i^{N-\Gamma}.$$

Now we get

$$\eta_{\rm gh} c_j(p) \eta_{\rm gh} = c_{-j}(p)$$

and

$$u^+ := \eta_{\rm gh} u^\dagger \eta_{\rm gh} = u, \qquad \tilde{u}^+ := \eta_{\rm gh} \tilde{u}^\dagger \eta_{\rm gh} = -\tilde{u},$$

together with

$$T_1^+ = T_1, \qquad Q^+ = Q.$$

An alternative definition for the ghost contribution in T_1 would be given by

$$\tfrac{1}{2} f^{abc} {:}A^a_\mu u^b \overleftrightarrow{\partial^\mu} \tilde{u}^c{:}(x) \quad \text{instead of} \quad f^{abc} {:}A^a_\mu u^b \partial^\mu \tilde{u}^c{:}(x).$$

The two forms differ by a pure divergence term plus a $Q_{\rm gh}$-coboundary, that is, a term of the form $[Q_{\rm gh}, K]_+$. Therefore the first one remains gauge invariant. The choice of it would allow the use of I as Krein operator, preserving all the good properties. The second one is employed partly for historical reasons.

To conclude, let me comment again on the different behaviour of the Poincaré group representation in the massive and the massless case. For the former, the representation is always unitary, and commutes with all charges $Q(A)$ and transformations $T(B)$. Therefore it is η-unitary as well. However, for the gauge potentials in the massless case the representation is not unitary, and η_A is introduced for reasons of covariance.

Bibliography

[1] C. N. Yang and R. L. Mills, Phys. Rev. **96** (1954) 191.
[2] L. O'Raifeartaigh, *The Dawning of Gauge Theory*, Princeton University Press, Princeton, NJ, 1997.
[3] R. Utiyama, Phys. Rev. **101** (1956) 1597.
[4] M. E. Peskin and D. V. Schroeder, *An introduction to quantum field theory*, Addison-Wesley, Reading, MA, 1995.
[5] P. H. Frampton, *Gauge field theories*, Wiley, New York, 2000.
[6] M. Chaichian and N. F. Nelipa, *Introduction to gauge field theories*, Springer, Berlin, 1984.
[7] J. F. Cariñena, 'Introducción a la teoría clásica de campos', unpublished notes, Zaragoza, 1995.
[8] S. Weinberg, *The quantum theory of fields* II, Cambridge University Press, Cambridge, 1996.
[9] C. Itzykson and J.-B. Zuber, *Quantum field theory*, McGraw-Hill, New York, 1980.
[10] E. Wigner, Ann. Math. **40** (1939) 149.
[11] S. Weinberg, *The quantum theory of fields* I, Cambridge University Press, Cambridge, 1995.
[12] D. Rivier, Helv. Phys. Acta **57** (1984) 577.

[13] T. A. Green and E. C. G. Stückelberg, Helv. Phys. Acta **24** (1951) 153.
[14] A. Petermann and E. C. G. Stückelberg, Helv. Phys. Acta **26** (1953) 499.
[15] E. C. G. Stückelberg and D. Rivier, Helv. Phys. Acta **23** (1950) 215.
[16] S. S. Schweber, *QED and the men who made it*, Princeton University Press, Princeton, NJ, 1994.
[17] E. C. G. Stückelberg, Helv. Phys. Acta **11** (1938) 225.
[18] W. Pauli, Rev. Mod. Phys. **13** (1941) 203.
[19] M. Dütsch and B. Schroer, J. Phys. A **33** (2000) 4317.
[20] H. Ruegg and M. Ruiz-Altaba, Int. J. Mod. Phys. A **19** (2004) 3265.
[21] A. A. Slavnov, Phys. Lett. B **620** (2005) 97.
[22] José M. Gracia-Bondía, 'Remarks on Noether's and Utiyama's paradigms', in *Fundamental Physics Meeting: Alberto Galindo*, R. F. Alvarez-Estrada, A. Dobado, L. A. Fernández, M. A. Martín Delgado and A. Muñoz Sudupe (eds.), Fundación BBVA, Madrid, 2004; pp. 283–290.
[23] T. Kunimasa and T. Goto, Prog. Theor. Phys. **37** (1967) 452.
[24] T. Sonoda and S. Y. Tsai, Prog. Theor. Phys. **71** (1984) 878.
[25] E. Fermi, Rev. Mod. Phys. **4** (1932) 125.
[26] L. Baulieu and J. Thierry-Mieg, Nucl. Phys. **B197** (1982) 477.
[27] R. Delbourgo, S. Twisk and G. Thompson, Int. J. Mod. Physics A **3** (1988) 435.
[28] R. P. Feynman, Acta Phys. Pol. **26** (1963) 697.
[29] F. Krahe, 'On the algebra of ghost fields', DIAS-STP-95-02 preprint, unpublished.
[30] G. Scharf, *Quantum gauge theories. A true ghost story*, Wiley, New York, 2001.
[31] T. Kugo and I. Ojima, Prog. Theor. Phys. Suppl. **66** (1979).
[32] S. S. Horuzhy and A. V. Voronin, Commun. Math. Phys. **123** (1989) 677.
[33] A. Galindo, Commun. Pure Appl. Math. **15** (1962) 423.
[34] L. Alvarez-Gaumé and L. Baulieu, Nucl. Phys. **B212** (1983) 255.

6

Large-N field theories and geometry

DAVID BERENSTEIN*

Abstract

This is a short introduction to the ideas of the AdS/CFT correspondence. In order to be self-contained, the chapter includes an introduction to the study of strings as geometric objects moving in spacetime and in particular their solvability in flat space. I also mention why strings give rise to a theory of gravity. D-branes are introduced as a collection of geometric objects where strings can end. The low-energy dynamics of a collection of D-branes is explored in two different ways, and this serves as a basis for a formulation of the AdS/CFT correspondence: an equivalence between a gravitational formulation of the dynamics and a gauge theory description. The problem of how to compare observables between the two formulations is presented, and some basic aspects of the representation theory of the superconformal group are explored, so that one can have tests of the AdS/CFT proposal.

6.1 Introduction

Roughly ten years ago, the AdS/CFT correspondence was formulated by Maldacena [20]. In its simplest example, the AdS/CFT correspondence states that a certain four-dimensional quantum field theory that is made from gauge fields and some matter content – a theory similar to the theory of strong, weak or electromagnetic interactions – is equivalent as quantum theory to type IIB string theory (as a theory of quantum gravity) on spacetimes that are asymptotic to a particular classical solution of type IIB string theory, namely $AdS_5 \times S^5$.

This correspondence has not been proved. This is in great part because we do not know what quantum gravity is. We believe that whatever it is, it will resemble

* Work supported in part by the DOE under grant DE-FG-02-91ER40618.

semiclassical gravity in the regimes where one is close to a classical solution with large radius of curvature. By comparison with these semiclassical calculations, there is overwhelming circumstantial evidence that the AdS/CFT correspondence conjecture is correct.

The purpose of this chapter is to give a basic introduction to the ideas of the AdS/CFT and to how one accumulates evidence in favor of such a strong conjecture. Obviously, if one learned how to cross systematically between the field theory and the quantum gravity side, many problems related to quantum gravity might find a solution in the dual field theory. Similarly, hard problems in quantum field theory might find a simple solution if studied gravitationally.

Finding a way of comparing sides – finding the AdS/CFT dictionary, so to speak – is a topic of research that has produced thousands of scientific papers and is an active area of current research in quantum field theory and string theory. Here we will just explore a tiny fraction of what is known. Unfortunately, there is no modern comprehensive review of the status of the field. An aggregate of most of the early evidence and literature of this correspondence can be found in [1]. A recent introductory set of notes is also available in [23].

The reader is expected to know some quantum field theory (at the very least the reader should be familiar with Feynman graphs and with elementary aspects of gauge theories) and some basic ideas of gravity: some knowledge of what a metric tensor is, and some basic aspects of coordinate changes and differential geometry. Although it will be easier to read this review if one already knows string theory, that is not required, and the basic ideas will also be introduced for completeness, to motivate the constructions of the AdS/CFT.

6.2 Strings: a geometric dynamical object

To understand the ideas of the AdS/CFT correspondence, it is important to understand first the notion of string theory as a theory of one-dimensional objects propagating on a fixed spacetime. The main idea is to introduce the smallest number of ingredients to obtain some sensible physical results and situations.

When we think of a string evolving in time, we imagine that we start with a configuration of some (oriented) loop in space, with some initial velocities, after which the loop will undergo motion given by some equations of motion. Usually in physical situations the equations of motion are derived from a least action principle. That is, the equations of motion are derived by performing a calculus of variations on a functional of the trajectory of the string. These trajectories describe a tubelike surface in spacetime. We call any such trajectory the worldsheet of the string. The allowed trajectories are those that extremize the action functional.

We want to have a theory with relativistic invariance, as this is a symmetry that we observe in nature. Thus spacetime should be endowed with a metric. For simplicity we will take this metric to be flat, but this is not required. Thus, at our disposal we have a metric tensor $g_{\mu\nu}$ that describes how we measure distance and time intervals. This will be fixed to a flat metric with a naturally adapted coordinate system $(t, \vec{x})$,

$$g_{\mu\nu} = -dt^2 + \sum_i (dx^i)^2. \tag{6.1}$$

With this information alone, we want to build the action functional describing the motion of the string in spacetime. We want this action to use only the preceding information: the trajectory that the string describes in spacetime (from some initial to some final configuration) and the metric tensor g. We also want the simplest possible action that we can write. In order to have dynamics depend on initial conditions alone, the action should have the property that it is local in time, described by some Lagrangian:

$$S = \int dt L, \tag{6.2}$$

where L depends on finitely many time derivatives.

Because it is local in time, the fact that Lorentz rotations mix time t and positions x indicates that the action should also be local in positions and therefore should be described by an integral over the worldsheet. Namely, we should have that

$$S = \int d^2\sigma \mathcal{L}, \tag{6.3}$$

where $\mathcal{L}$ is some Lagrangian density, and the coordinates σ (there are two of them, $\sigma^{1,2}$) are any convenient parametrization of the worldsheet of the string.

At this point, we are introducing a new ingredient. We are describing the worldsheet by some particular coordinate choice on its worldvolume, and from what we have written so far, this choice of coordinate frame might play an important role in the description of the motion of the string. The assumption that is made in string theory is that this coordinate choice is not physical, but just a mathematical device to label the points on the worldsheet.

Thus, the Lagrangian density $\mathcal{L}$ cannot depend on this choice, except as required so that the full action is independent of this coordinate choice. From this idea, it follows that $\mathcal{L}$ should be a density on the worldvolume built from the trajectory that the string describes, and it should depend only on $g_{\mu\nu}$ and the trajectory. This is because we do not want to introduce any additional structure unless it is absolutely necessary.

The tensor $g_{\mu\nu}$ is a tensor in spacetime, not on the worldsheet. We would like to build out of $g_{\mu\nu}$ a tensor field on the worldsheet itself. The natural way to do this is to use the induced metric tensor on the worldsheet, which we will denote by $g_{ind} \sim g_*$.

The simplest possible choice for the action is the volume of the worldsheet calculated in the metric g_*. This volume form is given by

$$\mathcal{L} \sim -T\sqrt{-\det g_*}. \tag{6.4}$$

We have introduced a numerical constant to restore proper units of action (energy times time). The number T is called the string tension. For convenience, we choose units where we set T to one. In principle, we could also consider other scalar tensors built out of g multiplied by the volume density. For example, the Ricci scalar would work. However, all of these are more complicated: one requires many more computational resources to calculate the Lagrangian density itself. It is also important to note that we have a minus sign in the definition of the square root. This follows from the fact that the metric that is induced on the worldsheet should be Lorentzian at every point of the worldsheet (the string evolves toward the future in all coordinate frames). Thus, the determinant that calculates the square of the volume form is negative. To have a real action, we need to take a square root of something positive. This condition encapsulates the idea that the string propagates to the future in all reference frames. Thus the total action is described by

$$S \sim -\int d^2\sigma\sqrt{-\det g_*}, \tag{6.5}$$

which is the Lorentzian area of the worldsheet. This action is called the Nambu–Goto action. The equations of motion that follow from it are very nonlinear. So far, the fact that g is flat has not been important. However, if we want to solve the equations of motion of the string, we need to make a choice of g, and this is where the simplest choice that we can make matters. From the physical standpoint, having flat space is a good approximation to the typical situations associated with accelerator experiments; thus studying strings in flat space is useful for that situation.

This is the starting point for the study of the bosonic string. All we have to do now is solve the dynamics and quantize the system.

Looking at the Nambu–Goto action, the presence of a square root complicates the analysis and makes the equations of motion very nonlinear. It also makes the process of quantization more difficult. It was found that quantization could be carried in the light-cone gauge, but manifest Lorentz invariance is lost. As a matter of fact, the quantum bosonic string theory can only be made to be Lorentz invariant if the spacetime dimension is equal to 26. This calculation is beyond the scope of the present review. The details can be found in standard textbooks on string theory [13,26].

6.2.1 The Polyakov string: more is the same

Given some of the conceptual difficulties in dealing with the Nambu–Goto string, it is interesting to ask if there is a better formulation of the Nambu–Goto string that makes quantization more straightforward and Lorentz covariant. Such a formulation is due to Polyakov. The main idea is to introduce an intrinsic metric on the worldsheet itself, which is independent of the induced metric g_*. We will call this metric $h_{\alpha\beta}$. It is added to what we already have. In the spirit that we have been working in, we are adding information that was not there before, so we are adding complications to the description of the string.

Given h and g_*, we can build a simple action for the embedding:

$$S = -\int \sqrt{-\det h}\ h^{\alpha\beta} g_{\alpha\beta *} \sim \int \sqrt{-\det h}\ \mathrm{Tr}(h^{-1} g_*). \tag{6.6}$$

If we want to be more explicit, we expand g_* in terms of the embedding functions, and we find that

$$S = \int \sqrt{-\det h}\ h^{\alpha\beta} \partial_\alpha X^\mu \partial_\beta X^\nu g_{\mu\nu}(X) \sim \int \sqrt{-\det h} h^{\alpha\beta} \partial_\alpha X^\mu \partial_\beta X^\mu, \tag{6.7}$$

where for the right-hand side we are using the simplifying assumption that the metric of the space in which the string is moving is flat.

This action is called the Polyakov action for the string. What we see is that in this more elaborate version of the string, the embedding functions X appear quadratically in the action. So if h is fixed, we have a quadratic action for scalar fields on the worldsheet (the coordinates X are just functions on the worldsheet and not tensors). A quadratic action for scalar fields is what we call a free field. Free fields are well understood, and it is easy to quantize them. So the advantage of the Polyakov action is that the process of quantization is much more transparent [27].

However, we have added the intrinsic metric h. If h is fixed a priori, the action we have is called a nonlinear σ-model (for g arbitrary). The idea of the Polyakov action is that h should be treated as just another dynamical variable. Having h dynamical means that we have to think of h as dynamical gravity in two dimensions. We could also add an Einstein–Hilbert term for the action in two dimensions. This would be an integral of the form

$$\kappa \int \sqrt{-\det h}\ R_h, \tag{6.8}$$

where R_h is the Ricci scalar associated to the metric h, and κ would play the role of Newton's constant. In two dimensions this integral is a topological quantity: it is the Euler number density. It can always be written locally as a total derivative, so the local variation of it with respect to h vanishes. Given a topology for a Riemann surface, the integral of $\sqrt{-\det h}\ R$ gives the genus of the surface times a normalization factor that depends on conventions for calculating the Ricci scalar.

Thus, when we apply a variational principle to the Polyakov action, we can ignore this term and focus on the Polyakov action (6.7). A quick calculation shows that the equations of motion of h state that up to a scalar factor, the worldsheet metric is proportional to the induced metric

$$h_{\alpha\beta} = \lambda(\sigma) g_{\alpha\beta *}, \tag{6.9}$$

where λ is undetermined. This might spell a problem, as we seem to have a degree of freedom that cannot be fixed by the equations of motion.

The crucial observation for the resolution of this puzzle is that the action (6.7) has more symmetry than just diffeomorphism invariance on the worldsheet. This insight is due to Polyakov.

If we take a metric h and consider the action for the metric $\tilde{h} = \exp(A(\sigma))h$ instead of h, we notice that

$$\sqrt{-\det(\tilde{h})} = \exp(A(\sigma))\sqrt{-\det(h)}, \tag{6.10}$$

whereas for the inverse metric $\tilde{h}^{-1}$ we have that

$$\tilde{h}^{-1} = \tilde{h}^{\alpha\beta} = \exp(-A(\sigma))h^{\alpha\beta}. \tag{6.11}$$

So when we consider the combination $\sqrt{-\det h}\ h^{-1}$, we have that

$$\left(\sqrt{-\det(\tilde{h})}\right)\tilde{h}^{\alpha\beta} = \left(\sqrt{-\det(h)}\right)h^{\alpha\beta}. \tag{6.12}$$

That is, the action and the dynamics are completely independent of these rescalings. This should be interpreted as an internal symmetry of the system. These rescalings of the metric are called Weyl rescalings. The Weyl invariance of the Polyakov action suggests that we have a redundancy of the description: we have functions that are undetermined by the equations of motion. This is a familiar property of gauge theories. The Weyl rescaling should be treated as a gauge symmetry and not a global symmetry of the system. What this means is that we are allowed to choose A to be whatever is most convenient.

For simplicity, from the solution of (6.9), we can rescale h so that it is exactly equal to the induced metric $h = g_*$. Because we solved for h algebraically, we can substitute it in the action and this process does not modify the equations of motion associated to the embedding itself. If we substitute this value of h in the action (6.7), we recover the action (6.5), so at the level of classical physics we find that the Polyakov action, with the gauged Weyl invariance, is completely equivalent as a dynamical system to the Nambu–Goto string. That is, the set of solutions of the equations of motion (the critical points of the action) of the two systems coincide exactly and describe the same object tracing the same trajectories in spacetime.

This process of adding fields to a theory without affecting the dynamics is called adding auxiliary fields. It is a trick commonly used in supersymmetric theories as well (see [30, 31] for example), where the main role auxiliary fields play is that they simplify the analysis if they are included.

6.3 Solving the Polyakov string

What we want to do now is solve the equations of motion for the string in flat space. This will also help us quantize the string. There are some tricky aspects to how to do this. One is that the metric h is Lorentzian, whereas for a random embedding the signature of g_* might be zero or two, depending on whether the worldsheet is embedded in spacetime along a proper timelike direction, or is embedded in a spacelike manner. This problem tells us that we cannot vary over all embeddings, but only those that satisfy the right conditions. Given a fixed worldsheet, this is easy to determine, but one might worry that some small variations of the fields are not allowed because they would violate this condition.

The other complication that we might worry about is that Lorentzian two-dimensional objects are more complicated than the Euclidean two-dimensional objects, which are just Riemann surfaces and are well understood.

Because Euclidean surfaces are simpler than their Lorentzian counterparts, we would like to analyze the same Polyakov action for Euclidean metrics and embeddings into an Euclidean space, as opposed to a Lorentzian space. The idea of why this might work is that as far as the variational equations of motion are concerned, these details do not affect the form of the resulting equations substantially (except for the obvious minus signs that have to be omitted).

The theory of two-dimensional metric surfaces up to conformal rescaling is mathematically well understood: it is characterized by Riemann surfaces and the theory of complex structures on them. The appearance of complex variables will also tell us that the problem we are studying is solved by analytic functions of complex variables, and these are also well understood and simple.

This procedure of going to the Euclidean setup is a kind of analytic continuation, similar to the familiar Wick rotation in quantum field theory that is used in the evaluation of Feynman diagrams. Unfortunately, it is not the case that any Lorentzian metric, can be analytically continued to a Riemannian metric, and we need to be careful about the way we do this. Because of the Weyl invariance this can be done locally, and to the extent that the solutions of the equations of motions are written in terms of analytic functions, the corresponding changes are reasonably easy to make. This process is sufficiently well understood that I can skip many details and just give the final answers.

I will, however, describe how to solve the problem, so that the reader can get acquainted with these techniques for use elsewhere.

6.3.1 The Euclidean setup

The idea is to analyze the Euclidean version of the Polyakov action.[1] This is given by

$$S \sim \int d^2\sigma \sqrt{\det(h)} h^{\alpha\beta} g_{\alpha\beta *}. \tag{6.13}$$

We understand this as an action for the embedding of the string worldsheet into a space that depends on some additional classical metric on the worldsheet. Because we care about classical solutions at this point, once we decide that we are looking at one solution in particular, we can make advantageous choices on that solution.

The main idea is to exploit the uniformization theorem for surfaces: any metric on a Riemann surface is (locally) conformally equivalent to a flat metric. Given a flat metric, with specially adapted coordinates σ^1, σ^2, we have

$$ds^2 = (d\sigma^1)^2 + (d\sigma^2), \tag{6.14}$$

and we can consider the complex coordinate

$$z = \frac{\sigma^1 + i\sigma^2}{\sqrt{2}} \tag{6.15}$$

plus its complex conjugate $\bar{z}$. This is a choice of gauge. Recasting the metric in such special coordinates can always be done locally. Globally one can show that the Riemann surface is endowed with a global complex structure in this way; thus one can talk about global holomorphic objects. We call such a local choice of coordinates and Weyl factor the conformal gauge.

In these complex coordinates, the metric takes the form

$$ds^2 = 2dzd\bar{z}. \tag{6.16}$$

Thus we find that in this coordinate system the Polyakov action is given by

$$S \sim \int d^2z\, \partial_z X^\mu \partial_{\bar{z}} X^\mu. \tag{6.17}$$

The equations of motion for X^μ, having fixed η, are simple. We have that

$$\partial_z \partial_{\bar{z}} X^\mu = 0. \tag{6.18}$$

[1] One should also consider that classical Euclidean solutions for field theories also describe tunneling processes in the WKB approximation in quantum mechanics, so one should take these continuations seriously.

We will simplify notation by using $\partial \sim \partial_z$ and $\bar{\partial} = \partial_{\bar{z}}$. Thus the equation (6.18) reads $\partial\bar{\partial}X^\mu = 0$. The equation of motion for X is that the Laplacian of X vanishes. Thus X is locally a harmonic function. For closed Riemann surfaces this would indicate that X is constant. However, in string theory we have to imagine that the Riemann surfaces we have correspond to setups for initial conditions. Thus the global Riemann surfaces will end on corresponding "initial" data and "final" data circles, and the functions X can be nontrivial.

If we have such a solution for X near a smooth point in the surface, then we can decompose X as

$$X^\mu(z, \bar{z}) \sim X^\mu_L(z) + X^\mu_R(\bar{z}). \tag{6.19}$$

These two pieces will be called the left and right movers of X^μ. These should be locally (anti)holomorphic in their domain of definition.[2] Also, if X is single valued, then so are ∂X and $\bar{\partial} X$. So the derivatives of X are globally (anti)holomorphic one-forms.

Because we made a choice of gauge for h that eliminated it completely from the Lagrangian in (6.17), we have assumed all along that we have a solution to the equations of motion for h; but we still need to verify it after solving the equations of motion for X. The consistency conditions for this to happen are called the Virasoro constraints. They tell us that given a harmonic set of functions X^μ, the following two equations must be satisfied:

$$T_{zz} = \partial X^\mu \partial X^\mu = 0, \tag{6.20}$$
$$T_{\bar{z}\bar{z}} = \bar{\partial} X^\mu \bar{\partial} X^\mu = 0. \tag{6.21}$$

These just state that the two components g_{zz*} and $g_{\bar{z}\bar{z}*}$ vanish, so that we indeed have a solution as described by Equation (6.9). The tensor T is also the stress–energy tensor of the fields X coupled to gravity. Here we find that two components of T need to vanish to have a solution of the equations of motion.

Thus, solving the equations is technically easy, but we have to verify that a given solution satisfies the constraints at the end.

6.3.2 Going back to real time: Fourier series and quantization

We now want to consider now how this solution of the string looks in a particular coordinate system. To describe an initial and a final shape for a string, we need a surface with two ends: a cylinder. We can imagine extending the cylinder infinitely far towards the future and the past in a real time situation, so the cylinder is represented by the complex plane z with identifications $z \equiv z + 2\pi$. Here it is

[2] The functions X^μ can be individually multivalued, so long as their sum is single valued on the worldsheet.

convenient to choose another coordinate w on $\mathbb{C}^*$, where $w = \exp(iz)$. In terms of w, we have that

$$\partial_w X^\mu \tag{6.22}$$

is holomorphic and single valued everywhere in the complex plane (except perhaps at the origin). Thus we can write ∂_w in terms of a Laurent series,

$$\partial_w X^\mu = \sum_{n=-\infty}^{\infty} \alpha_n w^n, \tag{6.23}$$

and similarly for the right movers. We can integrate these and put the left and right movers together (after transforming to the z variable) to find that

$$X^\mu(z, \bar{z}) = a^\mu z + b^\mu + \tilde{a}^\mu \bar{z} + \sum_{n\neq 0} \left(\frac{\alpha_n^\mu}{n} \exp(inz) + \frac{\tilde{\alpha}_n^\mu}{n} \exp(in\bar{z}) \right). \tag{6.24}$$

The requirement for X to be single valued forces us to have $a = -\tilde{a}$.

Now, let us split z into a real and an imaginary part, $z = \sigma + iT$. We will call T the Euclidean time. The metric (up to a rescaling factor) for the cylinder is given by $dT^2 + d\sigma^2$.

We notice that in this metric we can do an analytic continuation $T = i\tau$, and it becomes

$$dT^2 + d\sigma^2 \to -d\tau^2 + d\sigma^2. \tag{6.25}$$

This analytic continuation takes $z \to \sigma_- = \sigma - \tau$ and $\bar{z} \to \sigma + = \sigma + \tau$. In these equations the complex conjugate variables $z, \bar{z}$ become two independent real variables $\sigma_\pm$. We also carry out this continuation in the solution to the equations of motion and then impose that the X are real coordinates. This should be done after the analytic continuation.

The solution for the Lorentzian problem can be written in the form

$$X^\mu(\sigma, \tau) = p^\mu \tau + b^\mu + \sum_{n\neq 0} \left(\frac{\alpha_n^\mu}{n} \exp(in(\sigma - \tau)) + \frac{\tilde{\alpha}_n^\mu}{n} \exp(in(\sigma + \tau)) \right), \tag{6.26}$$

which we interpret by saying that we have waves moving to the left and to the right on the worldsheet, whose mode amplitudes are the α_n^μ. We have also relabeled the coefficient $a^\mu \sim -\tilde{a}^\mu$ as an equivalent variable p^μ. To this formal solution, we still need to add the Virasoro constraints, which tell us what solutions are acceptable. These end up ensuring that the string propagates forward in time in all possible coordinate systems.

This represents the complete solution of the bosonic string in flat space.

Now, the numbers p^μ denote the velocity of the string center of mass on all coordinates with respect to the parameter τ. One can show that the total momentum of the string is proportional to p^μ, so p^μ can be interpreted as the total momentum carried by the string.

If we integrate the Virasoro constraint on a slice of the string, we find that (schematically)

$$\int T_{++} = \int \partial_+ X^\mu \partial_+ X^\mu \sim \sum p^\mu p_\mu + \sum_{n\neq 0} g_{\mu\nu}\alpha_n^\mu \alpha_{-n}^\nu = 0. \tag{6.27}$$

The quantity $p^\mu p_\mu = -m^2$ is the relativistic mass of a particle. We see that the mass is determined by the amplitudes of oscillation of the different modes; indeed, the sum in (6.27) is the energy stored in the oscillators (remember that T is the stress energy tensor on the worldsheet, so the integral of T is the total energy). A similar statement holds for the left movers: $T_{--} = 0$ implies the same type of constraint for $\alpha \leftrightarrow \tilde{\alpha}$.

When we quantize the string, the coefficients of the general time-dependent solution are interpreted as quantum operators. Because solving the string is equivalent to solving a wave equation, the amplitudes of the modes of the wave end up giving us an infinite set of harmonic oscillators. This is the standard quantization of a free field on a circle. The coefficients of positive-frequency solutions are raising operators, and the coefficients for negative-frequency solutions are lowering operators. We can also read the energy of each oscillator from its frequency. The energy is given by an integer n times some standard frequency.

That means that once we take into account quantum corrections, the spectrum of values of $\sum_{n\neq 0} g_{\mu\nu}\alpha_n^\mu \alpha_{-n}^\nu$ consists basically of the set of integers, and the possible masses of strings are quantized (I will explain this in some more detail in the next subsection). In this way the string produces an infinite tower of unitary representations of the Lorentz group labeled by an integer N (the mass squared of the associated one-particle state). We call this integer number the level of the string state.

6.3.3 Massless particles and superstrings

The spectrum of the string is formally calculated by Equation (6.27). However, the infinite sum over n has operator ordering ambiguities (these are standard), and although the difference in energy between different levels is well defined, we still need to know the energy of the ground state. A derivation of the correct value of the energy for the ground state is beyond the scope of this paper. It is a standard calculation in string theory textbooks [13, 26]. For us, the important thing is that up to normalization constants

$$m^2 \sim (N-1). \tag{6.28}$$

The units of mass are determined by the string tension T and $\hbar$; the exact numbers depend on conventions. Here we are ignoring these and setting various coefficients to one.

The ground state is characterized by no oscillator excitations, $N = 0$. The mass squared of the associated particle is negative. Such a particle with negative mass squared is called a tachyon. In relativistic quantum field theories a tachyon signals an instability of the vacuum, not the presence of particles traveling faster than light. This means that the bosonic string is being expanded around an unstable vacuum and should not be used to describe nature, but instead as a model for strings in a slightly unphysical situation.

The next set of states correspond to $N = 1$. These are characterized by massless representations of the Lorentz group. Massless representations of a given spin have generally fewer polarizations than their massive counterparts. The missing polarizations usually signal some additional gauge symmetry of the system.

For the case at hand, the spectrum of massless particles gives rise to tensors with two indices that are transverse (more details can be found in [13, 26]) and a scalar field. These can be associated to spacetime tensors that describe the quantum fields that such particles generate. These are three different quantum fields

$$h_{\mu\nu}, B_{\mu\nu}, \phi, \tag{6.29}$$

a symmetric traceless two-tensor, an antisymmetric two-tensor potential and a scalar field. The field ϕ is called the dilaton, and its vacuum value describes the strength of the string interactions, $g_s \sim \exp(-\phi)$. The antisymmetric potential $B_{\mu\nu}$ is a generalized Maxwell field, with a gauge invariance of the type $B \to B + d\Lambda$, where Λ is a one form. The charged objects under B are the fundamental strings themselves. Finally, we have a symmetric traceless two-tensor, with a linearized gauge symmetry. This massless particle is of spin two. It is generally understood that consistent interactions of a massless spin two particle to matter require a conserved tensor to couple to. The symmetry requirements for such a consistent coupling lead to a generally covariant theory, and one then finds that the excitations associated to $h_{\mu\nu}$ are gravitational waves. A good discussion of these issues and their history can be found in [29]. In the case of string theory we obtain quanta of gravity in the flat background.

Standard calculations in string theory show that the conditions that make the string theory consistent quantum mechanically lead to Einstein's equations coupled to matter (represented by the dilaton and the antisymmetric tensor). In this way we find that the interacting quantum bosonic string requires (predicts) gravity. Beyond this analysis, for $N > 1$, we have a lot of different massive particles.

We are interested in what follows in understanding the low-energy limit of the string theory. This limit is characterized by only keeping the light particles (the massless particles in this case, as we have no other parameters that make some particles light compared to the others) and studying scattering process where the center-of-mass energy of the scattering problem is much smaller than the masses of the states that we are ignoring. This requirement guarantees that heavy particles cannot appear in the final products of the scattering process, as there is not enough energy to produce them. We therefore have to worry about a few states, and their interactions. These states, being massless, will not decay further once they are produced. This limit is a simplification that is justified by the fact that the scale characterizing fundamental string energies is high: we have not seen the characteristic spectrum of a string theory for fundamental particles in particle accelerators; thus in practice this is an experimental constraint.

The main technical problems with the bosonic string are two: first, we have the tachyon state, and this presents problems in that it is interpreted as an expansion around an unstable vacuum configuration that is largely irrelevant for physics. The second problem is with our understanding of the phenomenology of particle species. In nature it has been noted that particles can arise both in tensor (boson particles) and in spin representations of the Lorentz group (fermions). However, in the spectrum of the bosonic string there are no fermions. Thus the bosonic string fails to explain why an electron is both light and a fermion. In string theory at weak coupling the lightest particles that are not massless are string excitations.

A way to solve this problem of particle species was found by enlarging the symmetry group of the string worldsheet and requiring that it have supersymmetry. One could then also consistently remove the tachyon from the physical spectrum. The final result is a theory in ten spacetime dimensions (the extra symmetries of the string change the number from twenty-six to ten), and it has massless particles of spin 2 and massless particles of spin $3/2$. Their consistent coupling gives rise to a supergravity theory.

Beyond this, there are four closed supersymmetric string theories in ten dimensions, depending on various choices. We will be interested in the type II theories. These have two supersymmetries in ten dimensions (a total of 32 supercharges, transforming as two spinor representations of $SO(9, 1)$). Because supersymmetries in ten dimensions can be Majorana–Weyl, the supersymmetries carry a handedness (chirality). This has two possible values. The type IIA string theory has two supersymmetries of opposite chiralities, so it is not chiral. Its effective action is the dimensional reduction of eleven-dimensional supergravity on a circle. The type IIB string theory has two supersymmetries of the same chirality. The theory at low energies is given by type IIB supergravity.

Apart from the dilaton, the graviton and the antisymmetric $B_{\mu\nu}$ tensor potential, the type II theories also contain other generalized antisymmetric tensor potentials $A_{\mu\nu\ldots}$. These depend on the chirality of the string. For type IIA strings we get odd forms, and for type IIB string theory we get even forms:

$$\text{Type IIA:}\ \ A_\mu, A_{\mu\nu\rho}, A_{\mu\nu\rho\sigma\tau}, \dots, \tag{6.30}$$

$$\text{Type IIB:}\ \ A, A_{\mu\nu}, A_{\mu\nu\rho\sigma}, \dots. \tag{6.31}$$

These are also called Ramond–Ramond (RR) fields. The symmetry that these higher form fields have is similar to that of electromagnetism. Here, configurations where $A \to A + d\Lambda$ (in form notation) are to be considered equivalent by a gauge transformation.

Having these potentials suggests that there could be charged extended objects under their gauge symmetries. For such an extended object, their action would have a coupling of the form

$$\int_M A_*. \tag{6.32}$$

This generalizes an electric coupling of a particle of charge e to a potential $S_{em} \sim e\int A \sim e\int A_\mu dx^\mu$, where the integral is over the worldline trajectory of the particle: this is the pullback of the one-form A to the worldline of the particle. The generalization has us integrate the pullback of A to the worldvolume of the corresponding charged extended object. These possible extended objects are not the fundamental string and should be considered as nonperturbative excitations. Moreover, there is a self-duality constraint, so that the field strength of degree p is dual to the field strength of degree $10 - p$. Thus $F_p = dA_{p-1} \sim {}^*F_{10-p} = {}^*dA_{10-p-1}$. This is a situation where the objects charged under A_{p-1} are the electromagnetic duals of objects charged under A_{10-p-1}. For the special case $p = 4$, we have a self-dual field configuration.

One of the most important realizations of string theory is that such objects are tractable and can be studied in great detail [25]. They correspond to a very special class of defects called D-branes. We will now explore these objects.

6.4 Open strings and a stack of branes: setting up the AdS/CFT conjecture

The first thing we need to do is describe what a D-brane is. Technically, we think of a D-brane as a defect in spacetime. A D-brane modifies the dynamics of strings by allowing strings to end on it. Thus a D-brane begins its life as an abstract object that is described by the geometric locus where strings are allowed to end. At this level, it seems that we are adding these objects by hand to the theory of closed strings. We can add more than one D-brane to a given spacetime. Strings are allowed to have ends on different D-branes. The strings that have ends on a D-brane are called open strings. In order to have a simple setup, we require that the D-branes extend infinitely in time and that they do not move, and we also require that they be embedded as flat sheets in space (they are defined by hyperplanes containing time). This restriction is the same type of simplification associated to choosing flat space as the first place to start understanding strings. If we have more than one brane, we also require that they be parallel (and not moving with respect to each other).

Similarly to the case of closed strings, we can try to solve for the spectrum of the open strings, by using the same variational principle (the Polyakov action). This has to be supplemented with restricted variations at the boundary, so that the ends of the strings are located on the D-brane. The equations of motion of the string in the bulk are the same as before and can be solved similarly to those of the closed string. Our task is to find the relations between the coefficients of the formal solutions.

If a string has end points, we can choose these to be located at fixed values of the worldsheet coordinate σ that we introduced to find a mode expansion in Equation (6.26). The variation of the action given in the conformal gauge (Equation (6.17)) for a coordinate X^μ gives

$$\delta S \sim \int_R \nabla X^\mu \nabla \delta X^\mu \tag{6.33}$$

and needs to be integrated by parts in order to get unrestricted variations without derivatives. We then find that the variation to the action,

$$\delta S \sim \int_R (-\nabla^2 X^\mu)\delta X^\mu + \int_{\partial R} \delta X^\mu \cdot \nabla X^\mu, \tag{6.34}$$

has two contributions. The first one gives rise to the equations of motion. The second comes from the boundary of a region where we are doing the variation. We have two options to make the full variation of S vanish and to obtain a good variational principle. We can restrict ∇X^μ so that it vanishes, and therefore be allowed to treat δX^μ on the boundary as a free variation. This type of boundary condition for the preceding partial differential equation is a Neumann boundary condition. The other option is to restrict δX^μ so that it vanishes. This second boundary condition is a Dirichlet boundary condition. For the problem of the branes as described, we realize that the variations for the coordinates that lie along the branes have Neumann boundary conditions, whereas for those directions that are orthogonal to the brane we get Dirichlet boundary conditions (the end coordinates are fixed on the corresponding hyperplanes, so the corresponding $\delta X_\perp$ have to vanish; otherwise the end of the string would be separated from the brane). It is only when the branes correspond to flat hyperplanes that the boundary conditions are so simple.

Given these boundary conditions, it is straightforward to quantize the open string, and one finds a similar solution to the closed-string spectrum: In the open string the boundary conditions relate the coefficients α_n^μ and $\tilde{\alpha}_n^\mu$. If a string is suspended between two different branes at positions a, b, then the left value of a coordinate $X^\alpha = a$ is different from the value on the right end of the string, $X^\alpha = b$. This means that in the classical vacuum of the string there is a gradient, and this gives an additional contribution to the mass squared of the ground state that is proportional to $|a - b|^2$, namely

$$m^2 \sim \frac{1}{2}(N - 1) + \beta|a - b|^2. \tag{6.35}$$

The factor of $1/2$ in that equation is to indicate that the numbers one obtains for m^2 are half integers between the squared masses of the closed strings (these are integers in the convention of Equation (6.28)). The quantity β can be calculated precisely; we do not need the exact details.

Again, we notice that if the branes coincide, namely $a = b$, we have a tachyon state for $N = 0$. For $N = 1$ we now get a massless spin one particle A_μ. Just as the number of polarizations of a massless spin two tensor particle is smaller than for massive representations, so also for vector particles. These massless spin one particles must have a symmetry that removes the unphysical polarizations. This symmetry is the symmetry of a gauge theory. Thus the vector particle degrees of freedom should be interpreted as a connection on a principal bundle. The gauge theory for N D-branes is usually $U(N)$. The strings have labels that remind us on which D-branes they begin and end, and if we have N objects, the degrees of freedom can be accommodated in an $N \times N$ matrix.

Open strings interact by joining at their ends or by splitting at the D-brane. It is easy to show that the scattering amplitudes that one constructs for massless spin one particles coincide with those of the Yang–Mills perturbation theory in the low-energy limit (one only scatters massless particles with a small center-of-mass energy). Thus the effective action of the massless particles at low energies is identical to the Yang–Mills theory. The Yang–Mills theory is part of the fundamental structure of the standard model of particle physics. We see here that the D-branes, which are described as geometric objects, carry information about ordinary quantum field theories that we are interested in for describing physical phenomena that we observe in nature. Thus the D-brane setup geometrizes quantum field theories.

For example, we see that the D-brane positions must correspond to quantum fields. This is shown as follows: A D-brane in some location spontaneously breaks translation invariance in 26 dimensions. When one breaks a global symmetry, there is always an associated massless particle. Such particles are called Goldstone bosons. There is one such massless degree of freedom for each brane we have, because in perturbation theory the D-branes are treated as independent objects whose intrinsic degrees of freedom (strings with both ends on a given D-brane) are independent of the presence of the other D-branes.

When two D-branes coincide, we get extra massless particles, and they appear so that we get a 2×2 matrix of massless particles. However, from Equation (6.35) we notice that the vector particles become massive when the D-branes are separated. For this to happen, the vector particles need to acquire an extra polarization. This is provided by "eating" the Goldstone boson degrees of freedom for the off-diagonal entries of the matrix. This phenomenon is the standard description of spontaneous symmetry breaking in gauge theories: the Higgs mechanism. What we see from this example is that the Higgs mechanism can be understood geometrically, and that the masses of the associated W bosons (the massive off-diagonal vectors) can be computed by geometric manipulations.

Again, the presence of open string tachyons complicates the analysis, making the configuration around which we are expanding slightly unphysical. For the same reason as given previously, the analysis improves if we consider superstrings. For type IIA string theory, we find that the spectrum on a flat D-brane has no tachyons if the D-brane is a flat D0, D2, D4, D6 or D8 brane. A Dp-brane is a D-brane whose worldvolume has p spatial directions and one time direction. For the type IIB theory theory, the flat D-branes are tachyon free if we have D1, D3, D5, D7 or D9 branes. A pointlike D-instanton is also supersymmetric in type IIB string theory.

Indeed, the spectrum of particles on these special Dp-branes preserves half of the supersymmetries of the ten-dimensional type II theories in the absence of branes. Which half of the supersymmetries are preserved depends on the orientation of the brane. A brane with the opposite orientation will break the opposite half of the supersymmetries (we will call these antibranes). If we put a brane and an antibrane on top of each other, then there is a tachyon for the open strings stretching between them. If we put two D-branes of the same dimension on top of each other, we get no tachyon, and supersymmetry is preserved. Indeed, one can show that even if we separate a pair of parallel branes, they still preserve supersymmetry: the off-diagonal strings stretching between them also come in multiplets of the supersymmetry, as well as the interactions.

Thus, at the price of introducing supersymmetry, we can have a physical D-brane configuration that is stable and serves as a geometric toy model for a quantum field theory of particle physics. Complicating these configurations by using various branes, many people have tried to construct D-brane realizations of the standard model of particle physics (a recent review can be found in [10]).

6.4.1 The low-energy limit

Now, we want to take the low-energy limit of a collection of D3-branes in type IIB string theory stacked very near each other or on top of each other. We choose to work with D3-branes because those correspond to a field theory in four spacetime dimensions. This is the closest setup to physical reality in this simplified scenario.

We want to choose an energy scale E with respect to which we will measure all other energies. In natural units, this also corresponds to a unit of length $1/E$. We also have the string scale, which is characterized by the string tension T or the string scale ℓ_s, with $T \sim \ell_s^{-2}$. To take a low-energy limit, we want to consider processes at a center-of-mass energy of order E, and we want this to be small compared to the string scale. Namely, $E\ell_s = \epsilon \ll 1$. In the end we want to take the limit where $\epsilon \to 0$, if this is possible.

If the typical separation between the collection of branes is Δx, then the mass of the lowest-lying string modes between the branes is of order

$$m \sim \frac{|\Delta x|}{\ell_s^2}. \tag{6.36}$$

We want this quantity to be of the same order of magnitude as E. Thus, we should have that $|\Delta x| \sim E\ell_s^2 \sim \epsilon \ell_s$, so when we take $\epsilon \to 0$ we want to take the limit in such a way that the D-branes are very close to each other.

The other string states stretching between branes will have energies of order ℓ_s^{-1}, so they can be safely ignored, because they will not appear in the end products of any scattering process involving energies of order E.

We want to keep all of the open string states whose energy is allowed. This set includes only the lowest-lying states of the open strings – the massive W bosons, for example. We have a $U(N)$ Yang–Mills theory, and we have 16 supersymmetries (this corresponds to having four distinct spinor supersymmetries, also called $\mathcal{N} = 4$ supersymmetry in four dimensions). This tells us that the low-energy theory of open strings is characterized by the $\mathcal{N} = 4$ supersymmetric Yang–Mills theory, plus stringy corrections. These corrections have to be in the form of polynomials in powers of the energy, divided by the masses of the string particles. That is, the corrections are of order $\sum_{k>0} a_k(E\ell_s)^k$. Because we are taking ϵ to zero, in the low-energy limit the string corrections vanish, and the problem reduces to ordinary quantum field theory in four dimensions, with the requisite amounts of supersymmetry.

We now need to worry about the closed string states, because these also have massless particles (for example, the graviton). The closed string emission is also suppressed by powers of $E\ell_s = \epsilon$. This is typical of gravitational interactions, which are proportional to the energies of the objects that interact. So in the limit we are taking, the closed string degrees of freedom do not contribute, because it is extremely hard to produce them. When we take $\epsilon \to 0$, the production of closed strings goes to zero. We call such a type of limit a decoupling limit. We find in this way that the only degrees of freedom that survive are those of the ordinary $\mathcal{N} = 4$ SYM theory, with ordinary quantum field theory interactions.

The $\mathcal{N} = 4$ SYM theory is essentially unique [30]; thus, given the gauge group, the Lagrangian of the theory is characterized by a unique parameter: the coupling constant of the theory, g_{YM}. This is closely related to the open string coupling constant $g_{YM} \sim g_{open}$. The closed string coupling constant g_s is of order g_{open}^2.

The $\mathcal{N} = 4$ SYM theory is very special. It is superconformal [30] for any value of the coupling constant. This means that many quantities do not receive quantum field theory renormalizations, and the coupling constant of the theory does not run. This is often stated by saying that the beta function for the coupling constant vanishes ($\beta(g_{YM}) = 0$), or that the theory is at a fixed point of the renormalization group. If the distance between the branes is zero, then all the W bosons are massless, and the theory is nontrivial (the interaction strength does not vanish at small energies). This special vacuum has no scale associated with it, and corresponds to an exact superconformal vacuum. If the branes are separated, then there is a natural mass scale in the problem: the mass associated to the W bosons (the separation between the branes). The mass scale implies that superconformal invariance has been broken spontaneously.

6.4.2 A different point of view: gravity

We now want to think about this same system of D3-branes from a different point of view. We want to analyze it from the gravitational point of view. As we previously discussed, the D3-branes are defects in a theory of gravity. The fact that they are defects means that they should have a mass, or tension (energy per unit volume). The tension of the D3-branes is of order $1/g_s$ in string units [12, 25]. Thus the D3-branes weigh, and they bend spacetime. Because we know that we can put the D3-branes at finite distance from each other and that we preserve supersymmetry, this means that the D3-branes can stay at fixed distance from each other. In order to do this, the D3-branes must interact with other fields than just gravity (gravity is universally attractive), and the interactions must be carried by massless particles, because otherwise gravity would win at long distances. The other massless degrees of freedom are members of the supergravity theory, and the only option is the fields $A_{\mu\nu\rho\ldots}$. These are generalized electromagnetic fields. For parallel branes, these interactions must be repulsive, and they must balance gravity exactly. Thus, the branes are charged under these generalized potentials, and their charge must be related to their tension for this balance of forces to take place. If Q is the charge density and T the tension, we must have that

$$Q \simeq T. \tag{6.37}$$

Equalities between tension and charge of objects like these are usually a consequence of supersymmetry, and in this case they saturate an inequality of the form

$$T \geq Q. \tag{6.38}$$

These inequalities appear in two ways. In one form, they are a consequence of unitarity of the corresponding quantum mechanical theory in the presence of supersymmetry. The other way in which they appear is in the study of charged black holes that carry the corresponding charges. The inequality is then related to the absence of naked singularities in the gravitational theory.

Given that the D3-branes are massive, they deform the geometry of spacetime around them. This is because string theory is a theory of gravity, and massive objects necessarily distort the geometry of spacetime, as described by Einstein's equations for the corresponding theory.

Indeed, we can solve for the metric of the full ten-dimensional spacetime metric in the presence of the branes. This was first done in [17]. To understand the metric, we notice that we have rotational invariance in the directions transverse to the D-brane. Moreover, we have Lorentz invariance in the directions parallel to the D-brane. We can choose a coordinate system in spherical coordinates in the transverse directions, and a flat system of coordinates along the D-brane, so that everything can be written in terms of a radial variable alone. We define the radial variable so that the radial plus angular coordinates appear in the metric in the form $g(r)(dr^2 + r^2 d\Omega_5^2)$. Such a coordinate choice is always possible. The full solution for all the branes on top of each other is

$$ds^2 \sim f^{-1/2} dx_{||}^2 + f^{1/2}(dr^2 + r^2 d\Omega_5^2), \tag{6.39}$$

where

$$f = 1 + \frac{4\pi (N/g_s) g_s^2}{r^4} \sim 1 + \frac{4\pi N g_s}{r^4}. \tag{6.40}$$

The object f carries the information of the tension of the brane in its asymptotic form for $r \to \infty$. The mass density is of order N/g_s, whereas the gravitational constant is of order of the closed string coupling constant squared, g_s^2. Thus, $N g_s$ is the gravitational constant times the mass density of the D-branes.

The solution is characterized by a constant dilaton field (the strength of the string interactions does not depend on the location in the full spacetime). Thus, even near the brane, where spacetime is substantially curved, closed strings are weakly coupled in the full geometry.

The solution written down in [17] is similar to a black hole, but there is no horizon. It appears as an extremal limit of charged-brane black hole solutions. Also, the apparent singularity at $r = 0$ turns out to be a coordinate singularity and not a metric singularity.

We can actually see from the metric that the locus $r = 0$ is at infinite distance (we count this distance at constant-time slices in $x_{||}$). Thus the length of a curve starting at r and going to zero is given by

$$\lim_{a \to 0} \int_a^r f^{1/4}(r) dr \sim \lim_{a \to 0} \int_a \frac{dr}{r} \sim \lim_{a \to 0} \log(a) = \infty. \tag{6.41}$$

We also have a similar property to a horizon: In the region near $r = 0$ one can have arbitrarily large red shifts: we compare how proper time ticks at r with how it ticks at infinity. This is done by examining the ratio of $g_{tt}^{1/2}$ between r and infinity, which when r is small scales as

$$R(r) = \frac{f^{1/4}(r)}{f^{1/4}(\infty)} \sim \frac{1}{r}. \tag{6.42}$$

This is the amount of time that one second of proper time at r corresponds to in global time (the proper time at infinity). Thus, near the brane, time slows down arbitrarily, just as in the region near event horizons for a Schwarzschild black hole.

Because of large red shifts, placing a massive particle of mass m at rest near $r = 0$ will add a total amount of energy to the gravitational system of order $m/R(r)$. Thus, even for very massive states – let us say, heavy closed strings – if we place them sufficiently close to the horizon, the amount of energy that we have added to the system can be made arbitrarily small. Moreover, these particles cannot escape to infinity, as at infinity their energy would be of order m. We find that heavy massive particles very near the branes are bound to them and carry little energy. As a matter of fact, we can place the massive strings at sufficiently small values of r so that the energy of the massive string is less than E, as described in the previous subsection.

Thus, from the gravitational point of view we have a way to get a low-energy limit of physics near the stack of branes that looks completely different than the low-energy limit of the open strings.

We can expand the metric near $r = 0$, by noticing that in the function f, the term 1 becomes irrelevant. Thus, near $r \sim 0$, we can replace f by $4\pi N g_s/r^4$. With this substitution the metric becomes

$$ds^2 \sim R^2 \left(\frac{r^2}{R^4} dx_{||}^2 + \frac{dr^2}{r^2} + d\Omega_5^2 \right), \tag{6.43}$$

where R is given by $\sqrt{4\pi g_s N}$. If we introduce the variable $y = R^2/r$, we find that the metric takes the form

$$ds^2 = R^2 \left(\frac{dx_{||}^2 + dy^2}{y^2} + d\Omega_5^2 \right), \tag{6.44}$$

where R measures the radius of the metric in string units. The end metric is simple, and it is the product of a five-sphere of constant size and a homogeneous space, which is called five-dimensional anti-de Sitter space, AdS_5, in the physics literature. It is the analytic continuation of the homogeneous hyperbolic space of five dimensions to Lorentzian signature. The near-horizon metric of the D-brane system is thus $AdS_5 \times S^5$.

So far we have described the low-energy physics of the branes in two different ways: massless open strings stretching between the branes, and closed strings in the near-horizon geometry. The AdS/CFT conjecture [20] is that these two points of view on what constitutes the low-energy physics of the branes are completely equivalent to each other as quantum systems. When the conjecture was formulated, the evidence presented in favor of it was mainly the fact that one could calculate the absorption of massless closed strings on the D-branes by studying the production of open string states, and that this computation gave the same answer as the absorption of massless closed strings by the deformed geometry [14, 19].

Moreover, in [20] it was shown that both systems, the near-horizon geometry and the field theory, have the same symmetries: the isometries of the metric are identical to the global symmetries of the field theory. This is a requirement for the equivalence to be able to hold. I will explain this in more detail later on.

We now want to understand when to expect the gravity approximation to be good. This requires that strings interact weakly, so that the closed string coupling is tiny ($g_s \ll 1$). We also want the radius of curvature of the metric to be large in string units. Thus R should be large. Because $R \sim \sqrt{4\pi g_s N}$, we need $g_s N$ to be large. Because $g_s \sim g_{YM}^2$, in the field theory this requires that the quantity g_{YM} be small, but that the quantity $\lambda \sim g_{YM}^2 N$ be very large. Thus, N should be taken to be very large for gravitational physics to dominate the physics.

We see the appearance of a large number N of colors, and of the quantity $\lambda \sim g_{YM}^2 N$, also called the 't Hooft coupling constant [28], which needs to be large. The expansion around $N = \infty$ was introduced by 't Hooft by noting that the Feynman diagrams of all these theories are similar. He found that, if one carefully kept track of how they differ, the perturbation series for particles physics amplitudes looked like

$$A \sim \sum a_{n,g} (g_{YM}^2 N)^n (N^{-2g}), \tag{6.45}$$

where each graph could be interpreted as a skeleton of a Riemann surface. The quantity $2g$ is the genus of the corresponding surface, and n roughly counts the number of vertices in the diagram. For n large, we get fine meshings of a Riemann surface, and this suggests that if we can perform the sum over n exactly, at strong coupling the theory is dominated by infinite n. That is, the skeleton can be expected to form a continuous mesh on the Riemann surface. This is similar to how one would imagine a string theory to look, where the string coupling constant is roughly $g \sim 1/N^2$.

The strings of 't Hooft in the AdS/CFT proposal are supposed to be exactly the fundamental type IIB string in higher dimensions. The AdS/CFT setup was the first case where a precise string theory dual to a four-dimensional gauge theory was found, in the sense of 't Hooft. This knowledge makes the proposal more plausible.

6.5 Making sense of it all: observables in CFT and gravity

Now that we understand the basic premise of the AdS/CFT correspondence conjecture, we would like to make calculations both in field theory and in the string theory and to set up a comparison between them so that we can verify the conjecture. We need to ask about the nature of physical observables on both sides of the correspondence.

Let us begin with the field theory. The typical calculation that one does in field theory is to describe a scattering process. This is usually done by calculating the S-matrix of the theory: the probability amplitude for an initial state of m particles that are far apart from each other in the infinite past to evolve to a configuration of n particles that are far apart from each other in the infinite future.

However, this object does not exist in an interacting conformal field theory. The reason is that conformal field theories are theories of massless matter. In essence, one can radiate ultra-small amounts of energy on a single particle, and this would escape our detection in the laboratory. Such processes lead to infrared divergences from soft gluon and collinear amplitudes.

It is convenient to formulate the problem of observables in the conformal field theory in a different way. The idea is first to do a Wick rotation on a Euclidean field theory, after which we would analyze this theory by studying the Green's functions of allowed operators.

The simplest such operators are local gauge-invariant operators $\mathcal{O}(x)$. Thus, one would want to compute general quantum averages

$$\langle \mathcal{O}_1(x_1) \cdots \mathcal{O}_n(x_n) \rangle, \tag{6.46}$$

where the averages are analyzed on the vacuum state. The infrared divergences are eliminated because the operators are at finite distance from each other.

One would want to have a complete set of such Green's functions define the theory. One usually expresses this collection of numbers via a generating functional for Green's functions,

$$Z[\alpha] \sim \sum \frac{1}{a_1!} \left(\int \alpha_1(x) \mathcal{O}_1(x) \right)^{a_1} \cdots \frac{1}{a_k!} \left(\int \mathcal{O}_k(x_k) \right)^{a_k} \tag{6.47}$$

$$= \exp\left(\sum_k \int \alpha_k \mathcal{O}_k(x) \right). \tag{6.48}$$

This is to be interpreted as a formal series. The correlations are recovered by differentiation of Z with respect to α. The α variables are sources for the operators, and they can be thought of as generating infinitesimal deformations of the theory by the operators $\mathcal{O}$. Most of these cannot be integrated to a function.

In the gravitational theory, we have a different problem. The theory of gravity is invariant under general changes of coordinates. If we have a quantum state where many geometries are superposed, there is no invariant meaning of a position: there is no unique way to compare coordinates between different geometries. Thus, we require that the observables be diffeomorphism invariant.

Diffeomorphism-invariant observables are typically nonlocal – for example, integrals of densities over the whole space and time. These numbers can be compared between different geometries with ease.

In the classical limit, we would want to analyze semiclassical solutions to the gravity equations of motion against the $AdS_5 \times S^5$ background. Because we want to compare with the Euclidean quantum theory, we would need to solve the supergravity equation for the Euclidean signature. Thus, the AdS space becomes a global hyperbolic space (this can be represented as the Poincaré disk or the upper half plane in higher dimensions). Generally, one solves some elliptic differential equation problem on the space. If one insists that variations vanish at the boundary of the disk (that the solution is normalizable), one finds that there are no such solutions. Thus one needs to relax the requirement that the field variations in gravity vanish at the boundary of the disk. This can also be interpreted as having sources for gravitational fields on the boundary.

Indeed, the boundary is infinitely far away from any point of the interior. To bring infinity to finite distance, we need to perform a conformal rescaling of the metric. If we do this with the five-sphere as well, we find that the S^5 will shrink to zero size at the boundary. Thus the conformal boundary of Euclidean $AdS_5 \times S^5$ is a four-sphere (or $\mathbb{R}^4$). This is the same topological space as the space where we insert sources in the field theory.

Also, it is known that to leading order the closed strings couple to the field theory (open string) degrees of freedom via a disk diagram. This is approximately described by a local operator insertion at the position where the closed string is located. Thus, it seems natural to identify the sources in the field theory partition function with the boundary conditions of the gravity.

The proposed dictionary between the field theory and the gravitational dual is that [15, 32]

$$Z[\alpha] = Z[g]|_{\delta g(\infty)\sim\alpha}. \tag{6.49}$$

The left-hand side is the generating functional of correlators in the field theory. The right-hand side is the partition function of gravity subject to the modified boundary conditions. This second formulation is similar to an S-matrix: at infinity all points are at infinite distance from one another, so boundary conditions at different points in the boundary are essentially decoupled: they do not interact with each other. We still need to set up a precise dictionary between the boundary conditions and the sources if we want to compare the two sides.

This is what we hope to match precisely between the conformal field theory side and the gravitational side. Notice that the gravitational side is an on-shell description: we only allow configurations that satisfy the (possibly quantum-corrected) equations of motion of gravity.

6.6 The operator state correspondence, the superconformal group and unitary representations

So far I have described the observables in the conformal field theory in a very abstract form. If one is to have a quantum mechanical equivalence between gravity and the conformal field theory, one should also have an equivalence of the Hilbert space of states between the two formulations. Such Hilbert spaces are characterized by the unitary representations of the symmetry algebra. In essence, to test the correspondence we would need to show that the representations of the superconformal algebra are in one-to-one correspondence between the gravity side and the field theory side.

However, we have set up the problem in the conformal field theory in terms of operator insertions, a very algebraic point of view, rather than in terms of the Hilbert space of states of the conformal field theory. Fortunately, these two points of view are equivalent.

The main idea is that an operator insertion at the origin $\mathcal{O}(0)$ at the classical level acts as a source for the fields. That is, solutions of the field equations will have characteristic singularities as they approach the origin.

We can choose a radial coordinate system centered at $x = 0$. In such a coordinate system we have that

$$ds^2 = dr^2 + r^2 d\Omega_3^2 = r^2 \left(\frac{dr^2}{r^2} + d\Omega_3^2 \right), \tag{6.50}$$

where $d\Omega_3$ represents the full set of angular variables in four dimensions. Notice that the metric on flat space minus the origin is related by a rescaling to the metric on $S^3 \times \mathbb{R}$, if we define the Euclidean time $\tau = \log(r)$. The rescaled metric is

$$d\tau^2 + d\Omega_3^2. \tag{6.51}$$

Because we have a conformal field theory, the theory should only depend on the conformal class of the metric. When making a Weyl rescaling, we should be able to relate the theory on a background determined by some metric and its rescaled counterpart.

In doing this, the singularity of the solution of the equations of motion at the origin gets sent to a singularity of the solution of the equations of motion at $\tau \to -\infty$. This is, we have a boundary condition in the "infinite Euclidean past." In the realm of Euclidean field theories, this is just the definition of a quantum state of the system: a set of initial conditions. We can regulate this by describing the fields at finite τ and evolving the equations of motion from then on.

It is convenient to perform an analytic continuation $\tau = it$. In this situation, we have the Lorentzian metric

$$ds^2 = -dt^2 + d\Omega_3^2. \tag{6.52}$$

In this case, evolution in t is ordinary time evolution in quantum mechanics, and specifying the fields at some time t_0 is the classical description of a quantum state: a configuration of initial conditions. This isomorphism, which lets us think of local operator insertions as a set of initial conditions (that is, as a quantum state), is called the operator–state correspondence.

Also, if we consider the time evolution generated by ∂_t, this is related to the evolution according to the vector field $r\partial_r$. That is, the geometric interpretation of time evolution in t is the same as the evolution of an operator under rescalings of r. The problem we are interested in is how operator insertions behave under scale transformations.

Requiring these two evolutions to be related means that at the level of evolution equations

$$\partial_t |s\rangle = H|s\rangle = E_s|s\rangle, \tag{6.53}$$

$$r\partial_r \mathcal{O}_s = [\Delta, \mathcal{O}_s] = \delta_s \mathcal{O}_s \tag{6.54}$$

(that is, for evolution according to the Schrödinger equation), where we have states that are eigenvectors of the Hamiltonian with energy eigenvalues E_s, the latter need to be identified with the eigenvalues of the Heisenberg evolution of the corresponding operator under rescalings. Thus the dimension of the operator is identified with the energy of the state.

For this problem, it is convenient to give a description of the superconformal algebra for $\mathcal{N} = 4$ supersymmetries. We will classify the generators according to how they scale. The generators are given by the following diamond:

$$\begin{array}{ccccc} & & K_\mu & & \\ & S_\alpha & & \bar{S}_{\dot\alpha} & \\ \Delta & & M_{ij} & & R_{ab} \\ & Q_\alpha & & \bar{Q}_{\dot\alpha} & \\ & & P_\mu & & \end{array} \tag{6.55}$$

Here Δ is the generator of dilatations. P is the generator of translations in flat $\mathbb{R}^4$. The K are called special conformal generators. The $Q, \bar{Q}$ are the generators of supersymmetry, and the $S, \bar{S}$ are the special supersymmetry generators. They can be obtained from commutators of K with Q. Finally, M_{ij} are the rotations of the sphere, S^3. The generators R are global symmetries of the theory. They are usually called the R-charge of the theory. They generate an $SO(6) \sim SU(4)$ Lie algebra, and they correspond to a global symmetry of the field theory.

The commutator $[\Delta, X] = \alpha X$ determines the weight of X, α. This is usually called either the scaling dimension or the conformal weight of X. In the preceding paragraph, P has conformal weight 1 (scaling dimension 1), and $Q, \bar{Q}$ have conformal weight $1/2$, and Δ, $M_{ij} R_{ij}$ have conformal weight 0. In the field theory, Δ generates time translations and is therefore identified with the Hamiltonian for the theory on $S^3 \times \mathbb{R}$. It is a Hermitian operator. Taking complex conjugates, it is easy to see that $P^\dagger$ should have the opposite weight to P. There are two cases to consider.

If α is real, $P^\dagger$ has weight -1. Therefore, we have to make the following identifications in order to have a unitary representation of the algebra: we need that $K \sim P^\dagger$ and that $Q^\dagger \sim S$, $\bar{Q}^\dagger \sim \bar{S}$.

If instead we take the algebra on $\mathbb{R}^{3,1}$, then we take K, P, Δ to be Hermitian, and α is imaginary. $P^\dagger$ ends up having weight $-\bar{\alpha} = \alpha$. In this case $Q^\dagger \sim \bar{Q}$.

In the first case $P, Q, \bar{Q}$ act as some form of raising operator, and $K, S, \bar{S}$ act as lowering operators with respect to the Hamiltonian. If we have a well-defined ground state, it usually has minimal energy. Thus, the representation theory of the symmetry algebra will require that all states have positive energy.

Taking any allowed state and acting on it with lowering operators eventually gives us a lowest weight state that is annihilated by $S, \bar{S}, K$. Such states are called primary states. The full set of unitary representations are characterized by the primary states.

The Hilbert space is an infinite sum of these representations. If we want to match the field theory and the gravity theory, it is enough to show that we get the same primary states. Now, we need to find the corresponding Hilbert space on the gravitational side.

On that side we are not going to work on Euclidean space anymore. The correct coordinate system for the gravitational setup corresponds to global AdS coordinates. The Euclidean time is analytically continued to global time. The metric of AdS_5 in global coordinates is given by

$$ds^2 = -\cosh^2(\rho)dt^2 + d\rho^2 + \sinh^2(\rho)d\Omega_3^2. \tag{6.56}$$

We see that as we take $\rho \to \infty$, we can rescale the metric by $\exp(-2\rho)$, and in the boundary we get the metric of $S^3 \times \mathbb{R}$. The t variable in gravity corresponds to global time, and it is identified with the same t that we have in the conformal field theory description.

One should notice that it takes a finite amount of time for a light ray to go to the boundary and back. Thus anti-de Sitter space requires adding boundary conditions for the fields at infinity. This is the Lorentzian version of the problem in Euclidean gravity where we have to fix boundary conditions to describe the gravitational partition function.

Once we work in this Lorenzian setup, Einstein's equations are of a wave equation type. As opposed to the Euclidean setup, we can now have normalizable gravitational wave solutions of Einstein's equation propagating on $AdS_5 \times S^5$. This does not require singularities in the infinite past or infinite future, so long as we choose solutions with finite energy. The classification of linearized perturbations was done in [16, 18].

We can now at least attempt to match some representations between the gravity side of the correspondence and the quantum field theory. This will be the subject of the next section.

6.7 Matching of BPS representations

So far we have seen that both sides of the correspondence are characterized by a list of representations of the symmetry algebra. Unfortunately, the standard calculational tools of perturbation theory do not usually extrapolate from weak to strong coupling.

There are special classes of states where this extrapolation is more likely to be successful, and they are the cases where corrections vanish. In the case of supersymmetric theories, there are special collections of states where vanishing theorems prevent corrections from happening. The states that are protected by supersymmetry are characterized by saturating some inequality between various quantum numbers. This inequality is usually called the BPS inequality, after Bogomol'nyi, Prasad and Sommerfield. Such inequalities involve the energy and some conserved charge.

For the case that we are interested in, we want to find the states that preserve some large fraction of the supersymmetries. In particular, we want to study supersymmetric primary states as already defined, and match them. The primaries are characterized by being annihilated by S, K (the lowering operators cannot find a state of lower energy).

However, we also have to consider the Q, $\bar{Q}$ operators. If half of these annihilate a state, it is called a $1/2$ BPS state. Such states are the ones that we want to consider.

The BPS inequality will be obtained by requiring that $Q|s\rangle$ have positive norm. If it has zero norm, then it is annihilated by the corresponding Q operator. Because S is the adjoint of Q, we can use the commutation relations of the symmetry group to find that

$$\Delta \geq J, \tag{6.57}$$

where J is an $SO(2)$ generator of $SO(6)$ which leaves an unbroken $SO(4)$ symmetry.

So far, we have dealt with this problem abstractly. Now, we need to finally write down the degrees of freedom of $\mathcal{N} = 4$ SYM theory, so that we can understand what this means.

The field theory consists of gauge fields A_μ and their supersymmetric partners. These are four Weyl fermions ψ, and six real scalar fields ϕ. The fermions transform as a spinor representation of $SO(6)$, and the scalars transform as the defining representation of $SO(6)$, a vector.

In free field theory the scalar fields and the vector potential have dimension 1, whereas the fermions have dimension $3/2$. Because we will be analyzing the field theory in the free field limit, we will need to make use of this information to understand what is going on.

Finally, we need to put the field theory on an $S^3 \times \mathbb{R}$ geometry, as we have described. The action of the field theory in flat space (for the scalars ϕ) will be given by

$$\mathcal{S} \sim \frac{1}{2} \int d^4x \, \partial_\nu \phi^i \partial_\mu \phi^i \eta^{\mu\nu} + \text{interactions}. \tag{6.58}$$

When we do conformal transformations of the metric η, so that we have the metric on $S^3 \times \mathbb{R}$, we should generalize $\eta^{\mu\nu}$ to a general metric, so that we have

$$\mathcal{S} \sim \frac{1}{2} \int d^4x \sqrt{-g} \, \partial_\nu \phi^i \partial_\mu \phi^i g^{\mu\nu} + \text{interactions}. \tag{6.59}$$

However, under a Weyl rescaling of the metric $g \to \exp(2\sigma(x))g$ we find that in order for the action to be invariant, ϕ must transform as $\phi(x) \to \exp(-\sigma(x))\phi(x)$. From here, it follows that after this is done there are derivatives of σ appearing in the Lagrangian, and the action is not conformally invariant. It needs a correction to absorb these derivatives of σ. The correction is of the form

$$a \int \sqrt{-g} R (\phi^i)^2, \tag{6.60}$$

where R is the Ricci scalar curvature of the background metric. The particular value of a that one computes is a nonminimal coupling to a metric, which is called a conformally coupled scalar.

If we choose a sphere S^3 of radius one, we find that the full quadratic term in the action is

$$S = \frac{1}{2} \int_{S^3} dt [(\dot{\phi}^i)^2 - (\nabla \phi^i)^2 - (\phi^i)^2]. \tag{6.61}$$

This free field action is straightforward to quantize. We decompose the field ϕ into eigenvalues of the Laplacian of the three-sphere. For each such *spherical harmonic* one has an associated quantum oscillator. The energy of a quantum of each such oscillator will be equal to the frequency. All quantum states in the field theory can be represented by occupying finitely many oscillators. This is a Fock space.

The lowest-lying mode has a constant profile on the sphere. That mode has energy one. The higher spherical harmonics will end up having energy $n + 1$, where n is a nonzero integer.

With respect to the $SO(2) \subset SO(6)$ charge, four of the scalars are neutral, and two complex combinations $Z = \phi^1 + i\phi^2$, $\bar{Z} = \phi^1 - i\phi^2$ have charge $+1$ and -1, respectively.

We can do the same with the spinors and the gauge fields. The energies of the spinors will be $n + 3/2$, and those of the gauge fields we be $2 + n$, where n is a nonnegative integer.

If we want to saturate the inequality that the energy is equal to the R-charge, then the only degrees of freedom that have that property are the oscillators for the field Z, and not for $\bar{Z}$. These oscillators are matrix valued, because Z is an $N \times N$ matrix. We will call the corresponding raising operator $a_Z^\dagger \sim (a^\dagger)^i_j$.

We are still left with the problem of gauge invariance. In particular, we need to guarantee that the states are neutral with respect to gauge transformations. That is, we need to satisfy Gauss's law for the $SU(N)$ charges.

A generic state built by applying k oscillators $(a^\dagger)^i_j$ will transform into some complicated representation of $SU(N)$ with k upper indices and k lower indices. In order to build a gauge invariant, we need to contract these objects with some invariant tensor of $SU(N)$. There are only two algebraically independent such tensors: δ^i_j and $\epsilon_{a_1 \ldots a_N}$. The first one tells us that upper indices and lower indices transform as opposite representations. The ϵ tensor follows because the Lie algebra of $SU(N)$ preserves a complexified volume form (a holomorphic $(n,0)$-form).

If we have two ϵ tensors, one with upper and one with lower indices, they can be written in terms of δ^i_j, so all gauge-invariant states must follow from contractions between upper and lower indices. These can always be written as products of traces, by noticing that contractions between an upper index on one $a^\dagger$ and a lower index on another $a^\dagger$ can be written naturally in terms of matrix multiplication. Thus, the complete list of such states is

$$|s_1, \ldots s_k\rangle = \text{Tr}(a^\dagger)^{s_1} \ldots \text{Tr}((a^\dagger)^k)^{s_k}|0\rangle. \tag{6.62}$$

Their energies are $s_1 + 2s_2 + \cdots + ks_k$. This counting of energies looks the same as the energies one would have in a system with many quantum oscillators, of energies $1, 2, \ldots, k$ and occupation numbers $s_1, s_2, \ldots, s_k$.

One can check that this is approximately correct. Indeed, as long as we keep the total energy finite, and we take N large, toward infinity, the spectrum of states becomes an approximate Fock space where each trace is an independent oscillator. This means that the states with different occupation numbers become orthogonal in the limit when $N \to \infty$ (this has been discussed in some detail in [4]). The main tool one uses is the 't Hooft planar diagram counting [28]: one finds that the overlaps are suppressed by powers of $1/N$, but they depend also on the energies of the oscillators, so the approximation to a Fock space is worse as we take k large.

Indeed, if k is larger than N, we find that $\text{Tr}(Z^k)$ can be written as linear combinations of traces of powers of Z that are less than or equal to N, so one even fails to create new states by considering more complicated traces. However, one can find a complete set of orthogonal states [11].

All of these states that preserve this number of supersymmetries (half of them) are protected (preserved) in the passage from weak to strong coupling. This means that the counting of states and their energies will not change when we introduce interactions. This is a property of the half BPS unitary representations of the supersymmetry group: that they cannot combine with other unitary representations to form a generic non-supersymmetric representation. If one preserves fewer supersymmetries in the free field limit, this is not true any longer once one turns on the interactions.

On the gravity side, we can consider the spectrum of single-particle excitations that have $E = J$. The energy of the state is related to the time evolution in global coordinates. The J charge is related to motion on the sphere, as the $SO(6)$ symmetry group is the set of isometries of the five-sphere.

Here we find that for the BPS particles, their energy in AdS_5 is equal to their momentum on the S^5 sphere. In the limit where these particles are pointlike, we can interpret them in terms of geodesic motion on the full geometry. We find that for these particles the energy is equal to the momentum, so they are massless particles in ten dimensions and they must follow null geodesics.

In the type IIB gravity theory, all of the massless particles belong to the gravity multiplet. Moreover, all null geodesics on $AdS_5 \times S^5$ are conjugate to each other. What we have to show is that we can reproduce the integer label k. Semiclassically, the point particles will be moving on a great circle in S^5. This orbit is periodic, and therefore one should have some semiclassical quantization condition for quantization of angular momentum. This is roughly stated by saying that the phase change of the one-particle wave function on going around the sphere is a multiple of 2π. This restriction implies that the momentum is classified by an integer.

Thus, for each integer k, there is one such type of particles, and because these particles are bosons in gravity, we obtain a harmonic oscillator for each integer k. If we are careful with the spin of the particles, we end up with an exact match of the approximate Fock space of the field theory when the occupation numbers are small [32]. The $1/N$ corrections are interpreted as gravitational corrections in the ten-dimensional geometry, and it results that these gravitons interact with each other.

In essence, one can prove that this is enough to reproduce the spectrum of linearized fluctuations of supergravity on $AdS_5 \times S^5$ as computed in [16, 18].

6.8 Recent developments

Since the review [1] was written, the AdS/CFT ideas have developed much further, particularly in applying gravitational techniques to the study of various physical questions of field theories at strong coupling. These developments are too numerous to describe in detail. I will instead just describe some of the recent progress towards understanding the quantum field theory dynamics at strong coupling.

The first development for a truly stringy test of the AdS/CFT correspondence beyond supergravity arose in the work [8], where it was shown how to recover the spectrum of strings in a particular limit. Geometrically, the limit that one considers is that of a short string that is moving with high angular momentum on S^5. This particular class of string states stay close to a null geodesic, so one can expand the geometry around this null geodesic in a systematic way. This expansion leads to a plane wave background [9], and it was shown in [21] that the spectrum of the string on the maximally supersymmetric plane wave background can be solved for exactly. The work [8] showed how to recover the spectra of these strings in the field theory dual and gave a precise map for building the different string states. The field theory showed that the string was discretized to some type of space lattice, whereas the time was kept continuous.

This was further generalized in [22], where it was shown that the energies of single trace states for a particular scalar sector could be interpreted in terms of a spin chain model, and that moreover this model was integrable to one-loop order on the worldsheet. On the other hand, an integrable structure was also found when one studied classical string motion in $AdS_5 \times S^5$, in the work [3]. It was later shown that the full one-loop spectrum of planar energies was integrable [2]. Having integrability at weak and strong coupling was completely unexpected, and it is currently believed that one can extrapolate from weak to strong coupling by a one-parameter family of quantum integrable models. A review of some the literature on this subject can be found in [24]. This is now a rapidly advancing field, and the results that are being found look promising and have provided highly nontrivial tests that seem to indicate that this integrability property is true. The integrability property permits us in principle to give exact answers for the spectra of strings as a function of the coupling constant. Unfortunately, this only seems to work for the strings on $AdS_5 \times S^5$.

A different development has been realized in the work [5]. It has been shown there that if one tries to do an expansion at strong coupling of the field theory dynamics, then one can start by reconstructing the geometry of the S^5 in the field theory, and one has a way to calculate the emergence of geometry and strings from a very different point of view than the one that is suggested by integrability. Moreover, these techniques generalize to many other gauge–gravity dualities [6, 7] and seem to provide a handle on the strong coupling expansion of many different field theories with a unified framework. These works are giving powerful tests of the AdS/CFT correspondence and are complementary to the integrability program.

Bibliography

[1] O. Aharony, S. S. Gubser, J. M. Maldacena, H. Ooguri and Y. Oz, "Large *N* field theories, string theory and gravity," Phys. Rept. **323**, 183 (2000) [arXiv:hep-th/9905111].

[2] N. Beisert and M. Staudacher, Nucl. Phys. B **670**, 439 (2003) [arXiv:hep-th/0307042].

[3] I. Bena, J. Polchinski and R. Roiban, "Hidden symmetries of the AdS(5) × S**5 superstring," Phys. Rev. D **69**, 046002 (2004) [arXiv:hep-th/0305116].

[4] D. Berenstein, "Shape and holography: Studies of dual operators to giant gravitons," Nucl. Phys. B **675**, 179 (2003) [arXiv:hep-th/0306090].

[5] D. Berenstein, "Large *N* BPS states and emergent quantum gravity," JHEP **0601**, 125 (2006) [arXiv:hep-th/0507203].

[6] D. Berenstein, "Strings on conifolds from strong coupling dynamics, part I," arXiv:0710.2086 [hep-th].

[7] D. E. Berenstein and S. A. Hartnoll, "Strings on conifolds from strong coupling dynamics: quantitative results," arXiv:0711.3026 [hep-th].

[8] D. E. Berenstein, J. M. Maldacena and H. S. Nastase, "Strings in flat space and pp waves from $N = 4$ super Yang Mills," JHEP **0204**, 013 (2002) [arXiv:hep-th/0202021].
[9] M. Blau, J. Figueroa-O'Farrill, C. Hull and G. Papadopoulos, "Penrose limits and maximal supersymmetry," Class. Quant. Grav. **19**, L87 (2002) [arXiv:hep-th/0201081].
[10] R. Blumenhagen, M. Cvetic, P. Langacker and G. Shiu, "Toward realistic intersecting D-brane models," Ann. Rev. Nucl. Part. Sci. **55**, 71 (2005) [arXiv:hep-th/0502005].
[11] S. Corley, A. Jevicki and S. Ramgoolam, "Exact correlators of giant gravitons from dual $N = 4$ SYM theory," Adv. Theor. Math. Phys. **5**, 809 (2002) [arXiv:hep-th/0111222].
[12] J. Dai, R. G. Leigh and J. Polchinski, "New connections between string theories," Mod. Phys. Lett. A **4**, 2073 (1989).
[13] M. B. Green, J. H. Schwarz and E. Witten, "Superstring Theory. Vol. 1: Introduction," *Cambridge, UK: Cambridge Univ. Press (1987) (Cambridge Monographs on Mathematical Physics).*
[14] S. S. Gubser and I. R. Klebanov, "Absorption by branes and Schwinger terms in the world volume theory," Phys. Lett. B **413**, 41 (1997) [arXiv:hep-th/9708005].
[15] S. S. Gubser, I. R. Klebanov and A. M. Polyakov, "Gauge theory correlators from non-critical string theory," Phys. Lett. B **428**, 105 (1998) [arXiv:hep-th/9802109].
[16] M. Gunaydin and N. Marcus, "The spectrum of the S**5 compactification of the chiral $N = 2$, $D = 10$ supergravity and the unitary supermultiplets of U(2, 2/4)," Class. Quant. Grav. **2**, L11 (1985).
[17] G. T. Horowitz and A. Strominger, "Black strings and P-branes," Nucl. Phys. B **360**, 197 (1991).
[18] H. J. Kim, L. J. Romans and P. van Nieuwenhuizen, "The mass spectrum of chiral $N = 2$ $D = 10$ supergravity on S**5," Phys. Rev. D **32**, 389 (1985).
[19] I. R. Klebanov, "World-volume approach to absorption by non-dilatonic branes," Nucl. Phys. B **496**, 231 (1997) [arXiv:hep-th/9702076].
[20] J. M. Maldacena, "The large N limit of superconformal field theories and supergravity," Adv. Theor. Math. Phys. **2**, 231 (1998) [Int. J. Theor. Phys. **38**, 1113 (1999)] [arXiv:hep-th/9711200].
[21] R. R. Metsaev, "Type IIB Green–Schwarz superstring in plane wave Ramond–Ramond background," Nucl. Phys. B **625**, 70 (2002) [arXiv:hep-th/0112044].
[22] J. A. Minahan and K. Zarembo, "The Bethe-ansatz for $N = 4$ super Yang–Mills," JHEP **0303**, 013 (2003) [arXiv:hep-th/0212208].
[23] H. Nastase, "Introduction to AdS-CFT," arXiv:0712.0689 [hep-th].
[24] J. Plefka, "Spinning strings and integrable spin chains in the AdS/CFT correspondence," arXiv:hep-th/0507136.
[25] J. Polchinski, "Dirichlet-branes and Ramond–Ramond charges," Phys. Rev. Lett. **75**, 4724 (1995) [arXiv:hep-th/9510017].
[26] J. Polchinski, "String Theory. Vol. 1: An Introduction to the Bosonic String," *Cambridge, UK: Cambridge Univ. Press (1998).*
[27] A. M. Polyakov, "Quantum geometry of bosonic strings," Phys. Lett. B **103**, 207 (1981).
[28] G. 't Hooft, "A planar diagram theory for strong interactions," Nucl. Phys. B **72**, 461 (1974).
[29] R. M. Wald, "Spin-2 fields and general covariance," Phys. Rev. D **33**, 3613 (1986).

[30] P. C. West, *Introduction to Supersymmetry and Supergravity, Singapore: World Scientific (1986).*
[31] J. Wess and J. Bagger, "Supersymmetry and Supergravity," *Princeton, NJ: Princeton Univ. Press (1992).*
[32] E. Witten, "Anti-de Sitter space and holography," Adv. Theor. Math. Phys. **2**, 253 (1998) [arXiv:hep-th/9802150].

7

Functional renormalization group equations, asymptotic safety, and quantum Einstein gravity

MARTIN REUTER AND FRANK SAUERESSIG*

Abstract

This chapter provides a pedagogical introduction to a specific continuum implementation of the Wilsonian renormalization group, the effective average action. Its general properties and, in particular, its functional renormalization group equation are explained in a simple scalar setting. The approach is then applied to quantum Einstein gravity. The possibility of constructing a fundamental theory of quantum gravity in the framework of asymptotic safety is discussed, and the supporting evidence is summarized.

7.1 Introduction

After the introduction of a functional renormalization group equation for gravity [1], detailed investigations of the nonperturbative renormalization group (RG) behavior of quantum Einstein gravity (QEG) have become possible [1–16]. The exact RG equation underlying this approach defines a Wilsonian RG flow on a theory space which consists of all diffeomorphism-invariant functionals of the metric $g_{\mu\nu}$. The approach turned out to be ideal for investigating the asymptotic safety scenario in gravity [17–19], and, in fact, substantial evidence was found for the nonperturbative renormalizability of QEG. The theory emerging from this construction is not a quantization of classical general relativity. Instead, its bare action corresponds to a nontrivial fixed point of the RG flow and is a prediction therefore. Independent support for the asymptotic safety conjecture comes from a two-dimensional symmetry reduction of the gravitational path integral [20].

The approach of [1] employs the effective average action [21–24], which has crucial advantages over other continuum implementations of the Wilsonian RG

* F.S. is supported by the European Commission Marie Curie Fellowship no. MEIF-CT-2005-023966.

flow [25]. In particular, it is closely related to the standard effective action and defines a family of effective field theories $\{\Gamma_k[g_{\mu\nu}], 0 \leq k < \infty\}$ labeled by the coarse-graining scale k. The latter property opens the door to a rather direct extraction of physical information from the RG flow, at least in single-scale cases: If the physical process under consideration involves a single typical momentum scale p_0 only, it can be described by a tree-level evaluation of $\Gamma_k[g_{\mu\nu}]$, with $k = p_0$.[1]

The effective field theory techniques proved useful for an understanding of the scale-dependent geometry of the effective QEG spacetimes [26–28]. In particular it has been shown [3, 5, 28] that these spacetimes have fractal properties, with a fractal dimension of 2 at small, and 4 at large distances. The same dynamical dimensional reduction was also observed in numerical studies of Lorentzian dynamical triangulations [29–31]; in [32] Connes et al. speculated about its possible relevance to the noncommutative geometry of the standard model.

As for possible physics implications of the RG flow predicted by QEG, ideas from particle physics, in particular the *RG improvement*, have been employed in order to study the leading quantum gravity effects in black hole and cosmological spacetimes [33–43]. Among other results, it was found [33] that the quantum effects tend to decrease the Hawking temperature of black holes, and that their evaporation presumably stops completely once the black hole's mass is of the order of the Planck mass.

These notes are intended to provide the background necessary for understanding these developments. In the next section we introduce the general idea of the effective average action and its associated functional renormalization group equation (FRGE) by means of a simple scalar example [21, 23], before reviewing the corresponding construction for gravity [1] in Section 3. In all practical calculations based upon this approach which have been performed to date, the truncation of theory space has been used as a nonperturbative approximation scheme. In Section 3 we explain the general ideas and problems behind this method, and in Section 4 we illustrate it explicitly in a simple context, the so-called Einstein–Hilbert truncation. Section 5 introduces the concept of asymptotic safety, and Section 6 contains a summary of the results obtained using truncated flow equations, with an emphasis on the question whether there exists a nontrivial fixed point for the average action's RG flow. If so, QEG could be established as a fundamental theory of quantum gravity which is nonperturbatively renormalizable and asymptotically safe from unphysical divergences.

[1] The precision which can be achieved by this effective field theory description depends on the size of the fluctuations relative to mean values. If they turn out large, or if more than one scale is involved, it might be necessary to go beyond the tree-level analysis.

7.2 Introducing the effective average action

In this section we introduce the concept of the effective average action [21–24] in the simplest context: scalar field theory on flat d-dimensional Euclidean space $\mathbb{R}^d$.

7.2.1 The basic construction for scalar fields

We start by considering a single-component real scalar field $\chi\colon \mathbb{R}^d \to \mathbb{R}$ whose dynamics is governed by the bare action $S[\chi]$. Typically the functional S has the structure $S[\chi] = \int d^dx \left\{\frac{1}{2}(\partial_\mu \chi)^2 + \frac{1}{2}m^2\chi^2 + \text{interactions}\right\}$, but we shall not need to assume any specific form of S in the following. After coupling $\chi(x)$ to a source $J(x)$ we can write down an a-priori formal path integral representation for the generating functional of the connected Green functions: $W[J] = \ln \int \mathcal{D}\chi \exp\{-S[\chi] + \int d^dx\, \chi(x)J(x)\}$. By definition, the (conventional) effective action $\Gamma[\phi]$ is the Legendre transform of $W[J]$. It depends on the field expectation value $\phi \equiv \langle \chi \rangle = \delta W[J]/\delta J$ and generates all one-particle irreducible Green functions of the theory by multiple functional differentiation with respect to $\phi(x)$ and setting $\phi = \phi[J=0]$ thereafter. In order to make the functional integral well defined, a UV cutoff is needed; for example, one could replace $\mathbb{R}^d$ by a d-dimensional lattice $\mathbb{Z}^d$. The functional integral $\mathcal{D}\chi$ would then read $\prod_{x\in\mathbb{Z}^d} d\chi(x)$. In the following we implicitly assume such a UV regularization, but leave the details unspecified and use continuum notation for the fields and their Fourier transforms.

The construction of the effective average action [21] starts out from a modified form, $W_k[J]$, of the functional $W[J]$ which depends on a variable mass scale k. This scale is used to separate the Fourier modes of χ into *short wavelength* and *long wavelength*, depending on whether or not their momentum squared, $p^2 \equiv p_\mu p^\mu$, is larger or smaller than k^2. By construction, the modes with $p^2 > k^2$ contribute without any suppression to the functional integral defining $W_k[J]$, whereas those with $p^2 < k^2$ contribute only with a reduced weight or are suppressed altogether, depending on which variant of the formalism is used. The new functional $W_k[J]$ is obtained from the conventional one by adding a *cutoff action* $\Delta_k S[\chi]$ to the bare action $S[\chi]$:

$$\exp\{W_k[J]\} = \int \mathcal{D}\chi \exp\Big\{ - S[\chi] - \Delta_k S[\chi] + \int d^dx\, \chi(x)J(x)\Big\}. \tag{7.1}$$

The factor $\exp\{-\Delta_k S[\chi]\}$ serves the purpose of suppressing the IR modes having $p^2 < k^2$. In momentum space the cutoff action is taken to be of the form

$$\Delta_k S[\chi] \equiv \frac{1}{2}\int \frac{d^dp}{(2\pi)^d}\, \mathcal{R}_k(p^2)\, |\widehat{\chi}(p)|^2\,, \tag{7.2}$$

where $\widehat{\chi}(p) = \int d^d x\, \chi(x) \exp(-ipx)$ is the Fourier transform of $\chi(x)$. The precise shape of the function $\mathcal{R}_k(p^2)$ is arbitrary to some extent; what matters is its limiting behavior for $p^2 \gg k^2$ and $p^2 \ll k^2$ only. In the simplest case[2] we require that

$$\mathcal{R}_k(p^2) \approx \begin{cases} k^2 & \text{for } p^2 \ll k^2\,, \\ 0 & \text{for } p^2 \gg k^2\,. \end{cases} \tag{7.3}$$

The first condition leads to a suppression of the small-momentum modes by a soft masslike IR cutoff; the second guarantees that the large-momentum modes are integrated out in the usual way. Adding $\Delta_k S$ to the bare action $S[\chi]$ leads to

$$S + \Delta_k S = \frac{1}{2} \int \frac{d^d p}{(2\pi)^d} \Big[p^2 + m^2 + \mathcal{R}_k(p^2) \Big] |\widehat{\chi}(p)|^2 + \text{interactions.} \tag{7.4}$$

Obviously the cutoff function $\mathcal{R}_k(p^2)$ has the interpretation of a momentum-dependent mass squared which vanishes for $p^2 \gg k^2$ and assumes the constant value k^2 for $p^2 \ll k^2$. How $\mathcal{R}_k(p^2)$ is assumed to interpolate between these two regimes is a matter of calculational convenience. In practical calculations one often uses the exponential cutoff $\mathcal{R}_k(p^2) = p^2[\exp(p^2/k^2) - 1]^{-1}$, but many other choices are possible [23, 44]. One could also think of suppressing the $p^2 < k^2$ modes completely. This could be achieved by allowing $\mathcal{R}_k(p^2)$ to diverge for $p^2 \ll k^2$ so that $\exp\{-\Delta_k S[\chi]\} \to 0$ for modes with $p^2 \ll k^2$. Although this behavior of $\mathcal{R}_k(p^2)$ seems most natural from the viewpoint of a Kadanoff–Wilson-type coarse graining, its singular behavior makes the resulting generating functional problematic to deal with technically. For this reason, and because it still allows for the derivation of an exact RG equation, one usually prefers to work with a smooth cutoff satisfying (7.3). At the nonperturbative path integral level it suppresses the long-wavelength modes by a factor $\exp\{-\frac{1}{2}k^2 \int |\widehat{\chi}|^2\}$. In perturbation theory, according to Equation (7.4), the $\Delta_k S$ term leads to the modified propagator $[p^2 + m^2 + \mathcal{R}_k(p^2)]^{-1}$, which equals $[p^2 + m^2 + k^2]^{-1}$ for $p^2 \ll k^2$. Thus, when computing loops with this propagator, k^2 acts indeed as a conventional IR cutoff if $m^2 \ll k^2$. (It plays no role in the opposite limit $m^2 \gg k^2$, in which the physical particle mass cuts off the p-integration.) We note that by replacing p^2 with $-\partial^2$ in the argument of $\mathcal{R}_k(p^2)$ the cutoff action can be written in a way that makes no reference to the Fourier decomposition of χ:

$$\Delta_k S[\chi] = \frac{1}{2} \int d^d x\, \chi(x) \mathcal{R}_k(-\partial^2) \chi(x)\,. \tag{7.5}$$

The next steps toward the definition of the effective average action are similar to the usual procedure. One defines the (now k-dependent) field expectation value $\phi(x) \equiv \langle \chi(x) \rangle = \delta W_k[J]/\delta J(x)$, assumes that the functional relationship

[2] We shall discuss a slight generalization of these conditions at the end of this section.

$\phi = \phi[J]$ can be inverted to yield $J = J[\phi]$, and introduces the Legendre transform of W_k,

$$\widetilde{\Gamma}_k[\phi] \equiv \int d^d x \, J(x)\phi(x) - W_k[J] \,, \tag{7.6}$$

where $J = J[\phi]$. The actual effective average action, denoted by $\Gamma_k[\phi]$, is obtained from $\widetilde{\Gamma}_k$ by subtracting $\Delta_k S[\phi]$:

$$\Gamma_k[\phi] \equiv \widetilde{\Gamma}_k[\phi] - \frac{1}{2} \int dx \, \phi(x) \mathcal{R}_k(-\partial^2) \phi(x) \,. \tag{7.7}$$

The rationale for this definition becomes clear when we look at the list of properties enjoyed by the functional Γ_k:

7.2.1.1

The scale dependence of Γ_k is governed by the FRGE

$$k \frac{\partial}{\partial k} \Gamma_k[\phi] = \frac{1}{2} \mathrm{Tr}\Big[k \frac{\partial}{\partial k} \mathcal{R}_k \left(\Gamma_k^{(2)}[\phi] + \mathcal{R}_k \right)^{-1} \Big] \,. \tag{7.8}$$

Here the RHS uses a compact matrix notation. In a position space representation $\Gamma_k^{(2)}$ has the matrix elements $\Gamma_k^{(2)}(x, y) \equiv \delta^2 \Gamma_k / \delta\phi(x)\delta\phi(y)$, i.e., it is the Hessian of the average action, $\mathcal{R}_k(x, y) \equiv \mathcal{R}_k(-\partial_x^2)\delta(x - y)$, and the trace Tr corresponds to an integral $\int d^d x$. In (7.8) the implicit UV cutoff can be removed trivially. This is most easily seen in the momentum representation, where $k \frac{\partial}{\partial k} \mathcal{R}_k(p^2)$, considered as a function of p^2, is significantly different from zero only in the region around $p^2 = k^2$. Hence the trace receives contributions from a thin shell of momenta $p^2 \approx k^2$ only and is therefore well convergent both in the UV and in the IR.

The RHS of (7.8) can be rewritten in a style reminiscent of a one-loop expression:

$$k \frac{\partial}{\partial k} \Gamma_k[\phi] = \frac{1}{2} \frac{D}{D \ln k} \mathrm{Tr} \ln \Big(\Gamma_k^{(2)}[\phi] + \mathcal{R}_k \Big) \,. \tag{7.9}$$

Here the scale derivative $D/D \ln k$ acts only on the k-dependence of $\mathcal{R}_k$, not on $\Gamma_k^{(2)}$. The expression $\mathrm{Tr} \ln(\cdots) = \ln \det(\cdots)$ in (7.9) differs from a standard one-loop determinant in two ways: it contains the Hessian of the actual effective action rather than that of the bare action S, and it has a built-in IR regulator $\mathcal{R}_k$. These modifications make (7.9) an *exact* equation. In a sense, solving it amounts to solving the complete theory.

The derivation of (7.8) proceeds as follows [21]. Taking the k-derivative of (7.6) with (7.1) and (7.5) inserted, one finds

$$k \frac{\partial}{\partial k} \widetilde{\Gamma}_k[\phi] = \frac{1}{2} \int d^d x d^d y \, \langle \chi(x) \chi(y) \rangle \, k \frac{\partial}{\partial k} \mathcal{R}_k(x, y) \,, \tag{7.10}$$

with $\langle A\rangle \equiv e^{-W_k}\int\mathcal{D}\chi\, A\exp\{-S-\Delta_k S-\int J\phi\}$ defining the J- and k-dependent expectation values. Next it is convenient to introduce the connected two-point function $G_{xy}\equiv G(x,y)\equiv \delta^2 W_k[J]/\delta J(x)\delta J(y)$ and the Hessian of $\widetilde{\Gamma}_k$: $(\widetilde{\Gamma}_k^{(2)})_{xy}\equiv \delta^2\widetilde{\Gamma}_k[J]/\delta\phi(x)\delta\phi(y)$. Because W_k and $\widetilde{\Gamma}_k$ are related by a Legendre transformation, one shows in the usual way that G and $\widetilde{\Gamma}^{(2)}$ are mutually inverse matrices: $G\widetilde{\Gamma}^{(2)}=1$. Furthermore, taking two J-derivatives of (7.1), one obtains $\langle\chi(x)\chi(y)\rangle = G(x,y)+\phi(x)\phi(y)$. Substituting this expression for the two-point function into (7.10), we arrive at

$$\partial_t\widetilde{\Gamma}_k[\phi]=\frac{1}{2}\mathrm{Tr}[\partial_t\mathcal{R}_k G]+\frac{1}{2}\int d^dx\,\phi(x)\,\partial_t\mathcal{R}_k(-\partial^2)\,\phi(x)\,, \tag{7.11}$$

where $t\equiv\ln(k/k_0)$. In terms of Γ_k, the effective average action proper, this becomes $\partial_t\Gamma_k[\phi]=\frac{1}{2}\mathrm{Tr}[\partial_t\mathcal{R}_k G]$. The cancellation of the $\frac{1}{2}\int\phi\mathcal{R}_k\phi$ term is a first motivation for the definition (7.7), where this term is subtracted from the Legendre transform $\widetilde{\Gamma}_k$. The derivation is completed by noting that $G=[\widetilde{\Gamma}^{(2)}]^{-1}=(\Gamma_k^{(2)}+\mathcal{R}_k)^{-1}$, where the second equality follows by differentiating (7.7): $\Gamma_k^{(2)}=\widetilde{\Gamma}_k^{(2)}-\mathcal{R}_k$.

7.2.1.2

The effective average action satisfies the following integrodifferential equation:

$$\exp\{-\Gamma_k[\phi]\}=\int\mathcal{D}\chi\,\exp\Big\{-S[\chi]+\int d^dx\,(\chi-\phi)\frac{\delta\Gamma_k[\phi]}{\delta\phi}\Big\}$$
$$\times\exp\Big\{-\int d^dx\,(\chi-\phi)\mathcal{R}_k(-\partial^2)(\chi-\phi)\Big\}\,. \tag{7.12}$$

This equation is easily derived by combining Equations (7.1), (7.6), and (7.7), and by using the effective field equation $\delta\widetilde{\Gamma}_k/\delta\phi=J$, which is dual to $\delta W_k/\delta J=\phi$. (Note that it is $\widetilde{\Gamma}_k$ which appears here, not Γ_k.)

7.2.1.3

For $k\to 0$ the effective average action approaches the ordinary effective action ($\lim_{k\to 0}\Gamma_k=\Gamma$), and for $k\to\infty$ the bare action ($\Gamma_{k\to\infty}=S$). The $k\to 0$ limit is a consequence of (7.3); $\mathcal{R}_k(p^2)$ vanishes for all $p^2>0$ when $k\to 0$. The derivation of the $k\to\infty$ limit makes use of the integrodifferential equation (7.12). A formal version the argument is as follows. Because $\mathcal{R}_k(p^2)$ approaches k^2 for $k\to\infty$, the second exponential on the RHS of (7.12) becomes $\exp\{-k^2\int dx(\chi-\phi)^2\}$, which, up to a normalization factor, approaches a delta functional $\delta[\chi-\phi]$. The χ integration can be performed trivially then, and one ends up with $\lim_{k\to\infty}\Gamma_k[\phi]=S[\phi]$. In a more careful treatment [21] one shows that the saddle point approximation of the functional integral in (7.12) about the point $\chi=\phi$ becomes exact in the limit $k\to\infty$. As a result, $\lim_{k\to\infty}\Gamma_k$ and S differ at most by the infinite-mass limit of

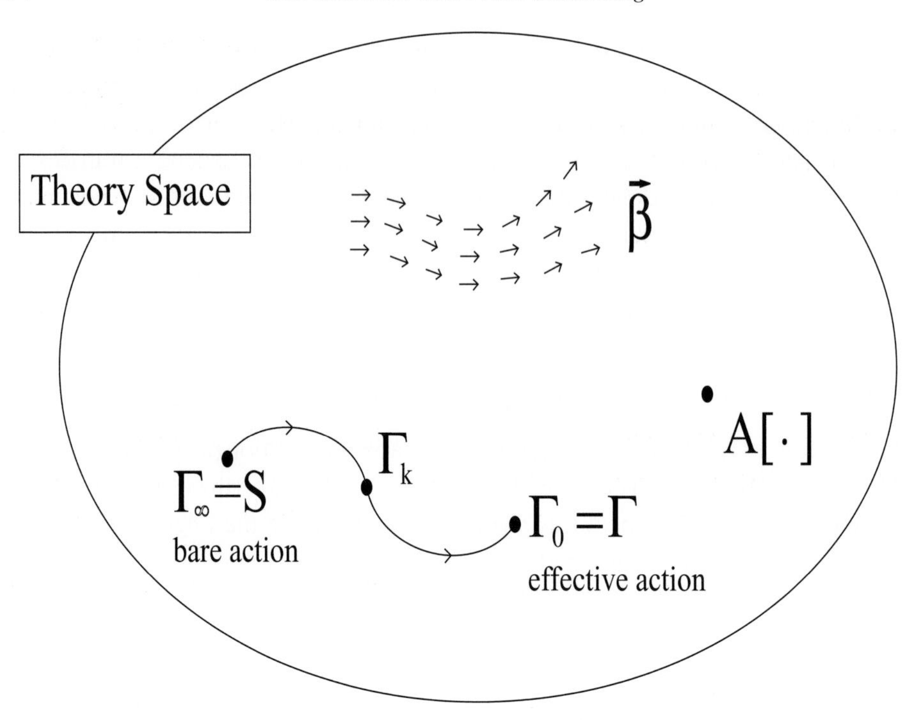

Fig. 7.1. The points of theory space are the action functionals $A[\,\cdot\,]$. The RG equation defines a vector field $\vec{\beta}$ on this space; its integral curves are the RG trajectories $k \mapsto \Gamma_k$. They start at the bare action S and end at the standard effective action Γ.

a one-loop determinant, which we suppress here because it plays no role in typical applications (see [45] for a more detailed discussion).

7.2.1.4

The FRGE (7.8) is independent of the bare action S, which enters only via the initial condition $\Gamma_\infty = S$. In the FRGE approach the calculation of the path integral for W_k is replaced by integrating the RG equation from $k = \infty$, where the initial condition $\Gamma_\infty = S$ is imposed, down to $k = 0$, where the effective average action equals the ordinary effective action Γ, the object which we actually would like to know.

7.2.2 Theory space

The arena in which the Wilsonian RG dynamics takes place is the *theory space*. Albeit a somewhat formal notion, it helps in visualizing various concepts related to functional renormalization group equations; see Figure 7.1. To describe it, we shall be slightly more general than in the previous subsection and consider an arbitrary set of fields $\phi(x)$. Then the corresponding theory space consists of all (action) functionals $A : \phi \mapsto A[\phi]$ depending on this set, possibly subject to certain symmetry requirements (a $\mathbb{Z}_2$-symmetry for a single scalar, or diffeomorphism invariance if ϕ denotes the spacetime metric, for instance). So the theory space $\{A[\,\cdot\,]\}$ is fixed once the field content and the symmetries are fixed. Let us assume we can find a set of *basis functionals* $\{P_\alpha[\,\cdot\,]\}$ such that every point of theory space has an expansion of the form [18]

$$A[\phi] = \sum_{\alpha=1}^{\infty} \bar{u}_\alpha \, P_\alpha[\phi]. \tag{7.13}$$

The basis $\{P_\alpha[\,\cdot\,]\}$ will include both local field monomials and nonlocal invariants, and we may use the *generalized couplings* $\{\bar{u}_\alpha, \alpha = 1, 2, \dots\}$ as local coordinates. More precisely, the theory space is coordinatized by the subset of *essential couplings*, i.e., those coordinates which cannot be absorbed by a field reparameterization.

Geometrically speaking, the FRGE for the effective average action, Equation (7.8) or its generalization for an arbitrary set of fields, defines a vector field $\vec{\beta}$ on theory space. The integral curves along this vector field are the *RG trajectories* $k \mapsto \Gamma_k$ parameterized by the scale k. They start, for $k \to \infty$, at the bare action S (up to the correction term mentioned earlier) and terminate at the ordinary effective action at $k = 0$. The natural orientation of the trajectories is from higher to lower scales k, the direction of increasing *coarse graining*. Expanding Γ_k as in (7.13),

$$\Gamma_k[\phi] = \sum_{\alpha=1}^{\infty} \bar{u}_\alpha(k) \, P_\alpha[\phi] \,, \tag{7.14}$$

the trajectory is described by infinitely many *running couplings* $\bar{u}_\alpha(k)$. Inserting (7.14) into the FRGE, we obtain a system of infinitely many coupled differential equations for the $\bar{u}_\alpha$s:

$$k\partial_k \, \bar{u}_\alpha(k) = \overline{\beta}_\alpha(\bar{u}_1, \bar{u}_2, \dots; k) \,, \quad \alpha = 1, 2, \dots . \tag{7.15}$$

Here the *beta functions* $\overline{\beta}_\alpha$ arise by expanding the trace on the RHS of the FRGE in terms of $\{P_\alpha[\,\cdot\,]\}$, i.e., $\frac{1}{2}\mathrm{Tr}\,[\cdots] = \sum_{\alpha=1}^{\infty} \overline{\beta}_\alpha(\bar{u}_1, \bar{u}_2, \dots; k) P_\alpha[\phi]$. The expansion coefficients $\overline{\beta}_\alpha$ have the interpretation of beta functions similar to those of perturbation theory, but not restricted to relevant couplings. In standard field theory jargon one would refer to $\bar{u}_\alpha(k = \infty)$ as the “bare” parameters and to $\bar{u}_\alpha(k = 0)$ as the “renormalized” or “dressed” parameters.

The notation with the bar on $\bar{u}_\alpha$ and $\overline{\beta}_\alpha$ is to indicate that we are still dealing with dimensionful couplings. Usually the flow equation is reexpressed in terms of the dimensionless couplings $u_\alpha \equiv k^{-d_\alpha} \bar{u}_\alpha$, where d_α is the canonical mass dimension of $\bar{u}_\alpha$. Correspondingly the essential u_α’s are used as coordinates of theory space.

The resulting RG equations

$$k\partial_k u_\alpha(k) = \beta_\alpha(u_1, u_2, \ldots) \tag{7.16}$$

are a coupled system of autonomous differential equations. The β_αs have no explicit k-dependence and define a *time-independent* vector field on theory space.

Figure 7.1 gives a schematic summary of the theory space and its structures. It should be kept in mind, though, that only the essential couplings are coordinates on theory space, and that Γ_∞ and S might differ by a simple, explicitly known functional.

7.2.3 Nonperturbative approximations through truncations

Up to this point our discussion has not involved any approximation. In practice, however, it is usually impossible to find exact solutions to the flow equation. As a way out, one could evaluate the trace on the RHS of the FRGE by expanding it with respect to some small coupling constant, for instance, thus recovering the familiar perturbative beta functions. A more interesting option, which gives rise to nonperturbative approximate solutions, is to truncate the theory space $\{A[\,\cdot\,]\}$. The basic idea is to project the RG flow onto a finite-dimensional subspace of theory space. The subspace should be chosen in such a way that the projected flow encapsulates the essential physical features of the exact flow on the full space.

Concretely the projection onto a truncation subspace is performed as follows. One makes an ansatz of the form $\Gamma_k[\phi] = \sum_{i=1}^{N} \bar{u}_i(k) P_i[\phi]$, where the k-independent functionals $\{P_i[\,\cdot\,], i = 1, \ldots, N\}$ form a *basis* on the subspace selected. For a scalar field, say, examples include pure potential terms $\int d^dx \phi^m(x)$, $\int d^dx \phi^n(x) \ln \phi^2(x), \ldots$, a standard kinetic term $\int d^dx (\partial\phi)^2$, higher-order derivative terms $\int d^dx\, \phi \left(\partial^2\right)^n \phi$, $\int d^dx\, f(\phi) \left(\partial^2\right)^n \phi \left(\partial^2\right)^m \phi, \ldots$, and nonlocal terms like $\int d^dx\, \phi \ln(-\partial^2)\phi, \ldots$. Even if $S = \Gamma_\infty$ is simple – a standard ϕ^4 action, say – the evolution from $k = \infty$ downwards will generate such terms, a priori constrained only by symmetry requirements. The difficult task in practical RG applications consists in selecting a set of P_is on the one hand is generic enough to allow for a sufficiently precise description of the physics one is interested in, and on the other hand is small enough to be computationally manageable.

The projected RG flow is described by a set of ordinary (if $N < \infty$) differential equations for the couplings $\bar{u}_i(k)$. They arise as follows. Let us assume we expand the ϕ-dependence of $\frac{1}{2}\mathrm{Tr}[\cdots]$ (with the ansatz for $\Gamma_k[\phi]$ inserted) in a basis $\{P_\alpha[\,\cdot\,]\}$ of the *full* theory space which contains the P_is spanning the truncated space as a subset:

$$\frac{1}{2}\mathrm{Tr}[\cdots] = \sum_{\alpha=1}^{\infty} \overline{\beta}_\alpha(\bar{u}_1, \ldots, \bar{u}_N; k)\, P_\alpha[\phi] = \sum_{i=1}^{N} \overline{\beta}_i(\bar{u}_1, \ldots, \bar{u}_N; k)\, P_i[\phi] + \text{rest}\,. \tag{7.17}$$

Here the "rest" contains all terms outside the truncated theory space; the approximation consists in neglecting precisely those terms. Thus, on equating (7.17) to the LHS of the flow equation, $\partial_t \Gamma_k = \sum_{i=1}^{N} \partial_t \bar{u}_i(k) P_i$, the linear independence of the P_is implies the coupled system of ordinary differential equations

$$\partial_t \bar{u}_i(k) = \overline{\beta}_i(\bar{u}_1, \ldots, \bar{u}_N; k)\,, \quad i = 1, \ldots, N\,. \tag{7.18}$$

Solving (7.18), one obtains an *approximation* to the exact RG trajectory projected onto the chosen subspace. Note that this approximate trajectory does not, in general, coincide with the projection of the exact trajectory, but if the subspace is well chosen, it will not be very different from it. In fact, the most nontrivial problem in using truncated flow equations is to find and justify a truncation subspace which should be as low-dimensional as possible to make the calculations feasible, but at the same time large enough to describe at least qualitatively the essential physics. We shall return to the issue of testing the quality of a given truncation later on.

As a simple example of a truncation we mention the *local potential approximation* [23]. The corresponding subspace consists of functionals containing a standard kinetic term plus arbitrary nonderivative terms:

$$\Gamma_k[\phi] \equiv \int d^d x \left\{ \frac{1}{2}(\partial\phi(x))^2 + U_k(\phi(x)) \right\}. \tag{7.19}$$

In this case N is infinite, the coordinates $\bar{u}_i$ on truncated theory space being the infinitely many parameters characterizing an arbitrary potential function $\phi \mapsto U(\phi)$. The infinitely many component equations (7.18) amount to a partial differential equation for the running potential $U_k(\phi)$. It is obtained by inserting (7.19) into the FRGE and projecting the trace onto functionals of the form (7.19). This is most easily done by inserting a *constant* field $\phi = \varphi = const$ into both sides of the equation, for this gives a nonvanishing value precisely to the nonderivative P_is. Because $\Gamma_k^{(2)} = -\partial^2 + U_k''(\varphi)$ with $U'' \equiv d^2U_k/d\phi^2$ has no explicit x-dependence, the trace is easily evaluated in momentum space. This leads to the following partial differential equation:

$$k\partial_k U_k(\varphi) = \frac{1}{2}\int \frac{d^d p}{(2\pi)^d} \frac{k\partial_k \mathcal{R}_k(p^2)}{p^2 + \mathcal{R}_k(p^2) + U_k''(\varphi)}\,. \tag{7.20}$$

It describes how the classical (or microscopic) potential $U_\infty = V_{\text{class}}$ evolves into the standard effective potential $U_0 = V_{\text{eff}}$. Remarkably, the limit $\lim_{k\to 0} U_k$ is automatically a convex function of φ, and there is no need to perform the Maxwell construction "by hand," in the case of spontaneous symmetry breaking. For a detailed discussion of this point we refer to [23].

One can continue the truncation process and make a specific ansatz for the φ-dependence of the running potential, $U_k(\varphi) = \frac{1}{2}\overline{m}(k)^2\varphi^2 + \frac{1}{12}\overline{\lambda}(k)\varphi^4$, say. Then, upon inserting $U_k''(\varphi) = \overline{m}(k)^2 + \overline{\lambda}(k)\varphi^2$ into the RHS of (7.19) and expanding to $O(\varphi^4)$, one can equate the coefficients of φ^2 and φ^4 to obtain the flow equations on a two-dimensional subspace: $k\partial_k\overline{m}^2 = \overline{\beta}_{\overline{m}^2}$, $k\partial_k\overline{\lambda} = \overline{\beta}_{\overline{\lambda}}$.

If one wants to go beyond the local potential approximation (7.19), the first step is to allow for a (ϕ-independent in the simplest case) wave function renormalization, i.e., a running prefactor of the kinetic term: $\Gamma_k = \int d^dx\,\{\frac{1}{2}Z_k(\partial\phi)^2 + U_k\}$. Using truncations of this type, one should employ a slightly different normalization of $\mathcal{R}_k(p^2)$, namely, $\mathcal{R}_k(p^2) \approx Z_k k^2$ for $p^2 \ll k^2$. Then $\mathcal{R}_k$ combines with $\Gamma_k^{(2)}$ to form the inverse propagator $\Gamma_k^{(2)} + \mathcal{R}_k = Z_k(p^2 + k^2) + \cdots$, as is necessary if the IR cutoff is to give rise to a mass squared of size k^2 rather than k^2/Z_k. In particular, in more complicated theories with more than one field, it is important that all fields be cut off at precisely the same k^2. This is achieved by a cutoff function of the form

$$\mathcal{R}_k(p^2) = \mathcal{Z}_k\, k^2\, R^{(0)}(p^2/k^2)\,, \tag{7.21}$$

where $R^{(0)}$ is normalized so that $R^{(0)}(0) = 1$ and $R^{(0)}(\infty) = 0$. In general the factor $\mathcal{Z}_k$ is a matrix in field space. In the sector of modes with inverse propagator $Z_k^{(i)}p^2 + \cdots$, this matrix is chosen diagonal with entries $\mathcal{Z}_k = Z_k^{(i)}$.

7.3 The effective average action for gravity

We saw that the FRGE of the effective average action does not depend on the bare action S. Given a theory space, the form of the FRGE and, as a result, the vector field $\vec{\beta}$ are completely fixed. To define a theory space $\{A[\,\cdot\,]\}$ one has to specify on which types of fields the functionals A are supposed to depend, and what their symmetries are. This is the only input data needed for finding the RG flow.

In the case of QEG the theory space consists, by definition, of functionals $A[g_{\mu\nu}]$ depending on a symmetric tensor field, the metric, in a diffeomorphism-invariant way. Unfortunately, it is not possible to straightforwardly apply the constructions of the previous section to this theory space. Diffeomorphism invariance leads to two types of complications one has to deal with [1].

The first one is not specific to the RG approach. It occurs already in the standard functional integral quantization of gauge or gravity theories, and is familiar from Yang–Mills theories. If one gauge-fixes the functional integral with an ordinary (covariant) gauge-fixing condition like $\partial^\mu A_\mu^a = 0$, couples the (non-abelian) gauge field A_μ^a to a source, and constructs the ordinary effective action, the resulting functional $\Gamma[A_\mu^a]$ is *not* invariant under the gauge transformations of A_μ^a,

$A_\mu^a \mapsto A_\mu^a + D_\mu^{ab}(A)\,\omega^b$. Only at the level of physical quantities constructed from $\Gamma[A_\mu^a]$ – S-matrix elements, for instance – is gauge invariance recovered.

The second problem is related to the fact that in a gauge theory a coarse graining based on a naive Fourier decomposition of $A_\mu^a(x)$ is not gauge covariant and hence not physical. In fact, if one were to gauge-transform a slowly varying $A_\mu^a(x)$ using a parameter function $\omega^a(x)$ with a fast x-variation, a gauge field with a fast x-variation would arise – which, however, still describes the same physics. In a nongauge theory the coarse graining is performed by expanding the field in terms of eigenfunctions of the (positive) operator $-\partial^2$ and declaring its eigenmodes to be of long or short wavelength depending on whether the corresponding eigenvalue p^2 is smaller or larger than a given k^2. In a gauge theory the best one can do in installing this procedure is to expand with respect to the *covariant* Laplacian or a similar operator, and then organize the modes according to the size of their eigenvalues. Though gauge covariant, this approach sacrifices to some extent the intuition of a Fourier coarse graining in terms of slow and fast modes. Analogous remarks apply to theories of gravity covariant under general coordinate transformations.

The key idea which led to a solution of both problems was the use of the background field method. In fact, it is well known [46, 47] that the background gauge-fixing method leads to an effective action which depends on its arguments in a gauge-invariant way. As it turned out [1, 22], this technique also lends itself to implementing a covariant IR cutoff, and it is at the core of the effective average action for Yang–Mills theories [22, 24] and for gravity [1]. In the following we briefly review the effective average action for gravity, which was introduced in [1].

The ultimate goal is to give meaning to an integral over "all" metrics $\gamma_{\mu\nu}$ of the form $\int \mathcal{D}\gamma_{\mu\nu} \exp\{-S[\gamma_{\mu\nu}] + \text{source terms}\}$ whose bare action $S[\gamma_{\mu\nu}]$ is invariant under general coordinate transformations,

$$\delta\gamma_{\mu\nu} = \mathcal{L}_v\gamma_{\mu\nu} \equiv v^\rho\partial_\rho\gamma_{\mu\nu} + \partial_\mu v^\rho\gamma_{\rho\nu} + \partial_\nu v^\rho\gamma_{\rho\mu}\,, \tag{7.22}$$

where $\mathcal{L}_v$ is the Lie derivative with respect to the vector field $v^\mu\partial_\mu$. To start with, we consider $\gamma_{\mu\nu}$ to be a Riemannian metric and assume that $S[\gamma_{\mu\nu}]$ is positive definite. Heading towards the background field formalism, the first step consists in decomposing the variable of integration according to $\gamma_{\mu\nu} = \bar{g}_{\mu\nu} + h_{\mu\nu}$, where $\bar{g}_{\mu\nu}$ is a fixed background metric. Note that we are not implying a perturbative expansion here; $h_{\mu\nu}$ is not supposed to be small in any sense. After the background split, the measure $\mathcal{D}\gamma_{\mu\nu}$ becomes $\mathcal{D}h_{\mu\nu}$ and the gauge transformations which we have to gauge-fix read

$$\delta h_{\mu\nu} = \mathcal{L}_v\gamma_{\mu\nu} = \mathcal{L}_v(\bar{g}_{\mu\nu} + h_{\mu\nu})\,, \quad \delta\bar{g}_{\mu\nu} = 0\,. \tag{7.23}$$

Picking an a priori arbitrary gauge-fixing condition $F_\mu(h;\bar{g}) = 0$, the Faddeev–Popov trick can be applied straightforwardly [46]. Upon including an IR cutoff as in the scalar case, we are led to the following k-dependent generating functional W_k for the connected Green functions:

$$\exp\Big\{W_k[t^{\mu\nu},\sigma^\mu,\bar{\sigma}_\mu;\bar{g}_{\mu\nu}]\Big\} = \int \mathcal{D}h_{\mu\nu}\mathcal{D}C^\mu\mathcal{D}\bar{C}_\mu \,\exp\Big\{-S[\bar{g}+h] - S_{\rm gf}[h;\bar{g}] \\ - S_{\rm gh}[h,C,\bar{C};\bar{g}] - \Delta_k S[h,C,\bar{C};\bar{g}] - S_{\rm source}\Big\}. \tag{7.24}$$

Here $S_{\rm gf}$ denotes the gauge-fixing term

$$S_{\rm gf}[h;\bar{g}] = \frac{1}{2\alpha}\int d^dx\sqrt{\bar{g}}\,\bar{g}^{\mu\nu}F_\mu F_\nu, \tag{7.25}$$

and $S_{\rm gh}$ is the action for the corresponding Faddeev–Popov ghosts C^μ and $\bar{C}_\mu$:

$$S_{\rm gh}[h,C,\bar{C};\bar{g}] = -\kappa^{-1}\int d^dx\,\bar{C}_\mu\,\bar{g}^{\mu\nu}\,\frac{\partial F_\nu}{\partial h_{\alpha\beta}}\,\mathcal{L}_C\left(\bar{g}_{\alpha\beta}+h_{\alpha\beta}\right). \tag{7.26}$$

The Faddeev–Popov action $S_{\rm gh}$ is obtained along the same lines as in Yang–Mills theory: one applies a gauge transformation (7.23) to F_μ and replaces the parameters v^μ by the ghost field C^μ. The integral over C^μ and $\bar{C}_\mu$ exponentiates the Faddeev–Popov determinant $\det[\delta F_\mu/\delta v^\nu]$. In (7.24) we coupled $h_{\mu\nu}$, C^μ, and $\bar{C}_\mu$ to sources $t^{\mu\nu}$, $\bar{\sigma}_\mu$, and σ^μ, respectively: $S_{\rm source} = -\int d^dx\,\sqrt{\bar{g}}\{t^{\mu\nu}h_{\mu\nu} + \bar{\sigma}_\mu C^\mu + \sigma^\mu\bar{C}_\mu\}$. The k- and source-dependent expectation values of $h_{\mu\nu}$, C^μ, and $\bar{C}_\mu$ are then given by

$$\bar{h}_{\mu\nu} = \frac{1}{\sqrt{\bar{g}}}\frac{\delta W_k}{\delta t^{\mu\nu}}, \qquad \xi^\mu = \frac{1}{\sqrt{\bar{g}}}\frac{\delta W_k}{\delta\bar{\sigma}_\mu}, \qquad \bar{\xi}_\mu = \frac{1}{\sqrt{\bar{g}}}\frac{\delta W_k}{\delta\sigma^\mu}. \tag{7.27}$$

As usual, we assume that one can invert the relations (7.27) and solve for the sources $(t^{\mu\nu},\,\sigma^\mu,\,\bar{\sigma}_\mu)$ as functionals of $(\bar{h}_{\mu\nu},\,\xi^\mu,\,\bar{\xi}_\mu)$ and, parametrically, of $\bar{g}_{\mu\nu}$. The Legendre transform $\widetilde{\Gamma}_k$ of W_k reads

$$\widetilde{\Gamma}_k[\bar{h},\xi,\bar{\xi};\bar{g}] = \int d^dx\,\sqrt{\bar{g}}\,\{t^{\mu\nu}\bar{h}_{\mu\nu} + \bar{\sigma}_\mu\xi^\mu + \sigma^\mu\bar{\xi}_\mu\} - W_k[t,\sigma,\bar{\sigma};\bar{g}]. \tag{7.28}$$

This functional inherits a parametric $\bar{g}_{\mu\nu}$-dependence from W_k.

As mentioned earlier for a generic gauge-fixing condition the Legendre transform (7.28) is not a diffeomorphism-invariant functional of its arguments, because the gauge breaking under the functional integral is communicated to $\widetilde{\Gamma}_k$ via the sources. Although $\widetilde{\Gamma}_k$ does indeed describe the correct on-shell physics satisfying all constraints coming from BRST invariance, it is not invariant off shell [46, 47]. The situation is different for the class of gauge-fixing conditions of the background type. Although – as any gauge-fixing condition must – they break the invariance under (7.23), they are chosen to be invariant under the so-called background gauge transformations

$$\delta h_{\mu\nu} = \mathcal{L}_v h_{\mu\nu}\,, \qquad \delta \bar{g}_{\mu\nu} = \mathcal{L}_v \bar{g}_{\mu\nu}\,. \tag{7.29}$$

The complete metric $\gamma_{\mu\nu} = g_{\mu\nu} + h_{\mu\nu}$ transforms as $\delta\gamma_{\mu\nu} = \mathcal{L}_v \gamma_{\mu\nu}$ both under (7.29) and under (7.23). The crucial difference is that the (*quantum*) gauge transformations (7.23) keep $\bar{g}_{\mu\nu}$ unchanged, so that the entire change of $\gamma_{\mu\nu}$ is ascribed to $h_{\mu\nu}$. This is the point of view one adopts in a standard perturbative calculation around flat space, where one fixes $\bar{g}_{\mu\nu} = \eta_{\mu\nu}$ and allows for no variation of the background. In the present construction, instead, we leave $\bar{g}_{\mu\nu}$ unspecified but insist on covariance under (7.29). This will lead to a completely background-covariant formulation.

Clearly there exist many possible gauge-fixing terms $S_{\text{gf}}[h;\bar{g}]$ of the form (7.25) which break (7.23) and are invariant under (7.29). A convenient choice which has been employed in practical calculations is the background version of the harmonic coordinate condition [46]:

$$F_\mu = \sqrt{2}\kappa \left[\delta^\beta_\mu \bar{g}^{\alpha\gamma}\bar{D}_\gamma - \frac{1}{2}\bar{g}^{\alpha\beta}\bar{D}_\mu\right] h_{\alpha\beta}\,. \tag{7.30}$$

The covariant derivative $\bar{D}_\mu$ involves the Christoffel symbols $\bar{\Gamma}^\rho_{\mu\nu}$ of the background metric. Note that (7.30) is linear in the quantum field $h_{\alpha\beta}$. On a flat background with $\bar{g}_{\mu\nu} = \eta_{\mu\nu}$ the condition $F_\mu = 0$ reduces to the familiar harmonic coordinate condition, $\partial^\mu h_{\mu\nu} = \frac{1}{2}\partial_\nu h_\mu^{\ \mu}$. In Equations (7.30) and (7.26) κ is an arbitrary constant with the dimension of a mass. We shall set $\kappa \equiv (32\pi\bar{G})^{-1/2}$ with $\bar{G}$ a constant reference value of Newton's constant. The ghost action for the gauge condition (7.30) reads

$$S_{\text{gh}}[h, C, \bar{C}; \bar{g}] = -\sqrt{2}\int d^d x\,\sqrt{\bar{g}}\,\bar{C}_\mu \mathcal{M}[g,\bar{g}]^\mu{}_\nu C^\nu \tag{7.31}$$

with the Faddeev–Popov operator

$$\mathcal{M}[g,\bar{g}]^\mu{}_\nu = \bar{g}^{\mu\rho}\bar{g}^{\sigma\lambda}\bar{D}_\lambda(g_{\rho\nu}D_\sigma + g_{\sigma\nu}D_\rho) - \bar{g}^{\rho\sigma}\bar{g}^{\mu\lambda}\bar{D}_\lambda g_{\sigma\nu}D_\rho\,. \tag{7.32}$$

It will prove crucial that for every background-type choice of F_μ, S_{gh} is invariant under (7.29) together with

$$\delta C^\mu = \mathcal{L}_v C^\mu\,, \qquad \delta\bar{C}_\mu = \mathcal{L}_v \bar{C}_\mu\,. \tag{7.33}$$

The essential piece in Equation (7.24) is the IR cutoff for the gravitational field $h_{\mu\nu}$ and for the ghosts. It is taken to be of the form

$$\Delta_k S = \frac{\kappa^2}{2}\int d^d x\,\sqrt{\bar{g}}\,h_{\mu\nu}\mathcal{R}_k^{\text{grav}}[\bar{g}]^{\mu\nu\rho\sigma}h_{\rho\sigma} + \sqrt{2}\int d^d x\,\sqrt{\bar{g}}\,\bar{C}_\mu \mathcal{R}_k^{\text{gh}}[\bar{g}]C^\mu\,. \tag{7.34}$$

The cutoff operators $\mathcal{R}_k^{\text{grav}}$ and $\mathcal{R}_k^{\text{gh}}$ serve the purpose of discriminating between high-momentum and low-momentum modes. Eigenmodes of $-\bar{D}^2$ with eigenvalues $p^2 \gg k^2$ are integrated out without any suppression, whereas modes with small eigenvalues $p^2 \ll k^2$ are suppressed. The operators $\mathcal{R}_k^{\text{grav}}$ and $\mathcal{R}_k^{\text{gh}}$ have the structure $\mathcal{R}_k[\bar{g}] = \mathcal{Z}_k k^2 R^{(0)}(-\bar{D}^2/k^2)$, where the dimensionless function $R^{(0)}$ interpolates between $R^{(0)}(0) = 1$ and $R^{(0)}(\infty) = 0$. A convenient choice is, e.g., the exponential cutoff $R^{(0)}(w) = w[\exp(w) - 1]^{-1}$, where $w = p^2/k^2$. The factors $\mathcal{Z}_k$ are different for the graviton and the ghost cutoff. For the ghost $\mathcal{Z}_k \equiv Z_k^{\text{gh}}$ is a pure number, whereas for the metric fluctuation $\mathcal{Z}_k \equiv \mathcal{Z}_k^{\text{grav}}$ is a tensor, constructed only from the background metric $\bar{g}_{\mu\nu}$, which must be fixed along the lines described at the end of Section 7.2.

A feature of $\Delta_k S$ which is essential from a practical point of view is that the modes of $h_{\mu\nu}$ and the ghosts are organized according to their eigenvalues with respect to the *background* Laplace operator $\bar{D}^2 = \bar{g}^{\mu\nu}\bar{D}_\mu\bar{D}_\nu$ rather than $D^2 = g^{\mu\nu}D_\mu D_\nu$, which would pertain to the full quantum metric $\bar{g}_{\mu\nu} + h_{\mu\nu}$. Using $\bar{D}^2$, the functional $\Delta_k S$ is quadratic in the quantum field $h_{\mu\nu}$, whereas it becomes extremely complicated if D^2 is used instead. The virtue of a quadratic $\Delta_k S$ is that it gives rise to a flow equation which contains *second* functional derivatives of Γ_k but no higher ones. The flow equations resulting from the cutoff operator D^2 are prohibitively complicated and can hardly be used for practical computations. A second property of $\Delta_k S$, which is crucial for our purposes, is that it is invariant under the background gauge transformations (7.29) with (7.34).

Having specified all the ingredients which enter the functional integral (7.24) for the generating functional W_k, we can write down the final definition of the effective average action Γ_k. It is obtained from the Legendre transform $\widetilde{\Gamma}_k$ by subtracting the cutoff action $\Delta_k S$ with the classical fields inserted:

$$\Gamma_k[\bar{h}, \xi, \bar{\xi}; \bar{g}] = \widetilde{\Gamma}_k[\bar{h}, \xi, \bar{\xi}; \bar{g}] - \Delta_k S[\bar{h}, \xi, \bar{\xi}; \bar{g}] \,. \tag{7.35}$$

It is convenient to define the expectation value of the quantum metric $\gamma_{\mu\nu}$,

$$g_{\mu\nu}(x) \equiv \bar{g}_{\mu\nu}(x) + \bar{h}_{\mu\nu}(x) \,, \tag{7.36}$$

and consider Γ_k as a functional of $g_{\mu\nu}$ rather than $\bar{h}_{\mu\nu}$:

$$\Gamma_k[g_{\mu\nu}, \bar{g}_{\mu\nu}, \xi^\mu, \bar{\xi}_\mu] \equiv \Gamma_k[g_{\mu\nu} - \bar{g}_{\mu\nu}, \xi^\mu, \bar{\xi}_\mu; \bar{g}_{\mu\nu}] \,. \tag{7.37}$$

So, what did we gain going through this seemingly complicated background field construction, eventually ending up with an action functional which depends on *two* metrics even? The main advantage of this setting is that the corresponding functionals $\widetilde{\Gamma}_k$, and as a result Γ_k, are invariant under general coordinate transformations where all its arguments transform as tensors of the corresponding rank:

$$\Gamma_k[\Phi + \mathcal{L}_v \Phi] = \Gamma_k[\Phi], \qquad \Phi \equiv \left\{ g_{\mu\nu}, \bar{g}_{\mu\nu}, \xi^\mu, \bar{\xi}_\mu \right\}. \tag{7.38}$$

Note that in (7.38), contrary to the quantum gauge transformation (7.23), also the background metric transforms as an ordinary tensor field: $\delta \bar{g}_{\mu\nu} = \mathcal{L}_v \bar{g}_{\mu\nu}$. Equation (7.38) is a consequence of

$$W_k\left[\mathcal{J} + \mathcal{L}_v \mathcal{J}\right] = W_k\left[\mathcal{J}\right], \qquad \mathcal{J} \equiv \left\{ t^{\mu\nu}, \sigma^\mu, \bar{\sigma}_\mu; \bar{g}_{\mu\nu} \right\}. \tag{7.39}$$

This invariance property follows from (7.24) if one performs a compensating transformation (7.29), (7.34) on the integration variables $h_{\mu\nu}$, C^μ, and $\bar{C}_\mu$ and uses the invariance of $S[\bar{g} + h]$, S_{gf}, S_{gh}, and $\Delta_k S$. At this point we assume that the functional measure in (7.24) is diffeomorphism invariant.

Because the $\mathcal{R}_k$s vanish for $k = 0$, the limit $k \to 0$ of $\Gamma_k[g_{\mu\nu}, \bar{g}_{\mu\nu}, \xi^\mu, \bar{\xi}_\mu]$ brings us back to the standard effective action functional, which still depends on two metrics, though. The ordinary effective action $\Gamma[g_{\mu\nu}]$ with one metric argument is obtained from this functional by setting $\bar{g}_{\mu\nu} = g_{\mu\nu}$, or equivalently $\bar{h}_{\mu\nu} = 0$ [46,47]:

$$\Gamma[g] \equiv \lim_{k\to 0} \Gamma_k[g, \bar{g} = g, \xi = 0, \bar{\xi} = 0] = \lim_{k\to 0} \Gamma_k[\bar{h} = 0, \xi = 0, \bar{\xi} = 0; g = \bar{g}]. \tag{7.40}$$

This equation brings about the magic property of the background field formalism: a priori the 1PI n-point functions of the metric are obtained by an n-fold functional differentiation of $\Gamma_0[\bar{h}, 0, 0; \bar{g}_{\mu\nu}]$ with respect to $\bar{h}_{\mu\nu}$. Thereby $\bar{g}_{\mu\nu}$ is kept fixed; it acts simply as an externally prescribed function which specifies the form of the gauge-fixing condition. Hence the functional Γ_0 and the resulting *off-shell* Green functions do depend on $\bar{g}_{\mu\nu}$, but the *on-shell* Green functions, related to observable scattering amplitudes, do not depend on $\bar{g}_{\mu\nu}$. In this respect $\bar{g}_{\mu\nu}$ plays a role similar to the gauge parameter α in the standard approach. Remarkably, the same on-shell Green functions can be obtained by differentiating the functional $\Gamma[g_{\mu\nu}]$ of (7.40) with respect to $g_{\mu\nu}$, or equivalently $\Gamma_0[\bar{h} = 0, \xi = 0, \bar{\xi} = 0; \bar{g} = g]$, with respect to its $\bar{g}$ argument. In this context, "on-shell" means that the metric satisfies the effective field equation $\delta\Gamma_0[g]/\delta g_{\mu\nu} = 0$.

With (7.40) and its k-dependent counterpart

$$\bar{\Gamma}_k[g_{\mu\nu}] \equiv \Gamma_k[g_{\mu\nu}, g_{\mu\nu}, 0, 0], \tag{7.41}$$

we succeeded in constructing a diffeomorphism-invariant generating functional for gravity: thanks to (7.38), $\Gamma[g_{\mu\nu}]$ and $\bar{\Gamma}_k[g_{\mu\nu}]$ are invariant under general coordinate transformations $\delta g_{\mu\nu} = \mathcal{L}_v g_{\mu\nu}$. However, there is a price to be paid for their invariance: the simplified functional $\bar{\Gamma}_k[g_{\mu\nu}]$ does not satisfy an exact RG equation, basically because it contains insufficient information. The actual RG evolution has to be performed at the level of the functional $\Gamma_k[g, \bar{g}, \xi, \bar{\xi}\,]$. Only *after* the evolution may one set $\bar{g} = g$, $\xi = 0$, $\bar{\xi} = 0$. As a result, the actual theory space of QEG, $\{A[g, \bar{g}, \xi, \bar{\xi}\,]\}$, consists of functionals of all four variables, $g_{\mu\nu}$, $\bar{g}_{\mu\nu}$, ξ^μ, $\bar{\xi}_\mu$, subject to the invariance condition (7.38).

The derivation of the FRGE for Γ_k is analogous to the scalar case. Following exactly the same steps, one arrives at

$$\begin{aligned}\partial_t \Gamma_k[\bar{h}, \xi, \bar{\xi}; \bar{g}] =& \frac{1}{2}\mathrm{Tr}\left[\left(\Gamma_k^{(2)} + \widehat{\mathcal{R}}_k\right)^{-1}_{\bar{h}\bar{h}} \left(\partial_t \widehat{\mathcal{R}}_k\right)_{\bar{h}\bar{h}}\right] \\ & - \frac{1}{2}\mathrm{Tr}\left[\left\{\left(\Gamma_k^{(2)} + \widehat{\mathcal{R}}_k\right)^{-1}_{\bar{\xi}\xi} - \left(\Gamma_k^{(2)} + \widehat{\mathcal{R}}_k\right)^{-1}_{\xi\bar{\xi}}\right\} \left(\partial_t \widehat{\mathcal{R}}_k\right)_{\bar{\xi}\xi}\right].\end{aligned} \tag{7.42}$$

Here $\Gamma_k^{(2)}$ denotes the Hessian of Γ_k with respect to the dynamical fields $\bar{h}$, ξ, $\bar{\xi}$ at fixed $\bar{g}$. It is a block matrix labeled by the fields $\varphi_i \equiv \{\bar{h}_{\mu\nu}, \xi^\mu, \bar{\xi}_\mu\}$:

$$\Gamma_k^{(2)ij}(x, y) \equiv \frac{1}{\sqrt{\bar{g}(x)\bar{g}(y)}} \frac{\delta^2 \Gamma_k}{\delta\varphi_i(x)\delta\varphi_j(y)}. \tag{7.43}$$

(In the ghost sector the derivatives are understood as left derivatives.) Likewise, $\widehat{\mathcal{R}}_k$ is a block-diagonal matrix with entries $(\widehat{\mathcal{R}}_k)^{\mu\nu\rho\sigma}_{\bar{h}\bar{h}} \equiv \kappa^2(\mathcal{R}_k^{\mathrm{grav}}[\bar{g}])^{\mu\nu\rho\sigma}$ and $\widehat{\mathcal{R}}_{\bar{\xi}\xi} = \sqrt{2}\mathcal{R}_k^{\mathrm{gh}}[\bar{g}]$. Evaluating the trace in the position representation includes an integration $\int d^d x \sqrt{\bar{g}(x)}$ involving the background volume element. For any cutoff which is qualitatively similar to the exponential cutoff, the traces on the RHS of Equation (7.42) are well convergent, in both the IR and the UV. By virtue of the factor $\partial_t \widehat{\mathcal{R}}_k$, the dominant contributions come from a narrow band of generalized momenta centered around k. Large momenta are exponentially suppressed.

Besides the FRGE, the effective average action also satisfies an exact integro-differential equation similar to (7.12) in the scalar case. By the same argument as there, it can be used to find the $k \to \infty$ limit of the average action:

$$\Gamma_{k\to\infty}[\bar{h}, \xi, \bar{\xi}; \bar{g}] = S[\bar{g} + \bar{h}] + S_{\mathrm{gf}}[\bar{h}; \bar{g}] + S_{\mathrm{gh}}[\bar{h}, \xi, \bar{\xi}; \bar{g}]. \tag{7.44}$$

Note that the initial value $\Gamma_{k\to\infty}$ includes the gauge-fixing and ghost actions. At the level of the functional $\bar{\Gamma}_k[g]$, Equation (7.44) boils down to $\bar{\Gamma}_{k\to\infty}[g] = S[g]$. However, as $\Gamma_k^{(2)}$ involves derivatives with respect to $\bar{h}_{\mu\nu}$ (or equivalently $g_{\mu\nu}$) at fixed $\bar{g}_{\mu\nu}$, it is clear that the evolution cannot be formulated entirely in terms of $\bar{\Gamma}_k$ alone.

The background gauge invariance of Γ_k, expressed in Equation (7.38), is of enormous practical importance. It implies that if the initial functional does not contain non-invariant terms, the flow will not generate such terms. Very often this reduces the number of terms to be retained in a reliable truncation ansatz quite considerably. Nevertheless, even if the initial action is simple, the RG flow will generate all sorts of local and nonlocal terms in Γ_k which are consistent with the symmetries.

Let us close this section by remarking that, at least formally, the construction of the effective average action can be repeated for Lorentzian signature metrics. In this case one deals with oscillating exponentials e^{iS}, and for arguments like the one leading to (7.44) one has to employ the Riemann–Lebesgue lemma. Apart from the obvious substitutions $\Gamma_k \to -i\Gamma_k$, $\mathcal{R}_k \to -i\mathcal{R}_k$, the evolution equation remains unaltered.

7.4 Truncated flow equations

Solving the FRGE (7.42) subject to the initial condition (7.44) is equivalent to (and in practice as difficult as) calculating the original functional integral over $\gamma_{\mu\nu}$. It is therefore important to devise efficient approximation methods. The truncation of theory space is the one which makes maximum use of the FRGE reformulation of the quantum field theory problem at hand.

As for the flow on the theory space $\{A[g,\bar{g},\xi,\bar{\xi}]\}$, a still very general truncation consists in neglecting the evolution of the ghost action by making the ansatz

$$\Gamma_k[g,\bar{g},\xi,\bar{\xi}] = \bar{\Gamma}_k[g] + \widehat{\Gamma}_k[g,\bar{g}] + S_{\rm gf}[g-\bar{g};\bar{g}] + S_{\rm gh}[g-\bar{g},\xi,\bar{\xi};\bar{g}]\,, \quad (7.45)$$

where we have extracted the classical $S_{\rm gf}$ and $S_{\rm gh}$ from Γ_k. The remaining functional depends on both $g_{\mu\nu}$ and $\bar{g}_{\mu\nu}$. It is further decomposed as $\bar{\Gamma}_k + \widehat{\Gamma}_k$ where $\bar{\Gamma}_k$ is defined as in (7.41) and $\widehat{\Gamma}_k$ contains the deviations for $\bar{g} \neq g$. Hence, by definition, $\widehat{\Gamma}_k[g,g] = 0$, and $\widehat{\Gamma}_k$ contains in particular quantum corrections to the gauge fixing term which vanishes for $\bar{g} = g$, too. This ansatz satisfies the initial condition (7.44) if

$$\bar{\Gamma}_{k\to\infty} = S \qquad \text{and} \qquad \widehat{\Gamma}_{k\to\infty} = 0\,. \quad (7.46)$$

Inserting (7.45) into the exact FRGE (7.42), one obtains an evolution equation on the truncated space $\{A[g,\bar{g}]\}$:

$$\begin{aligned}\partial_t \Gamma_k[g,\bar{g}] = {} & \frac{1}{2}\mathrm{Tr}\left[\left(\kappa^{-2}\Gamma_k^{(2)}[g,\bar{g}] + \mathcal{R}_k^{\rm grav}[\bar{g}]\right)^{-1} \partial_t \mathcal{R}_k^{\rm grav}[\bar{g}]\right] \\ & - \mathrm{Tr}\left[\left(-\mathcal{M}[g,\bar{g}] + \mathcal{R}_k^{\rm gh}[\bar{g}]\right)^{-1} \partial_t \mathcal{R}_k^{\rm gh}[\bar{g}]\right]. \end{aligned} \quad (7.47)$$

This equation evolves the functional

$$\Gamma_k[g,\bar{g}] \equiv \bar{\Gamma}_k[g] + S_{\rm gf}[g-\bar{g};\bar{g}] + \widehat{\Gamma}_k[g,\bar{g}]\,. \quad (7.48)$$

Here $\Gamma_k^{(2)}$ denotes the Hessian of $\Gamma_k[g,\bar{g}]$ with respect to $g_{\mu\nu}$ at fixed $\bar{g}_{\mu\nu}$.

The truncation ansatz (7.45) is still too general for practical calculations to be easily possible. The first truncation for which the RG flow has been found [1] is the *Einstein–Hilbert truncation*, which retains in $\bar{\Gamma}_k[g]$ only the terms $\int d^dx\sqrt{g}$ and $\int d^dx\sqrt{g}R$, already present in the in the classical action, with k-dependent coupling constants, and includes only the wave function renormalization in $\widehat{\Gamma}_k$:

$$\Gamma_k[g,\bar{g}] = 2\kappa^2 Z_{Nk} \int d^dx\,\sqrt{g}\left\{-R(g) + 2\bar{\lambda}_k\right\} + \frac{Z_{Nk}}{2\alpha}\int d^dx\,\sqrt{\bar{g}}\,\bar{g}^{\mu\nu}F_\mu F_\nu\,. \tag{7.49}$$

In this case the truncation subspace is two-dimensional. The ansatz (7.49) contains two free functions of the scale, the running cosmological constant $\bar{\lambda}_k$, and Z_{Nk} or, equivalently, the running Newton constant $G_k \equiv \bar{G}/Z_{Nk}$. Here $\bar{G}$ is a fixed constant, and $\kappa \equiv (32\pi\bar{G})^{-1/2}$. As for the gauge-fixing term, F_μ is given by Equation (7.30) with $\bar{h}_{\mu\nu} \equiv g_{\mu\nu} - \bar{g}_{\mu\nu}$ replacing $h_{\mu\nu}$; it vanishes for $g = \bar{g}$. The ansatz (7.49) has the general structure of (7.45) with $\widehat{\Gamma}_k = (Z_{Nk} - 1)S_{\text{gf}}$. Within the Einstein–Hilbert approximation the gauge-fixing parameter α is kept constant. Here we shall set $\alpha = 1$ and comment on generalizations later on.

Upon inserting the ansatz (7.49) into the flow equation (7.47), it boils down to a system of two ordinary differential equations for Z_{Nk} and $\bar{\lambda}_k$. Their derivation is rather technical, so we shall focus on the conceptual aspects here. In order to find $\partial_t Z_{Nk}$ and $\partial_t\bar{\lambda}_k$ it is sufficient to consider (7.47) for $g_{\mu\nu} = \bar{g}_{\mu\nu}$. In this case the LHS of the flow equation becomes $2\kappa^2\int d^dx\sqrt{g}[-R(g)\partial_t Z_{Nk} + 2\partial_t(Z_{Nk}\bar{\lambda}_k)]$. The RHS is assumed to admit an expansion in terms of invariants $P_i[g_{\mu\nu}]$. In the Einstein–Hilbert truncation only two of them, $\int d^dx\sqrt{g}$ and $\int d^dx\sqrt{g}R$, need to be retained. They can be extracted from the traces in (7.47) by standard derivative expansion techniques. On equating the result to the LHS and comparing the coefficients of $\int d^dx\sqrt{g}$ and $\int d^dx\sqrt{g}R$, a pair of coupled differential equations for Z_{Nk} and $\bar{\lambda}_k$ arises. It is important to note that, on the RHS, we may set $g_{\mu\nu} = \bar{g}_{\mu\nu}$ only *after* the functional derivatives of $\Gamma_k^{(2)}$ have been obtained, for they must be taken at fixed $\bar{g}_{\mu\nu}$.

In principle this calculation can be performed without ever considering any specific metric $g_{\mu\nu} = \bar{g}_{\mu\nu}$. This reflects the fact that the approach is background covariant. The RG flow is universal in the sense that it does not depend on any specific metric. In this respect gravity is not different from the more traditional applications of the renormalization group: the RG flow in the Ising universality class, say, has nothing to do with any specific spin configuration; it rather reflects the statistical properties of many such configurations.

Although there is no conceptual necessity to fix the background metric, it nevertheless is sometimes advantageous from a computational point of view to pick a specific class of backgrounds. Leaving $\bar{g}_{\mu\nu}$ completely general, the calculation of the functional traces is usually hard work. In principle there exist well-known derivative expansion and heat kernel techniques, which could be used for this purpose, but their application is usually an extremely lengthy and tedious task. Moreover, typically the operators $\Gamma_k^{(2)}$ and $\mathcal{R}_k$ are of a complicated nonstandard type, so that no efficient use of the tabulated Seeley coefficients can be made. However, often calculations of this type simplify if one can assume that $g_{\mu\nu} = \bar{g}_{\mu\nu}$ has specific properties. Because the beta functions are background independent, we may therefore restrict $\bar{g}_{\mu\nu}$ to lie in a conveniently chosen class of geometries which is still general enough to disentangle the invariants retained and at the same time simplifies the calculation.

For the Einstein–Hilbert truncation the most efficient choice is a family of d-spheres $S^d(r)$, labeled by their radius r. For those geometries, $D_\alpha R_{\mu\nu\rho\sigma} = 0$, so they give a vanishing value to all invariants constructed from $g = \bar{g}$ containing covariant derivatives acting on curvature tensors. What remains (among the local invariants) are terms of the form $\int \sqrt{g} P(R)$, where P is a polynomial in the Riemann tensor with arbitrary index contractions. To linear order in the (contractions of the) Riemann tensor the two invariants relevant for the Einstein–Hilbert truncation are discriminated by the S^d metrics as the latter scale differently with the radius of the sphere: $\int \sqrt{g} \sim r^d$, $\int \sqrt{g} R \sim r^{d-2}$. Thus, in order to compute the beta functions of $\bar{\lambda}_k$ and Z_{Nk} it is sufficient to insert an S^d metric with arbitrary r and to compare the coefficients of r^d and r^{d-2}. If one wants to do better and include the three quadratic invariants $\int R_{\mu\nu\rho\sigma} R^{\mu\nu\rho\sigma}$, $\int R_{\mu\nu} R^{\mu\nu}$, and $\int R^2$, the family $S^d(r)$ is not general enough to separate them; all scale like r^{d-4} with the radius.

Under the trace we need the operator $\Gamma_k^{(2)}[\bar{h}; \bar{g}]$. It is most easily calculated by Taylor expanding the truncation ansatz, $\Gamma_k[\bar{g} + \bar{h}, \bar{g}] = \Gamma_k[\bar{g}, \bar{g}] + O(\bar{h}) + \Gamma_k^{\text{quad}}[\bar{h}; \bar{g}] + O(\bar{h}^3)$, and stripping off the two $\bar{h}$s from the quadratic term, $\Gamma_k^{\text{quad}} = \frac{1}{2} \int \bar{h} \Gamma_k^{(2)} \bar{h}$. For $\bar{g}_{\mu\nu}$ the metric on $S^d(r)$ one obtains

$$\Gamma_k^{\text{quad}}[\bar{h}; \bar{g}] = \frac{1}{2} Z_{Nk} \kappa^2 \int d^d x \left\{ \widehat{h}_{\mu\nu} \left[-\bar{D}^2 - 2\bar{\lambda}_k + C_T \bar{R} \right] \widehat{h}^{\mu\nu} \right.$$
$$\left. - \left(\frac{d-2}{2d} \right) \phi \left[-\bar{D}^2 - 2\bar{\lambda}_k + C_S \bar{R} \right] \phi \right\}, \tag{7.50}$$

with $C_T \equiv (d(d-3)+4)/(d(d-1))$, $C_S \equiv (d-4)/d$. In order to partially diagonalize this quadratic form, $\bar{h}_{\mu\nu}$ has been decomposed into a traceless part $\widehat{h}_{\mu\nu}$ and the trace part proportional to ϕ: $\bar{h}_{\mu\nu} = \widehat{h}_{\mu\nu} + d^{-1} \bar{g}_{\mu\nu} \phi$, $\bar{g}^{\mu\nu} \widehat{h}_{\mu\nu} = 0$. Further, $\bar{D}^2 = \bar{g}^{\mu\nu} \bar{D}_\mu \bar{D}_\nu$ is the covariant Laplace operator corresponding to the background geometry, and $\bar{R} = d(d-1)/r^2$ is the numerical value of the curvature scalar on $S^d(r)$.

At this point we can fix the constants $\mathcal{Z}_k$ which appear in the cutoff operators $\mathcal{R}_k^{\text{grav}}$ and $\mathcal{R}_k^{\text{gh}}$ of (7.34). They should be adjusted in such a way that for every low-momentum mode the cutoff combines with the kinetic term of this mode to form $-\bar{D}^2 + k^2$ times a constant. Looking at (7.50), we see that the respective kinetic terms for $\widehat{h}_{\mu\nu}$ and ϕ differ by a factor of $-(d-2)/2d$. This suggests the following choice:

$$\left(\mathcal{Z}_k^{\text{grav}}\right)^{\mu\nu\rho\sigma} = \left[\left(\mathbb{1} - P_\phi\right)^{\mu\nu\rho\sigma} - \frac{d-2}{2d} P_\phi^{\mu\nu\rho\sigma}\right] Z_{Nk} \,. \tag{7.51}$$

Here $(P_\phi)_{\mu\nu}{}^{\rho\sigma} = d^{-1}\bar{g}_{\mu\nu}\bar{g}^{\rho\sigma}$ is the projector on the trace part of the metric. For the traceless tensor, (7.51) gives $\mathcal{Z}_k^{\text{grav}} = Z_{Nk}\mathbb{1}$, and for ϕ the different relative normalization is taken into account. (See [1] for a detailed discussion of the subtleties related to this choice.) Thus we obtain in the $\widehat{h}$ and the ϕ sector, respectively,

$$\begin{aligned} \left(\kappa^{-2}\Gamma_k^{(2)}[g,g] + \mathcal{R}_k^{\text{grav}}\right)_{\widehat{h}\widehat{h}} &= Z_{Nk}\left[-D^2 + k^2 R^{(0)}(-D^2/k^2) - 2\bar{\lambda}_k + C_T R\right], \\ \left(\kappa^{-2}\Gamma_k^{(2)}[g,g] + \mathcal{R}_k^{\text{grav}}\right)_{\phi\phi} & \\ = -\frac{d-2}{2d} Z_{Nk}\left[-D^2 + k^2 R^{(0)}(-D^2/k^2) - 2\bar{\lambda}_k + C_S R\right]. \end{aligned} \tag{7.52}$$

From now on we may set $\bar{g} = g$, and for simplicity we have omitted the bars from the metric and the curvature. Inasmuch as we did not take into account any renormalization effects in the ghost action, we set $Z_k^{\text{gh}} \equiv 1$ in $\mathcal{R}_k^{\text{gh}}$ and obtain

$$-\mathcal{M} + \mathcal{R}_k^{\text{gh}} = -D^2 + k^2 R^{(0)}(-D^2/k^2) + C_V R\,, \tag{7.53}$$

with $C_V \equiv -1/d$. At this point the operator under the first trace on the RHS of (7.47) has become block diagonal, with the $\widehat{h}\widehat{h}$ and $\phi\phi$ blocks given by (7.52). Both block operators are expressible in terms of the Laplacian D^2, in the former case acting on traceless symmetric tensor fields, in the latter on scalars. The second trace in (7.47) stems from the ghosts; it contains (7.53) with D^2 acting on vector fields.

It is now a matter of straightforward algebra to compute the first two terms in the derivative expansion of those traces, proportional to $\int d^d x\sqrt{g} \sim r^d$ and $\int d^d x\sqrt{g}R \sim r^{d-2}$. Considering the trace of an arbitrary function of the Laplacian, $W(-D^2)$, the expansion up to second derivatives of the metric is given by

$$\begin{aligned} \text{Tr}[W(-D^2)] = (4\pi)^{-d/2}\text{tr}(I)\Big\{ & Q_{d/2}[W]\int d^d x\,\sqrt{g} \\ & + \frac{1}{6}Q_{d/2-1}[W]\int d^d x\,\sqrt{g}R + O(R^2)\Big\}. \end{aligned} \tag{7.54}$$

The Q_n's are defined as

$$Q_n[W] = \frac{1}{\Gamma(n)} \int_0^\infty dz\, z^{n-1} W(z) \tag{7.55}$$

for $n > 0$, and $Q_0[W] = W(0)$ for $n = 0$. The trace $\mathrm{tr}(I)$ counts the number of independent field components. It equals 1, d, and $(d-1)(d+2)/2$ for scalars, vectors, and symmetric traceless tensors, respectively. The expansion (7.54) is easily derived using standard heat kernel and Mellin transform techniques [1].

Using (7.54) it is easy to calculate the traces in (7.47) and to obtain the RG equations in the form $\partial_t Z_{Nk} = \cdots$ and $\partial_t (Z_{Nk}\bar{\lambda}_k) = \cdots$. We shall not display them here, because it is more convenient to rewrite them in terms of the dimensionless running cosmological constant and Newton constant, respectively:

$$\lambda_k \equiv k^{-2}\bar{\lambda}_k\,, \qquad g_k \equiv k^{d-2} G_k \equiv k^{d-2} Z_{Nk}^{-1} \bar{G}\,. \tag{7.56}$$

Recall that the dimensionful running Newton constant is given by $G_k = Z_{Nk}^{-1}\bar{G}$. In terms of the dimensionless couplings g and λ the RG equations become a system of autonomous differential equations:

$$\begin{aligned} \partial_t g_k &= \left[d - 2 + \eta_N(g_k, \lambda_k)\right] g_k \equiv \beta_g(g_k, \lambda_k)\,, \\ \partial_t \lambda_k &= \beta_\lambda(g_k, \lambda_k)\,. \end{aligned} \tag{7.57}$$

Here $\eta_N \equiv -\partial_t \ln Z_{Nk}$ is the anomalous dimension of the operator $\sqrt{g}R$,

$$\eta_N(g_k, \lambda_k) = \frac{g_k\, B_1(\lambda_k)}{1 - g_k\, B_2(\lambda_k)}\,, \tag{7.58}$$

with the following functions of λ_k:

$$\begin{aligned} B_1(\lambda_k) \equiv \frac{1}{3}(4\pi)^{1-d/2} \Big[& d(d+1)\Phi^1_{d/2-1}(-2\lambda_k) - 6d(d-1)\Phi^2_{d/2}(-2\lambda_k), \\ & -4d\,\Phi^1_{d/2-1}(0) - 24\Phi^2_{d/2}(0)\Big], \end{aligned} \tag{7.59}$$

$$B_2(\lambda_k) \equiv -\frac{1}{6}(4\pi)^{1-d/2}\left[d(d+1)\widetilde{\Phi}^1_{d/2-1}(-2\lambda_k) - 6d(d-1)\widetilde{\Phi}^2_{d/2}(-2\lambda_k)\right].$$

The beta function for λ is given by a similar expression:

$$\begin{aligned} \beta_\lambda(g_k, \lambda_k) = -(2 - \eta_N)\lambda_k + \frac{1}{2} g_k (4\pi)^{1-d/2}\Big[& 2d(d+1)\Phi^1_{d/2}(-2\lambda_k) \\ & - 8d\,\Phi^1_{d/2}(0) - d(d+1)\eta_N \widetilde{\Phi}^1_{d/2}(-2\lambda_k)\Big]. \end{aligned} \tag{7.60}$$

The *threshold functions* Φ and $\widetilde{\Phi}$ appearing in (7.59) and (7.60) are certain integrals involving the normalized cutoff function $R^{(0)}$:

$$\begin{aligned}
\Phi_n^p(w) &\equiv \frac{1}{\Gamma(n)} \int_0^\infty dz\, z^{n-1} \frac{R^{(0)}(z) - z R^{(0)\prime}(z)}{[z + R^{(0)}(z) + w]^p}\,, \\
\widetilde{\Phi}_n^p(w) &\equiv \frac{1}{\Gamma(n)} \int_0^\infty dz\, z^{n-1} \frac{R^{(0)}(z)}{[z + R^{(0)}(z) + w]^p}\,.
\end{aligned} \tag{7.61}$$

They are defined for positive integers p, and $n > 0$.

With the derivation of the system (7.57) we managed to find an approximation to a two-dimensional projection of the RG flow. Its properties, and in particular the domain of applicability and reliability of the Einstein–Hilbert truncation, will be discussed in the following section.

Although there are (a few) aspects of the truncated RG flow which are independent of the cutoff scheme, i.e., independent of the function $R^{(0)}$, the explicit solution of the flow equation requires a specific choice of this function. As we discussed already, the normalized cutoff function $R^{(0)}(w)$, $w = p^2/k^2$, describes the shape of $\mathcal{R}_k(p^2)$ in the transition region, where it interpolates between the prescribed behavior for $p^2 \ll k^2$ and $p^2 \gg k^2$, respectively, and is therefore referred to as the *shape function*. In the literature various forms of $R^{(0)}$'s have been employed. Easy to handle, but disadvantageous for high-precision calculations, is the sharp cutoff [4] defined by $\mathcal{R}_k(p^2) = \lim_{\hat{R}\to\infty} \hat{R}\,\theta(1 - p^2/k^2)$, where the limit is to be taken after the p^2 integration. This cutoff allows for an evaluation of the Φ and $\widetilde{\Phi}$ integrals in closed form. Taking $d = 4$ as an example, the equations (7.57) boil down to the following simple system of equations:[3]

$$\partial_t \lambda_k = -(2 - \eta_N)\lambda_k - \frac{g_k}{\pi}\left[5\ln(1 - 2\lambda_k) - 2\zeta(3) + \frac{5}{2}\eta_N\right], \tag{7.62a}$$

$$\partial_t g_k = (2 + \eta_N)\, g_k, \tag{7.62b}$$

$$\eta_N = -\frac{2\, g_k}{6\pi + 5\, g_k}\left[\frac{18}{1 - 2\lambda_k} + 5\ln(1 - 2\lambda_k) - \zeta(2) + 6\right]. \tag{7.62c}$$

Also, the *optimized cutoff* [44] with $R^{(0)}(w) = (1 - w)\theta(1 - w)$ allows for an analytic evaluation of the integrals [14]. In order to check the scheme (in)dependence of the results it is desirable to perform the calculation for a whole class of $R^{(0)}$s. For this purpose the following one-parameter family of exponential cutoffs has been used [3, 5, 8]:

$$R^{(0)}(w; s) = \frac{s w}{e^{s w} - 1}\,. \tag{7.63}$$

[3] To be precise, (7.62c) corresponds to the sharp cutoff with $s = 1$; see [4].

The precise form of the cutoff is controlled by the *shape parameter* s. For $s = 1$, (7.63) coincides with the standard exponential cutoff. The exponential cutoffs are suitable for precision calculations, but the price to be paid is that their Φ and $\widetilde{\Phi}$ integrals can be evaluated only numerically. The same is true for a one-parameter family of shape functions with compact support, which was used in [3, 5].

We have illustrated the general ideas and constructions underlying gravitational RG flows by means of the simplest example, the Einstein–Hilbert truncation. In the literature various extensions have been investigated. The derivation and analysis of these more general flow equations, corresponding to higher-dimensional truncation subspaces, is an extremely complex and calculationally demanding problem in general. For this reason we cannot go into the technical details here and just mention some further developments.

7.4.1

The natural next step beyond the Einstein–Hilbert truncation consists in generalizing the functional $\bar{\Gamma}_k[g]$, while keeping the gauge fixing and ghost sector classical, as in (7.45). During the RG evolution the flow generates all possible diffeomorphism-invariant terms in $\bar{\Gamma}_k[g]$ which one can construct from $g_{\mu\nu}$. Both local and nonlocal terms are induced. The local invariants contain strings of curvature tensors and covariant derivatives acting upon them, with any number of tensors and derivatives, and of all possible index structures. The first truncation of this class which has been worked out completely [5, 6] is the R^2 *truncation* defined by (7.45) with the same $\widehat{\Gamma}_k$ as before, and the curvature-squared action

$$\bar{\Gamma}_k[g] = \int d^d x \sqrt{g} \Big\{ (16\pi G_k)^{-1} [-R(g) + 2\bar{\lambda}_k] + \bar{\beta}_k R^2(g) \Big\} . \qquad (7.64)$$

In this case the truncated theory space is three-dimensional. Its natural (dimensionless) coordinates are (g, λ, β), where $\beta_k \equiv k^{4-d} \bar{\beta}_k$, and g and λ are defined in (7.56). Even though (7.64) contains only one additional invariant, the derivation of the corresponding RG equations is far more complicated than in the Einstein–Hilbert case. We shall summarize the results obtained with (7.64) [5, 6] in Section 7.6.

7.4.2

As for generalizing the ghost sector of the truncation beyond (7.45), no results are available yet, but there is a partial result concerning the gauge-fixing term. Even if one makes the ansatz (7.49) for $\Gamma_k[g, \bar{g}]$ in which the gauge-fixing term has the classical (or, more appropriately, bare) structure, one should treat its prefactor as a running coupling: $\alpha = \alpha_k$. The beta function of α has not been determined yet from the FRGE, but there is a simple argument which allows us to bypass this calculation.

In nonperturbative Yang–Mills theory and in perturbative quantum gravity, $\alpha = \alpha_k = 0$ is known to be a fixed point for the α evolution. The following reasoning suggests that the same is true within the nonperturbative FRGE approach to gravity. In the standard functional integral the limit $\alpha \to 0$ corresponds to a sharp implementation of the gauge-fixing condition, i.e., $\exp(-S_{\text{gf}})$ becomes proportional to $\delta[F_\mu]$. The domain of the $\int \mathcal{D}h_{\mu\nu}$ integration consists of those $h_{\mu\nu}$s which satisfy the gauge-fixing condition exactly, $F_\mu = 0$. Adding the IR cutoff at k amounts to suppressing some of the $h_{\mu\nu}$ modes while retaining the others. But because all of them satisfy $F_\mu = 0$, a variation of k cannot change the domain of the $h_{\mu\nu}$ integration. The delta functional $\delta[F_\mu]$ continues to be present for any value of k if it was there originally. As a consequence, α vanishes for all k, i.e., $\alpha = 0$ is a fixed point of the α evolution [48].

Thus we can mimic the dynamical treatment of a running α by setting the gauge-fixing parameter to the constant value $\alpha = 0$. The calculation for $\alpha = 0$ is more complicated than at $\alpha = 1$, but for the Einstein–Hilbert truncation the α-dependence of β_g and β_λ, for arbitrary constant α, has been found in [3,49]. The R^2 truncations could be analyzed only in the simple $\alpha = 1$ gauge, but the results from the Einstein–Hilbert truncation suggest the UV quantities of interest do not change much between $\alpha = 0$ and $\alpha = 1$ [3,5].

7.4.3

Up to now we have considered pure gravity. For the general formalism, the inclusion of matter fields is straightforward. The structure of the flow equation remains unaltered, except that now $\Gamma_k^{(2)}$ and $\mathcal{R}_k$ are operators on the larger Hilbert space of both gravity and matter fluctuations. In practice, however, the derivation of the projected RG equations can be quite a formidable task, the difficult part being the decoupling of the various modes (diagonalization of $\Gamma_k^{(2)}$), which in most calculational schemes is necessary for the computation of the functional traces. Various matter systems, both interacting and noninteracting (apart from their interaction with gravity) have been studied in the literature [2,50,51]. A rather detailed analysis has been performed by Percacci et al. In [2,12] arbitrary multiplets of free (massless) fields with spin 0, 1/2, 1, and 3/2 were included. In [12] an interacting scalar theory coupled to gravity in the Einstein–Hilbert approximation was analyzed, and a possible solution to the triviality and the hierarchy problem [16] was proposed in this context.

7.4.4

Finally we mention another generalization of the simplest case that we have reviewed, which is of a more technical nature [3]. In order to facilitate the calculation of the functional traces it is helpful to employ a transverse traceless (TT) decomposition of the metric: $h_{\mu\nu} = h^T_{\mu\nu} + \bar{D}_\mu V_\nu + \bar{D}_\nu V_\mu + \bar{D}_\mu \bar{D}_\nu \sigma - d^{-1}\bar{g}_{\mu\nu}\bar{D}^2\sigma + d^{-1}\bar{g}_{\mu\nu}\phi$. Here $h^T_{\mu\nu}$ is a transverse traceless tensor, V_μ a transverse vector, and σ and ϕ are scalars. In this framework it is natural to formulate the cutoff in terms of the component fields appearing in the TT decomposition: $\Delta_k S \sim \int h^T_{\mu\nu}\mathcal{R}_k h^{T\mu\nu} + \int V_\mu \mathcal{R}_k V^\mu + \cdots$. This cutoff is referred to as a cutoff of *type B*, in contradistinction to the *type A* cutoff described in Section 7.3: $\Delta_k S \sim \int h_{\mu\nu}\mathcal{R}_k h^{\mu\nu}$. Because covariant derivatives do not commute, the two cutoffs are not exactly equal even if they contain the same shape function. Thus, comparing type A and type B cutoffs is an additional possibility for checking scheme (in)dependence [3, 5].

7.5 Asymptotic Safety

In intuitive terms, the basic idea of asymptotic safety can be understood as follows. The boundary of theory space depicted in Figure 7.1 is meant to separate points with coordinates $\{u_\alpha, \alpha = 1, 2, \ldots\}$ with all the essential couplings u_α well defined, from points with undefined, divergent couplings. The basic task of renormalization theory consists in constructing an *infinitely long* RG trajectory which lies entirely within this theory space, i.e., a trajectory which does not leave theory space (that is, develops divergences) either in the UV limit $k \to \infty$ or in the IR limit $k \to 0$. Every such trajectory defines one possible quantum theory.

The idea of asymptotic safety is to perform the UV limit $k \to \infty$ at a fixed point $\{u^*_\alpha, \alpha = 1, 2, \ldots\} \equiv u^*$ of the RG flow. The fixed point is a zero of the vector field $\vec{\beta} \equiv (\beta_\alpha)$, i.e., $\beta_\alpha(u^*) = 0$ for all $\alpha = 1, 2, \ldots$. The RG trajectories, solutions of $k\partial_k u_\alpha(k) = \beta_\alpha(u(k))$, have a low "velocity" near or a fixed point because the β_αs are small there and directly at the fixed point the running stops completely. As a result, one can "use up" an infinite amount of RG time near or at the fixed point if one bases the quantum theory on a trajectory which runs into such a fixed point for $k \to \infty$. This is the key idea of asymptotic safety: If in the UV limit the trajectory ends at a fixed point, an *inner point* of theory space giving rise to a well-behaved action functional, we can be sure that, for $k \to \infty$, the trajectory does not escape from theory space, i.e., does not develop pathological properties such as divergent couplings. For $k \to \infty$ the resulting quantum theory is *asymptotically safe* from unphysical divergences. In the context of gravity, Weinberg [17] proposed to use a non-Gaussian fixed point (NGFP) for letting $k \to \infty$. By definition, not all of its coordinates u^*_α vanish.[4]

Recall from Section 2.2 that the coordinates u_α are the *dimensionless* essential couplings related to the dimensionful ones $\bar{u}_\alpha$ by $u_\alpha \equiv k^{-d_\alpha} \bar{u}_\alpha$. Hence the running of the $\bar{u}$'s is given by

$$\bar{u}_\alpha(k) = k^{d_\alpha}\, u_\alpha(k)\,. \tag{7.65}$$

Therefore, even directly at a NGFP where $u_\alpha(k) \equiv u^*_\alpha$, the dimensionful couplings keep running according to a power law involving their canonical dimensions d_α:

$$\bar{u}_\alpha(k) = u^*_\alpha\, k^{d_\alpha}\,. \tag{7.66}$$

Furthermore, nonessential dimensionless couplings are not required to attain fixed-point values.

Given a NGFP, an important concept is its *UV critical hypersurface* $\mathcal{S}_{\rm UV}$, or synonymously, its *unstable manifold*. By definition, it consists of all points of theory space which are pulled into the NGFP by the inverse RG flow, i.e., for *increasing* k. Its dimensionality $\dim(\mathcal{S}_{\rm UV}) \equiv \Delta_{\rm UV}$ is given by the number of attractive (for *increasing* cutoff k) directions in the space of couplings.

Writing the RG equations as $k\,\partial_k u_\alpha = \beta_\alpha(u_1, u_2, \ldots)$, the linearized flow near the fixed point is governed by the Jacobi matrix $\mathbf{B} = (B_{\alpha\gamma})$, $B_{\alpha\gamma} \equiv \partial_\gamma \beta_\alpha(u^*)$:

$$k\,\partial_k\, u_\alpha(k) = \sum_\gamma B_{\alpha\gamma}\, \left(u_\gamma(k) - u^*_\gamma\right). \tag{7.67}$$

The general solution to this equation reads

$$u_\alpha(k) = u^*_\alpha + \sum_I C_I\, V^I_\alpha \left(\frac{k_0}{k}\right)^{\theta_I}, \tag{7.68}$$

where the V^Is are the right eigenvectors of $\mathbf{B}$ with eigenvalues $-\theta_I$, i.e., $\sum_\gamma B_{\alpha\gamma}\, V^I_\gamma = -\theta_I\, V^I_\alpha$. Inasmuch as $\mathbf{B}$ is not symmetric in general, the θ_I's are not guaranteed to be real. We assume that the eigenvectors form a complete system, though. Furthermore, k_0 is a fixed reference scale, and the C_I's are constants of integration.

If $u_\alpha(k)$ is to describe a trajectory in $\mathcal{S}_{\rm UV}$, then $u_\alpha(k)$ must approach u^*_α in the limit $k \to \infty$, and therefore we must set $C_I = 0$ for all I with $\operatorname{Re}\theta_I < 0$. Hence the dimensionality $\Delta_{\rm UV}$ equals the number of $\mathbf{B}$-eigenvalues with a negative real part, i.e., the number of θ_Is with $\operatorname{Re}\theta_I > 0$. The corresponding eigenvectors span the tangent space to $\mathcal{S}_{\rm UV}$ at the NGFP.

[4] In contrast, $u^*_\alpha = 0\,\forall\alpha = 1, 2, \ldots$ is a so-called Gaussian fixed point (GFP). In a sense, standard perturbation theory takes the $k \to \infty$ limit at the GFP; see [18] for a detailed discussion.

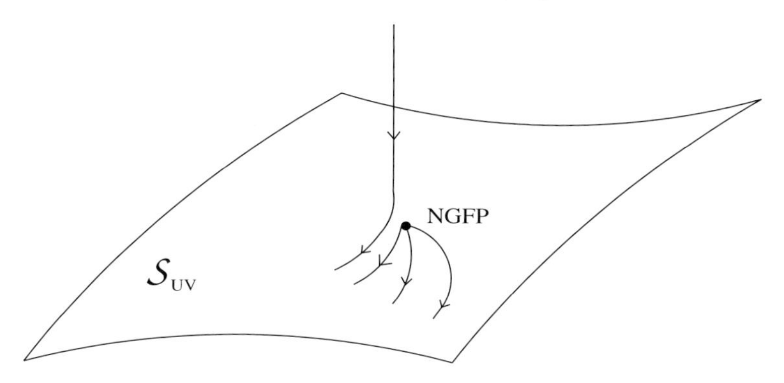

Fig. 7.2. Schematic picture of the UV critical hypersurface $\mathcal{S}_{\mathrm{UV}}$ of the NGFP. It is spanned by RG trajectories emanating from the NGFP as the RG scale k is lowered. Trajectories not in the surface are attracted towards $\mathcal{S}_{\mathrm{UV}}$ as k decreases. (The arrows point in the direction of decreasing k, from the UV to the IR.)

If $u_\alpha(k)$ describes a generic trajectory with all C_I nonzero and we *lower* the cutoff, only Δ_{UV} *relevant* parameters corresponding to the eigendirections tangent to $\mathcal{S}_{\mathrm{UV}}$ grow ($\mathrm{Re}\,\theta_I > 0$), while the remaining *irrelevant* couplings pertaining to the eigendirections normal to $\mathcal{S}_{\mathrm{UV}}$ decrease ($\mathrm{Re}\,\theta_I < 0$). Thus near the NGFP a generic trajectory is attracted towards $\mathcal{S}_{\mathrm{UV}}$; see Figure 7.2.

Coming back to the asymptotic safety construction, let us now use this fixed point in order to take the limit $k \to \infty$. The trajectories which define an infinite-cutoff limit for QEG are special in that all irrelevant couplings are set to zero: $C_I = 0$ if $\mathrm{Re}\,\theta_I < 0$. These conditions place the trajectory exactly on $\mathcal{S}_{\mathrm{UV}}$. There is a Δ_{UV}-parameter family of such trajectories, and the experiment must decide which one is realized in nature. Therefore the predictive power of the theory increases with decreasing dimensionality of $\mathcal{S}_{\mathrm{UV}}$, i.e., number of UV attractive eigendirections of the NGFP. (If $\Delta_{\mathrm{UV}} < \infty$, the quantum field theory thus constructed is comparable to and as predictive as a perturbatively renormalizable model with Δ_{UV} *renormalizable* couplings, i.e., couplings relevant at the GFP.)

The quantities θ_I are referred to as critical exponents, because when the renormalization group is applied to critical phenomena (second-order phase transitions) the traditionally defined critical exponents are related to the θ_Is in a simple way [23]. In fact, one of the early successes of the RG ideas was an explanation of the universality properties of critical phenomena, i.e., the fact that systems at the critical point seem to "forget" the precise form of their microdynamics and just depend on the universality class, characterized by a set of critical exponents, they belong to.

In the present context, "universality" means that certain very special quantities related to the RG flow are independent of the precise form of the cutoff and, in particular, its shape function $R^{(0)}$. Universal quantities are potentially measurable or at least closely related to observables. The θ_Is are examples of universal quantities; the coordinates of the fixed point, u_α^*, are not, even in an exact calculation. Quantities independently known to be universal provide an important tool for testing the reliability or accuracy of *approximate* RG calculations and of truncations in particular. Because they are known to be $R^{(0)}$ independent in an exact treatment, we can determine the degree of their $R^{(0)}$ dependence within the truncation and use it as a measure for the quality of the truncated calculation.

For a more detailed and formal discussion of asymptotic safety and, in particular, its relation to perturbation theory we refer to the review [18].

7.6 Average Action approach to Asymptotic Safety

Our discussion of the asymptotic safety construction in the previous section was at the level of the exact (untruncated) RG flow. In this section we are going to implement these ideas in the context of explicitly computable approximate RG flows on truncated theory spaces. We shall mostly concentrate on the Einstein–Hilbert (R) truncation and the R^2 truncation of pure gravity in $d = 4$. The corresponding d-dimensional flow equations were derived in [1] and [5], respectively.

7.6.1 The phase portrait of the Einstein–Hilbert truncation

In [4] the RG equations (7.57) implied by the Einstein–Hilbert truncation were analyzed in detail, using both analytical and numerical methods. In particular, all RG trajectories of this system were classified, and examples were computed numerically. The most important classes of trajectories in the phase portrait on the g–λ plane are shown in Figure 7.3. The trajectories were obtained by numerically solving the system (7.62) for a sharp cutoff; using a smooth one, all qualitative features remain unchanged. The RG flow is found to be dominated by two fixed points (g^*, λ^*): the GFP at $g^* = \lambda^* = 0$, and a NGFP with $g^* > 0$ and $\lambda^* > 0$. There are three classes of trajectories emanating from the NGFP: trajectories of types Ia and IIIa run toward negative and positive cosmological constants, respectively, and the single trajectory of type IIa (*separatrix*) hits the GFP for $k \to 0$. The high-momentum properties of QEG are governed by the NGFP; for $k \to \infty$, in Figure 7.3 all RG trajectories on the half plane $g > 0$ run into this point. Note that near the NGFP the dimensionful Newton constant vanishes for $k \to \infty$ according to $G_k \equiv g_k/k^2 \approx g^*/k^2 \to 0$, whereas the cosmological constant diverges: $\bar{\lambda}_k \equiv \lambda_k k^2 \approx \lambda^* k^2 \to \infty$.

So, the Einstein–Hilbert truncation does indeed predict the existence of a NGFP with exactly the properties needed for the asymptotic safety construction. Clearly the crucial question to be analyzed now is whether the NGFP found is the projection

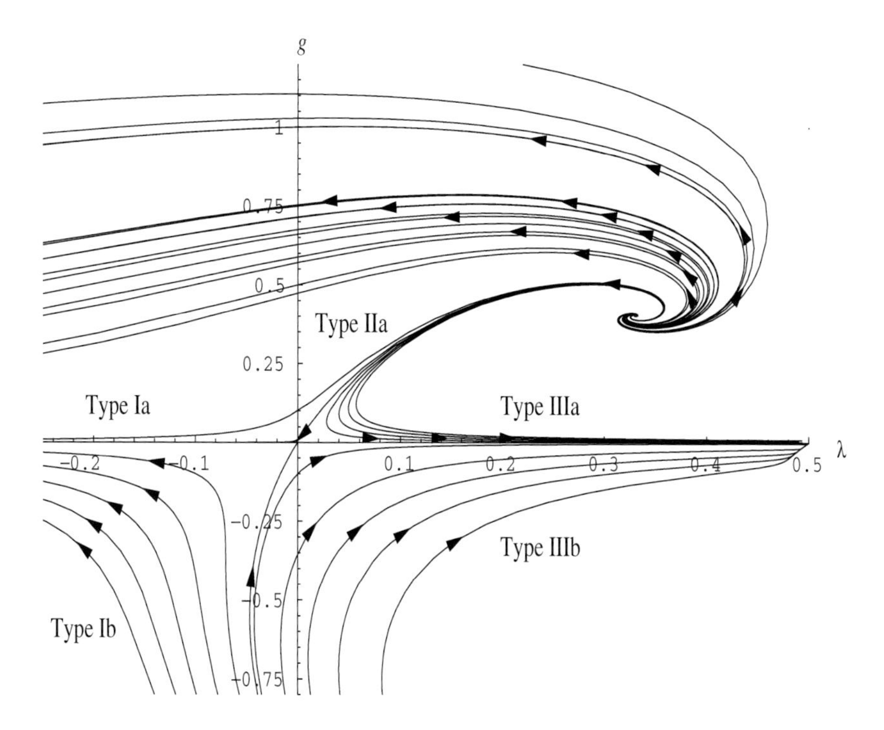

Fig. 7.3. RG flow in the g–λ plane. The arrows point in the direction of increasing coarse graining, i.e., of decreasing k. (From [4].)

of a fixed point in the exact theory on the untruncated theory space or whether it is merely the artifact of an insufficient approximation.

7.6.2 Testing the Einstein–Hilbert truncation

We mentioned already that the residual $R^{(0)}$ dependence of universal quantities is a measure for the quality of a truncation. This test has been applied to the Einstein–Hilbert truncation in [3, 8]. We shall display the results in the next subsection. In accordance with the general theory, the coordinates of the fixed point (g^*, λ^*) are not universal. However, it can be argued that they should give rise to a universal combination, the product $g^*\lambda^*$, which can be measured in principle [3]. Although k and, at a fixed value of k, G_k and $\bar{\lambda}_k$ cannot be measured separately, we may invert the function $k \mapsto G_k$ and insert the result $k = k(G)$ into $\bar{\lambda}_k$. This leads to an in principle experimentally testable relationship $\bar{\lambda} = \bar{\lambda}(G)$ between Newton's constant and the cosmological constant. Here $\bar{\lambda}$ and G should be determined in experiments involving similar scales. In the fixed-point regime this relationship reads $\bar{\lambda}(G) = g^*\lambda^*/G$. So, even if this is quite difficult in practice, one can determine the product $g^*\lambda^*$ experimentally. As a consequence, in any reliable calculation $g^*\lambda^*$ should be approximately $R^{(0)}$ independent.

The ultimate justification of a given truncation consists in checking that if one adds further terms to it, its physical predictions remain robust. The first step towards testing the robustness of the Einstein–Hilbert truncation near the NGFP against the inclusion of other invariants has been taken in [5, 6], where the R^2 truncation of Equation (7.64) has been analyzed. The corresponding beta functions for the three generalized couplings g, λ, and β have been derived, but they are too complicated to be reproduced here. Suffice it to say that on the three-dimensional (g, λ, β) space, too, a NGFP has been found which generalizes the one from the pure R calculation. This allows for a comparison of the fixed-point results for the R^2 and the Einstein–Hilbert truncation, and for a check of the approximate $R^{(0)}$ independence of universal quantities in the three-dimensional setting. For the Einstein–Hilbert truncation the universality analysis has been performed for an arbitrary constant gauge parameter α, including the "physical" value $\alpha = 0$ [3]. Because of its algebraic complexity, the R^2 analysis [5] was carried out in the simpler $\alpha = 1$ gauge.

7.6.3 Evidence for Asymptotic Safety

We now summarize the results concerning the NGFP which were obtained with the R (Sections 7.6.3.1–7.6.3.5) and R^2 truncations (Sections 7.3.1.6–7.3.1.9), respectively [3–6]. All properties mentioned in the following are independent pieces of evidence pointing in the direction that QEG is indeed asymptotically safe in four dimensions. Except for Section 7.6.3.5, all results refer to $d = 4$.

7.6.3.1 Universal existence

Both for type A and for type B cutoffs, the non-Gaussian fixed point exists for all shape functions $R^{(0)}$. (This generalizes earlier results in [8].) It seems impossible to find an admissible cutoff that destroys the fixed point in $d = 4$. This result is highly nontrivial in that in higher dimensions ($d \gtrsim 5$) the existence of the NGFP depends on the cutoff chosen [4].

7.6.3.2 Positive Newton constant

Although the position of the fixed point is scheme dependent, all cutoffs yield positive values of g^* and λ^*. A negative g^* might have been problematic for stability reasons, but there is no mechanism in the flow equation which would exclude it on general grounds.

7.6.3.3 Stability

For any cutoff employed, the NGFP is found to be UV attractive in both directions of the λ–g plane. Linearizing the flow equation according to Equation (7.67), we obtain a pair of complex conjugate critical exponents $\theta_1 = \theta_2^*$ with positive real part θ' and imaginary parts $\pm\theta''$. In terms of $t = \ln(k/k_0)$ the general solution to the linearized flow equations reads

$$(\lambda_k, g_k)^{\mathbf{T}} = (\lambda^*, g^*)^{\mathbf{T}} + 2\Big\{\big[\mathrm{Re}\, C \, \cos(\theta'' t) + \mathrm{Im}\, C \, \sin(\theta'' t)\big] \mathrm{Re}\, V + \big[\mathrm{Re}\, C \, \sin(\theta'' t) - \mathrm{Im}\, C \, \cos(\theta'' t)\big] \mathrm{Im}\, V \Big\} e^{-\theta' t} \tag{7.69}$$

with $C \equiv C_1 = (C_2)^*$ an arbitrary complex number and $V \equiv V^1 = (V^2)^*$ the right eigenvector of $\mathbf{B}$ with eigenvalue $-\theta_1 = -\theta_2^*$. Equation (7.67) implies that, due to the positivity of θ', all trajectories hit the fixed point as t is sent to infinity. The nonvanishing imaginary part θ'' has no effect on the stability. However, it influences the shape of the trajectories that spiral into the fixed point for $k \to \infty$. Thus, the fixed point has the stability properties needed in the asymptotic safety scenario.

Solving the full, nonlinear flow equations [4] shows that the asymptotic scaling region where the linearization (7.69) is valid extends from $k = \infty$ down to about $k \approx m_{\mathrm{Pl}}$ with the Planck mass defined as $m_{\mathrm{Pl}} \equiv G_0^{-1/2}$. Here m_{Pl} plays a role similar to Λ_{QCD} in QCD: it marks the lower boundary of the asymptotic scaling region. We set $k_0 \equiv m_{\mathrm{Pl}}$ so that the asymptotic scaling regime extends from about $t = 0$ to $t = \infty$.

7.6.3.4 *Scheme and gauge dependence*

Analyzing the cutoff scheme dependence of θ', θ'', and $g^*\lambda^*$ as a measure for the reliability of the truncation, the critical exponents were found to be reasonably constant, within about a factor of 2. For $\alpha = 1$ and $\alpha = 0$, for instance, they assume values in the ranges ($1.4 \lesssim \theta' \lesssim 1.8$, $2.3 \lesssim \theta'' \lesssim 4$) and ($1.7 \lesssim \theta' \lesssim 2.1$, $2.5 \lesssim \theta'' \lesssim 5$), respectively. The universality properties of the product $g^*\lambda^*$ are even more impressive. Despite the rather strong scheme dependence of g^* and λ^* separately, their product has almost no visible s-dependence for not too small values of s. Its value is

$$g^*\lambda^* \approx \begin{cases} 0.12 & \text{for } \alpha = 1, \\ 0.14 & \text{for } \alpha = 0. \end{cases} \tag{7.70}$$

The differences between the "physical" (fixed-point) value of the gauge parameter, $\alpha = 0$, and the technically more convenient $\alpha = 1$ are at the level of about 10 to 20 percent.

7.6.3.5 *Higher and lower dimensions*

The beta functions implied by the FRGE are continuous functions of the spacetime dimensionality, and it is instructive to analyze them for $d \neq 4$. In [1] it has been shown that for $d = 2 + \epsilon$, $|\epsilon| \ll 1$, the FRGE reproduces Weinberg's [17] fixed

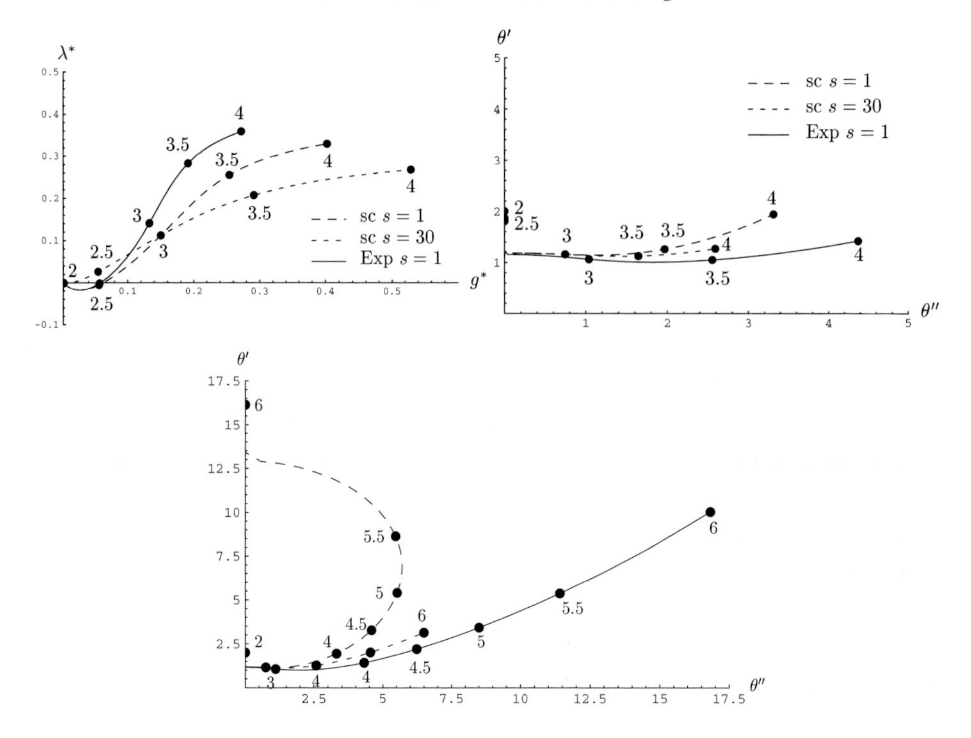

Fig. 7.4. Comparison of λ^*, g^*, θ', and θ'' for different cutoff functions in dependence on the dimension d. Two versions of the sharp cutoff (sc) and the exponential cutoff with $s = 1$ (Exp) have been employed. The upper line shows that for $2 + \epsilon \leq d \leq 4$ the cutoff scheme dependence of the results is rather small. The lower diagram shows that increasing d beyond about 5 leads to a significant difference in the results for θ', θ'' obtained with the different cutoff schemes. (From [4].)

point for Newton's constant, $g^* = \frac{3}{38}\epsilon$, and also supplies a corresponding fixed-point value for the cosmological constant, $\lambda^* = -\frac{3}{38}\Phi_1^1(0)\epsilon$, with the threshold function given in (7.61). For arbitrary d and a generic cutoff the RG flow is quantitatively similar to the four-dimensional one for all d smaller than a certain critical dimension d_{crit}, above which the existence or nonexistence of the NGFP becomes cutoff dependent. The critical dimension is scheme dependent, but for any admissible cutoff it lies well above $d = 4$. As d approaches d_{crit} from below, the scheme dependence of the universal quantities increases drastically, indicating that the R truncation becomes insufficient near d_{crit}.

In Figure 7.4 we show the d-dependence of g^*, λ^*, θ', and θ'' for two versions of the sharp cutoff (with $s = 1$ and $s = 30$, respectively) and for the exponential cutoff with $s = 1$. For $2 + \epsilon \leq d \leq 4$ the scheme dependence of the critical exponents is rather weak; it becomes appreciable only near $d \approx 6$ [4]. Figure 7.4 suggests that the Einstein–Hilbert truncation at $d = 4$ performs almost as well as near $d = 2$. Its

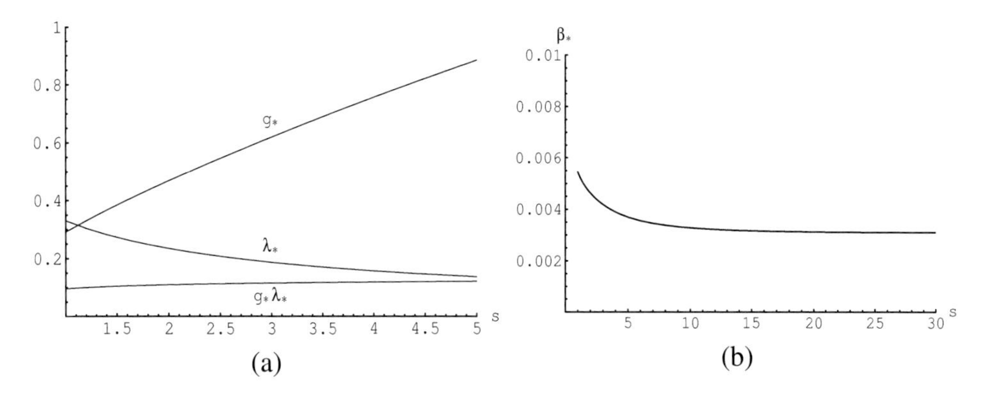

Fig. 7.5. (a) g^*, λ^*, and $g^*\lambda^*$ as functions of s for $1 \le s \le 5$, and (b) β^* as a function of s for $1 \le s \le 30$, using the family of exponential shape functions (7.63). (From [6].)

validity can be extended toward larger dimensionalities by optimizing the shape function [14].

7.6.3.6 *Position of the fixed point* (R^2)

Also with the generalized truncation the NGFP is found to exist for all admissible cutoffs. Figure 7.5 shows its coordinates $(\lambda^*, g^*, \beta^*)$ for the family of shape functions (7.63) and the type B cutoff. For every shape parameter s, the values of λ^* and g^* are almost the same as those obtained with the Einstein–Hilbert truncation. In particular, the product $g^*\lambda^*$ is constant with high accuracy. For $s = 1$, for instance, one obtains $(\lambda^*, g^*) = (0.348, 0.272)$ from the Einstein–Hilbert truncation and $(\lambda^*, g^*, \beta^*) = (0.330, 0.292, 0.005)$ from the generalized truncation. It is quite remarkable that β^* is always significantly smaller than λ^* and g^*. Within the limited precision of our calculation this means that in the three-dimensional parameter space the fixed point practically lies on the λ–g plane with $\beta = 0$, i.e., in the parameter space of the pure Einstein–Hilbert truncation.

7.6.3.7 *Eigenvalues and -vectors* (R^2)

The NGFP of the R^2 truncation proves to be UV attractive in each of the three directions of the (λ, g, β) space for all cutoffs used. The linearized flow in its vicinity is always governed by a pair of complex conjugate critical exponents $\theta_1 = \theta' + \mathrm{i}\theta'' = \theta_2^*$ with $\theta' > 0$ and a single real, positive critical exponent $\theta_3 > 0$. It may be expressed as

$$(\lambda_k, g_k, \beta_k)^{\mathrm{T}} = \left(\lambda^*, g^*, \beta^*\right)^{\mathrm{T}} + 2\Big\{\left[\mathrm{Re}\, C \cos\left(\theta'' t\right) + \mathrm{Im}\, C \sin\left(\theta'' t\right)\right] \mathrm{Re}\, V + \left[\mathrm{Re}\, C \sin\left(\theta'' t\right) - \mathrm{Im}\, C \cos\left(\theta'' t\right)\right] \mathrm{Im}\, V\Big\}\, e^{-\theta' t} + C_3 V^3 e^{-\theta_3 t} \tag{7.71}$$

with arbitrary complex $C \equiv C_1 = (C_2)^*$ and real C_3, and with $V \equiv V^1 = (V^2)^*$ and V^3 the right eigenvectors of the stability matrix $(B_{ij})_{i,j\in\{\lambda,g,\beta\}}$ with eigenvalues $-\theta_1 = -\theta_2^*$ and $-\theta_3$, respectively. Clearly the conditions for UV stability are $\theta' > 0$ and $\theta_3 > 0$. They are indeed satisfied for all cutoffs. For the exponential shape function with $s = 1$, for instance, we find $\theta' = 2.15$, $\theta'' = 3.79$, $\theta_3 = 28.8$, and $\mathrm{Re}\, V = (-0.164, 0.753, -0.008)^{\mathbf{T}}$, $\mathrm{Im}\, V = (0.64, 0, -0.01)^{\mathbf{T}}$, $V^3 = -(0.92, 0.39, 0.04)^{\mathbf{T}}$. (The vectors are normalized so that $\|V\| = \|V^3\| = 1$.) The trajectories (7.71) comprise three independent normal modes with amplitudes proportional to $\mathrm{Re}\, C$, $\mathrm{Im} C$, and C_3, respectively. The first two are again of the spiral type; the third is a straight line.

For any cutoff, the numerical results have several quite remarkable properties. They all indicate that, close to the NGFP, the RG flow is rather well approximated by the pure Einstein–Hilbert truncation.

(a) The β-components of $\mathrm{Re} V$ and $\mathrm{Im} V$ are tiny. Hence these two vectors span a plane which virtually coincides with the g–λ subspace at $\beta = 0$, i.e., with the parameter space of the Einstein–Hilbert truncation. As a consequence, the $\mathrm{Re} C$ and $\mathrm{Im} C$ normal modes are essentially the same trajectories as the old normal modes already found without the R^2 term. Also, the corresponding θ' and θ'' values coincide within the scheme dependence.

(b) The new eigenvalue θ_3 introduced by the R^2 term is significantly larger than θ'. When a trajectory approaches the fixed point from below ($t \to \infty$), the old normal modes $\propto \mathrm{Re}\, C, \mathrm{Im}\, C$ are proportional to $\exp(-\theta' t)$, but the new one is proportional to $\exp(-\theta_3 t)$, so that it decays much quicker. For every trajectory running into the fixed point, i.e., for every set of constants $(\mathrm{Re}\, C, \mathrm{Im}\, C, C_3)$, we find therefore that once t is sufficiently large, the trajectory lies entirely in the $\mathrm{Re}\, V$–$\mathrm{Im}\, V$ subspace – for practical purposes, the $\beta = 0$ plane. Due to the large value of θ_3, the new scaling field is very relevant. However, when we start at the fixed point ($t = \infty$) and lower t, it is only at the low energy scale $k \approx m_{\mathrm{Pl}}$ ($t \approx 0$) that $\exp(-\theta_3 t)$ reaches unity, and only there, i.e., far away from the fixed point, does the new scaling field start growing rapidly.

(c) Inasmuch as the matrix $\mathbf{B}$ is not symmetric, its eigenvectors have no reason to be orthogonal. In fact, one finds that V^3 lies almost in the $\mathrm{Re} V$–$\mathrm{Im} V$ plane. For the angles between the eigenvectors given, we obtain $\sphericalangle(\mathrm{Re} V, \mathrm{Im} V) = 102.3^\circ$, $\sphericalangle(\mathrm{Re} V, V^3) = 100.7^\circ$, $\sphericalangle(\mathrm{Im} V, V^3) = 156.7^\circ$. Their sum is 359.7°, which confirms that $\mathrm{Re} V$, $\mathrm{Im} V$, and V^3 are almost coplanar. This implies that when we lower t and move away from the fixed point so that the V^3 scaling field starts growing, it is again predominantly the $\int d^d x\, \sqrt{g}$ and $\int d^d x\, \sqrt{g} R$ invariants which get excited, but not $\int d^d x\, \sqrt{g} R^2$ in the first place.

Summarizing the preceding three points, we can say that close to the fixed point the RG flow seems to be essentially two-dimensional, and that this two-dimensional flow is well approximated by the RG equations of the Einstein–Hilbert truncation.

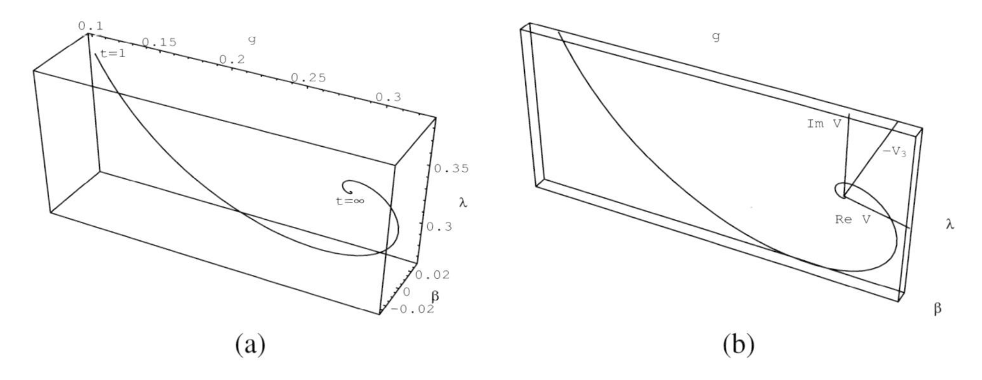

Fig. 7.6. Trajectory of the linearized flow equation obtained from the R^2 truncation for $1 \leq t = \ln(k/k_0) < \infty$. In (b) we depict the eigendirections and the "box" to which the trajectory is confined. (From [6].)

In Figure 7.6 we show a typical trajectory, which has all three normal modes excited with equal strength ($\mathrm{Re}\, C = \mathrm{Im}\, C = 1/\sqrt{2}$, $C_3 = 1$). All the way down from $k = \infty$ to about $k = m_{\mathrm{Pl}}$ it is confined to a thin box surrounding the $\beta = 0$ plane.

7.6.3.8 Scheme dependence (R^2)

The scheme dependence of the critical exponents and of the product $g^*\lambda^*$ turns out to be of the same order of magnitude as in the case of the Einstein–Hilbert truncation. Figure 7.7 shows the cutoff dependence of the critical exponents, using the family of shape functions (7.63). For the cutoffs employed, θ' and θ'' assume values in the ranges $2.1 \lesssim \theta' \lesssim 3.4$ and $3.1 \lesssim \theta'' \lesssim 4.3$, respectively. Whereas the scheme dependence of θ'' is weaker than in the case of the Einstein–Hilbert truncation, one finds that that of θ' it is slightly stronger. The exponent θ_3 suffers from relatively strong variations as the cutoff is changed ($8.4 \lesssim \theta_3 \lesssim 28.8$), but it is always significantly larger than θ'. The product $g^*\lambda^*$ again exhibits an extremely weak scheme dependence. Figure 7.5(a) displays $g^*\lambda^*$ as a function of s. It is impressive to see how the cutoff dependences of g^* and λ^* cancel almost perfectly. Figure 7.5(a) suggests the universal value $g^*\lambda^* \approx 0.14$. Comparing this value with those obtained from the Einstein–Hilbert truncation, we find that it differs slightly from the one based upon the same gauge $\alpha = 1$. The deviation is of the same size as the difference between the $\alpha = 0$ and the $\alpha = 1$ results of the Einstein–Hilbert truncation.

As for the universality of the critical exponents, we emphasize that the qualitative properties just listed ($\theta', \theta_3 > 0$, $\theta_3 \gg \theta'$, etc.) are obtained for all cutoffs. The θ's have a much stronger scheme dependence than $g^*\lambda^*$, however. This is most

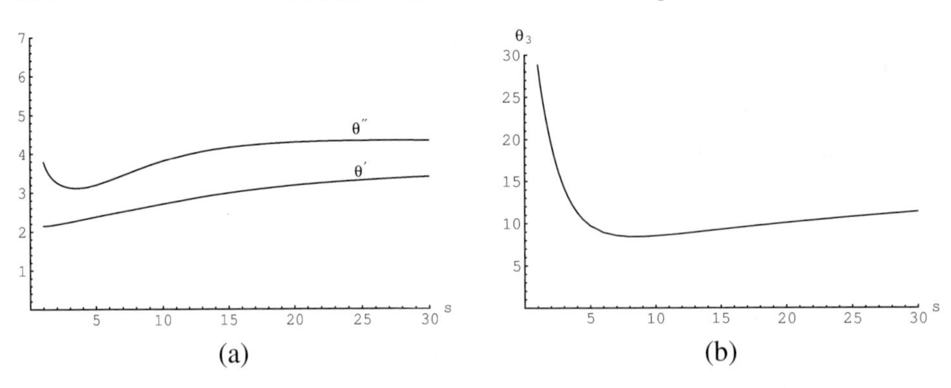

Fig. 7.7. (a) $\theta' = \mathrm{Re}\,\theta_1$ and $\theta'' = \mathrm{Im}\,\theta_1$, and (b) θ_3 as functions of s, using the family of exponential shape functions (7.63). (From [5].)

probably due to neglecting further relevant operators in the truncation, so that the **B**-matrix we are diagonalizing is still too small.

7.6.3.9 Dimensionality of $\mathcal{S}_{\mathrm{UV}}$

According to the canonical dimensional analysis, the (curvature)^n invariants in four dimensions are classically marginal for $n = 2$ and irrelevant for $n > 2$. The results for θ_3 indicate that there are large nonclassical contributions, so that there might be relevant operators perhaps even beyond $n = 2$. With the present approach it is clearly not possible to determine their number Δ_{UV}. However, as it is hardly conceivable that the quantum effects change the signs of arbitrarily large (negative) classical scaling dimensions, Δ_{UV} should be finite [17].

A first confirmation of this picture comes from the R^2 calculation, which has also been performed at $d = 2 + \varepsilon$, where, at least canonically, the dimensional count is shifted by two units. In this case we find indeed that the third scaling field is irrelevant: $\theta_3 < 0$. Therefore the dimensionality of $\mathcal{S}_{\mathrm{UV}}$ could be as small as $\Delta_{\mathrm{UV}} = 2$ (but this is not a proof, of course). If so, the quantum theory would be characterized by only two free parameters: the renormalized Newton constant and the cosmological constant.

7.7 Discussion and conclusion

On the basis of the preceding results we believe that the non-Gaussian fixed point occurring in the Einstein–Hilbert truncation is not a truncation artifact, but rather the projection of a fixed point in the exact theory space. The fixed point and all its qualitative properties are stable against variations of the cutoff and the inclusion of a further invariant in the truncation. It is particularly remarkable that within the scheme dependence the additional R^2 term has essentially no impact on the fixed point. We interpret the results and their mutual consistency as quite nontrivial indications supporting the conjecture that four-dimensional QEG indeed possesses a RG fixed point with precisely the properties needed for its nonperturbative renormalizability and asymptotic safety.

Recently this picture has been beautifully confirmed by Codello, Percacci, and Rahmede [52], who, for $d = 4$, considered truncations of the form

$$\bar{\Gamma}_k[g] = \int d^4x \sqrt{g} \sum_{n=0}^{N} \bar{u}_n(k)\, R^n \,. \tag{7.72}$$

In the most advanced case the highest power of the curvature scalar was as large as $N = 7$. An important result obtained with these truncations is that, going beyond the R^2 truncation, the new eigendirections at the NGFP are all UV repulsive ($\mathrm{Re}\,\theta_I < 0$), indicating that Δ_{UV} is indeed likely to be a small finite number. On increasing the order N of the curvature polynomial, the values of the universal quantities show a certain degree of convergence; in particular, $g^*\lambda^*$ agrees with the Einstein–Hilbert result (7.70) to within 10 or 20 percent for any $N = 2, \ldots, 7$. It is amazing how well the RG flow near the NGFP is approximated by the Einstein–Hilbert truncation; the reason for this is not yet fully understood.

In this chapter we have focused on the average action approach to QEG. For a detailed discussion, including evidence for asymptotic safety from other approaches, we refer to [18].

Before closing, some further comments might be helpful here.

(1) The construction of an effective average action for gravity as introduced in [1] represents a *background-independent* approach to quantum gravity. Somewhat paradoxically, this background independence is achieved by means of the background field formalism: One fixes an arbitrary background, quantizes the fluctuation field in this background, and afterwards adjusts $\bar{g}_{\mu\nu}$ in such a way that the expectation value of the fluctuation vanishes: $\bar{h}_{\mu\nu} = 0$. In this way the background gets fixed dynamically.

(2) The combination of the effective average action with the background field method has been successfully tested within conventional field theory. In QED and Yang–Mills type gauge theories it reproduces the known results and extends them into the nonperturbative domain [22, 24].

(3) The coexistence of asymptotic safety and perturbative nonrenormalizability is well understood. In particular, upon fixing $\bar{g}_{\mu\nu} = \eta_{\mu\nu}$ and expanding the trace on its RHS in powers of G, the FRGE reproduces the divergences of perturbation theory; see [18] for a detailed discussion of this point.

(4) It is to be emphasized that in the average action framework the RG flow, i.e., the vector field $\vec{\beta}$, is completely determined once a theory space is fixed. As a consequence, the choice of theory space determines the set of fixed points Γ^* at which asymptotically safe theories can be defined. Therefore, in the asymptotic safety scenario the bare

action $S = \Gamma^*$ is a *prediction* of the theory, rather than an ad hoc postulate as usual in quantum field theory. (Ambiguities could arise only if there is more than one suitable NGFP.)

(5) According to the results available to date, the Einstein–Hilbert action of classical general relativity seems not to play any distinguished role in the asymptotic safety context, at least not at the conceptual level. The only known NGFP on the theory space of QEG has the structure $\Gamma^* =$ (Einstein–Hilbert action) $+$ more, where "more" stands for both local and nonlocal corrections. So it seems that the Einstein–Hilbert action is only an approximation to the true fixed-point action, albeit an approximation that was found to be rather reliable for many purposes.

(6) Any quantum theory of gravity must reproduce the successes of classical general relativity. As for QEG, it cannot be expected that this will happen for all RG trajectories in $\mathcal{S}_{\rm UV}$, but it should happen for some or at least one of them. Within the Einstein–Hilbert truncation it has been shown [41] that there actually do exist trajectories (of type IIIa) which have an extended classical regime and are consistent with all observations.

(7) In the classical regime mentioned, the spacetime geometry is nondynamical to a good approximation. In this regime the familiar methods of quantum field theory in curved classical spacetimes apply, and it is clear therefore that effects such as Hawking radiation or cosmological particle production are reproduced by the general framework of QEG with matter.

(8) On coupling free massless matter fields to gravity, it turned out [12] that the fixed point continues to exist under weak conditions concerning the number of various types of matter fields (scalars, fermions, etc.). No fine tuning with respect to the matter multiplets is necessary. In particular, asymptotic safety does not seem to require any special constraints or symmetries among the matter fields such as supersymmetry, for instance.

(9) Because the NGFP seems to exist already in pure gravity, it is likely that a widespread prejudice about gravity may be incorrect: its quantization seems not to require any kind of unification with the other fundamental interactions.

Given that by now the asymptotic safety of QEG hardly can be questioned, future work will have to focus on its physics implications. The effective average action is an ideal framework for investigations of this sort in that, contrary to other exact RG schemes, it provides a family of scale-dependent *effective* (rather than *bare*) actions, $\{\Gamma_k[\,\cdot\,], 0 \leq k < \infty\}$. Dealing with phenomena involving typical scales k, a tree-level evaluation of Γ_k is sufficient for finding the leading quantum gravity effects. The investigations already performed in this direction employed the following methods:

(a) *RG improvement*: In [33] and [35], respectively, a first study of the asymptotic safety-based phenomenology of black hole and cosmological spacetimes has been carried out by RG-improving the classical field equations or their solutions. Thereby k is identified with a fixed, geometrically motivated scale. Using the same method, modified dispersion relations of point particles were discussed in [42].

(b) *Scale-dependent geometry:* In the spirit of the gravitational average action, a spacetime manifold can be visualized as a fixed differentiable manifold equipped with infinitely many metric structures $\{\langle g_{\mu\nu}\rangle_k, 0 \leq k < \infty\}$ where $\langle g_{\mu\nu}\rangle_k$ is a solution to the effective field equation implied by Γ_k. Comparably to the situation in fractal geometry, the metric, and therefore all distances, depend on the resolution of the experiment by means of which spacetime is probed. A general discussion of the geometrical issues involved (scale-dependent diffeomorphisms, symmetries, causal structures, etc.) was given in [27], and in [26] these ideas were applied to show that QEG can generate a minimum length dynamically. In [3,5] it has been pointed out that the QEG spacetimes should have fractal properties, with a fractal dimension equal to 4 on macroscopic and 2 on microscopic scales. This picture was confirmed by the computation of their spectral dimension in [28]. Quite remarkably, the same dynamical dimensional reduction from 4 to 2 has also been observed in Monte Carlo simulations using the causal triangulation approach [29–31]. It is therefore intriguing to speculate that this discrete approach and the gravitational average action actually describe the same underlying theory.

Bibliography

[1] M. Reuter, Phys. Rev. D 57 (1998) 971, hep-th/9605030.
[2] D. Dou and R. Percacci, Class. Quant. Grav. 15 (1998) 3449.
[3] O. Lauscher and M. Reuter, Phys. Rev. D 65 (2002) 025013, hep-th/0108040.
[4] M. Reuter and F. Saueressig, Phys. Rev. D 65 (2002) 065016, hep-th/0110054.
[5] O. Lauscher and M. Reuter, Phys. Rev. D 66 (2002) 025026, hep-th/0205062.
[6] O. Lauscher and M. Reuter, Class. Quant. Grav. 19 (2002) 483, hep-th/0110021.
[7] O. Lauscher and M. Reuter, Int. J. Mod. Phys. A 17 (2002) 993, hep-th/0112089.
[8] W. Souma, Prog. Theor. Phys. 102 (1999) 181.
[9] M. Reuter and F. Saueressig, Phys. Rev. D 66 (2002) 125001, hep-th/0206145; Fortschr. Phys. 52 (2004) 650, hep-th/0311056.
[10] A. Bonanno and M. Reuter, JHEP 02 (2005) 035, hep-th/0410191.
[11] For a review see O. Lauscher and M. Reuter, in *Quantum Gravity*, B. Fauser, J. Tolksdorf, and E. Zeidler (Eds.), Birkhäuser, Basel, 2007, hep-th/0511260.
[12] R. Percacci and D. Perini, Phys. Rev. D 67 (2003) 081503; Phys. Rev. D 68 (2003) 044018; D. Perini, Nucl. Phys. Proc. Suppl. 127 C (2004) 185.
[13] A. Codello and R. Percacci, Phys. Rev. Lett. 97 (2006) 221301.
[14] D. Litim, Phys. Rev. Lett. 92 (2004) 201301; AIP Conf. Proc. 841 (2006) 322. P. Fischer and D. Litim, Phys. Lett. B 638 (2006) 497; AIP Conf. Proc. 861 (2006) 336.
[15] R. Percacci and D. Perini, Class. Quant. Grav. 21 (2004) 5035.
[16] R. Percacci, J. Phys. A 40 (2007) 4895.
[17] S. Weinberg, in *General Relativity, an Einstein Centenary Survey*, S.W. Hawking and W. Israel (Eds.), Cambridge University Press, 1979; S. Weinberg, hep-th/9702027.
[18] For a comprehensive review of asymptotic safety in gravity see M. Niedermaier and M. Reuter, Living Rev. Relativity 9 (2006) 5.

[19] For a general introduction see C. Kiefer, *Quantum Gravity*, Second Edition, Oxford Science Publications, Oxford, 2007; C. Rovelli, *Quantum Gravity*, Cambridge University Press, Cambridge, 2004.
[20] P. Forgács and M. Niedermaier, hep-th/0207028; M. Niedermaier, JHEP 12 (2002) 066; Nucl. Phys. B 673 (2003) 131; gr-qc/0610018.
[21] C. Wetterich, Phys. Lett. B 301 (1993) 90.
[22] M. Reuter and C. Wetterich, Nucl. Phys. B 417 (1994) 181, Nucl. Phys. B 427 (1994) 291, Nucl. Phys. B 391 (1993) 147, Nucl. Phys. B 408 (1993) 91; M. Reuter, Phys. Rev. D 53 (1996) 4430, Mod. Phys. Lett. A 12 (1997) 2777.
[23] For a review see J. Berges, N. Tetradis, and C. Wetterich, Phys. Rep. 363 (2002) 223; C. Wetterich, Int. J. Mod. Phys. A 16 (2001) 1951.
[24] For reviews of the effective average action in Yang–Mills theory see M. Reuter, hep-th/9602012; J. Pawlowski, hep-th/0512261; H. Gies, hep-ph/0611146.
[25] For reviews see C. Bagnuls and C. Bervillier, Phys. Rep. 348 (2001) 91; T.R. Morris, Prog. Theor. Phys. Suppl. 131 (1998) 395; J. Polonyi, Central Eur. J. Phys. 1 (2004) 1.
[26] M. Reuter and J. Schwindt, JHEP 01 (2006) 070, hep-th/0511021.
[27] M. Reuter and J. Schwindt, JHEP 01 (2007) 049, hep-th/0611294.
[28] O. Lauscher and M. Reuter, JHEP 10 (2005) 050, hep-th/0508202.
[29] J. Ambjørn, J. Jurkiewicz, and R. Loll, Phys. Rev. Lett. 93 (2004) 131301.
[30] J. Ambjørn, J. Jurkiewicz, and R. Loll, Phys. Lett. B 607 (2005) 205.
[31] J. Ambjørn, J. Jurkiewicz, and R. Loll, Phys. Rev. Lett. 95 (2005) 171301; Phys. Rev. D 72 (2005) 064014; Contemp. Phys. 47 (2006) 103.
[32] A. Connes, JHEP 11 (2006) 081; A.H. Chamseddine, A. Connes, and M. Marcolli, hep-th/0610241.
[33] A. Bonanno and M. Reuter, Phys. Rev. D 62 (2000) 043008, hep-th/0002196; Phys. Rev. D 73 (2006) 083005, hep-th/0602159; Phys. Rev. D 60 (1999) 084011, gr-qc/9811026.
[34] M. Reuter and E. Tuiran, hep-th/0612037.
[35] A. Bonanno and M. Reuter, Phys. Rev. D 65 (2002) 043508, hep-th/0106133; M. Reuter and F. Saueressig, JCAP 09 (2005) 012, hep-th/0507167.
[36] A. Bonanno and M. Reuter, Phys. Lett. B 527 (2002) 9, astro-ph/0106468; Int. J. Mod. Phys. D 13 (2004) 107, astro-ph/0210472.
[37] E. Bentivegna, A. Bonanno, and M. Reuter, JCAP 01 (2004) 001, astro-ph/0303150.
[38] A. Bonanno, G. Esposito, and C. Rubano, Gen. Rel. Grav. 35 (2003) 1899; Class. Quant. Grav. 21 (2004) 5005; A. Bonanno, G. Esposito, C. Rubano, and P. Scudellaro, Class. Quant. Grav. 23 (2006) 3103 and 24 (2007) 1443.
[39] M. Reuter and H. Weyer, Phys. Rev. D 69 (2004) 104022, hep-th/0311196.
[40] M. Reuter and H. Weyer, Phys. Rev. D 70 (2004) 124028, hep-th/0410117.
[41] M. Reuter and H. Weyer, JCAP 12 (2004) 001, hep-th/0410119.
[42] F. Girelli, S. Liberati, R. Percacci, C. Rahmede, gr-qc/0607030.
[43] J. Moffat, JCAP 05 (2005) 003, astro-ph/0412195; J.R. Brownstein and J. Moffat, Astrophys. J. 636 (2006) 721; Mon. Not. Roy. Astron. Soc. 367 (2006) 527.
[44] D. Litim, Phys. Lett. B 486 (2000) 92; Phys. Rev. D 64 (2001) 105007; Int. J. Mod. Phys. A 16 (2001) 2081.
[45] M. Reuter and C. Wetterich, Nucl. Phys. B 506 (1997) 483, hep-th/9605039.
[46] L. F. Abbott, Nucl. Phys. B 185 (1981) 189; B. S. DeWitt, Phys. Rev. 162 (1967) 1195; M. T. Grisaru, P. van Nieuwenhuizen and C. C. Wu, Phys. Rev. D 12 (1975)

3203; D. M. Capper, J. J. Dulwich, and M. Ramon Medrano, Nucl. Phys. B 254 (1985) 737; S. L. Adler, Rev. Mod. Phys. 54 (1982) 729.
[47] For an introduction see M. Böhm, A. Denner, and H. Joos, *Gauge Theories of the Strong and Electroweak Interactions*, Teubner, Stuttgart, 2001.
[48] D. Litim and J. Pawlowski, Phys. Lett. B 435 (1998) 181.
[49] S. Falkenberg and S. D. Odintsov, Int. J. Mod. Phys. A 13 (1998) 607.
[50] L. N. Granda, Europhys. Lett. 42 (1998) 487.
[51] G. O. Pires, Int. J. Mod. Phys. A 13 (1998) 5425.
[52] A. Codello, R. Percacci, and C. Rahmede, arXiv:0705.1769 [hep-th].

8

When is a differentiable manifold the boundary of an orbifold?

ANDRÉS ANGEL*

Abstract

The aim of this short chapter is to review some classical results on cobordism of manifolds and discuss recent extensions of this theory to orbifolds. In particular, I present an answer to the question "When is a differentiable manifold the boundary of an orbifold?" in the oriented case and in the unoriented case when we restrict to isotropy groups of odd order.

Introduction

When is a differentiable manifold the boundary of another differentiable manifold? This question was answered by Thom [14] in the 1950s; the necessary and sufficient condition is the vanishing of certain characteristic numbers, invariants defined by evaluating characteristic classes of the tangent bundle on the fundamental class of the manifold. His proof is one of the cornerstones of algebraic topology [2] and shows the powerful tools that homotopy theory gives to the study of the geometry of manifolds.

Orbifolds, originally introduced as V-manifolds by Satake [12], and so named by Thurston [15], are useful generalizations of manifolds: locally they look like the quotient of Euclidean space by the action of a finite group. Their study lies at the intersection of many different areas of mathematics, and they appear naturally in many situations such as moduli problems, noncommutative geometry and

* The author would like to thank Bernardo Uribe and Ernesto Lupercio for their interest and helpful conversations about this work, and the organizers of the 2007 summer school "Geometrical and Topological Methods in Quantum Field Theory" for a great event and the opportunity to present some of the results mentioned in this work. Some of the results presented here are part of the author's dissertation under the direction of Professor Ralph Cohen, to whom the author owes thanks for many hours of guidance and suggestions. The author wants to thank also Kimberly Druschel for making available to him a copy of her thesis, which has been a constant source of ideas.

This paper is dedicated on her birthday to Nora Raggio, for her support and love.

foliation theory. The local character of the definition of orbifolds allows many constructions that can be applied to manifolds to be extended to orbifolds, and it is natural to ask, *When is an orbifold the boundary of another orbifold?* Key ingredients of Thom's proof do not seem to extend naturally to orbifolds, and the interplay between geometry and homotopy theory, in this case, is still not understood.

A complete answer to this question is still open, but some partial results are known. In the oriented case Druschel [8] defines generalized Pontrjagin numbers that determine when a multiple of an oriented orbifold is a boundary. In [4] the author provides similar characteristic numbers that determine when an orbifold with isotropy groups of odd order is a boundary.

The purpose of this short chapter is to present an answer to the more restricted question, *When is a differentiable manifold the boundary of an orbifold?* In this case the constructions in [8] and [4] take a simpler form, and the following answer can be given:

An oriented manifold is the boundary of an oriented orbifold precisely when all Pontrjagin numbers vanish.

A differentiable manifold is the boundary of an orbifold with only odd singularities precisely when all Stiefel–Whitney numbers vanish.

This paper is expository and is organized as follows. In Section 1, I review some of the classical theory for manifolds, and state the results of Thom [14] and Wall [16] on the cobordism rings. In Section 2, I discuss the necessary background on orbifolds. Section 3 is the main part of this chapter; it gives a proof that if an oriented differentiable manifold is the boundary of an oriented orbifold, then some multiple of the manifold bounds. This easily follows from the existence of a rational fundamental class for oriented orbifolds and the fact that Pontrjagin classes determine the cobordism class of a manifold up to torsion.

Also I present a proof that if a manifold is the boundary of an orbifold with only odd singularities, then it is actually the boundary of a manifold. This also follows from the existence of $\mathbb{Z}_2$-fundamental classes for this type of orbifolds, but instead I present a direct geometric proof that constructs the bounding manifold out of the orbifold.

Throughout this chapter *manifold* will mean compact differentiable manifold.

8.1 Cobordism of manifolds

In this section I present the basic results on the classical theory of cobordism of manifolds. I give the definitions of the cobordism groups and characteristic numbers and state the main results of Thom and Wall that led to a description of the cobordism ring of manifolds. They provide a complete answer to the question:

When is a manifold the boundary of another manifold? The main references for this section are [13], [10] and [16].

8.1.1 Unoriented cobordism

We say that two closed n-dimensional manifolds M_1^n, M_2^n are cobordant if there exists an $n+1$-dimensional manifold with boundary W^{n+1} such that the boundary is the disjoint union of M_1^n and M_2^n:

$$\partial W^{n+1} = M_1^n \sqcup M_2^n.$$

The cobordism relation is an equivalence relation on the class of differentiable manifolds. We denote by $[M]$ the corresponding equivalence class, and denote by $\mathfrak{N}_n$ the set of equivalence classes of n-dimensional manifolds. This set can be endowed with a group operation given by disjoint union and the empty set as identity. We call $\mathfrak{N}_n$ the *cobordism group* of n-dimensional manifolds. This is the central object of our study, and for its determination more structure on these groups turns out to be of utmost importance.

The Cartesian product of manifolds induces a ring structure on the graded vector space

$$\mathfrak{N}_* = \bigoplus_n \mathfrak{N}_n.$$

Observe that $\mathfrak{N}_*$ is a $\mathbb{Z}_2$-algebra, i.e., every element has order two, because two copies of a manifold M are the boundary of the manifold $M \times I$.

Remark 8.1.1 We have translated the question *"When is a manifold M the boundary of another manifold?"* into the algebraic task "Determine when the class $[M]$ is zero in the cobordism ring $\mathfrak{N}_*$."

In [14] Thom completely determined this algebra: $\mathfrak{N}_*$ is the polynomial algebra over $\mathbb{Z}_2$ generated by elements in dimensions not of the form $2^j - 1$. He provided complete invariants that determine when two manifolds are cobordant.

8.1.2 Characteristic numbers

I will now introduce invariants of the cobordism class of a manifold. These invariants are defined in terms of the differentiable structure of the manifold. Given an n-dimensional differentiable manifold M, the tangent bundle $TM \to M$ is a rank n real vector bundle over M. To such a vector bundle you can assign certain cohomology classes $\omega_i(TM) \in H^i(M; \mathbb{Z}_2)$, called Stiefel–Whitney classes [10]. Here $H^i(M; \mathbb{Z}_2)$ denotes the singular cohomology of M with $\mathbb{Z}_2$ coefficients.

Recall that any closed n-dimensional manifold has a $\mathbb{Z}_2$-fundamental class, i.e., a top homology class that restricts at any point $x \in M$ to the nonzero class of

$$H_n(M, M - \{x\}; \mathbb{Z}_2) \cong \mathbb{Z}_2.$$

Now, given any partition $I = (i_1, \dots, i_m)$ of n, the cup product

$$\omega_I(TM) = \omega_{i_1}(TM) \cup \cdots \cup \omega_{i_m}(TM)$$

is a top cohomology class. By evaluating on the fundamental class of M, we obtain elements of $\mathbb{Z}_2$ that are called the *Stiefel–Whitney numbers* associated to the partition I,

$$\langle \omega_I(TM), [M] \rangle \in \mathbb{Z}_2.$$

Here $[M]$ denotes the fundamental class, and $\langle\,,\rangle$ is the Kronecker pairing between singular cohomology and homology. These numbers turn out to be invariants of the cobordism class. To see this, it is enough to prove that if M is the boundary of W, then all its Stiefel–Whitney numbers are zero. Suppose that $M^n = \partial W^{n+1}$, and let $I = (i_1, \dots, i_m)$ be a partition of n. Denote by $\iota : M \to W$ the inclusion of M into W as the boundary. By the collaring theorem there is a neighborhood of M in W that is diffeomorphic to $M \times [0, 1)$. Therefore $\iota^*(TW) = TM \oplus \mathbb{R}$, the sum of the tangent bundle with a trivial one, and we have

$$\begin{aligned}
\langle \omega_I(TM), [M] \rangle &= \langle \omega_I(TM \oplus \mathbb{R}), [M] \rangle && \text{by stability} \\
&= \langle \omega_I(\iota^*(TW)), [M] \rangle && \text{by the preceding remark} \\
&= \langle \iota^*\omega_I(TW), [M] \rangle && \text{by naturality} \\
&= \langle \omega_I(TW), \iota_*[M] \rangle && \text{by functoriality of the pairing.}
\end{aligned}$$

Now, in the long exact sequence, on homology with $\mathbb{Z}_2$-coefficients, of the pair (W, M),

$$\cdots \to H_{n+1}(M) \xrightarrow{\iota_*} H_{n+1}(W) \to H_{n+1}(W, M) \xrightarrow{\partial} H_n(M) \xrightarrow{\iota_*} H_n(W) \to \cdots,$$

we have that $\iota_*\partial = 0$; but W is a manifold with boundary, and it has a fundamental class $[W] \in H_{n+1}(W, M)$ such that $\partial[W] = [M]$; therefore $\iota_*([M]) = 0$. This shows that the Stiefel–Whitney numbers are zero.

Example 8.1.2 For real projective spaces (see [10])

$$\omega_I(T\mathbb{RP}^n) = \binom{n+1}{i_1} \cdots \binom{n+1}{i_k} \quad \text{mod } 2,$$

and in particular for the trivial partition of $n = 2k$, $I = (2k)$, we have

$$\omega_{(2k)}(T\mathbb{RP}^{2k}) \neq 0,$$

i.e., $\mathbb{RP}^{2k}$ is not the boundary of another manifold.

Thom's proof starts by realizing that the cobordism groups are the stable homotopy groups of a spectrum, now called a Thom spectrum and denoted by MO. Thus the problem of determining when a manifold is the boundary of another manifold has been translated into a homotopy theory problem: determine when a map from a sphere into the spectrum MO is null-homotopic.

The calculation of the stable homotopy groups of MO is still not trivial and was accomplished by Thom in [14]. A corollary of his work is that unorientably, a manifold bounds if and only if all its Stiefel–Whitney numbers are zero. His proof also determines completely the cobordism ring: it is the polynomial algebra $\mathfrak{N}_* \cong \mathbb{Z}_2[x_i \mid i \neq 2^j - 1]$. On even dimensions, the generators can be taken to be the real projective spaces; odd-dimensional generators were given by Dold [6] soon after Thom's paper.

8.1.3 Oriented cobordism

Observe that an orientation of a manifold with boundary induces an orientation of the boundary on choosing a normal unit vector field on the boundary. By always considering the outward normal vector, one makes the boundary of an oriented manifold also an oriented manifold; therefore it is possible to talk also about cobordism of such objects. We say closed oriented n-dimensional manifolds M_1^n, M_2^n are cobordant if there exists an $n+1$-dimensional oriented manifold with boundary W^{n+1} whose boundary is the disjoint union of M_1^n and $-M_2^n$:

$$\partial W^{n+1} = M_1^n \cup -M_2^n,$$

where $-M_2^n$ is just M_2^n with the reverse orientation.

Just as before, cobordism is an equivalence relation on the class of n-dimensional oriented manifolds; the set of equivalence classes is denoted by Ω_n, and is a group under disjoint union. The graded vector space

$$\Omega_* = \bigoplus_n \Omega_n$$

is a graded commutative ring, the oriented cobordism ring. Given an oriented manifold, we will use the linear structure on the tangent bundle to define invariants that determine the oriented cobordism class, at least up to torsion. They are defined in similar way to the Stiefel–Whitney numbers, but now the homology and cohomology are with integer coefficients.

Recall that an oriented closed n-dimensional manifold has a fundamental class $[M] \in H_n(M;\mathbb{Z})$, that is, a top homology class that restricts at each point to the class

$$H_n(M, M - \{x\}; \mathbb{Z}) \cong \mathbb{Z}$$

given by the orientation.

To the tangent bundle $TM \to M$, we can associate certain characteristic classes $p_i(TM) \in H^{4i}(M;\mathbb{Z})$ called Pontrjagin classes; see [11] and [10]. To define invariants of the cobordism class we evaluate these classes at the fundamental class.

Given any partition $I = (i_1, \dots, i_m)$ of n, the cup product $p_I(TM) = p_{i_1} \cup \dots \cup p_{i_m}(TM)$ is an n-dimensional cohomology class that, when we evaluate on the fundamental class

$$\langle p_I(TM), [M] \rangle,$$

gives an integer. The same proof as with Stiefel–Whitney classes shows that these are also cobordism invariants. Note that for dimensional reasons these numbers are always zero unless n is divisible by four.

Example 8.1.3 For even complex projective spaces (see [10])

$$p_I(T\mathbb{CP}^{2n}) = \binom{2n+1}{i_1} \cdots \binom{2n+1}{i_k}$$

if $i_1 + \cdots i_k = n$, and zero otherwise. In particular, they are not the boundary of an oriented manifold.

In the same manner, we can identify the oriented cobordism ring with the stable homotopy groups of another spectrum, now denoted MSO. In this case the homotopy problem turns out to be even harder, but Thom managed to calculate the stable homotopy groups of MSO after tensoring with the rationals, thereby giving a complete description of the ring $\Omega_* \otimes \mathbb{Q}$. It is a polynomial algebra over the rationals generated by classes on dimensions that are multiples of 4. Generators can be taken to be the complex projective spaces on even dimensions,

$$\mathbb{Q}[\mathbb{CP}^2, \mathbb{CP}^4, \dots].$$

Thom's proof shows also that Pontrjagin numbers completely determine an element of $\Omega_* \otimes \mathbb{Q}$. Therefore a multiple of a manifold bounds if and only if all its Pontrjagin numbers vanish.

Later work of Milnor [10] and Wall [16] settled the complete calculation of the torsion, and gave an algebraic description of the oriented cobordism ring. The main results contained in those papers can be summarized as follows:

- Ω_* has no odd torsion.
- Ω_* has no element of order 4.
- The torsion-free part of Ω_* is a polynomial ring.
- Two oriented manifolds are cobordant if and only if their Pontrjagin and Stiefel–Whitney numbers are the same.

Therefore,

An oriented manifold is the boundary of an oriented manifold precisely when all Pontrjagin and Stiefel–Whitney numbers vanish.

8.2 Orbifolds

Orbifolds were first introduced by Satake [12] in the fifties as generalizations of smooth manifolds that allow mild singularities. Since then, orbifolds have became a subject of study of their own (see [1], for example), and their use has transcended mathematics: nowadays orbifolds are used in string theory and crystallography. In this section we follow the classical perspective on orbifolds by defining orbifold charts and atlases akin to the way manifolds are defined. We define the analogs of vector bundles and fiber bundles, and discuss orientations and fundamental classes. The main references for this section are [12], [1], and [9].

8.2.1 Charts

Let X be a paracompact Hausdorff topological space. Now we introduce the analog of charts for manifolds: an *n-dimensional orbifold chart* on X is a triple $(\overline{U}, G, U)$ where $\overline{U}$ is a connected manifold, G is a finite group acting on $\overline{U}$, and U is an open subset of X, homeomorphic to $\overline{U}/G$.

To be able to glue charts, we need to specify when two charts are compatible. First, an *embedding of charts* $(\overline{U}, G, U) \hookrightarrow (\overline{V}, H, V)$ is a differentiable embedding $\overline{U} \hookrightarrow \overline{V}$ that is equivariant with respect to a monomorphism $G \hookrightarrow H$ that preserves the kernel of the actions. Two charts $(\overline{U}, G, U) \hookrightarrow (\overline{V}, H, V)$ are *compatible* if for every point in $U \cap V$, there exists a chart $(\overline{W}, K, W)$ with embeddings of charts:

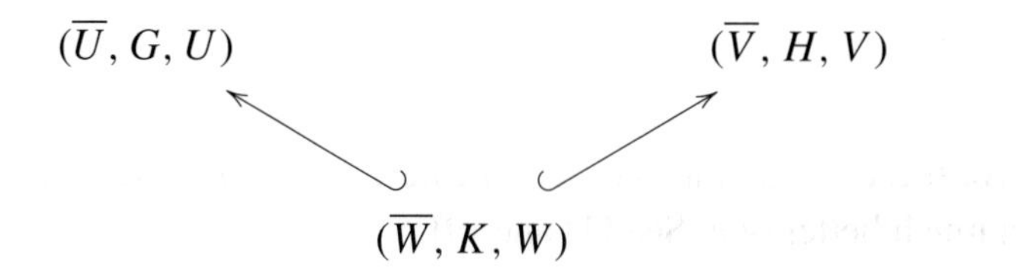

As with manifolds, an *orbifold atlas* on X is a family of charts $(\overline{U}_\alpha, G_\alpha, U_\alpha)$ that is compatible and covers X. An *orbifold structure* on X is just an equivalence class of orbifold atlases, where two atlases are equivalent if there is a zigzag of common refinements. We will denote an orbifold structure on X by calligraphic letters, like $\mathcal{X}$, and the topological space X will be called the underlying space and will be denoted by $|\mathcal{X}|$.

Example 8.2.1 A manifold M with an action of a finite group G gives rise to an orbifold structure that we will denote by $[M/G]$. It has an atlas with only one chart. In a more general vein, if a compact Lie group acts differentiably on a manifold with finite stabilizers, then the quotient space can be endowed with an orbifold structure. Charts can be constructed with the slice theorem for differentiable actions.

Example 8.2.2 For a finite group G, the orbifold $[*/G]$ is a zero-dimensional orbifold. The underlying space has only one point.

Given a point $x \in X$, take a chart $(\overline{U}, G, U)$ around x, and let $\overline{x} \in \overline{U}$ be a lift of x. Then we define the *isotropy group* of x to be

$$G_x = \{g \in G \mid g\overline{x} = \overline{x}\}.$$

This group is well defined up to isomorphism, and the restriction of the action of the isotropy group on any chart gives a well-defined representation of the isotropy group that we call the *local representation* at x. A point is called *nonsingular* if G_x is trivial, and *singular* otherwise. The set $\Sigma\mathcal{X} = \{x \in X \mid G_x \neq \{1\}\}$ is called the *singular locus* of X. In general the singular locus is not an orbifold.

Remark 8.2.3 The isotropies give a stratification of $|\mathcal{X}|$.

Example 8.2.4 Consider the action of S^1 on $\mathbb{C}^2 - \{0\}$ given by $\lambda(z_1, z_2) = (\lambda^2 z, \lambda z)$. It is a an action with finite stabilizers, and therefore the underlying space, which topologically is just a sphere, has an orbifold structure coming from this action. It has one singular point with isotropy group $\mathbb{Z}_2$, but it can be seen that is not the quotient of a manifold by a finite group (even though it is a quotient by a compact Lie group).

We say that an orbifold is *effective* if we can find an atlas for which all the local groups act effectively. On the other side, we say that an orbifold is *purely ineffective* if all the local groups act trivially.

Remark 8.2.5 Extra care should be taken when working with orbifolds that are not effective; for these our definition is not the best one, and the language of groupoids provides a much better one. See [1] and [9].

8.2.2 *Orbibundles*

A *vector orbibundle* $\mathcal{V}$ over an orbifold $\mathcal{Q}$ is an orbifold $\mathcal{V}$ that has an atlas of the form $(\overline{U}_\alpha \times \mathbb{R}^n, G_\alpha)$, where $(\overline{U}_\alpha, G_\alpha)$ is an atlas for $\mathcal{Q}$, all the actions are linear on the second component, and we require also the embeddings between the charts to be linear on the second component.

We have an induced map on the underlying spaces $p : |\mathcal{W}| \to |\mathcal{Q}|$, which in general is not a vector bundle. The fibers of a point are topological spaces homeomorphic to $\mathbb{R}^n/G_x$.

Example 8.2.6 If G is a finite group, and M is a manifold with an action of G, then a G-bundle E gives an orbivector bundle $[E/G]$ over $[M/G]$. In particular, for a representation V of G, we have an orbivector bundle $[V/G]$ over $[*/G]$.

In general, given a manifold F and a group acting effectively on F, we define a *fiber orbibundle* with structure group G in a similar way, by requiring an atlas of the form $(\overline{U}_\alpha \times F, G_\alpha)$, where the actions and embeddings on the second component are through elements of G. For example, given an orbivector bundle $\mathcal{V}$, we can form the projectivization $\mathbb{RP}(\mathcal{V})$, and the charts are of the form $(\overline{U}_\alpha \times \mathbb{RP}^{n-1}, G_\alpha)$. Note that the projectivization can have more singularities coming from elements fixing setwise (and not necessarily pointwise) a line.

Example 8.2.7 Let ρ be the representation of $\mathbb{Z}_2$ on $\mathbb{R}^2$ given by multiplying by -1. The linear orbifold $[\mathbb{R}^2/\mathbb{Z}_2]$ has only one singular point. The projectivization $\mathbb{RP}(\rho \oplus \mathbb{R})$ is the orbifold $[\mathbb{RP}^2/\mathbb{Z}_2]$. Its singular locus consists of one point and a copy of S^1. The local representation at the isolated point is ρ, and the local representation at points on S^1 is given by complex conjugation.

A *local orientation* of an orbifold is a choice of an orientation at each point that makes the action of G_x orientation preserving; this induces orientations on all smaller charts. This is equivalent to identifying G_x with a subgroup of $SO(n)$. As with manifolds, an *orientation of an orbifold* is just a choice of local orientations such that the transition functions are orientation preserving. We say that an orbifold is *locally orientable* if all the local groups act preserving the orientation.

Locally orientable orbifolds share many of the properties that manifolds have. In particular, a locally orientable n-dimensional orbifold is a *rational homology manifold*, i.e., for every $x \in |\mathcal{X}|$,

$$H_n(|\mathcal{X}|, |\mathcal{X}| - \{x\}; \mathbb{Q}) = \mathbb{Q}.$$

Therefore a closed oriented orbifold of dimension n has a fundamental class in $H_n(|\mathcal{X}|; \mathbb{Q})$. This is not true when working with integer coefficients and not even with $\mathbb{Z}_2$ coefficients; the order of the groups and the local actions can introduce torsion in $H_n(|\mathcal{X}|, |\mathcal{X}| - \{x\}; \mathbb{Z})$.

To an orbifold $\mathcal{X}$ we can associate a topological space $B\mathcal{X}$ called the *classifying space of the orbifold*; see [1,9]. For example, for a global quotient $\mathcal{X} = [M/G]$

with G a finite group acting on M, $B\mathcal{X}$ is homotopy equivalent to the homotopy orbit space $EG \times_G M$. A key feature of this classifying space construction is that if $\mathcal{V}$ is an orbibundle on $\mathcal{X}$, then $B\mathcal{V} \to B\mathcal{X}$ is an honest vector bundle, and therefore we can talk about characteristic classes, now elements of $H^*(B\mathcal{X})$.

To an orbifold $\mathcal{X}$ we have associated two topological spaces, the underlying space $|\mathcal{X}|$ and the classifying space $B\mathcal{X}$; there is a map $B\mathcal{X} \to |\mathcal{X}|$, but in general this map is far from being a homotopy equivalence.

Example 8.2.8 For a finite group G, $B[*/G]$ is homotopy equivalent to BG, the classifying space for principal G-bundles [10], but $|[*/G]|$ is just a point.

Over the rationals these spaces are the same, at least homologically, i.e.,

$$H_*(|\mathcal{X}|; \mathbb{Q}) \cong H_*(B\mathcal{X}; \mathbb{Q}).$$

Remark 8.2.9 Therefore for an oriented orbifold we can talk at the same time about a fundamental class and characteristic classes, at least rationally.

8.3 Cobordism of orbifolds

In her thesis, Druschel [7] started the study of orbifold cobordism by introducing a complete set of invariants that determine the oriented cobordism class up to torsion. To study the torsion, Druschel [7] considers cobordism with restrictions on the set of local groups and how they fit into a commutative diagram to show that every two- and three-dimensional effective oriented orbifold bounds.

By restricting the type of singularities that we allow, we get different cobordism groups, and the study of how these groups relate when we allow more singularities is the main theme of [7] and [5], where the interplay between different cobordism groups is codified algebraically by a commutative diagram and a spectral sequence, respectively. Druschel's result on two- and three-dimensional orbifolds is explained in [5] as the collapse of this spectral sequence.

Generalizing classical constructions from equivariant cobordism, such as families of subgroups and fixed point homomorphisms, in [4] a framework to study cobordism groups with restricted singularities is introduced. For example, restricting the isotropy to groups of odd order, we get a nontrivial cobordism ring $\mathfrak{N}^{odd}_{*,orb}$, for which a complete description in terms of bordism theory is given.

In this section I present a complete answer to the question "When is a differentiable manifold the boundary of an orbifold?" for orbifolds with isotropy groups of odd order and oriented orbifolds. The proofs are self-contained and are adaptations of the main techniques of [8] and [4].

8.3.1 Orbifolds with boundary

As with manifolds, we can talk about *orbifolds with boundary*. Now the charts correspond to open sets in $\mathbb{R}^{n-1} \times [0, \infty)$ with a finite group acting linearly on them. An orbifold has a well-defined boundary, found just by taking the restriction of the action to the boundary on each chart. But some care should be taken: being a point on the boundary is not a condition that can be checked topologically on the underlying space. For manifolds the boundary was just the set of points for which

$$H_n(M, M - \{x\}) = 0.$$

Example 8.3.1 Consider the interval $I = [-1, 1]$ with the action of $\mathbb{Z}_2$ taken as multiplication by -1. The quotient $[I/\mathbb{Z}_2]$ is an orbifold with boundary; the underlying space is an interval, but the boundary is just one point.

Example 8.3.2 Consider S^1 the complex numbers of norm 1, and the $\mathbb{Z}_2$ action given by complex conjugation. The quotient $[S^1/\mathbb{Z}_2]$ is an orbifold without boundary, but the underlying space is homeomorphic to a closed interval.

The assiduous reader will realize that in both examples the action is non-orientable and has codimension-one fixed points. As before, we can say that two orbifolds $\mathcal{Q}_1, \mathcal{Q}_2$ are cobordant if there exists an $n+1$-dimensional orbifold with boundary $\mathcal{W}^{n+1}$ such that

$$\partial \mathcal{W}^{n+1} = \mathcal{Q}_1 \sqcup \mathcal{Q}_2.$$

Without restricting the types of orbifolds that we allow, this definition is vacuous: every orbifold is the boundary of another orbifold. The boundary of the orbifold $[I/\mathbb{Z}_2]$ is only one point. From this easy observation it follows that all orbifolds are boundaries.

Remark 8.3.3 $\mathcal{Q} \times [I/\mathbb{Z}_2]$ is an orbifold with boundary precisely $\mathcal{Q}$.

Because any manifold is naturally an orbifold, we have a well-defined map $\mathfrak{N}_* \to \mathfrak{N}^{odd}_{*,orb}$ and $\Omega_* \to \Omega_{*,orb}$, and the question "When is a differentiable manifold the boundary of an orbifold?" can be recast algebraically.

Remark 8.3.4 Determine the kernels of the map $\mathfrak{N}_* \to \mathfrak{N}^{odd}_{*,orb}$ and $\Omega_* \to \Omega_{*,orb}$.

8.3.2 The unoriented case

As we have seen before, if we impose no restriction on the type of orbifolds that we allow, any orbifold is the boundary of another orbifold: simply consider $\mathcal{X} \times [I/\mathbb{Z}_2]$. A good class of orbifolds for which the theory is not trivial is the class of orbifolds with only odd singularities. For this class of orbifolds, in [4] it is shown that the long exact sequences that appear when we increase the allowed singularities split. The splitting has a clear geometric construction: it is the blowup along the singular set.

Let us see now that if a manifold is the boundary of an orbifold with only odd singularities, then it is actually the boundary of a manifold. The proof will give a way to construct the bounding manifold out of the orbifold.

Suppose that M is an n-dimensional manifold that is the boundary of an $n+1$-dimensional orbifold $\mathcal{W}$ with isotropies of odd order only. Inasmuch as $\mathcal{W}$ is compact, let H be a group with isotropy of maximal order. The points of $\mathcal{W}$ that have isotropy H form a suborbifold, called the H-singular set:

$$\mathcal{W}^H = \{x \in |\mathcal{W}| \mid G_x \cong H\}.$$

$\mathcal{W}^H$ is a purely ineffective orbifold; on each connected component the action of H on a local chart is a fixed representation of H. Because $\mathcal{W}^H$ is a suborbifold of $\mathcal{W}$, there is a normal orbibundle $\nu \to \mathcal{W}^H$. For orbifolds the tubular neighborhood theorem also holds, and therefore we can identify a tubular neighborhood with the total space of the normal orbibundle.

Consider $\mathbb{RP}(\nu \oplus \mathbb{R})$, the projectivization of the orbibundle $\nu \oplus \mathbb{R}$. This also can be thought of as the orbibundle that one gets by identifying antipodal points on the disk orbibundle of $\nu \oplus \mathbb{R}$.

Because we are assuming that all groups are of odd order, an element fixes a line setwise if and only if it fixes the line pointwise. Therefore $\mathbb{RP}(\nu \oplus \mathbb{R})$ has the same isotropy as $\mathcal{W}$.

The singular sets of $\mathbb{RP}(\nu \oplus \mathbb{R})$ and $\mathcal{W}$ can be identified, as well as their respective tubular neighborhoods. We can take the connected sum of $\mathcal{W}$ and $\mathbb{RP}(\nu \oplus \mathbb{R})$ along $\mathcal{W}^H$; the resulting orbifold has strictly less isotropy groups. Because the boundary of $\mathcal{W}$ is a manifold and $\mathbb{RP}(\nu \oplus \mathbb{R})$ has no boundary, this process does not alter the boundary. We can iterate this construction, and because in each step we are reducing the number of singularities, this process is finite. At the end we have a manifold whose boundary is precisely M. Therefore

Theorem 8.3.5 *The kernel of the map* $\mathfrak{N}_* \to \mathfrak{N}^{odd}_{*,orb}$ *is zero.*

8.3.3 The oriented case

As with manifolds, we can also talk about cobordism between oriented orbifolds. The cobordism ring of oriented orbifolds, $\Omega_{*,orb}$, was studied in [8], where $\Omega_{*,orb} \otimes \mathbb{Q}$ was determined.

Let us proceed now and see that if an oriented manifold bounds an oriented orbifold, then actually some multiple of it bounds a manifold (i.e., is a torsion element of Ω_*). This now easily follows from the existence of fundamental classes for oriented orbifolds (over the rationals) and Pontrjagin classes for orbibundles.

Suppose M^n is an oriented manifold that is the boundary of an oriented orbifold $\mathcal{W}$. We want to see that the Pontrjagin numbers of M are zero, which shows that M is a torsion class of Ω_n. The proof is a slight modification of the one given before for the vanishing of the Stiefel–Whitney numbers.

Denote by $\iota : M \to \mathcal{W}$ the inclusion. This map induces a map between the classifying spaces $B\iota : BM \to B\mathcal{W}$. Because M is a manifold, the projection $BM \to M$ is a homotopy equivalence that has the property that $p^*(TM) \cong BTM$. Denote by $[BM] = p_*^{-1}([M])$ the preimage of the fundamental class of M.

Also, $\mathcal{W}$ is an orbifold and therefore has a tangent orbibundle $T\mathcal{W}$. This is not a vector bundle, but $BT\mathcal{W} \to B\mathcal{W}$ is an honest vector bundle, and even more, $B\iota^*(BT\mathcal{W}) = BTM \oplus \mathbb{R}$. We have

$$
\begin{aligned}
\langle p_I(TM), [M]\rangle &= \langle p_I(TM), p_*[BM]\rangle \\
&= \langle p^* p_I(TM), [BM]\rangle && \text{by functoriality} \\
&= \langle p_I(p^*TM), [BM]\rangle && \text{by naturality} \\
&= \langle p_I(BTM), [BM]\rangle \\
&= \langle p_I(BTM \oplus \mathbb{R}), [BM]\rangle && \text{by stability} \\
&= \langle p_I(B\iota^*(BT\mathcal{W})), [BM]\rangle && \text{by the preceding remark} \\
&= \langle B\iota^* p_I(TW), [BM]\rangle && \text{by naturality} \\
&= \langle p_I(TW), B\iota_*[BM]\rangle && \text{by functoriality.}
\end{aligned}
$$

As before, by looking at the long exact sequence of the pair $(B\mathcal{W}, BM)$ we see that $B\iota_*[BM] = 0$. Therefore all the Pontrjagin numbers of M are zero, i.e., M is a torsion element of Ω_*.

In [3], Anderson showed that torsion elements of Ω_* may be represented as sums of classes of manifolds of the form

$$V^n = \mathbb{RP}(\lambda \oplus \mathbb{R}^{2k+1}),$$

the projectivization of the direct sum of a line bundle λ and a trivial $(2k + 1)$-bundle. These manifolds have orientation-reversing involutions coming from the bundle involution on λ.

Remark 8.3.6 If an orbifold $\mathcal{X}$ admits an orientation-reversing involution (actually, any orientation-reversing periodic map), then it is the boundary of an oriented orbifold. Just consider the orbifold $[\mathcal{X} \times I]/\mathbb{Z}_2$, where $\mathbb{Z}_2$ acts on $\mathcal{X}$ by the orientation-reversing involution, and on $I = [-1, 1]$ by multiplying by -1.

By the preceding remark and Anderson's result, all the torsion elements of Ω_* are the boundaries of oriented orbifolds.

Theorem 8.3.7 *The kernel of the map $\Omega_* \to \Omega_{*,orb}$ is precisely the torsion of Ω_**

The idea of using orientation-reversing involutions to construct bounding orbifolds is the key geometric argument used in [7] and [5], where it is used to show that all two- and three-dimensional oriented orbifolds bound.

Bibliography

[1] A. Adem, J. Leida, Y. Ruan. *Orbifolds and Stringy Topology*, Cambridge University Press (2007).

[2] M. Atiyah. *The impact of Thom's cobordism theory*, Bull. Amer. Math. Soc. (N.S.) **41** (2004), 337–340.

[3] P. Anderson. *Cobordism classes of squares of orientable manifolds*, Ann. of Math., **83** (1966), 47–53.

[4] A. Angel. *Cobordism of Orbifolds*, Stanford University Ph.D. thesis (2008).

[5] A. Angel. *A spectral sequence for orbifold cobordism.* In Algebraic Topology: Old and New, M. M. Postnikov Memorial Conference. Polish Academy of Sciences, Institute of Mathematics, Banach Center Publ. 85, Warsaw (2009), 141–154.

[6] A. Dold. *Erzeugende der Thomschen Algebra* $\mathfrak{N}$. Math. Zeit. **65** (1956), 25–35.

[7] K. S. Druschel. *The cobordism of oriented three dimensional orbifolds*, Pacific J. Math. **193** (2000), 45–55.

[8] K. S. Druschel. *Oriented orbifold cobordism*, Pacific J. Math **164** (1994), 299–319.

[9] I. Moerdijk. *Orbifolds as groupoids: an introduction*, in *Orbifolds in Mathematics and Physics* (Madison, WI), 2001; *Contemp. Math.* **310**, Amer. Math. Soc., Providence, RI (2002), pp. 205–222.

[10] J. Milnor. *On the cobordism ring* Ω_*, Notices Amer. Math. Soc. **5** (1968), 457.

[11] L. S. Pontrjagin. *Characteristic cycles of differentiable manifolds,* Amer. Math. Soc. Transl. (1) **7** (1962), 279–331; Mat. Sb. **21** (1947), 233–284.

[12] I. Satake. *On a generalization of the notion of manifold*, Proc. Nat. Acad. Sci. **42**, (1956), 359–363.

[13] R. E. Stong. *Notes on Cobordism Theory*, Princeton University Press, Princeton (1968).

[14] R. Thom. *Quelques propriétés globales des variétés différentiables*, Comment. Math. Helv. **28** (1954), 17–86.

[15] W. Thurston. *The geometry and topology of* 3*-manifolds*, Lectures Notes, Princeton University (1978).

[16] C. T. C. Wall. *Determination of the cobordism ring*, Ann. of Math. **72** (1960), 292–231.

9

Canonical group quantization, rotation generators, and quantum indistinguishability

CARLOS BENAVIDES AND ANDRÉS REYES-LEGA*

Abstract

Using the method of canonical group quantization, we construct the angular momentum operators associated to configuration spaces with the topology of (i) a sphere and (ii) a projective plane. In the first case, the angular momentum operators derived this way are the quantum version of Poincaré's vector, i.e., the physically correct angular momentum operators for an electron coupled to the field of a magnetic monopole. In the second case, the operators one obtains represent the angular momentum operators of a system of two indistinguishable spin zero quantum particles in three spatial dimensions. The relevance of the proposed formalism for the progress in our understanding of the spin–statistics connection in nonrelativistic quantum mechanics is discussed.

9.1 Introduction

The connection between the spin of quantum particles and the statistics they obey is a remarkable example of a simply stated physical fact without the recognition of which many physical phenomena (ranging from the stability of matter and the electronic configuration of atoms to Bose–Einstein condensation and superconductivity) would not have an explanation. Nevertheless, the simplicity of the assertion "integer spin particles obey Bose statistics, and half-integer spin particles obey Fermi statistics" stands in bold contrast to its intricate physical origin. Indeed, Pauli's proof of the *spin–statistics theorem* [Pau40] (improving on earlier work by Fierz [Fie39]), showed that the spin–statistics connection was deeply rooted in relativistic quantum field theory. The path to a rigorous proof of this theorem (from the mathematical point of view) was a long one and involved the efforts of many people (see, e.g., the book by Duck and Sudarshan [DS98b]). The

* Financial support from Universidad de los Andes is gratefully acknowledged.

modern proof of the theorem, in the framework of the general theory of quantum fields, is described in the book by Streater and Wightman [SW00], where many references to original sources are given. There are also treatments of the theorem within the algebraic approach to quantum field theory [Haa96]. An approach within Lagrangian field theory (based on earlier work by Schwinger), which makes use of Lorentz invariance, but in a restricted sense, was pioneered by Sudarshan [Sud75, DS98a]. Nowadays, the spin–statistics theorem stands as a well-established result of theoretical physics.

In spite of all of these triumphs, many authors have been of the opinion that there might be alternative ways to prove the spin–statistics theorem, in a way that does not use the whole machinery of relativistic quantum field theory. This "belief" in a nonrelativistic proof of the theorem has its origin (presumably) in the realization that the topology of the underlying structures of a quantum theory (symmetry groups, configuration spaces, gauge potentials, etc.) may lead to the explanation and clarification of many features of the theory. For instance, from the work of Schulman [Sch68] it became clear that the path integral approach to quantization had to be modified if it was to be applied to a multiply connected configuration space.

This led Laidlaw and DeWitt [LD71] to study the path integral quantization of the configuration space of N indistinguishable (spinless) particles in $\mathbb{R}^3$. They arrived at the conclusion that there were exactly two inequivalent quantizations of such a system, one leading to Fermi statistics, the other leading to Bose statistics. The lesson was: If the indistinguishability of particles is taken into account *before* quantizing, then the Fermi–Bose alternative emerges (in three spatial dimensions) as a consequence of the nontrivial topology of the configuration space. In this sense, one can dispense with the symmetrization postulate, if quantum indistinguishability is taken into account right from the beginning.

Parallel to these developments was the work on quantization of nonlinear field configurations by Finkelstein and Rubinstein [FR68], where a general relation between kink exchange and rotations was established, using homotopy arguments, that resembled the connection between spin and statistics (exchange of particles produces a phase $(-1)^{2S}$ in the wave function, the same effect that a rotation through 2π has on the wave function of a single particle of spin S).

Leinaas and Myrheim [LM77] reformulated the problem studied by Laidlaw and DeWitt in a language close to that of fiber bundles, obtaining the same results for three spatial dimensions (the Fermi–Bose alternative). In addition, they found that in two spatial dimensions, the possible statistics were given not by a sign, but by a phase factor (the so-called anyon statistics [Wil82]).

Since then, a considerable amount of work has been devoted to attempts at alternative, nonrelativistic proofs of the spin–statistics theorem. An approach where pair creation and annihilation are incorporated indirectly in the topology of the configuration space was proposed in [BDG+90].

Perhaps one of the most interesting and influential proposals that have been put forward in recent times is that of Berry and Robbins [BR97]. It is based on a generalization of Leinaas and Myrheim's work in which the spin degrees of freedom are included in the treatment of two indistinguishable quantum particles. Although it does not lead to a new proof of the theorem [BR00], it has inspired new developments, both in mathematics and in physics. For instance, the recent work of Atiyah and coworkers on configuration spaces [Ati01, AS02, AB02] was motivated by the technical difficulties that appear when one tries to generalize the Berry–Robbins construction to the N-particle case. Of more relevance for physics, their work seems to have given new impetus to the nonrelativistic spin–statistics issue (see, for example, [Ana02, Pes03b, AM03, Pes03a, SD03, Kuc04, CJ04]). It has also led to several questions that, in the opinion of the authors, deserve attention.

So, in order to advance in our understanding of the problem, it is necessary to first settle those issues. One of them – a crucial aspect of the Berry–Robbins approach – is the imposition of single-valuedness on the wave function. This condition has been studied in detail by the second-named author and collaborators [PPRS04], arriving at the conclusion that the single-valuedness condition is inconsistent with the assumption that the wave function is a section of a vector bundle over the *physical* configuration space. The global approach proposed in [PPRS04] also allows one to explain why the proof presented in [Pes03b] fails [RL].

Another point, which will be the topic of this paper, has to do with the rotational properties of a quantum system of indistinguishable particles. Recent work by Kuckert shows that it is possible to characterize the connection between spin and statistics in terms of a unitary equivalence between the angular momentum operator of a single-particle system and the angular momentum operator of a two-particle system, both operators being restricted to suitable domains [Kuc04]. We believe that a detailed analysis of that equivalence, which takes fully into account the topology of the problem, could lead to interesting results. For this reason, in this paper we will construct the angular momentum operators for a system of two indistinguishable particles of spin zero, using Isham's canonical group quantization [Ish84]. We shall see how, using Isham's method, we obtain structures ($SU(2)$ equivariance) that are already present in the Berry–Robbins construction, though not explicitly. This is interesting, because one of the advantages of the spin basis of Berry and Robbins (Schwinger construction) is that it allows for explicit computations. Thus, we expect that the Berry–Robbins construction, suitably reinterpreted (as proposed in [PPRS04]), may in fact lead to an advance in our understanding of the spin–statistics connection.

Let us finish this introduction with a description of the contents of this chapter. In Section 9.2 we briefly review Isham's canonical group quantization method. In Section 9.3 – as an example illustrating the dependence of quantum observables on the topology of the configuration space – we then construct, using Isham's method, the angular momentum operators for an electron coupled to the field of a magnetic monopole. In Section 9.4 we consider a system of two indistinguishable, spin zero particles. Again using Isham's method, we construct the corresponding angular momentum operators. The chapter finishes with some remarks and conclusions on Section 9.5.

9.2 Canonical group quantization

Quantization of a classical system described by means of a symplectic manifold (M, ω) involves the construction of a Hilbert space $\mathcal{H}$ and of a quantization map "$\hat{}$" allowing one to replace classical observables f (that is, smooth, real-valued functions on M) by self-adjoint operators $\hat{f}$ acting on $\mathcal{H}$. The quantization map is required to be real, linear and injective, and should map constant functions to multiples of the identity operator. Additionally, the Poisson bracket of two classical observables must be mapped to the commutator of the corresponding quantum observables (Dirac's quantization conditions). It is well known that such a full quantization (which includes an additional irreducibility requirement) is, in general, not implementable (Van Hove's theorem). Nevertheless, there are several quantization methods that allow one to pick a subalgebra of the Poisson algebra $(C^\infty(M), \{\,,\,\})$ and to map it homomorphically to an algebra of operators, satisfying physically and mathematically reasonable conditions. One of them, widely known, is geometric quantization [Woo80]. In this section we will briefly review a scheme developed by Isham [Ish84], the method of canonical group quantization. It has some similarities with geometric quantization and also uses some of the techniques developed by Mackey [Mac68] and Kirillov [Kir76].

The starting point of Isham's approach is the observation that, behind the usual quantum theory of a scalar particle on $\mathbb{R}^n$, where the canonical commutation relations (CCR)

$$\left[\hat{q}^i, \hat{p}_j\right] = i\hbar\delta^i_j, \qquad \left[\hat{q}^i, \hat{q}^j\right] = 0 = \left[\hat{p}_i, \hat{p}_j\right] \tag{9.1}$$

are satisfied, there is a group acting on the classical phase space of the theory by symplectic, transitive and effective transformations. In fact, regarding $\mathbb{R}^n \times \mathbb{R}^n$ as an additive group, we see that the action defined by

$$\begin{aligned} (\mathbb{R}^n \times \mathbb{R}^n) \times T^*\mathbb{R}^n &\longrightarrow T^*\mathbb{R}^n, \\ ((a, b), (q, p)) &\longmapsto (q - a, p + b) \end{aligned} \tag{9.2}$$

has the properties mentioned. That (9.1) and (9.2) have something in common can be seen if one considers the exponentiated (Weyl) form of the CCR. In fact, defining unitary operators $U(a)$ and $V(b)$ by

$$U(a) := e^{-ia\hat{p}}, \qquad V(b) := e^{-ib\hat{q}}, \tag{9.3}$$

one easily checks that the position and momentum operators transform according to

$$\begin{aligned} U(a)\hat{q}U(a)^{-1} &= \hat{q} - \hbar a, \\ V(b)\hat{p}V(b)^{-1} &= \hat{p} + \hbar b. \end{aligned} \tag{9.4}$$

On a general configuration space $\mathcal{Q}$, there are no a-priori given position and momentum operators. For example, the natural choice for the position operator on $\mathcal{Q} = S^1$ is the angle variable, which, as is well known, cannot be used as the basis for a quantum theory on S^1 [Kas06]. In such cases, a good starting point is the consideration of the symmetry groups of the classical configuration space. Once the appropriate group, the *canonical group* $\mathcal{C}$, has been identified, the construction of the corresponding quantum theory proceeds by studying the unitary, irreducible representations of the group. One then sees that in the particular case of $\mathcal{Q} = \mathbb{R}^n$ the CCR (9.1) arise as the unique (by virtue of the Stone–von Neumann theorem) solution of a purely geometric problem: the operators (12.3) provide an irreducible unitary representation of the unique simply connected Lie group the Lie algebra of which is a central extension of the Lie algebra of the group $\mathcal{G} = (\mathbb{R}^n \times \mathbb{R}^n, +)$. So, in this special case, the canonical group $\mathcal{C}$ turns out to be the Heisenberg group.

Keeping these preliminary remarks in mind, let us proceed to describe the general scheme. It is based on a careful analysis of the following diagram:

$$\begin{array}{ccccccccc} 0 & \longrightarrow & \mathbb{R} & \longrightarrow & C^\infty(M,\mathbb{R}) & \stackrel{\jmath}{\longrightarrow} & \mathrm{HamVF}(M) & \longrightarrow & 0. \\ & & & & \nwarrow & & \uparrow \gamma & & \\ & & & & & P & \mathcal{L}(\mathcal{G}) & & \end{array} \tag{9.5}$$

The meaning of the different terms appearing in (9.5) is the following:

- M is a symplectic manifold. We are mainly interested in the case where it is a phase space, of the form $M = T^*\mathcal{Q}$, with $\mathcal{Q}$ a homogeneous space.
- $\jmath$ is the map that assigns to each function f on phase space (the negative of) its Hamiltonian vector field. Following the notation in [Ish84], we shall write $\jmath(f) = -\xi_f$. The kernel of $\jmath$ is the set of constant functions on phase space, thus making the first row of the diagram a short exact sequence.

- $\mathcal{G}$ is a Lie group, acting by symplectic transformations on M. The Lie algebra of $\mathcal{G}$ will be denoted $\mathcal{L}(\mathcal{G})$.
- The map $\gamma : \mathcal{L}(\mathcal{G}) \to \mathrm{HamVF}(M)$ is the Lie algebra homomorphism induced by the $\mathcal{G}$-action.
- Once the appropriate $\mathcal{G}$-action has been found (certain requirements must be met), one looks for a linear map $P : \mathcal{L}(\mathcal{G}) \to C^\infty(M, \mathbb{R})$ that is also a Lie algebra homomorphism.

The idea of the quantization scheme is the following. Let us assume that P maps $\mathcal{L}(\mathcal{G})$ isomorphically onto some Lie subalgebra of $(C^\infty(M, \mathbb{R}), \{\,,\,\})$. In this case one can define a quantization map by fixing a representation U of the group and assigning to each function lying in the image of P the self-adjoint generator obtained from U by means of P^{-1}. The existence of a map P with the desired properties is not something obvious. There are obstructions coming from the fact that the map P determines a class in the second cohomology group of $\mathcal{L}(\mathcal{G})$ (with values in $\mathbb{R}$). Of course, there might be many $\mathcal{G}$-actions on M that could be considered. But the restriction will be imposed that the diagram (9.5) must be commutative. The reason for the imposition of this restriction is that, given a (finite-dimensional) Lie subalgebra $\mathfrak{h}$ of $C^\infty(M, \mathbb{R})$, the Hamiltonian vector field that a function $f \in \mathfrak{h}$ generates, ξ_f, gives place to a one-parameter group, acting by symplectic transformations on M. If all these vector fields are complete, their one-parameter groups will generate a group $\mathcal{G}$ of symplectic transformations, and, if the mapping sending $\mathfrak{h}$ into the set of Hamiltonian vector fields is injective, we obtain a Lie algebra isomorphism $\mathfrak{h} \cong \mathcal{L}(\mathcal{G})$. On the other hand, given a symplectic action of a Lie group $\mathcal{G}$ on M, there is a naturally induced map $\gamma : \mathcal{L}(\mathcal{G}) \to \mathfrak{X}(M)$. It is only if $\gamma(A)$ is a Hamiltonian vector field that we can assign a function on phase space to the Lie algebra element A. For this reason, the requirement that the image of γ lie in $\mathrm{HamVF}(M)$ must be imposed.[1] The idea is, therefore, to try to reverse this procedure: starting with a group $\mathcal{G}$ of symplectic transformations, we seek a kind of inverse to the map $\jmath$. More precisely, we look for a Lie algebra homomorphism P such that $\jmath \circ P = \gamma$. In other words, P must be a linear map satisfying

$$\{P(A), P(B)\} = P([A, B]) \tag{9.6}$$

and

$$\gamma(A) = -\xi_{P(A)} \tag{9.7}$$

for all A and B in $\mathcal{L}(\mathcal{G})$.

Because every exact sequence of vector spaces splits, there is no difficulty in finding a linear map P such that the diagram commutes. The problem lies in (9.6). The condition (9.7) fixes $P(A)$ only up to a constant (because $\ker \jmath = \mathbb{R}$), and in

[1] This is automatically satisfied if $H^1(M; \mathbb{R}) = 0$ or if $\mathcal{G}$ is semisimple.

some cases it is possible to adjust these constants so as to satisfy (9.6). But this is only possible if the cocycle defined by

$$z(A, B) := \{P(A), P(B)\} - P([A, B]) \tag{9.8}$$

is also a coboundary. We thus see how the obstruction is measured by the second cohomology group of $\mathcal{L}(\mathcal{G})$. In case the cocycle cannot be made to vanish by a redefinition of P, a central extension of $\mathcal{L}(\mathcal{G})$ by $\mathbb{R}$ can be used to construct the desired map. As mentioned, this is precisely the way in which the Heisenberg group (and with it the CCR) arises from the action (9.2).

Once the appropriate canonical group $\mathcal{C}$ has been found,[2] a quantization map can be defined by assigning to each element $P(A) \in \mathrm{Im}P \subseteq C^\infty(M, \mathbb{R})$ the self-adjoint generator corresponding to A induced by a unitary, irreducible representation of the canonical group. Because there may be inequivalent representations of the canonical group, we may also obtain different, inequivalent quantizations of the same classical system. The general scheme can thus be divided in two main steps:

(i) Find the canonical group $\mathcal{C}$.
(ii) Study the irreducible, unitary representations of the canonical group.

In the particular case where $M = T^*\mathcal{Q}$, there is a natural place to start the search for the canonical group, and it turns out that the representations can be constructed using Mackey's theory of induced representations. When M is the cotangent bundle of some configuration space $\mathcal{Q}$, then every diffeomorphism on it induces a symplectic transformation, given by the pullback operation on the bundle. Additionally, the exterior differential of any smooth function on $T^*\mathcal{Q}$ induces a canonical transformation, by translations along the fibers. Because none of these actions is transitive, it is necessary to consider both of them. The natural combination of these operations can be regarded as coming from the group action ρ defined by ($[h] \in C^\infty(\mathcal{Q}, \mathbb{R})/\mathbb{R}$, $\phi \in \mathrm{Diff}\mathcal{Q}$ and $l \in T_q^*\mathcal{Q}$):

$$\rho_{([h],\phi)}(l) := \phi^{-1*}(l) - (dh)_{\phi(q)}, \tag{9.9}$$

provided the set $C^\infty(\mathcal{Q}, \mathbb{R})/\mathbb{R} \times \mathrm{Diff}\mathcal{Q}$ is endowed with the structure of a semidirect product. That is, $C^\infty(\mathcal{Q}, \mathbb{R})/\mathbb{R} \rtimes \mathrm{Diff}\mathcal{Q}$ is the group with elements of the form $([h], \phi) \in C^\infty(\mathcal{Q}, \mathbb{R})/\mathbb{R} \times \mathrm{Diff}\mathcal{Q}$ and with product

$$([h_2], \phi_2) \cdot ([h_1], \phi_1) = ([h_2] + [h_1 \circ \phi_2^{-1}], \phi_2 \circ \phi_1). \tag{9.10}$$

Thus, for $M = T^*\mathcal{Q}$, step (i) reduces to the problem of finding a suitable finite-dimensional subspace W of $C^\infty(\mathcal{Q}, \mathbb{R})/\mathbb{R}$ and a suitable finite-dimensional subgroup G of $\mathrm{Diff}\mathcal{Q}$. The group $\mathcal{G}$ of the diagram (9.5) will then be given by $W \rtimes G$.

[2] In some cases it is given by $\mathcal{G}$; in others, it will be a Lie group whose Lie algebra is the central extension of $\mathcal{L}(\mathcal{G})$.

At this point, we refer the reader to Isham's article [Ish84] for a thorough discussion of the method. For the applications that will be presented in the next two sections, it will be enough to briefly comment on how the vector space W and the group G make their appearance in the still more special case in which $\mathcal{Q}$ is a homogeneous space. A brief discussion of the way in which the representations are constructed in this case will also be presented at the end of this section.

Assume that $R : G \to \mathrm{GL}(W)$ is a representation of a Lie group on a real, finite-dimensional vector space W. Then, a contragredient representation R^* is naturally induced on W^*, by duality. It is defined as follows ($g \in G$, $u \in W$ and $\varphi \in W^*$):

$$\left(R^*(g)\varphi\right)(u) := \varphi\left(R(g)u\right). \tag{9.11}$$

Regarding W as a configuration space, we have $T^*W \cong W \times W^*$. Using the representation R^*, one can construct the semidirect product $W^* \rtimes G$ in the usual way. It is then possible to define a left action of $\mathcal{G} := W^* \rtimes G$ on T^*W, by setting

$$l_{(\varphi',g)}(u,\varphi) := \left(R(g)u,\, R^*(g^{-1})\varphi - \varphi'\right). \tag{9.12}$$

An element φ of the dual space W^* can be naturally regarded as a function $f^\varphi \in C^\infty(W,\mathbb{R})$ by setting $f^\varphi(u) := \varphi(u)$. The map P is then naturally given by ($\tilde{A} \equiv (\varphi, A)$)

$$\begin{aligned} P : \mathcal{L}(W^* \rtimes G) &\longrightarrow C^\infty(T^*W,\mathbb{R}), \\ \tilde{A} &\longmapsto P(\tilde{A}) : (u,\psi) \mapsto \psi\left(R(A)u\right) + \varphi(u). \end{aligned} \tag{9.13}$$

As explained in detail in [Ish84], all properties that the diagram (9.5) must satisfy are fulfilled in this case, with the exception that the $\mathcal{G}$-action is not transitive. This problem can be solved by restricting the action to a G-orbit of W, say $\mathcal{O}_v$, for some $v \in W$. This leads us directly to configuration spaces of the form $\mathcal{Q} = G/H$ (if $\mathcal{Q} = \mathcal{O}_v$, then H is the little group of v). The action (9.12), as well as the map (9.13), can then be restricted to $G/H \cong \mathcal{O}_v \subseteq W$, and one can show that (9.12) is exactly of the form (9.9).

Thus, starting with a homogeneous space of the form G/H, one has to find a vector space W on which G acts, and such that G/H is a G-orbit. In this case, the canonical group can be chosen as $\mathcal{C} \equiv \mathcal{G} := W^* \rtimes G$. The unitary irreducible representations of this group can be constructed using Mackey's theory of induced representations. Generally, the resulting representation space will be the space of square-integrable sections of a vector bundle E over $\mathcal{Q}' = G/H$, constructed as an associated bundle to the principal bundle $G \to G/H$, by means of an irreducible unitary representation of H. Here, the subgroup H is regarded as the isotropy group of a previously chosen element in the character group of W^*, $\mathrm{Char}(W^*)$.[3]

[3] Hence, $\mathcal{Q}'$ is a G-orbit in $\mathrm{Char}(W^*)$. In the examples we are interested in, these orbits coincide with the G-orbits in W^* and we can identify them, i.e., $\mathcal{Q}' \cong \mathcal{Q}$.

Integration of sections is carried out using the Hermitian structure of the vector bundle and a G-quasi-invariant measure μ on configuration space. The operators giving the representation of the subgroup G of $\mathcal{C}$ are constructed using a lift $l^{\uparrow}$ of the G-action l on $\mathcal{Q}'$ to the corresponding vector bundle. This lift is naturally induced by the right action of G on the principal bundle. We are thus naturally led to consider G-vector bundles over $\mathcal{Q}'$. Recall that a G-vector bundle is a vector bundle (with total space E) over a G-space $\mathcal{Q}$, together with a lift $l^{\uparrow}$, i.e., a G-action on E which is linear on the fibers and such that the following diagram commutes ($g \in G$):

$$\begin{array}{ccc} E & \xrightarrow{l_g^{\uparrow}} & E \\ {\scriptstyle\pi}\big\downarrow & & {\scriptstyle\pi}\big\downarrow \\ \mathcal{Q}' & \xrightarrow{l_g} & \mathcal{Q}'. \end{array} \tag{9.14}$$

If Ψ is a section of the bundle (i.e., a *wave function*), then the unitary operator $U(g)$ acts on it as follows:

$$(U(g)\Psi)(x) := \sqrt{\frac{d\mu_g}{d_\mu}(x)}\, l_g^{\uparrow}\Psi(g^{-1}\cdot x), \tag{9.15}$$

where $d\mu_g/d\mu$ is the Radon–Nikodym derivative of μ_g with respect to μ. This G-representation can be extended to the whole group $\mathcal{C}$ as follows (recall that x is an element in a G-orbit of $\mathrm{Char}(W^*)$):

$$(V(\varphi)\Psi)(x) := x(\varphi)\Psi(x). \tag{9.16}$$

The infinitesimal version of these relations gives place to the corresponding self-adjoint generators, of which the angular momentum operators of a particle whose configuration space is a sphere are one example, to which we now turn our attention.

9.3 Magnetic monopole

9.3.1 The classical problem

In this section, we explore a simple but fundamental example: the problem of a point electric charge coupled to the the (external) magnetic field of a fixed magnetic monopole. As is well known, the importance of this problem lies in the fact that, in order for the quantum problem to be consistent, the electric charge of the particle must be quantized [Dir31].

Classically, the dynamics of a particle of mass m and charge e coupled to the field produced by a magnetic monopole of strength g can be described by the

Lagrangian

$$L(q, \dot{q}) = \frac{1}{2}m\dot{q}^2 + \frac{e}{c}\dot{q} \cdot A(q), \tag{9.17}$$

where $q = (q_1, q_2, q_3)$ denotes the position of the particle. The vector potential A must be chosen in such a way that its curl gives a radial field. If g denotes the magnetic "charge," then (using the notation $r = \|q\|$ and $\hat{r} = q/r$) we require

$$B := \nabla \times A \stackrel{!}{=} g\frac{\hat{r}}{r^2}. \tag{9.18}$$

This condition cannot be satisfied using a global gauge potential A. Thus, the Lagrangian (9.17) is only locally defined. It is possible to give a global description of this problem, in the Lagrangian setting, but the introduction of additional structures is necessary [ZSN+83]. For our purposes, the local description will be sufficient. We therefore introduce the following local expressions for the gauge potential:

$$\begin{aligned} A^N(q) &:= \frac{g}{r(r+q_3)}(-q_2, q_1, 0), \\ A^S(q) &:= \frac{g}{r(r-q_3)}(q_2, -q_1, 0). \end{aligned} \tag{9.19}$$

Using the general form of Noether's theorem,[4] one can show that there are three conserved quantities, related to the action of the rotation group on the configuration space. Because the Lagrangian is only locally defined, one has to compute the conserved quantities using the two expressions for the gauge potential, A^N and A^S. The conserved quantities obtained using A^N and A^S turn out to be the same, up to an irrelevant constant term. They can be combined into a single vector

$$J = mq \times \dot{q} - \frac{eg}{c}\hat{r}, \tag{9.20}$$

which is to be interpreted as the *angular momentum vector* of the particle. In fact, working in the Hamiltonian formalism (still in local coordinates), one obtains the following expression for J:

$$J = q \times \left(p - \frac{e}{c}A^N\right) - \frac{eg}{c}\hat{r} = L - \frac{eg}{c}K^N, \tag{9.21}$$

where

$$\begin{aligned} L &:= q \times p, \\ K^N &:= \frac{1}{g}(q \times A^N) + \hat{r} = \frac{q - r\hat{z}}{r - z}. \end{aligned} \tag{9.22}$$

[4] The general form of the theorem guarantees the existence of a conserved quantity whenever the Lagrangian is invariant under a one-parameter group of transformations *up to a gauge transformation*.

Equation (12.13) can be used in order to compute the Poisson brackets of the components of J. The result is $\{J_i, J_j\} = -\varepsilon_{ijk} J_k$. Thus, the components of J satisfy angular momentum commutation relations and are to be regarded as giving the correct expression for the angular momentum of the particle:

$$J = L - \frac{eg}{c} K. \tag{9.23}$$

In spite of the fact that the description of this system can only be given in local terms, the angular momentum is a well-defined, global function. But, as is well known, the situation changes drastically when we consider the quantum version of the problem. There are different ways to analyze it, all yielding the same result: the wave function for an electron coupled to the field of a magnetic monopole is a section of a line bundle over the configuration space. The topology of this bundle is characterized by an integer number n that relates magnetic and electric charge, giving rise to Dirac's famous result:

$$eg/c = \frac{n}{2}\hbar. \tag{9.24}$$

Because the wave function is a section in some bundle, the corresponding angular momentum operators must be maps from the space of sections to itself. A physically motivated and detailed analysis of this problem, involving the construction of the angular momentum operators, can be found in [BL81]. There, the form of the angular momentum operators is guessed from the classical expression, leading to an operator of the form $L - \mu K$, where $\mu = eg/\hbar c$. The quantization condition (9.24) arises from a consistency requirement on the theory.[5] In the next section we will arrive at the same result by applying the canonical group quantization method to the magnetic monopole problem.

9.3.2 The quantum problem

The configuration space for the monopole problem is given by $\mathbb{R}^3 \setminus \{0\}$. Because the monopole field is spherically symmetric and we are only interested in the rotational properties of the system, we can regard the sphere S^2 as the configuration space on which the magnetic monopole problem is defined. Moreover, because the sphere is a deformation retract of $\mathbb{R}^3 \setminus \{0\}$, the topological effects produced by both spaces in the quantum theory are the same.

In order to quantize, we want to think of the configuration space as a homogeneous space. We choose the description of the sphere as the quotient $SU(2)/U(1)$.

[5] This comes from the fact that the wave functions, as well as the angular momentum operators, are defined only locally. The consistency requirement imposed is that expectation values of the quantum operators, computed using the different local expressions, must coincide in the overlap regions.

In this case, the canonical group is given by $\mathcal{C} = (\mathbb{R}^3)^* \rtimes SU(2)$. Because we are only interested in obtaining the angular momentum operators, we only need to construct the U operators, as defined in (9.15). The Jacobian factor $d\mu_g/d\mu$ is equal to one in this case, because the measure is $SU(2)$-invariant. Hence, all we have to do is to choose an irreducible unitary representation of $U(1)$ in order to construct a vector bundle associated to the principal bundle $SU(2) \to SU(2)/U(1)$. The lift $l^\uparrow$ is naturally induced by the group product in $SU(2)$, as explained in the following.

Let

$$\begin{aligned} \mathcal{U}_n : U(1) &\longrightarrow \mathrm{Gl}(\mathbb{C}), \\ e^{i\phi} &\longmapsto \mathcal{U}_n(e^{i\phi}) := e^{in\phi} \end{aligned} \tag{9.25}$$

denote one of the unitary representations of $U(1)$ on $\mathbb{C}$, labeled by an integer n. The elements of the associated bundle $\mathcal{L}_n := SU(2) \times_{\mathcal{U}_n} \mathbb{C}$ are equivalence classes of the form $[(p, v)]$, with $p \in SU(2)$ and $v \in \mathbb{C}$. The equivalence relation is

$$(p, v) \sim (p \cdot \lambda, \mathcal{U}_n(\lambda^{-1})v). \tag{9.26}$$

Here, $U(1)$ is regarded as the subgroup of $SU(2)$ consisting of all diagonal matrices of the form $\mathrm{diag}(\lambda, \bar{\lambda})$, with $\|\lambda\| = 1$. If we adopt the convention of denoting the elements of $SU(2)$ by tuples (z_0, z_1) that represent matrices of the form

$$\begin{pmatrix} z_0 & \bar{z}_1 \\ -z_1 & \bar{z}_0 \end{pmatrix}, \tag{9.27}$$

then the right action of $U(1)$ on $SU(2)$, which is given by

$$\begin{pmatrix} z_0 & \bar{z}_1 \\ -z_1 & \bar{z}_0 \end{pmatrix} \longmapsto \begin{pmatrix} z_0 & \bar{z}_1 \\ -z_1 & \bar{z}_0 \end{pmatrix} \begin{pmatrix} \lambda & 0 \\ 0 & \bar{\lambda} \end{pmatrix} = \begin{pmatrix} \lambda z_0 & (\overline{\lambda z_1}) \\ -(\lambda z_1) & (\overline{\lambda z_0}) \end{pmatrix}, \tag{9.28}$$

can be equivalently expressed as

$$(z_0, z_1) \longmapsto (z_0, z_1) \cdot \lambda = (\lambda z_0, \lambda z_1). \tag{9.29}$$

We will use these conventions in order to identify the bundle $SU(2) \to SU(2)/U(1)$ with the Hopf fibration $S^3 \to S^2$, when appropriate. If in addition we consider the equivalence of S^2 with $\mathbb{C}P^1$, we can regard the projection $\pi : SU(2) \to SU(2)/U(1)$ as the map $\pi((z_0, z_1)) = [z_0 : z_1]$. Thus, the left action of $SU(2)$ on $S^2 \cong \mathbb{C}P^1$ takes the following form ($g = (\alpha, \beta)$):

$$\begin{aligned} l : SU(2) \times \mathbb{C}P^1 &\longrightarrow \mathbb{C}P^1, \\ (g, [z_0 : z_1]) &\longmapsto l_g([z_0 : z_1]) = \left[\alpha z_0 - \bar{\beta} z_1 : \beta z_0 + \bar{\alpha} z_1\right]. \end{aligned} \tag{9.30}$$

The left action of $SU(2)$ on itself given by the group product allows one to lift the action l to the bundle $\mathcal{L}_n$. It is given by the following expression

($g, p \in SU(2), v \in \mathbb{C}$):

$$l_g^{\uparrow}([(p, v)]) := [(gp, v)]\,. \tag{9.31}$$

The action of the angular momentum operators on wave functions can then be obtained from the infinitesimal version of (9.15).

Given that these operators act on the space of global sections of the bundle, it is necessary, in order to be able to compare with the expressions known from the physics literature, to obtain local expressions. Therefore, we will construct local trivializations for the bundle $\mathcal{L}_n$ and will then compute the action of the infinitesimal generators, using local sections.

9.3.3 Local description of $\mathcal{L}_n$

The total space of the line bundle $\mathcal{L}_n = SU(2) \times_{\mathcal{U}_n} \mathbb{C}$ consists of equivalence classes of the form $[((z_0, z_1), v)]$, with $(z_0, z_1) \in SU(2)$ and $v \in \mathbb{C}$. The projection is the map $\pi_n : \mathcal{L}_n \to S^2 \cong \mathbb{C}P^1$ given by $\pi_n([((z_0, z_1), v)]) := [z_0 : z_1]$. In order to construct local trivializations for this bundle, we start by defining local charts, as follows.

Set

$$U_N = S^2 \setminus \{N\} \qquad \text{(sphere with north pole removed)},$$
$$U_S = S^2 \setminus \{S\} \qquad \text{(sphere with south pole removed)}.$$

We define local charts using stereographic projections onto the complex plane. Let us denote the local coordinates as follows:

$$\begin{aligned} z : U_N &\longrightarrow \mathbb{C}, \\ x &\longmapsto z(x) \end{aligned} \tag{9.32}$$

(stereographic projection from the north pole) and

$$\begin{aligned} \zeta : U_S &\longrightarrow \mathbb{C}, \\ x &\longmapsto \zeta(x) \end{aligned} \tag{9.33}$$

(stereographic projection from the south pole). Notice that if on $\mathbb{C}P^1$ we set $U_0 := \{[z_0 : z_1] \mid z_1 \neq 0\}$ and $U_1 := \{[z_0 : z_1] \mid z_0 \neq 0\}$, then we can define local charts that coincide with z and ζ through the equivalence $S^2 \cong \mathbb{C}P^1$, as follows:

$$\begin{aligned} U_0 &\longrightarrow \mathbb{C}, \\ [z_0 : z_1] &\longmapsto z := \frac{z_0}{z_1} \end{aligned} \tag{9.34}$$

and

$$\begin{aligned} U_1 &\longrightarrow \mathbb{C}, \\ [z_0 : z_1] &\longmapsto \zeta := \frac{z_1}{z_0}. \end{aligned} \tag{9.35}$$

Hence, U_0 can be identified with U_N, and U_1 with U_S. It will be convenient to keep in mind that if x is a point in the sphere with polar coordinates (θ, φ), then

$$z(x) = \frac{e^{i\varphi} \sin\theta}{1 - \cos\theta} \quad \text{and} \quad \zeta(x) = \frac{e^{-i\varphi} \sin\theta}{1 + \cos\theta}. \tag{9.36}$$

Local trivializations for the bundle $\mathcal{L}_n$ can be defined in the following way.

Using the notation $g \equiv (z_0, z_1) \in SU(2)$, set

$$\begin{aligned} \varphi_N : \pi_n^{-1}(U_N) &\longrightarrow U_N \times \mathbb{C}, \\ [(g, v)] &\longmapsto \left([z_0 : z_1], \left(\frac{z_1}{|z_1|} \right)^n v \right) \end{aligned} \tag{9.37}$$

and

$$\begin{aligned} \varphi_S : \pi_n^{-1}(U_S) &\to U_S \times \mathbb{C}, \\ [(g, v)] &\longmapsto \left([z_0 : z_1], \left(\frac{z_0}{|z_0|} \right)^n v \right). \end{aligned} \tag{9.38}$$

As can be easily checked, these maps are well defined, and provide local homeomorphisms. From these local trivializations we obtain, for the transition function g_{SN},

$$(\varphi_S \circ \varphi_N^{-1})([z_0 : z_1], w) = \left([z_0 : z_1], \left(\frac{z}{|z|} \right)^n w \right), \tag{9.39}$$

that is, $g_{SN}([z_0 : z_1]) = (z/|z|)^n$. From this we see that the first Chern number of $\mathcal{L}_n$ is n. This means that an integer number that at first was chosen to (partially) label a representation of the canonical group also determines the topology of the bundle where the space of physical states is defined.

9.3.4 Construction of the angular momentum operators

Recall that the lifting $l^\uparrow$ of the $SU(2)$ action on the sphere to $\mathcal{L}_n$ is induced by the corresponding lifting on the principal bundle. Therefore, $\mathcal{L}_n$ has the structure of a homogeneous $SU(2)$ bundle:

$$\begin{array}{ccc} SU(2) \times_{\mathcal{U}_n} \mathbb{C} & \xrightarrow{l_g^\uparrow} & SU(2) \times_{\mathcal{U}_n} \mathbb{C} \\ \pi_n \downarrow & & \pi_n \downarrow \\ S^2 & \xrightarrow{l_g} & S^2, \end{array} \tag{9.40}$$

where $l_g[g'] := [gg']$ and $l_g^\uparrow([g', v]) = [gg', v]$.

What we want to do now is to use the local trivializations φ_N and φ_S to obtain a local version of (9.40). Using the map φ_N, we can obtain a local expression for $l_g^\uparrow$.

The corresponding map will be denoted σ_g:

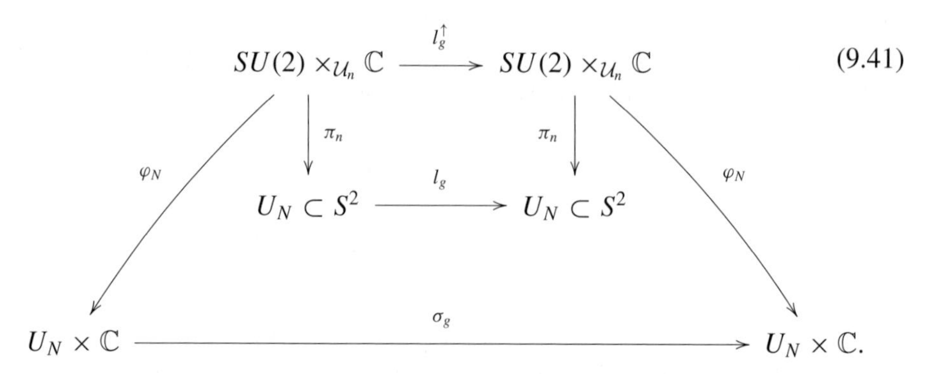

(9.41)

According to the diagram, we have $\sigma_g = \varphi_N \circ l_g^{\uparrow} \circ \varphi_N^{-1}$. Off course, this map is well defined only for elements $g \in SU(2)$ such that $l_g(U_N) \subset U_N$. Because we are interested in the infinitesimal generators of the group action, we will only consider group elements near the identity, so that this condition will always be satisfied.

Thus, for $g = (\alpha, \beta) \in SU(2)$ and $([z_0 : z_1], w) \in U_N \times \mathbb{C}$ we obtain

$$\begin{aligned}
\sigma_g\left(([z_0 : z_1], w)\right) &= (\varphi_N \circ l_g^{\uparrow} \circ \varphi_N^{-1})([z_0 : z_1], w) \\
&\overset{(9.37)}{=} (\varphi_N \circ l_g^{\uparrow})\left(\left[\left((z_0, z_1), (z_1/|z_1|)^{-n}\, w\right)\right]\right) \\
&\overset{(9.31)}{=} \varphi_N\left(\left[\left((\alpha, \beta)\cdot(z_0, z_1), (z_1/|z_1|)^{-n}\, w\right)\right]\right) \\
&\overset{(9.37),(9.30)}{=} \left([z_0' : z_1'], \left(\frac{\beta z + \bar{\alpha}}{|\beta z + \bar{\alpha}|}\right)^n w\right),
\end{aligned} \tag{9.42}$$

where $z_0' = \alpha z_0 - \bar{\beta} z_1$ and $z_1' = \beta z_0 + \bar{\alpha} z_1$. Let $s : S^2 \to \mathcal{L}_n$ be a section of the bundle $\mathcal{L}_n$. Using the local trivializations, we get local sections ($\gamma = N, S$)

$$\begin{aligned}
s_\gamma : U_\gamma &\longrightarrow U_\gamma \times \mathbb{C}, \\
x &\longmapsto \varphi_\gamma \circ s.
\end{aligned} \tag{9.43}$$

These are necessarily of the form $s_\gamma(x) = (x, |\psi_\gamma(x)\rangle)$, with $x \mapsto |\psi_\gamma(x)\rangle$ a complex-valued function defined on U_γ. The local version of (9.15) is, for $\gamma = N$ and $g = (\alpha, \beta)$,

$$\begin{aligned}
(U_{\text{loc}}(g) s_N)(x) &= \sigma_g\left(s_N(g^{-1}\cdot x)\right) = \sigma_g\left((g^{-1}\cdot x, |\psi_N(g^{-1}\cdot x)\rangle)\right) \\
&= \left(x, \left(\frac{\beta z(x) + \bar{\alpha}}{|\beta z(x) + \bar{\alpha}|}\right)^n |\psi_N(g^{-1}\cdot x)\rangle\right) \\
&=: \left(x, \omega(x, g)\, |\psi_N(g^{-1}\cdot x)\rangle\right),
\end{aligned} \tag{9.44}$$

with $\omega(x, g)$ defined through the last equality. In order to find local expressions for the infinitesimal generators, we introduce, for each generator, an appropriate parametrization $t \mapsto g(t)$. The corresponding generators are then defined by their action on local sections ($s_N \mapsto J s_N$) in the following way:

$$(J s_N)(x) := i \frac{d}{dt}\bigg|_{t=0} (U_{\text{loc}}(g(t)) s_N)(x). \tag{9.45}$$

The general form of the generator will be $J = \tilde{L} + \tilde{\omega}$, where $\tilde{\omega}$ is an x-dependent factor and $\tilde{L}$ a differential operator. This can be seen from

$$\begin{aligned} & i \frac{d}{dt}\bigg|_{t=0} \left(\omega(x, g(t)) \left|\psi_N(g(t)^{-1} \cdot x)\right\rangle\right) \\ &= \underbrace{\left(i \frac{d}{dt}\bigg|_{t=0} \omega(x, g(t))\right)}_{=\tilde{\omega}(x)} |\psi_N(x)\rangle + \underbrace{i \frac{d}{dt}\bigg|_{t=0} \left|\psi_N(g(t)^{-1} \cdot x)\right\rangle}_{=\tilde{L}|\psi_N(x)\rangle}. \end{aligned} \tag{9.46}$$

The generator for rotations around the z axis is obtained by putting $g(t) = (\alpha(t), \beta(t)) = (e^{it/2}, 0)$. In this case. $\tilde{L} = \hat{L}_z$, the third component of the usual (orbital) angular momentum operator. For $\tilde{\omega}$ we obtain

$$\tilde{\omega}_z(x) = i \frac{d\omega}{dt}\bigg|_{t=0} = \frac{n}{2}.$$

Thus,

$$\hat{J}_z^N = \hat{L}_z + \frac{n}{2}.$$

For rotation around the y axis we put $\alpha(t) = \cos(t/2)$ and $\beta(t) = \sin(t/2)$. This leads to

$$\begin{aligned} \tilde{\omega}_y(x) &= i \frac{d\omega(x, g(t))}{dt}\bigg|_{t=0} = i \frac{d}{dt}\bigg|_{t=0} \left(\frac{\cos t/2 + \sin t/2\, z(x)}{\cos t/2 + \sin t/2\, \overline{z(x)}}\right)^{n/2} \\ &= -\frac{n}{4}\left(z(x) - \overline{z(x)}\right) = \frac{n}{2} \frac{\sin\theta \sin\varphi}{1 - \cos\theta}. \end{aligned} \tag{9.47}$$

Here again we have $\tilde{L} = \hat{L}_y$, with $\hat{L}_y$ the second component of the usual (orbital) angular momentum operator. It follows that

$$J_y^N = L_y - \frac{n}{2} \frac{y}{1 - z}. \tag{9.48}$$

Using the commutation relations, we obtain, for the remaining generator,

$$J_x^N = L_x - \frac{n}{2} \frac{x}{1 - z}. \tag{9.49}$$

Writing $J^N = (J_x^N, J_y^N, J_z^N)$, we can express the result of the previous computations as follows:

$$J^N = L - \frac{n}{2} K^N. \tag{9.50}$$

Here, L represents the usual orbital angular momentum operator, and K^N is given by (9.22) (here it is regarded as a multiplication operator). The result for the local operator J^S is obtained in the same way.

Comparing with the classical expression (9.23), we see that the condition $\mu = n/2$ must be imposed in order to obtain a consistent quantum theory. This is, in fact, an expression of the quantization of the electric charge, obtained by the canonical group quantization method. Notice that here we are only considering the *kinematical* part of the problem.

9.4 Rotation generators for indistinguishable particles

9.4.1 Configuration space

The configuration space for a system of N indistinguishable, noncolliding particles in $\mathbb{R}^3$ is defined as

$$Q_N = \tilde{Q}_N / S_N, \tag{9.51}$$

where

$$\tilde{Q}_N = \left\{ (r_1, \ldots, r_N) \in \mathbb{R}^{3N} \mid r_i \neq r_j \text{ whenever } i \neq j \right\}, \tag{9.52}$$

with S_N denoting the permutation group. We are interested in the case $N = 2$, for which we have the following homeomorphism:

$$Q_2 \cong \mathbb{R}^3 \times \mathbb{R}_+ \times \mathbb{R}P^2. \tag{9.53}$$

Here, the projective space $\mathbb{R}P^2$ is obtained, through identification of exchanged configurations, from the sphere consisting of all normalized relative position vectors. Because we are only interested in topological effects, we regard $\mathbb{R}P^2$ as the configuration space for this problem.

It is well known that the quotient map $S^2 \to \mathbb{R}P^2$ gives rise to a $\mathbb{Z}_2$-bundle structure and also that there are two inequivalent (scalar) quantizations on $\mathbb{R}P^2$, determined by the characters of the fundamental group $\pi_1(\mathbb{R}P^2) \cong \mathbb{Z}_2$. Because our aim is to construct the infinitesimal generators of rotations for this problem, it will be convenient to describe the configuration space both as the quotient $S^2/\mathbb{Z}_2$ and as a homogeneous space, of the form $SU(2)/H$. Setting

$$H := \left\{ \begin{pmatrix} \lambda & 0 \\ 0 & \bar{\lambda} \end{pmatrix}, \begin{pmatrix} 0 & \bar{\lambda} \\ -\lambda & 0 \end{pmatrix} \mid |\lambda|^2 = 1 \right\}, \tag{9.54}$$

one can show that the space of right orbits of H on $SU(2)$ is homeomorphic to $\mathbb{R}P^2$. We will denote the orbits of this action as $[[z_0 : z_1]]$, where

$$[z_0 : z_1] \in \mathbb{C}P^1 \cong S^2.$$

9.4.2 Construction of the angular momentum operators

The construction is similar to the one presented in the previous section. Because the configuration space is of the form $SU(2)/H$, we start by considering unitary representations of the group H. In one complex dimension, we only have two possibilities, given by the trivial representation (boson statistics) and by

$$\begin{aligned} \kappa : H &\longrightarrow \mathrm{Gl}(\mathbb{C}), \\ \begin{pmatrix} \lambda & 0 \\ 0 & \bar{\lambda} \end{pmatrix} &\longmapsto 1, \\ \begin{pmatrix} 0 & \bar{\lambda} \\ -\lambda & 0 \end{pmatrix} &\longmapsto -1. \end{aligned} \tag{9.55}$$

From now on, we will only consider this representation, which is the one giving rise to Fermi statistics (i.e., wave functions for scalar particles violating the spin–statistics connection). The total space of the line bundle $SU(2) \times_\kappa \mathbb{C}$ associated to the principal bundle $SU(2) \to SU(2)/H$ is the space $\{[(g, v)] \mid g \in SU(2) \text{ and } v \in \mathbb{C}\}$ of equivalence classes defined by the equivalence relation $(g, v) \sim (gh, \kappa(h^{-1})v)$. The projection is given by

$$\pi_\kappa\left([(z_0, z_1), v]\right) = [[z_0 : z_1]]. \tag{9.56}$$

The action of the rotation group $SU(2)$ on the configuration space is the one naturally induced by the action on the sphere. That is, for $g = (\alpha, \beta) \in SU(2)$ and $p = [[z_0 : z_1]] \in SU(2)/H$, we have

$$l_g(p) = [[\alpha z_0 - \bar{\beta} z_1 : \beta z_0 + \bar{\alpha} z_1]]. \tag{9.57}$$

As in the previous section, the action can be lifted to the total space of the bundle, by setting

$$l_g^\uparrow\left([(z_0, z_1), v]\right) = \left[(g(z_0, z_1), v)\right]. \tag{9.58}$$

In the example of the magnetic monopole we had to introduce local trivializations in order to obtain the known expressions for the angular momentum operators. For the case of indistinguishable particles that we are considering in this section, our purpose is to establish a bridge between our formalism and the one presented in [BR97]. The latter does not make explicit use of vector bundles. Instead, it uses a position-dependent spin basis. The spin basis vectors are actually sections of a trivial bundle on the sphere, but their transformation properties allow one to regard wave functions constructed from them as sections on a bundle over the physical configuration space.[6] So, in order to establish this connection between the two formalisms, we will construct an explicit isomorphism between the bundle $SU(2) \times_\kappa \mathbb{C}$ and a line subbundle $\mathcal{L}_-$ of the trivial bundle $\mathbb{R}P^2 \times \mathbb{C}^3 \to \mathbb{R}P^2$, as follows.

Let us regard the projective plane as the quotient $S^2/\mathbb{Z}_2$. Then, points on it are equivalence classes of the form $[x] = \{x, -x\}$, where $x = (x_1, x_2, x_3) \in S^2$. With this, the following open cover can be defined ($\alpha = 1, 2, 3$):

$$U_\alpha = \{[x] \in \mathbb{R}P^2 \mid x_\alpha \neq 0\}. \tag{9.59}$$

Let us now define a line bundle $\mathcal{L}_-$ (a subbundle of the trivial bundle $\mathbb{R}P^2 \times \mathbb{C}^3$) as follows. The total space of the bundle is given by the following set:

$$\left\{([x], \lambda\, |\phi(x)\rangle) \in \mathbb{R}P^2 \times \mathbb{C}^3 \mid \lambda \in \mathbb{C} \,\text{and}\, x \in [x]\right\}, \tag{9.60}$$

where

$$\begin{aligned} |\phi(-)\rangle : S^2 &\longrightarrow \mathbb{C}^3, \\ x &\longmapsto |\phi(x)\rangle \end{aligned} \tag{9.61}$$

is *any* map from S^2 to $\mathbb{C}^3$ satisfying the following conditions:

(i) It is smooth.
(ii) $|\phi(x)\rangle \neq 0$ for all $x \in S^2$.
(iii) $|\phi(-x)\rangle = -|\phi(x)\rangle$ for all $x \in S^2$.

The bundle projection is defined through $\pi\left(([x], \lambda\, |\phi(x)\rangle)\right) = [x]$. According to (9.60), an element in the total space of $\mathcal{L}_-$ is given by a tuple of the form $([x], \lambda|\phi(x)\rangle)$. Notice that there is some ambiguity in this expression, for a representative x is being explicitly used. However, there is no problem if one realizes that a choice of representative $x \in [x]$ uniquely fixes the value of λ. Assuming that the representative x has been chosen, and that to it corresponds the scalar λ, then from property (iii) it follows that the other choice of representative, $-x$, forces the value of the scalar to be $-\lambda$. An alternative way to define the bundle is by saying that the fiber over the point $[x]$ is the subset $\{[x]\} \times V_{[x]}$ of $\mathbb{R}P^2 \times \mathbb{C}^3$, where $V_{[x]}$ is the vector space generated by the vector $|\phi(x)\rangle \in \mathbb{C}^3$. Local trivializations for

[6] There are certain subtleties involved in this identification, which have been discussed in [Rey06].

$\mathcal{L}_-$ are given by ($\alpha = 1, 2, 3$)

$$\begin{aligned} \varphi_\alpha : \pi^{-1}(U_\alpha) &\longrightarrow U_\alpha \times \mathbb{C}, \\ ([x], \lambda\, |\phi(x)\rangle) &\longmapsto ([x], \operatorname{sign}(x_\alpha)\lambda)\,. \end{aligned} \tag{9.62}$$

They give rise to the following transition functions:

$$\begin{aligned} g_{\alpha\beta} : U_\alpha \cap U_\alpha &\longrightarrow \mathbb{Z}_2 \leqslant U(1), \\ [x] &\longmapsto g_{\alpha\beta}([x]) = \operatorname{sign}(x_\alpha x_\beta). \end{aligned} \tag{9.63}$$

Yet another point of view is provided by the Serre–Swan equivalence of bundles and modules: Given a (normalized) map $|\phi(-)\rangle$ satisfying properties (i)–(iii), it can be shown that the projector $p : [x] \mapsto |\phi(x)\rangle\langle\phi(x)|$ gives place to a finitely generated projective module $p(\mathcal{A}_+^3)$ over the algebra $\mathcal{A}_+$ of complex, continuous even functions over the sphere [Pas01, PPRS04]. This module is isomorphic to the module of sections on the bundle $\mathcal{L}_-$.

If $g = (z_0, z_1) \in SU(2)$ and $v \in \mathbb{C}$, then $\pi_\kappa([(g, v)]) = [[z_0 : z_1]]$ is a point in $SU(2)/H$. Let $x(g)$ denote the point in S^2 obtained from g through the quotient map $SU(2) \to SU(2)/U(1)$, and let $[x(g)]$ denote the corresponding equivalence class, with respect to the quotient map $S^2 \to S^2/\mathbb{Z}_2$. Then it is clear that $\pi_\kappa([(g, v)]) = [x(g)]$, independently of the chosen g. This fact allows us to construct the following map between the total spaces of $SU(2) \times_\kappa \mathbb{C}$ and $\mathcal{L}_-$:

$$\begin{aligned} \Phi : SU(2) \times_\kappa \mathbb{C} &\longrightarrow \mathcal{L}_-, \\ [(g, v)] &\longmapsto ([x(g)]\,, v\, |\phi(x(g))\rangle)\,. \end{aligned} \tag{9.64}$$

It is easy to check that this map is well defined and that, in fact, it provides a bundle isomorphism. Thus, we obtain an induced lift on the bundle $\mathcal{L}_-$, as indicated in the following diagram:

$$\begin{array}{ccc} \mathcal{L}_- & \overset{\tau_g}{\dashrightarrow} & \mathcal{L}_- \\ {\scriptstyle \Phi^{-1}}\big\downarrow & & \big\uparrow{\scriptstyle \Phi} \\ SU(2) \times_{\mathcal{U}_n} \mathbb{C} & \xrightarrow{\;l_g^\uparrow\;} & SU(2) \times_{\mathcal{U}_n} \mathbb{C} \\ \big\downarrow{\scriptstyle \pi_\kappa} & & {\scriptstyle \pi_\kappa}\big\downarrow \\ \mathbb{R}P^2 & \xrightarrow{\;l_g\;} & \mathbb{R}P^2. \end{array} \tag{9.65}$$

From $\tau_g = \Phi \circ l_g^{\uparrow} \circ \Phi^{-1}$ we get, for $g = (\alpha, \beta)$,

$$\begin{aligned}\tau_g\left([x], \lambda\,|\phi(x)\rangle\right) &= (\Phi \circ l_g^{\uparrow} \circ \Phi^{-1})\left([x], \lambda\,|\phi(x)\rangle\right)\\ &= (\Phi \circ l_g^{\uparrow})\,[((z_0, z_1), \lambda)]\\ &= \Phi\,[((\alpha, \beta)\cdot(z_0, z_1), \lambda)]\\ &= \left(\left[x(z_0', z_1')\right], \lambda\,\left|\phi(x(z_0', z_1'))\right\rangle\right), \end{aligned} \tag{9.66}$$

where, as in the previous section, $z_0' = \alpha z_0 - \bar{\beta} z_1$ and $z_1' = \beta z_0 + \bar{\alpha} z_1$. Here, (z_0, z_1) is chosen in such a way that $x \equiv x(z_0, z_1) = [[z_0 : z_1]]$.

Now, notice that a smooth section on $\mathcal{L}_-$ can always be written in the form $\Psi([x]) = ([x], a(x)\,|\phi(x)\rangle)$, with $a : S^2 \to \mathbb{C}$ a smooth *antisymmetric* function. Such a section transforms under the action of $SU(2)$ in the following way:

$$\begin{aligned}(U(g)\Psi)([x]) := \tau_g(\Psi(g^{-1}\cdot[x])) &= \tau_g([g^{-1}\cdot x], a(g^{-1}\cdot x)|\phi(g^{-1}\cdot x)\rangle)\\ &= \left([x]\,, a(g^{-1}\cdot x)\,|\phi(x)\rangle\right). \end{aligned} \tag{9.67}$$

From this we immediately see that the infinitesimal generators J_i are given by

$$(J_i\Psi)([x]) = ([x]\,, (L_i a)(x)\,|\phi(x)\rangle)\,, \tag{9.68}$$

where L_i is the usual (orbital) angular momentum operator.

9.5 Conclusions

The generally accepted (relativistic) quantum field theory proof of the spin–statistics theorem is perhaps one of the most interesting results of the general theory of quantum fields, and there might be no apparent reason for trying to look for a different proof. But, as a brief look at the current literature on the subject will show, the interest in the problem of the spin–statistics connection in nonrelativistic quantum mechanics has increased in the last years.

One reason might be that there is the opinion that nonrelativistic quantum mechanics describes, without relativity, an astonishing number of physical phenomena. As it is a theory that stands on a firm mathematical foundation, one would like to be able to obtain the physically correct spin–statistics connection without having to draw a theorem from another theory (which, anyway, is more fundamental).

Another motivation might be the study of the spin–statistics connection in different contexts: quantum gravity, quantum field theory on noncommutative spaces, etc. In any case, in contrast to the opinion of many authors, our interest is not so much to find a *simple* proof of the theorem, or even one which does not use relativistic invariance, but rather to understand the connection from a different point of view.

Just the fact that the Fermi–Bose alternative can be obtained as a consequence of the topology of the configuration space is a quite remarkable result. But if, in the end, it turns out that the connection has something to do with topology or geometry, one should not expect to obtain an understanding of it without using the tools of those disciplines. The approach that we are presenting here, which in some aspects is a continuation of [PPRS04], has the purpose of establishing a bridge between the proposed mathematical–physical framework and the current literature on the subject. We believe that a clear formulation of the problem in mathematical terms might help in providing a firm foundation to many works where interesting physical ideas have been put forward and in establishing a link between them.

The main result of the present paper is the construction of the angular momentum operators for a system of two indistinguishable particles obeying fermionic statistics. Taking into account the equivalence $\Gamma(\mathcal{L}_-) \cong \mathcal{A}_-$ [Pas01, Rey06], we see from (9.68) not only that sections on $\mathcal{L}_-$ can be isomorphically mapped to antisymmetric functions over S^2, but also that the generators of rotations, obtained here by means of a well-defined quantization map, correspond to the *usual* angular momentum operators. Thus, whereas it is true that by taking seriously into account the indistinguishability of quantum particles we are forced to consider nontrivial geometric–topological structures, at the end we see that all these structures can be mapped isomorphically to the ones that we are familiar with. One could argue that this only means we have not won anything. On the contrary, we believe that taking these structures into account could eventually lead to an advance in our understanding of the subject. In particular, we believe that it would be a fruitful idea to obtain a global version of the theorem proven in [Kuc04], using the tools discussed in the present paper.

Bibliography

[AB02] M. Atiyah and R. Bielawski. Nahm's equations, configurations spaces and flag manifolds. *Bull. Brazilian Math. Soc.,* 33:157–166, 2002.

[AM03] R. E. Allen and A. R. Mondragon. Comment on "spin and statistics in nonrelativistic quantum mechanics: The spin-zero case." *Phys. Rev. A,* 68:046101, 2003.

[Ana02] Charis Anastopoulos. Spin–statistics theorem and geometric quantization. *Int. J. Mod. Phys. A,* 19:655–676, 2002.

[AS02] M. Atiyah and P. Sutcliffe. The geometry of point particles. *Proc. R. Soc. London A,* 458:1089–1115, 2002.

[Ati01] M. Atiyah. Configurations of points. *Phil. Trans. Roy. Soc. Lond. A359,* 359(1784):1375–1387, 2001.

[BDG+90] A. P. Balachandran, A. Daughton, Z. C. Gu, G. Marmo, R. D. Sorkin, and A. M. Srivastava. A topological spin–statistics theorem or a use of the antiparticle. *Mod. Phys. Lett. A,* 5:1575–1585, 1990.

[BL81] L. C. Biedenharn and J. D. Louck. The Racah–Wigner algebra in quantum theory. In G. C. Rota, editor, *Encyclopedia of Mathematics and its Applications, Vol. 9.* Addison-Wesley, 1981.

[BR97] M. V. Berry and J. M. Robbins. Indistinguishability for quantum particles: spin, statistics and the geometric phase. *Proc. R. Soc. London A,* 453:1771–1790, 1997.

[BR00] M. V. Berry and J. M. Robbins. Quantum indistinguishability: alternative constructions of the transported basis. *J. Phys. A,* 33:L207–L214, 2000.

[CJ04] Dariusz Chruscinski and Andrzej Jamiolkowski. *Geometric Phases in Classical and Quantum Mechanics.* Birkhäuser, Boston, 2004.

[Dir31] P. A. M. Dirac. Local observables and particle statistics II. *Proc. Roy. Soc.* A, 133:60, 1931.

[DS98a] I. Duck and E. C. G. Sudarshan. Toward an understanding of the spin–statistics theorem. *Am. J. Phys.,* 66:284–303, 1998.

[DS98b] Ian Duck and E. C. G. Sudarshan. *Pauli and the Spin–Statistics Theorem.* World Scientific, 1998.

[Fie39] M. Fierz. On the relativistic theory of force free particles of arbitrary spin. *Helv. Phys. Acta,* 12:3–37, 1939.

[FR68] David Finkelstein and Julio Rubinstein. Connection between spin, statistics, and kinks. *J. Math. Phys.,* 9:1762–1779, 1968.

[Haa96] Rudolf Haag. *Local Quantum Physics.* Texts and Monographs in Physics. Springer Verlag, 2nd edition, 1996.

[Ish84] C. J. Isham, Topological and global aspects of quantum theory. In Bryce S. DeWitt and Raymond Stora, editors, *Relativity, Groups and Topology II,* pages 1059–1290, Amsterdam, Kluwer Academic Publishers, 1984.

[Kas06] H. A. Kastrup. Quantization of the canonically conjugate pair angle and orbital angular momentum. *Phys. Rev. A*, 73:052104, 2006.

[Kir76] A. A. Kirillov. *Elements of the Theory of Representations.* Springer Verlag, 1976.

[Kuc04] B. Kuckert. Spin and statistics in nonrelativistic quantum mechanics, I. *Phys. Lett. A*, 322:47–53, 2004.

[LD71] Michael G. G. Laidlaw and Cecile Morette DeWitt. Feynman functional integrals for systems of indistinguishable particles. *Phys. Rev. D,* 3:1375–1378, 1971.

[LM77] J. M. Leinaas and J. Myrheim. On the theory of identical particles. *Nuovo Cimento*, 37B:1–23, 1977.

[Mac68] G. W. Mackey. *Induced Representations and Quantum Mechanics*. W. A. Benjamin, New York, 1968.

[Pas01] M. Paschke. *Von nichtkommutativen Geometrien, ihren Symmetrien und etwas Hochenergiephysik*. PhD thesis, Institut für Physik, Universität Mainz, 2001.

[Pau40] W. Pauli. The connection between spin and statistics. *Phys. Rev.*, 58:716–722, 1940.

[Pes03a] Murray Peshkin. Reply to comment on spin and statistics in nonrelativistic quantum mechanics: The spin-zero case. *Phys. Rev. A*, 68:046102, 2003.

[Pes03b] Murray Peshkin. Spin and statistics in nonrelativistic quantum mechanics: The spin-zero case. *Phys. Rev. A*, 67:042102, 2003.

[PPRS04] N. A. Papadopoulos, M. Paschke, A. Reyes, and F. Scheck. The spin–statistics relation in nonrelativistic quantum mechanics and projective modules. *Ann. Math. Blaise-Pascal*, 11(2):205–220, 2004.

[Rey06] A. Reyes. *On the Geometry of the Spin–Statistics Connection in Quantum Mechanics*. PhD Dissertation, Mainz, 2006.

[RL] A. F. Reyes-Lega. On the geometry of quantum indistinguishability. In preparation.

[Sch68] Lawrence Schulman. A path integral for spin. *Phys. Rev.*, 5:1558–1569, 1968.

[SD03] E. C. G. Sudarshan and I. M. Duck. What price the spin–statistics theorem? *Pramana J. Phys.*, 61:1–9, 2003.

[Sud75] E. C. G. Sudarshan. Relation between spin and statistics. *Stat. Phys. Suppl.: J. Indian Inst. Sci.*, June:123–137, 1975.

[SW00] R. F. Streater and A. S. Wightman. *Spin, Statistics and All That*. Princeton Univ. Press, Princeton, 2000.

[Wil82] Frank Wilczek. Quantum mechanics of fractional-spin particles. *Phys. Rev. Lett.*, 49:957–959, 1982.

[Woo80] N. Woodhouse. *Geometric Quantization*. Clarendon Press, Oxford, 1980.

[ZSN+83] F. Zaccaria, E. C. G. Sudarshan, J. S. Nilsson, N. Mukunda, G. Marmo, and A. P. Balachandran. Universal unfolding of Hamiltonian systems: From symplectic structure to fiber bundles. *Phys. Rev. D*, 27:2327–2340, 1983.

10

Conserved currents in Kähler manifolds

JAIME R. CAMACARO AND JUAN CARLOS MORENO

Abstract

We revisit some properties of Lie algebroids, Lie bialgebroids and bidifferential calculi, in order to stress the relation of the Lie algebroid structure with the constructions of hierarchies of conserved currents and show the existence of a particular class of conserved currents. We give some examples related to Kähler and generalized Kähler structures.

10.1 Introduction

In a recent paper [1] Dimakis and Müller-Hoissen showed how to generate conservations laws in completely integrable systems by making use of a bidifferential calculus. More recently, Crampin et al. proved that the approach of Dimakis and Müller-Hoissen was related to the standard approach using bi-Hamiltonian structures of the Poisson–Nijenhuis type [2]. These results were extended in [3], where the Poisson–Nijenhuis case was discussed in a detailed way. Some interesting remarks on the so-called gauged bidifferential calculus by Dimakis and Müller-Hoissen were also given. A detailed description of geometrical properties and a generalization to Lie algebroids were given respectively in [4, 5]. Apart from the previous citations, one can find of interest the relation of the bidifferential calculi and the non-Noether symmetries (see, e.g., [6, 7]), and the application of the bidifferential calculi and Lie algebroids to specific integrable systems (see, e.g., [8, 9]).

As was pointed out in [10], the concepts of recursion operator and hierarchies of conservation laws are related to the Lenard chains, and also the algebraic action of the Nijenhuis operator is related to the Poisson–Nijenhuis structure. Here we will use the relation between the Dimakis and Müller-Hoissen constructions and the Lie algebroid and bialgebroid structures established by means of the recursion operators and their differential action, studied in [5], to show the role played by the

Lie algebroids as a bridge between the algebraic action and the differential action of the recursion operator. Finally, we show the existence of a new subsequence of conserved currents in the construction of Crampin et al., generated by means of the hierarchies of symplectic forms.

10.2 Lie algebroids

In this section we recall some facts about Lie algebroids and their representation as an exterior differential algebra.

Definition 10.2.1 A Lie algebroid is a vector bundle $\tau : \mathcal{A} \to M$ together with:

- a Lie bracket $[\cdot, \cdot]_{\mathcal{A}}$ on the space $\Gamma(\mathcal{A})$ of sections of $\mathcal{A}$,
- a vector bundle map $\rho : \mathcal{A} \to TM$ over the identity, called the anchor, such that the induced map $\rho : \Gamma(\mathcal{A}) \to \Gamma(TM)$ is a Lie algebra morphism,
- the identity $[X, fY]_{\mathcal{A}} = f[X, Y]_{\mathcal{A}} + (\rho(X)f)Y$ holding for each pair $X, Y \in \Gamma(\mathcal{A})$ and $f \in C^\infty(M)$.

If we consider coordinates $(x^1, \ldots, x^n)$ on a local chart $U \subset M$ and a basis $\{e_\alpha \mid \alpha = 1, \ldots, r\}$ of local sections of $\tau : \mathcal{A}\big|_U \to U$, then the expressions for the Lie bracket and the anchor are

$$[e_\alpha, e_\beta]_{\mathcal{A}} = \sum_{\gamma=1}^{r} C_{\alpha\beta}{}^{\gamma}(x) e_\gamma \,, \qquad \alpha, \beta = 1, \ldots r \,, \tag{10.1}$$

and

$$\rho(e_\alpha) = \sum_{i=1}^{n} a^i{}_\alpha(x) \frac{\partial}{\partial x^i} \,, \qquad \alpha = 1, \ldots r \,, \tag{10.2}$$

respectively, where $C_{\alpha\beta}{}^{\gamma}(x)$ and $a^i{}_\alpha(x)$ are the structural functions of the Lie algebroid.

Examples of Lie algebroids are the following:

(i) Every finite-dimensional Lie algebra $\mathfrak{g}$, regarded as a vector bundle over a single point. Sections are elements of $\mathfrak{g}$, the Lie bracket is that of $\mathfrak{g}$, and the anchor map is identically zero.

(ii) The tangent bundle $\tau_M : TM \to M$ with the usual bracket on vector fields and with anchor the identity map $\mathbb{I}_{TM}$ on TM.

We define the graded exterior algebra of a Lie algebroid $(\mathcal{A}, \rho, [\cdot, \cdot]_{\mathcal{A}})$ via the exterior vector bundle $\bigwedge^\bullet \mathcal{A}$. The members of $\Gamma(\bigwedge^\bullet \mathcal{A})$ are called $\mathcal{A}$-vector fields. Sections of the dual bundle $\mathcal{A}^*$ are called $\mathcal{A}$ – 1-forms. Similarly, sections $\Gamma(\bigwedge^\bullet \mathcal{A}^*)$ of $\bigwedge^\bullet \mathcal{A}^*$ are called $\mathcal{A}$-forms. The bundle $\mathcal{A}^*$ is endowed with a nilpotent differential operator $d_{\mathcal{A}}$ of degree 1, $d_{\mathcal{A}} : \Gamma(\bigwedge^k \mathcal{A}^*) \to \Gamma(\bigwedge^{k+1} \mathcal{A}^*)$, given by

$$d_{\mathcal{A}}\vartheta(X_1, \ldots, X_i, \ldots, X_{k+1}) = \sum_i (-1)^{i+1} \rho(X_i)\vartheta(X_1, \ldots, \hat{X}_i, \ldots, X_{k+1}) + \sum_{i<j} (-1)^{i+j} \vartheta([X_i, X_j]_{\mathcal{A}}, X_1, \ldots, \hat{X}_i, \ldots, \hat{X}_j, \ldots, X_{k+1}).$$

This exterior differential algebra is specially interesting in that it is equivalent to the Lie algebroid structure in $\mathcal{A}$, with the structural maps ρ and $[\cdot, \cdot]_{\mathcal{A}}$ defined as follows:

$$\rho(X)f = \langle d_{\mathcal{A}} f, X \rangle,$$
$$\langle \theta, [X, Y] \rangle = \rho(X)(\langle \theta, Y \rangle) - \rho(Y)(\langle \theta, X \rangle) - d_{\mathcal{A}}\theta(X, Y)$$

for $X, Y \in \Gamma(\mathcal{A})$, $f \in C^\infty(M)$ and $\theta \in \Gamma(\mathcal{A}^*)$.

Another important fact about Lie algebroids is the one–one correspondence between a Lie algebroid structure in a vector bundle $\mathcal{A}$ and a Poisson structure in the dual bundle $\mathcal{A}^*$ whose linear functions form a Lie subalgebra (see, e.g., [3]).

Moreover, associated to a Lie algebroid structure there is also a graded Lie bracket

$$[\cdot, \cdot]_{(\mathcal{A})} : \Gamma(\textstyle\bigwedge^\bullet \mathcal{A}) \times \Gamma(\bigwedge^\bullet \mathcal{A}) \to \Gamma(\bigwedge^\bullet \mathcal{A}),$$

which constitutes, together with the exterior product, the well-known Gerstenhaber algebra of the Lie algebroid, also called an odd Poisson bracket or Schouten bracket (see, e.g., [3]). If the Gerstenhaber bracket is generated by a linear operator $\partial : \Gamma(\bigwedge^\bullet \mathcal{A}) \to \Gamma(\bigwedge^\bullet \mathcal{A})$ of degree -1 such that $\partial^2 = 0$, and

$$[X, Y]_{(\mathcal{A})} = (-1)^k \left(\partial(X \wedge Y) - \partial X \wedge Y - (-1)^k X \wedge \partial Y \right)$$

for every $X \in \Gamma(\bigwedge^k \mathcal{A})$, $Y \in \Gamma(\bigwedge^\bullet \mathcal{A})$, then we have an Batalin–Vilkovisky (BV) algebra (see, e.g., [12]).

In general, it is possible to define a mapping $d_X : \Gamma(\bigwedge^p \mathcal{A}^*) \to \Gamma(\bigwedge^{p+1} \mathcal{A}^*)$ for each $X \in \Gamma(\bigwedge^k \mathcal{A})$ by

$$d_X\vartheta = i_X d\vartheta + (-1)^{|k|} d i_X \vartheta,$$

where i_X is the inner contraction with X.

A type of Lie algebroids, remarkable because of its relations with integrable systems, are the Lie bialgebroids, introduced by Kossmann–Schwarzbach in [14]:

Definition 10.2.2 A Lie bialgebroid is a pair $((\mathcal{A}, \rho, [\cdot, \cdot]_{\mathcal{A}}), (\mathcal{A}^*, \rho^*, [\cdot, \cdot]_{\mathcal{A}^*}))$ of Lie algebroids, where $\mathcal{A}^*$ is the dual bundle of $\mathcal{A}$, such that the differential $d_{\mathcal{A}}$ is a derivation of the Gerstenhaber algebra $(\Gamma(\bigwedge^\bullet \mathcal{A}^*), \wedge, [\cdot, \cdot]_{\mathcal{A}^*})$, and the differential $d_{\mathcal{A}^*}$ is a derivation of the Gerstenhaber algebra $(\Gamma(\bigwedge^\bullet \mathcal{A}), \wedge, [\cdot, \cdot]_{\mathcal{A}})$.

The bialgebroid structure is specially interesting because of its relation with the bi-Hamiltonian structure, via the well-known relation with Poisson–Nijenhuis (PN) structures [5, 14].

10.3 Bidifferential calculi, PN structures and Lie algebroids

We begin this section using the theory of contraction of a Lie algebroid structure via a Nijenhuis tensor (see, e.g., [15, 19]) in order to reproduce the bidifferential calculi developed by Dimakis and Müller-Hoissen and the bi-Hamiltonian structures of Crampin et al.

Consider a manifold M endowed with a (1, 1) tensor field N. We say that N is a Nijenhuis tensor if the corresponding Nijenhuis torsion $T(N)$ defined by

$$T(N)(X, Y) = [N(X), N(Y)] - N([N(X), Y] + [X, N(Y)]) + N^2([X, Y])$$

for every $X, Y \in \Gamma(TM)$ vanishes. By means of N, it's possible to define on TM an alternative Lie algebroid structure (see [15, 16]) with bracket

$$[\cdot, \cdot]_N = [N(\cdot), \cdot] + [\cdot, N(\cdot)] - N([\cdot, \cdot]) \tag{10.3}$$

such that for every $X, Y \in \Gamma(M)$ and $\theta \in \Gamma(T^*M)$,

$$\begin{aligned}\langle \theta, [X, Y]_N \rangle &= N(X)\langle \theta, Y\rangle - N(Y)\langle \theta, X\rangle + X\langle \theta, NY\rangle - Y\langle \theta, NX\rangle \\ &\quad - \langle \theta, N([X, Y])\rangle - d\theta(NX, Y) - d\theta(X, NY).\end{aligned}$$

The corresponding anchor is $\rho = \widehat{N} : TM \to TM$, defined by contraction of N with vector fields. Using the graded commutator $[\cdot, \cdot]_{\mathbf{D}}$ on $Der\Gamma(\bigwedge^{\bullet} T^*M) = \bigoplus_k Der_k\Gamma(\bigwedge^{\bullet} T^*M)$ of all graded derivations, we write the corresponding differential operator on the new Lie algebroid structure $(TM, N, [\cdot, \cdot]_N)$ by

$$d_N = [i_N, d]_{\mathbf{D}}, \tag{10.4}$$

where d is the de Rham differential and i_N is a derivation of degree 0, defined by

$$i_N\, \vartheta\, (X_1, \ldots, X_p) = \sum_{i=1}^{n} \vartheta(X_1, \ldots, N(X_i), \ldots, X_p),$$

for any differential form $\vartheta \in \Gamma\left(\bigwedge^p T^*M\right)$.

It's straightforward to check that the two exterior differential operators d and d_N acting over $\Gamma(\bigwedge^{\bullet} T^*M)$, satisfy

$$d^2 = d_N^2 = 0$$

and

$$[d, d_N]_{\mathbf{D}} = 0.$$

Then the relations between d and d_N correspond to the classical theory of bidifferential calculi of Frölicher and Nijenhuis. Notice, the condition $d_N^2 = 0$ is equivalent to the analogous condition for $d_{\mathcal{A}}$, i.e., the differential operator d_N defines a Lie algebroid structure with bracket and anchor defined by

$$\begin{aligned} \langle \theta, [X, Y]_N \rangle &= N(X)(\langle \theta, Y \rangle) - N(Y)(\langle \theta, X \rangle) - d_N\theta(X, Y), \\ N(X)f &= \langle d_N f, X \rangle \end{aligned}$$

for every $X, Y \in \Gamma(TM)$ and $f \in C^\infty(M)$. Moreover, we have the following identity:

$$\begin{aligned} d_N^2\theta(X, Y) &= N(X)\langle d_N\theta, Y \rangle - N(Y)\langle d_N\theta, X \rangle - \langle d_N\theta, [X, Y]_N \rangle \\ &= N(X)\langle \theta, N(Y) \rangle - N(Y)\langle \theta, N(X) \rangle - \langle \theta, N([X, Y]_N) \rangle \\ &= \langle \theta, [N(X), N(Y)] - N([X, Y]_N) \rangle = 0, \end{aligned}$$

and therefore

$$N([X, Y]_N) = [N(X), N(Y)]$$

if and only if $T(N)(X, Y) = 0$. So d_N defines an alternative Lie algebroid structure corresponding to the contraction by N of the original Lie algebroid structure.

The action of d_N on $f \in C^\infty(M)$ and $\vartheta \in \Gamma\left(\bigwedge^\bullet T^*M\right)$ is given by

$$\begin{aligned} d_N f &= N^* df, \\ d_N \vartheta &= i_N(d\vartheta) - d(i_N\vartheta), \\ d_N(df) &= d(N^* df). \end{aligned}$$

The structure obtained in this way is the action of the Nijenhuis tensor N on forms, as was defined in [10].

Proposition 10.3.1 *Let M be a differentiable manifold. Given a* $(1, 1)$ *tensor field N, the triple* $(\Gamma(\bigwedge^\bullet T^*M), d, d_N)$ *is a bidifferential calculus of Frölicher and Nijenhuis if and only if the triple* $(TM, \widehat{N}, [\cdot, \cdot]_N)$ *defines a Lie algebroid structure on M.*

Continuing with the construction of the hierarchy of conserved currents, we find that the construction of Dimakis and Müller-Hoissen can be applied: starting under the original but more restrictive condition $d_N\chi^{(0)} = 0$ or with the less restrictive and more familiar condition $dd_N\chi^{(0)} = -d_N d\chi^{(0)} = 0$ for $\chi^{(0)} \in \Gamma(\bigwedge^{s-1} T^*B)$, $s \geq 1$ [1,2], it is possible to construct inductively a sequence of forms $\chi^{(m)}$, with $m = 0, 1, 2, \ldots$, satisfying

$$d\chi^{(m+1)} = d_N\chi^{(m)}.$$

Here the Lie algebroid structure is acting as a bridge between the algebraic structure of the Nijenhuis tensor and its differential part, by means of a contraction of the Lie algebroid bracket and the operator d_N. As we will see, the Lie algebroid and bialgebroid structures are a fundamental link between the algebraic action by contractions and the action on differential forms of the Nijenhuis tensor, in addition to the relation of these two actions with the Lenard chains (for details see e.g., [10, 11, 17]).

Now consider a Poisson–Nijenhuis manifold (M, P, N), that is, a Poisson manifold (M, P) endowed with a Nijenhuis tensor which satisfies the compatibility condition

$$[NP, NP]_S = [P, NP]_S = 0,$$

where $[\cdot, \cdot]_S$ is the Schouten bracket. So the linear combination of this Poisson structures is again a Poisson structure, defining in this way a bi-Hamiltonian system.

It is well known that (M, P, N) being a Poisson–Nijenhuis manifold is equivalent to the pair (TM, T^*M) being a Lie bialgebroid [14].

The Poisson bivector P defines a map $\widehat{P} : T^*M \to TM$ by contraction with covectors. So, we can endow the vector bundle $\pi_M : T^*M \to M$ with a Lie algebroid structure defined by the anchor map $\rho = \widehat{P}$ and Lie bracket on 1-forms given by

$$[\theta, \eta]^P = \mathcal{L}_{\widehat{P}(\theta)}\eta - \mathcal{L}_{\widehat{P}(\eta)}\theta - d(P(\theta, \eta)),$$

where $\mathcal{L}_X$ denotes the Lie derivative. The exterior differential in this case is the Lichnerowicz–Poisson differential $d_P = [P, \cdot]_S$. As we already pointed out, a Nijenhuis structure gives a new Lie algebroid structure on TM by means of the Lie bracket (10.3), the anchor map $\rho = \widehat{N}$, and the exterior differential operator d_N. Using both structures, we find again the bidifferential calculus defined by the operators d and d_N, which now can be endowed with a bi-Hamiltonian system related to the Poisson–Nijenhuis structure of the base manifold.

If we assume a nondegenerate Poisson structure on M – i.e., there is a symplectic structure ω_0 which satisfies

$$\omega_0(N(X), Y) = \omega_0(X, N(Y)) \quad \text{for all } X, Y \in \Gamma\left(\bigwedge TM\right)$$

– then there is a 2-form ω_1 defined by $\omega_1 = \omega_0(N(X), Y)$. Moreover, if $d\omega_1 = 0$, we can define a new Poisson structure P_1. In this way, as was stated in [2, 3], the sequence of currents obtained by the bidifferential construction are in involution with respect to both Poisson structures. So we recover all the elements of Cramping et al. and of Dimakis and Müller-Hoissen in the context of contracted Lie algebroids (Frölicher–Nijenhuis calculus), Notice that the bidifferential structure is in one-to-one correspondence with the Nijenhuis tensor.

All the preceding construction is made possible by the property

$$\omega_0(N(X), Y) = \omega_0(X, N(Y)).$$

Now let us consider the case where the compatibility condition between the symplectic form and the Nijenhuis tensor is

$$\omega_0(N(X), Y) = -\omega_0(X, N(Y)),$$

which is a kind of Kähler relation between the Nijenhuis operator and the symplectic form.

There is an alternative construction to generate a sequence of conserved currents in involution. This construction is based on the powers of the Nijenhuis tensor N, i.e., N^k with k an integer. Recall that if N is a Nijenhuis tensor, then N^k is again a Nijenhuis tensor. This kind of structures is specially interesting in that it defines the so-called hierarchies of 2-forms, by (see, e.g., [17])

$$\omega_k(X, Y) = \omega_0(X, N^k(Y)).$$

Theorem 10.3.2 *Let (M, ω) be a simply connected manifold endowed with a non-degenerate symplectic structure, P be the Poisson bivector field associated to ω, and N be a Nijenhuis tensor satisfying*

$$\omega(NX, Y) = -\omega(X, NY) \quad forevery\ X, Y \in \Gamma\left(\bigwedge TM\right).$$

Then, given $\chi^0 \in C^\infty(M)$ satisfying $dd_N\chi^{(0)} = d_N d\chi^{(0)} = 0$, there exists a sequence $\{\chi^{(k)} \in C^\infty(M)\}_{k=0}^{\infty}$ such that

$$P(d\chi^{(m)}, d\chi^{(n)}) = 0$$

for every $m, n \in \mathbb{N}$ satisfying $m - n = 2k$, with $k = 0, 1, 2, 3, \ldots$.

Proof For $\chi^{(0)} \in C^\infty(M)$ we consider

$$j^{(1)} = d_N\chi^{(0)} \in \Gamma\left(\bigwedge T^*M\right), \tag{10.5}$$

which satisfies

$$dj^{(1)} = dd_N\chi^{(0)} = -d_N d\chi^{(0)} = 0$$

as a consequence of $d_N d\chi^{(0)} = 0$. Then, from the Poincaré lemma, there exists $\chi^{(1)} \in C^\infty(M)$ such that

$$j^{(1)} = d\chi^{(1)}. \tag{10.6}$$

Now, applying $i_N|_{\Gamma(\wedge^1 T^*M)} : \Gamma\left(\bigwedge^\bullet T^*M\right) \to \Gamma\left(\bigwedge^\bullet T^*M\right)$ and equation (10.6), we define

$$\tilde{j}^{(1)} = i_N j^{(1)} = d_N\chi^{(1)} \in \Gamma\left(\bigwedge T^*M\right).$$

Because $i_N \circ i_N = i_{N^2}$ on $\Gamma\left(\bigwedge T^*M\right)$, it follows from (10.5) that

$$\tilde{j}^{(1)} = i_N j^{(1)} = i_N d_N \chi^{(0)} = d_{N^2} \chi^{(0)}.$$

Then, using $d_N = i_N \circ d - d \circ i_N$ and $d_N^2 = d^2 = 0$, we immediately find

$$d(\tilde{j}^{(1)}) = -d_N j^{(1)} + i_N d j^{(1)} = -d_N^2 \chi^{(0)} = 0,$$

that is, the 1-form $\tilde{j}^{(1)}$ is d-closed. In general, given a nonzero $\chi^{(m-1)} \in C^\infty(M)$, we define $j^{(m)} \in \Gamma\left(\bigwedge T^*M\right)$ satisfying

$$dj^{(m)} = 0, \qquad j^{(m)} = d_N \chi^{(m-1)},$$

and consider $\tilde{j}(m) = i_N j^{(m)} = d_{N^2} \chi^{(m-1)} \in \Gamma\left(\bigwedge T^*M\right)$. Because $dj^{(m)} = 0$, then by the Poincaré lemma there exists $\chi^{(m)} \in C^\infty(M)$ for which $j^{(m)} = d\chi^{(m)}$ and $i_N j^{(m)} = d_N \chi^{(m)}$. Therefore, using again the identities $d_N = i_N \circ d - d \circ i_N$ and $d_N^2 = d^2 = 0$, it follows that

$$d(\tilde{j}^{(1)}) = -d_N j^{(m)} + i_N d j^{(m)} = -d_N^2 \chi^{(m-1)} + i_N (d^2 \chi^{(m)}) = 0.$$

So $j^{(m+1)} = d_N \chi^{(m)}$ and $\tilde{j}^{(m+1)} = i_N j^{(m+1)} = d_{N^2} \chi^{(m+1)}$ are 1-forms satisfying

$$dj^{(m+1)} = dd_N \chi^{(m)} = -d_N d\chi^{(m)} = -d_N j^{(m)} = -d_N^2 \chi^{(m-1)} = 0,$$

and therefore

$$d(i_N j^{(m+1)}) = -d_N j^{(m+1)} + i_N (dj^{m+1}) = -d_N^2 \chi^{(m)} = 0.$$

In this way we obtain sequences of d-closed forms $\left\{j^{(k)} \in \Gamma\left(\bigwedge T^*M\right)\right\}_{k=1}^\infty$, $\left\{\tilde{j}^{(k)} \in \Gamma\left(\bigwedge T^*M\right)\right\}_{k=1}^\infty$ satisfying

$$j^{(k+1)} = d\chi^{(k+1)} = d_N \chi^{(k)}, \qquad \tilde{j}^{(k+1)} = d_N \chi^{(k+1)} = d_{N^2} \chi^{(k)}. \tag{10.7}$$

Finally,

$$d\chi^{(k+2)} = d_{N^2} \chi^{(k)}.$$

From the preceding construction it is straightforward that, for $k \geq 0$,

$$
\begin{aligned}
P(d\chi^{(m)}, d\chi^{(n)}) &= P(d\chi^{(n+2k)}, d\chi^{(n)}) \\
&= P(d_{N^2}\chi^{(n+2k-2)}, d\chi^{(n)}) \\
&= P(d\chi^{(n+2k-2)}, d_{N^2}\chi^{(n)}) \\
&= P(d\chi^{(n+2k-2)}, d\chi^{(n+2)}) \\
&\;\;\vdots \\
&= P(d_{N^2}\chi^{(n+2k-2k)}, d\chi^{(n+2(k-1))}) \\
&= P(d\chi^{(n+2k-2k)}, d_{N^2}\chi^{(n+2(k-1))}) \\
&= P(d\chi^{(n+2k-2k)}, d\chi^{(n+2(k-1)+2)}) \\
&= P(d\chi^{(n)}, \chi^{(m)}).
\end{aligned}
$$

In view of the skew symmetry of P,

$$P(d\chi^{(m)}, d\chi^{(n)}) = 0. \qquad \square$$

So even in the case where the condition

$$\omega(NX, Y) = -\omega(X, NY),$$

is satisfied, one can generate a sequence of conserved currents in involution. The relation that plays the principal role in this construction is

$$\omega(N^2(X), Y) = \omega(X,\ N^2(Y)),$$

in view of which it is straightforward to notice that there is a bi-Hamiltonian structure associated to this construction, related to the preceding hierarchy of symplectic forms. In particular, for the element with $k = 2$ we have

$$\omega_2(X, Y) = \omega_0(X, N^2(Y)).$$

Moreover, we can build a new sequence of conserved currents on the bidifferential $(\Gamma(\bigwedge^{\bullet} T^*M), d,\ d_{N^2})$, using the original sequence generated by $\chi^{(0)}$ on $(\Gamma(\bigwedge^{\bullet} T^*M), d,\ d_N)$. It is worthy of notice that this new sequence it is not in a direct relation with the original one, because of the Kähler-like relation between N and ω.

Thinking in terms of the Lie algebroid structures, we obtain a second contraction of the original structure on TM. Notice that N^2 is again a Nijenhuis tensor, because $d_{N^2}^2 = 0$, so there is a third Lie algebroid structure on TM. There is no reason to stop the number contractions at $k = 2$, so one can think of a hierarchy of contracted Lie algebroids with the same characteristics of the hierarchies of N.

10.4 Examples

Example 10.4.1 A Kähler manifold is the simplest case in which a subsequence of conserved currents can be expected [20].

Let M be a be a real manifold of dimension $2n$, endowed with a complex structure, namely, a $(1, 1)$ tensor field J with vanishing Nijenhuis torsion and satisfying $J^2 = -\mathbb{I}_{TM}$. Assuming that M admits an Hermitian structure for which the fundamental 2-form

$$\omega(X, Y) = g(J(X), Y) \quad \text{for all } X, Y \in \Gamma(TM)$$

is *closed*, we obtain a Kähler manifold. In this manifold we have for each pair $X, Y \in \Gamma(TM)$ the following relation between J and ω:

$$\omega(J(X), Y) = -\omega(X, J(Y)).$$

Now, given $\chi^{(0)} \in C^\infty(M)$ with $dd_J\chi^{(0)} = 0$, we can build, by Theorem 10.3.2, a sequence $\{\chi^{(k)} \in C^\infty(M)\}_{k=0}^\infty$ where

$$d\chi^{(k+2)} = -d\chi^{(k)}.$$

Then, we find a trivial sequence of function in involution with the Poisson structure defined by the 2-closed form ω.

Example 10.4.2 For a manifold M of dimension $2n$, a generalized Kähler structure on it is a triple (g, J_+, J_-) consisting of a Riemannian metric and two g-compatible complex structures $J_\pm$ such that $J_+J_- = J_-J_+$, and the fundamental 2-forms

$$\omega_\pm(X, Y) = g(J_\pm(X), Y) \quad \text{for all } X, Y \in \Gamma\,(TM)$$

satisfy the following integrability relations:

$$d_{J_+}\omega_+ + d_{J_-}\omega_- = 0, \quad dd_{J_\pm}\omega_\pm = 0$$

(see, e.g., [21]). It is possible to consider the case where $J_- = \pm J$ and $J_+ = J$, but, following the construction in [21], we will consider only the case where $J_+ \neq \pm J_-$. In this case we see that for each $X, Y \in \Gamma\,(TM)$,

$$\omega_\pm(J_\mp(X), Y) = -\omega_\pm(X, J_\mp(Y)).$$

Associated to the generalized Kähler structure there are two alternative Lie algebroids, which are determined by the bidifferential calculi $(\Gamma(\bigwedge^\bullet T^*M), d, d_{J_\pm})$. Moreover, starting from the 2-forms $\chi_\pm^{(0)} = \omega_\pm \in \Gamma(\bigwedge^2 T^*M)$ and following the original construction in [1], we find the sequences $\{\chi_+^{(k)} \in \Gamma(T^*M)\}_{k=0}^\infty$ and $\{\chi_-^{(k)} \in \Gamma(T^*M)\}_{k=0}^\infty$ such that

$$d\chi_+^{(k+1)} = d_{J_+}\chi_+^{(k)}, \quad d\chi_-^{(k+1)} = d_{J_-}\chi_-^{(k)}.$$

Example 10.4.3 Continuing with the generalized complex structure, we restrict to the submanifold B where the fundamental 2-forms are closed. Then, the complex structures J_+ and J_- are parallel with the Levi–Civita connections determined by g, namely, ∇_+ and ∇_-, respectively:

$$\nabla_\pm J_\pm = 0.$$

So we see that for every $X, Y \in \Gamma(TB)$,

$$J_+([X, Y]_{J_-}) + J_-([X, Y]_{J_+}) = [J_+(X), J_-(Y)] + [J_-(X), J_+(Y)];$$

therefore, $T(J_+ + J_-)(X, Y) = 0$ (see, e.g., [23]). Moreover, considering that $J_+J_- = J_-J_+$, we have the following relation between the 2-forms $\omega_\pm$ and the Nijenhuis operator $J_+ + J_-$:

$$\omega_\pm\left((J_+ + J_-)(X),\ Y\right) = -\omega_\pm\left(X,\ (J_+ + J_-)(Y)\right),$$

Then $J_+ + J_-$ is a Nijenhuis operator satisfying a Kähler-like condition. Now, instead of taking the same steps as in the previous examples, we will take advantage of the fact that

$$(J_+ + J_-)^2 = 2J_+J_- - 2\mathbb{I}_{TB}.$$

By Theorem 10.3.2, we find that, given $\chi^{(0)} \in C^\infty(B)$ satisfying

$$d_{J_+ + J_-} d\chi^{(0)} = -dd_{J_+ + J_-}\chi^{(0)} = 0,$$

there exists a sequence $\{\chi^{(k)} \in C^\infty(B)\}_{k=0}^\infty$ such that, if $P_\pm$ denote the Poisson bivectors corresponding to $\omega_\pm$, then

$$P_\pm(d\chi^{(m)}, d\chi^{(n)}) = 0$$

if $m - n$ is even.

10.5 Final Comments

Further research is necessary to gain a better understanding of the meaning and applications of this construction. For example, would be enlightening to consider different useful variations of the bidifferential calculi and their applications to integrable systems, as well as possible applications in sigma models, where it becomes an important task to consider Kähler manifolds, hyper-Kähler manifolds and $N = 1, 2$ manifolds [18].

The construction carried out in this chapter could be extended to generalized structures. For a first look at the algebraic construction see [19]. For applications see [21] and [22]; as they show, the contraction of a Courant algebroid by a Nijenhuis tensor is possible under some conditions, which include generalized structures. We are looking forward to generalizations of the bidifferential calculi having in this case the backing of the bialgebroids and Courant algebroids.

Bibliography

[1] A. Dimakis and F. Müller-Hoissen, *Bi-differential calculi and integrable models*, J. Phys. A: Math. Gen. **33** (2000), 957–974.
[2] M. Crampin, W. Sarlet and G. Thompson, *Bi-differential calculi and bi-Hamiltonian systems*, J. Phys. A: Math. Gen. **33** (2000), L177–L180.
[3] M. Crampin, W. Sarlet and G. Thompson, *Bi-differential calculi, bi-Hamiltonian systems and conformal Killing tensors*, J. Phys. A: Math. Gen. **33** (2000), 8755–8770.
[4] P. Guha, *Bidifferential calculi, bicomplex structure and its application to bihamiltonian systems*, Int. J. Geom. Methods Mod. Phys. **3** (2006), 209–232.
[5] J. Camacaro and J. F. Cariñena, *Alternative Lie algebroid structures and bidifferential calculi*. Applied Differential Geometry and Mechanics, ed. W. Sarlet and F. Cantrijn (Gent, Academia Press, 2003).
[6] G. Chavchanidze, *Remark on non-Noether symmetries and bidifferential calculi*. Preprint arXiv:math-ph/0105037.
[7] G. Chavchanidze, *Non-Noether symmetries and their influence on phase space geometry*, J. Geom. Phys. **48** (2003), 190–202.
[8] A. Meucci, *Compatible Lie algebroids and the periodic Toda lattice*, J. Geom. Phys. **35** (2000), 273–287.
[9] P. Lorenzoni and F. Magri, *A cohomological construction of integrable hierarchies of hydrodynamic type*, Int. Math. Res. Not. **34** (2005), 2087–2100.
[10] F. Magri, *Lenard chains for classical integrable systems*, Theor. Math. Phys. **137** (2003), 1716–1722.
[11] P. Guha, *A note on bidifferential calculi and bihamiltonian systems*, Arch. Math. (Brno) **40** (2004), 17–22.
[12] Y. Kosmann-Schwarzbach, *Exact Gerstenhaber algebras and Lie bialgebroids*, Act. Appl. Math. **41** (1995), 153–165.
[13] A. C. da Silva and A. Weinstein, *Lectures on geometrical models for noncommutative algebra* (Amer. Math. Soc., Providence, RI, 1999).
[14] Y. Kosmann-Schwarzbach, *The Lie bialgebroid of a Poisson–Nijenhuis manifold*, Lett. Math. Phys. **38** (1996), 421–428.
[15] J. F. Cariñena, J. Grabowski and G. Marmo, *Contractions: Nijenhuis and Saletan tensors for general algebraic structures*, J. Phys. A: Math. Gen. **34** (2001), 3769–3789.
[16] J. Grabowski and P. Urbański, *Lie algebroids and Poisson–Nijenhuis structures*, Rep. Math. Phys. **40** (1997), 195–208.
[17] I. Dorfman, *Dirac structures and integrability of nonlinear evolution equations*, (John Willey & Sons, 1993).
[18] M. Carvalho, J. Helayël-Neto and M. De Oliveira *Locally product structures and supersymmetry*, Lett. Math. Phys. **64** (2003), 93–104.
[19] J. F. Cariñena, J. Grabowski and G. Marmo, *Courant algebroid and Lie bialgebroid contractions*, J. Phys. A: Math. Gen. **37** (2004), 5189–5202.
[20] K. Yano and M. Kon, *Structures on manifolds*, Series in Pure Mathematics 3 (World Scientific, 1984).

[21] V. Apostolov and M. Gualtieri, *Generalized Kähler manifolds, commuting complex structures, and split tangent bundle*, Commun. Math. Phys. **271** (2007), 561–575.
[22] N. Hitchin, *Instantons, Poisson structures and generalized Kähler geometry*, Commun. Math. Phys. **265** (2006), 131–164.
[23] J. F. Cariñena, J. Grabowski and G. Marmo, *Quantum bi-Hamiltonian system*, Int. J. Mod. Phys. **A15** (2000), 4797–4810.

11

A symmetrized canonical determinant on odd-class pseudodifferential operators

MARIE-FRANÇOISE OUEDRAOGO*

Abstract

Inspired by Braverman's symmetrized determinant, I introduce a symmetrized logarithm $\log^{\text{sym}}$ *for certain elliptic pseudodifferential operators. The symmetrized logarithm of an operator lies in the odd class whenever the operator does. Using the canonical trace extended to log-polyhomogeneous pseudodifferential operators, I define an associated canonical symmetrized determinant* DET^{sym} *on odd-class classical elliptic operators in odd dimensions:* $\text{DET}^{\text{sym}} = \exp \circ \text{TR} \circ \log^{\text{sym}}$, *which provides a canonical description of Braverman's symmetrized determinant. Using the cyclicity of the canonical trace on odd-class operators, one then easily infers multiplicative properties of this canonical symmetrized determinant.*

Introduction

The ordinary determinant on matrices, $\det = \exp \circ \text{tr} \circ \log$, extends to certain elliptic pseudodifferential operators (ψDOs) via the ζ-determinant $\det_\zeta$, which is used in physics to compute partition functions. But $\det_\zeta$ is not multiplicative in general, i.e., $\det_\zeta(AB) \neq \det_\zeta(A)\det_\zeta(B)$. The multiplicativity of the ordinary determinant on matrices follows from the cyclicity of the trace combined with a Campbell–Hausdorff formula: $\det(AB) = \exp \text{tr}(\log(AB)) = \exp(\text{tr}(\log A) + \text{tr}(\log B)) = \det(A)\det(B)$. Our aim is to mimic this construction using an extension of the canonical trace TR introduced by Kontsevich and Vishik [KV] (defined on noninteger-order ψDOs) to an appropriate class of elliptic ψDOs, stable under logarithms, on which TR is cyclic. A natural class is that of odd-class ψDOs [Gr, KV] in odd dimensions. However, it is not stable under logarithms; nevertheless, logarithms of even-order odd-class elliptic ψDOs lie in the odd class. Inspired by Braverman

* I am grateful to Maxim Braverman for his helpful comments on a preliminary version of this article.

[B], we introduce a symmetrized logarithm:

$$\log_\theta^{\mathrm{sym}} A = \frac{1}{2}\left(\log_\theta A + \log_{\theta - a\pi} A\right),$$

which lies in the odd class whenever A does, independently of the order (Proposition 11.3.1). Here θ and $\theta - a\pi$ are both Agmon angles of A, and a is its order. It is therefore natural to define (Definition 11.5.1)

$$\mathrm{DET}_\theta^{\mathrm{sym}} A = \exp\left(\mathrm{TR} \log_\theta^{\mathrm{sym}} A\right)$$

on a set of odd-class elliptic ψDOs in odd dimensions, provided TR extends to symmetrized logarithms of odd-class elliptic ψDOs in odd dimensions. This holds because $\log_\theta^{\mathrm{sym}} A$ lies in the odd class according to results in [PS], where it was shown that TR extends to the algebra of odd-class log-polyhomogeneous operators in odd dimensions. Here we moreover observe that all operators in this algebra are of the type

$$A = \sum A_l \left(\log_\alpha^{\mathrm{sym}} Q\right)^l,$$

where A_l is an odd-class classical ψDO and Q is an odd-class admissible ψDO with positive order and Agmon angle α (Proposition 11.3.3).

The symmetrized canonical determinant $\mathrm{DET}^{\mathrm{sym}}$ coincides with the symmetrized determinant defined by Braverman [B], and it coincides with $\mathrm{DET} = \exp \circ \mathrm{TR} \circ \log$, which arises in [PS] (Theorem 2.11) in the case of an even-order odd-class operator in odd dimensions, which in turn coincides with the ζ-determinant. Just like the ζ-determinant, the symmetrized canonical determinant depends on the spectral cut, but, as already observed by Braverman, this dependence is weaker than for the ζ-determinant (Proposition 11.5.5). The fact that $\mathrm{DET}^{\mathrm{sym}}$ is multiplicative on commuting operators is an easy consequence of the Campbell–Hausdorff formula (see Proposition 11.5.6). As already observed by Braverman [B], the multiplicative property actually holds for noncommuting operators under some natural restrictions on the spectral cuts; here this multiplicative property arises as a straightforward consequence of the traciality of the canonical trace on odd-class operators in odd dimensions (see Proposition 11.5.8).

This chapter offers a summary of some results of my PhD thesis [Ou] in preparation.

11.1 Log-polyhomogeneous operators

Let us first recall some basic concepts related to pseudodifferential operators and their symbols. Let U be an open subset of $\mathbb{R}^n$. Given $m \in \mathbb{C}$, the space of symbols

$S^m(U)$ consists of functions $\sigma(x,\xi)$ in $C^\infty(U \times \mathbb{R}^n)$ such that for any compact subset K of U and any two multiindices $\alpha = (\alpha_1, \ldots, \alpha_n) \in \mathbb{N}^n$, $\beta = (\beta_1, \ldots, \beta_n) \in \mathbb{N}^n$ there exists a constant $C_{K\alpha\beta}$ satisfying, for all $(x,\xi) \in K \times \mathbb{R}^n$,

$$|\partial_x^\alpha \partial_\xi^\beta \sigma(x,\xi)| \le C_{K\alpha\beta}(1+|\xi|)^{\mathrm{Re}(m)-|\beta|}, \tag{11.1}$$

where $\mathrm{Re}(m)$ is the real part of m, and $|\beta| = \beta_1 + \cdots + \beta_n$.

We denote by $S(U) := \langle \bigcup_{m\in\mathbb{C}} S^m(U)\rangle$ the algebra generated by all symbols on U for the star product defined as follows: if $\sigma_1 \in S^{m_1}(U)$ and $\sigma_2 \in S^{m_2}(U)$,

$$\sigma_1 \star \sigma_2(x,\xi) \sim \sum_{\alpha\in\mathbb{N}^n} \frac{(-i)^{|\alpha|}}{\alpha!} \partial_\xi^\alpha \sigma_1(x,\xi) \partial_x^\alpha \sigma_2(x,\xi),$$

i.e., for any integer $N \ge 1$ we have

$$\sigma_1 \star \sigma_2(x,\xi) - \sum_{|\alpha|<N} \frac{(-i)^{|\alpha|}}{\alpha!} \partial_\xi^\alpha \sigma_1(x,\xi) \partial_x^\alpha \sigma_2(x,\xi) \in S^{m_1+m_2-N}(U).$$

A symbol $\sigma \in S^m(U)$ is *classical* of order m if there is an asymptotic expansion

$$\sigma(x,\xi) \sim \sum_{j=0}^{\infty} \psi(\xi)\,\sigma_{m-j}(x,\xi),$$

where $\sigma_{m-j}(x,\xi)$ is a positively homogeneous function on $C^\infty(U \times (\mathbb{R}^n - \{0\}))$ of degree $m-j$, i.e., $\sigma_{m-j}(x,t\xi) = t^{m-j}\sigma_{m-j}(x,\xi)$ for all $t \in \mathbb{R}^+$. Here $\psi \in C^\infty(\mathbb{R}^n)$ is any cut-off function which vanishes for $|\xi| \le \frac{1}{2}$ and such that $\psi(\xi) = 1$ for $|\xi| \ge 1$.

Let $CS^m(U)$ denote the subset of classical symbols of order m, and let $CS(U) = \langle \bigcup_{m\in\mathbb{C}} CS^m(U)\rangle$ denote the algebra generated by all classical symbols on U.

A symbol $\sigma \in S^{m+\epsilon}(U)$ for any $\epsilon > 0$ is *log-polyhomogeneous* of order m and log degree k (see [L]) if its has an asymptotic expansion of the form

$$\sigma(x,\xi) \sim \sum_{j=0}^{\infty}\sum_{l=0}^{k} \psi(\xi)\,\sigma_{m-j,l}(x,\xi)\,\log^l|\xi|, \qquad (x,\xi) \in T^\star U, \quad m \in \mathbb{C}, \tag{11.2}$$

where k is a nonnegative integer and every $\sigma_{m-j,l}$, $l = 0, \ldots, k$, is positively homogeneous of degree $m-j$. We denote the set of log-polyhomogeneous symbols of order m and log degree k by $CS^{m,k}(U)$, and we set

$$CS^{\star,k}(U) = \bigcup_{m\in\mathbb{C}} CS^{m,k}(U), \qquad CS^{\star,\star}(U) = \left\langle \bigcup_{k\ge 0} CS^{\star,k}(U)\right\rangle.$$

Note that $CS^{\star,0}(U) = CS(U)$. For a vector space V, we set

$$CS^m(U, V) = CS^m(U) \otimes \mathrm{End}(V), \qquad CS^{\star,k}(U, V) = CS^{\star,k}(U) \otimes \mathrm{End}(V).$$

Let $\pi : E \to M$ be a finite-rank Hermitian vector bundle over a smooth closed Riemannian manifold M of dimension n. A ψDO A acting on the space $C^\infty(M, E)$ of smooth sections of E is classical of order a [is log-polyhomogeneous of order a and log degree k] if in any local trivialization $E_{|U} \simeq U \times V$, its symbol $\sigma(A)^U$ is classical of order a, i.e., $\sigma(A)^U \in CS^a(U, V)$ [respectively, log-polyhomogeneous of order a and log degree k, i.e., $\sigma(A)^U \in CS^{a,k}(U, V)$]. We denote the set of classical operators of order a [log-polyhomogeneous operators of order a and log degree k] by $C\ell^a(M, E)$ [$C\ell^{a,k}(M, E)$], and we set

$$C\ell(M, E) = \left\langle \bigcup_{a \in \mathbb{C}} C\ell^a(M, E) \right\rangle, \quad C\ell^{\star,k}(M, E) = \bigcup_{a \in \mathbb{C}} C\ell^{a,k}(M, E),$$

$$C\ell^{\star,\star}(M, E) = \left\langle \bigcup_{k \geq 0} C\ell^{\star,k}(M, E) \right\rangle,$$

where, as before, $\langle S \rangle$ stands for the algebra generated by the set S. Here the product is the ordinary composition. Note that $C\ell^a(M, E) = C\ell^{a,0}(M, E)$.

Let us finally recall the notion of odd-class symbols and operators.

Definition 11.1.1 [PS] A log-polyhomogeneous symbol (11.2) with *integer* order m is of *odd class* if for each $j \geq 0$, for $l = 0, \ldots, k$,

$$\sigma_{m-j,l}(x, -\xi) = (-1)^{m-j} \sigma_{m-j,l}(x, \xi) \quad \text{for } |\xi| \geq 1. \tag{11.3}$$

A log-polyhomogeneous operator A will be said to be of odd class if in any local trivialization it is of odd class. We denote by $C\ell^{a,k}_{(-1)}(M, E)$ the set of odd-class log-polyhomogeneous operators of order a and log degree k, and by $C\ell^{\star,\star}_{(-1)}(M, E)$ the set of odd-class log-polyhomogeneous operators.

Example 11.1.2

(i) All differential operators and their parametrices belong to $C\ell^{\star,\star}_{(-1)}(M, E)$.
(ii) The operator $(1 + \Delta)^{-1}$, where Δ is the Laplacian, is of odd class.

Proposition 11.1.3 [PS] *(See also* [Ou].*)* $C\ell^{\star,\star}_{(-1)}(M, E)$ *is an algebra.*

11.2 Holomorphic families of log-polyhomogeneous operators

Let us recall the notion of holomorphic family of symbols and operators first introduced in the classical case by Guillemin [G] under the name of gauged symbols, later popularized by Kontsevich and Vishik [KV] and generalized in [PS] to the log-polyhomogeneous case.

Definition 11.2.1 Let k be a nonnegative integer, and Ω a domain of $\mathbb{C}$. A family $(\sigma(z))_{z\in\Omega} \subset CS^{\alpha(z),k}(U)$ of log-polyhomogeneous symbols is *holomorphic* if:

(i) The order $\alpha(z)$ of $\sigma(z)$ is holomorphic on Ω.

(ii) For $(x,\xi) \in U \times \mathbb{R}^n$, the function $z \to \sigma(z)(x,\xi)$ is holomorphic as a function in $C^\infty(\Omega \times U \times \mathbb{R}^n)$, and for each z in Ω,

$$\sigma(z)(x,\xi) \sim \sum_{j=0}^{\infty} \chi(\xi)\sigma_{\alpha(z)-j}(z)(x,\xi)$$

lies in $CS^{\alpha(z),k}(U)$ for some cut-off function χ.

(iii) For any integer, $N \geq 1$, the remainder

$$\sigma_N(z)(x,\xi) := \sigma(z)(x,\xi) - \sum_{j=0}^{N-1} \chi(\xi)\sigma_{\alpha(z)-j}(z)(x,\xi)$$

is holomorphic in $z \in \Omega$ as an element of $C^\infty(\Omega \times U \times \mathbb{R}^n)$, and its lth derivative $\sigma_N^{(l)}(z)(x,\xi) := \partial_z^l(\sigma_N(z)(x,\xi))$ lies in $S^{\mathrm{Re}(\alpha(z))-N+\epsilon}(U)$ for all $\epsilon > 0$ locally uniformly on Ω, i.e., the lth derivative $\partial_z^k \sigma_{(N)}(z)$ satisfies a uniform estimate (11.1) with respect to z on compact subsets in Ω.

Proposition 11.2.2 [PS] *Let k be a nonnegative integer, and let Ω be a domain of $\mathbb{C}$. If $\sigma(z) \in CS^{\alpha(z),k}(U)$ is a holomorphic family, then so is each derivative*

$$\sigma^{(l)}(z)(x,\xi) := \partial_z^l(\sigma(z)(x,\xi)) \in CS^{\alpha(z),k+l}(U).$$

Precisely, $\sigma^{(l)}(z)(x,\xi)$ has an asymptotic expansion

$$\sigma^{(l)}(z)(x,\xi) \sim \sum_{j\geq 0} \sigma^{(l)}(z)_{\alpha(z)-j}(x,\xi).$$

Now we recall the notion of holomorphic log-polyhomogeneous operators.

Definition 11.2.3 A family $(A(z))_{z\in\Omega}$ of log-polyhomogeneous operators on $C^\infty(M,E)$ with distribution kernels $(x,y) \mapsto K_{A(z)}(x,y)$ is holomorphic if:

(i) The order $\alpha(z)$ of $A(z)$ is holomorphic in z.

(ii) In any local trivialization of E, we can write $A(z)$ in the form $A(z) = Op(\sigma(z)) + R(z)$, for some holomorphic family of symbols $(\sigma(z))_{z\in\Omega}$ and some holomorphic family $(R(z))_{z\in\Omega}$ of smoothing operators, i.e., given by a holomorphic family of smooth Schwartz kernels.

(iii) The (smooth) restrictions of the distribution kernels $K_{A(z)}$ to the complement of the diagonal $\Delta \subset M \times M$ form a holomorphic family with respect to the topology given by the uniform convergence in all derivatives on compact subsets of $M \times M - \Delta$.

Let us recall the construction of complex powers of elliptic operators, which provide classical examples of holomorphic families of operators. An angle θ is called a *principal angle* of a ψDO $A \in C\ell(M, E)$ if for every $(x, \xi) \in T^*M - \{0\}$, the leading symbol $\sigma^L(A)(x, \xi)$ of A has no eigenvalues on the ray $L_\theta = \{re^{i\theta}, r \geq 0\}$; in that case A is elliptic and its spectrum is discrete. We call a ψDO $A \in C\ell(M, E)$ *admissible* with spectral cut (or *Agmon angle*) θ if A has principal angle θ and the spectrum of A does not meet L_θ. In that case, A is invertible elliptic. Let $A \in C\ell(M, E)$ be admissible with spectral cut θ and nonnegative order a. For $\mathrm{Re}(z) < 0$, the complex power A_θ^z is defined by the Cauchy integral [Se]:

$$A_\theta^z = \frac{i}{2\pi} \int_{\Gamma_{r,\theta}} \lambda_\theta^z (A - \lambda)^{-1} \, d\lambda,$$

where $\lambda_\theta^z = |\lambda|^z e^{iz(\arg\lambda)}$ with $\theta \leq \arg\lambda < \theta + 2\pi$. Here $\Gamma_{r,\theta} = \Gamma_{r,\theta}^1 \cup \Gamma_{r,\theta}^2 \cup \Gamma_{r,\theta}^3$, where $\Gamma_{r,\theta}^1 = \{\rho e^{i\theta}, \infty > \rho \geq r\}$, $\Gamma_{r,\theta}^2 = \{\rho e^{i(\theta - 2\pi)}, \infty > \rho \geq r\}$, and $\Gamma_{r,\theta}^3 = \{re^{it}, \theta - 2\pi \leq t < \theta\}$, is a contour along the ray L_θ around the spectrum of A; and r is any small positive real number such that $\Gamma_{r,\theta} \cap Sp(A) = \emptyset$. The operator A_θ^z is classical of order az; the homogeneous components of its symbol are

$$\sigma_{az-j}(A_\theta^z)(x, \xi) = \frac{i}{2\pi} \int_{\Gamma_\theta} \lambda_\theta^z \, q_{-a-j}(x, \xi, \lambda) \, d\lambda$$

with q_{-a-j} the homogeneous components of the resolvent $(A - \lambda I)^{-1}$. The definition of complex powers can be extended to the whole complex plane by setting $A_\theta^z := A^k A_\theta^{z-k}$ for $k \in \mathbb{N}$ and $\mathrm{Re}(z) < k$; this definition is independent of the choice of k and preserves the usual properties: $A_\theta^{z_1} A_\theta^{z_2} = A_\theta^{z_1+z_2}$, $A_\theta^k = A^k$ for $k \in \mathbb{Z}$.

The family $(A_\theta^z)_{z \in \mathbb{C}}$ is a holomorphic family of ψDOs, and the logarithm of A is defined in terms of the derivative at $z = 0$ of this complex power [Se]:

$$\log_\theta(A) = \partial_z A_\theta^z|_{z=0}.$$

In some local chart [Se], the symbol of $\log_\theta(A)$ admits an asymptotic expansion:

$$\sigma(\log_\theta(A))(x, \xi) \sim a \log|\xi| I + \sum_{j=0}^{\infty} \sigma_{-j}(\log_\theta(A))(x, \xi),$$

where a denotes the order of A. Then $\log_\theta(A)$ is a log-polyhomogeneous operator.

11.3 Symmetrized logarithms

Inspired by Braverman [B], we introduce the symmetrized logarithm of an odd-class admissible ψDO. Let $C\ell_{(-1)}^a(M, E)$ denote the set of odd-class classical operators of order a, and let $A \in C\ell_{(-1)}^a(M, E)$ be an admissible ψDO which admits Agmon angles θ and $\theta - a\pi$. The *symmetrized logarithm* of A is defined by

$$\log_{\theta}^{\mathrm{sym}} A := \frac{1}{2}\left(\log_{\theta} A + \log_{\theta - a\pi} A\right). \tag{11.4}$$

As already pointed out by Braverman (see [B], Corollary 2.14), symmetrized logarithms of odd-class operators lie in the odd class.

Proposition 11.3.1 *The symmetrized logarithm of $A \in C\ell^{a}_{(-1)}(M, E)$ with Agmon angles θ and $\theta - a\pi$ is an odd-class log-polyhomogeneous operator.*

Proof The proof is the same as in [B] (see also [P]). □

This confirms known results in the case of even-order operators [D].

Corollary 11.3.2 *If $A \in C\ell^{a}_{(-1)}(M, E)$ admits an Agmon angle θ and if a is even, then* $\log_{\theta}(A)$ *is an odd-class polyhomogeneous operator.*

Let us now observe that all (odd-class) log-polyhomogeneous operators can be written as finite linear combinations of products of (odd-class) classical operators and symmetrized logarithms.

Proposition 11.3.3 *Let $Q \in C\ell_{(-1)}(M, E)$ be any admissible odd-class classical ψDO with positive order q and Agmon angle α, $\alpha - q\pi$. Then*

$$\text{(i)}\quad Cl^{\star,k}(M, E) = \bigoplus_{l=0}^{k} C\ell(M, E)\,(\log_{\alpha}^{\mathrm{sym}} Q)^{l} = (\log_{\alpha}^{\mathrm{sym}} Q)^{l} \bigoplus_{l=0}^{k} C\ell(M, E),$$

$$\text{(ii)}\quad C\ell^{\star,k}_{(-1)}(M, E) = (\log_{\alpha}^{\mathrm{sym}} Q)^{l} \bigoplus_{l=0}^{k} C\ell_{(-1)}(M, E) = \bigoplus_{l=0}^{k} C\ell_{(-1)}(M, E)\,(\log_{\alpha}^{\mathrm{sym}} Q)^{l}.$$

Remark 11.3.4 $\log_{\alpha}^{\mathrm{sym}} Q$ can be replaced by $\log_{\alpha} Q$ in (i), as already observed in [D]. For the proof, see [Ou].

11.4 The canonical trace extended to odd-class operators in odd dimension

An operator $A \in C\ell(M, E)$ of order $< -n$ is trace-class. In this case, the trace of A, which is finite, is given by $\mathrm{tr}(A) := \int_M dx \int_{T_x^*M} \mathrm{tr}_x\,(\sigma_A(x, \xi))\, đ\xi$, where tr_x is the fibrewise trace and where $đ\xi := \frac{1}{(2\pi)^n}\, d\xi$ with $d\xi$ the ordinary Lebesgue measure on $T_x^*M \simeq \mathbb{R}^n$. Kontsevich and Vishik [KV] introduced the *canonical trace* TR for a noninteger-order classical ψDO A:

$$\mathrm{TR}(A) := \int_M \mathrm{TR}_x(A)\, dx = \int_M \left(⨍_{T_x^*M} \mathrm{tr}_x\,(\sigma_A(x, \xi))\, đ\xi \right) dx,$$

where $⨍_{T_x^*M} \mathrm{tr}_x\,(\sigma_A(x,\xi))\, đ\xi := \mathrm{fp}_{R\to\infty} \int_{B_x(0,R)} \mathrm{tr}_x\,(\sigma_A(x,\xi))\, đ\xi$. Here $B_x(0,R)$ is a ball of radius R. It further extends to noninteger-order log-polyhomogeneous operators using the same formula [L]. In contrast, the *noncommutative residue* of a classical ψDO A is defined by [G, W] (see also [K])

$$\mathrm{res}(A) := \int_M \left(\int_{S_x^*M} \mathrm{tr}_x\,((\sigma_{-n}(A)(x,\xi)))\, đ\xi \right) dx.$$

From this formula, it is easy to see that the noncommutative residue vanishes on ψDOs of order strictly less than $-n$ and also on ψDOs of noninteger order.

Let $A \in C\ell^{\star,k}(M,E)$ be a log-polyhomogeneous operator with the asymptotic expansion (11.2) in local coordinates. We set, by extension, for any $x \in M$ (see [PS]),

$$\mathrm{res}_{x,l}(A) := \int_{S_x^*M} \mathrm{tr}_x\,\left((\sigma_{-n,l}(A)(x,\xi))\right) đ\xi \quad \text{for } l \leq k.$$

Proposition 11.4.1 *Assume that M is an odd-dimensional manifold. Let $A \in C\ell^{\star,k}_{(-1)}(M,E)$. Then $\mathrm{res}_{x,l}(A)$ vanishes for $l = 0,\ldots,k$; hence*

$$\mathrm{res}_l(A) := \int_M \mathrm{res}_{x,l}(A)(x,\xi)\, dx = 0$$

and

$$\mathrm{res}(A) = \int_M \left(\int_{S_x^*M} \mathrm{tr}_x\,((\sigma_{-n}(A)(x,\xi)))\, đ\xi \right) dx = \mathrm{res}_0(A) = 0.$$

Proof Let $A \in C\ell^{\star,k}_{(-1)}(M,E)$ be an odd-class operator, so its symbol verifies $\sigma_{-n,l}(A)(x,-\xi) = (-1)^n \sigma_{-n,l}(A)(x,\xi)$. The dimension of M is odd, and we integrate over the unit sphere S_x^*M the odd function $\sigma_{-n,l}(A)(x,\xi)$. The results follow. □

Given an admissible operator $Q \in C\ell(M,E)$ with positive order q and spectral cut α, and given $A \in C\ell(M,E)$, the map $z \mapsto \mathrm{TR}\left(A\, Q_\alpha^{-z}\right)$ is meromorphic with simple poles, and the Q-weighted trace of A is defined by

$$\mathrm{tr}_\alpha^Q(A) := \mathrm{fp}_{z=0} \mathrm{TR}\left(A\, Q_\alpha^{-z}\right),$$

where $\mathrm{fp}_{z=0}$ denotes the constant term in the Laurent expansion. This definition of Q-weighted trace is extended in [D] to logarithms of elliptic admissible ψDOs by picking out the constant term of the meromorphic map $z \mapsto \mathrm{TR}(\log_\theta A\, Q^{-z})$ (which can have double poles).

Let us recall a result which extends results of [PS] to the log-polyhomogeneous case (this is unpublished work by the authors of [PS], which can be found in [P]).

Theorem 11.4.2 *Let $A(z) \in C\ell^{\alpha(z),k}(M, E)$ be a holomorphic family of log-polyhomogeneous operators parametrized by $z \in \Omega$, a domain of $\mathbb{C}$. Then for any $z_0 \in \Omega$ such that $\alpha'(z_0) \neq 0$ we have*

$$\mathrm{fp}_{z=z_0}\mathrm{TR}(A(z)) = \int_M dx \left(\mathrm{TR}_x(A(z_0)) + \sum_{l=0}^{k} \frac{(-1)^{l+1}}{(\alpha'(z_0))^{l+1}} \mathrm{res}_{x,l}(A^{(l+1)}(z_0)) \right).$$

Applying this theorem to $A(z) = AQ_\alpha^{-z}$, where $A \in C\ell^{\star,\star}(M, E)$, Q is admissible with Agmon angle α, $z = 0$, and $A(0) = A$, we obtain the following formula for the weighted trace of a log-polyhomogeneous operator [PS]:

$$\mathrm{tr}_\alpha^Q(A) = \int_M dx \left(\mathrm{TR}_x(A) + \sum_{l=0}^{k} \frac{1}{(-q)^{l+1}} \mathrm{res}_{x,l}\left(A (\log_\alpha Q)^{l+1} \right) \right). \tag{11.5}$$

Remark 11.4.3 If A is a classical pseudodifferential operator, (11.5) gives back the formula

$$\mathrm{tr}_\alpha^Q(A) = \int_M dx \left(\mathrm{TR}_x(A) - \frac{1}{q} \mathrm{res}_x(A \log_\alpha Q) \right).$$

Kontsevich and Vishik observed in [KV] that in odd dimensions, the weighted trace $\mathrm{tr}^Q(A)$ is independent of the choice of the even-order odd-class weight Q whenever A lies in the odd-class. Alternatively, on the basis of defect formulae in [PS] one can show that the canonical trace TR extends to odd-class operators in odd dimensions ([PS, Corollary 2.7]; see also [P]); it was actually shown in [MSS] that it is the unique extension to odd-class operators of the ordinary L^2 trace. The following theorem extends some of these facts to log-polyhomogeneous operators.

Theorem 11.4.4 *Assume that the dimension of M is odd. Let $A(z) \in C\ell^{\alpha(z),k}(M, E)$ be a holomorphic family of log-polyhomogeneous operators with $\alpha'(0) \neq 0$ such that $A(0) = A$ and $\forall j \geq 0$, $A^{(j)}(0)$ lies in the odd class. Then:*

(i) $\mathrm{TR}_x(A)dx = \left(⨍_{T_x^*M} \mathrm{tr}_x\left(\sigma(A)(x,\xi)\right) đ\xi \right) dx$ *defines a global density on M, so that* $\mathrm{TR}(A) = \int_M \mathrm{TR}_x(A)dx$ *is well defined.*
(ii) $\mathrm{fp}_{z=0}\mathrm{TR}(A(z)) = \lim_{z\to 0} \mathrm{TR}(A(z)) = \mathrm{TR}(A)$.

Proof Following the line of proof of Theorem 3 of [P], let us assume that $A(z) \in C\ell^{\alpha(z),k}(M, E)$. We know that $\forall j \geq 0$, $A^{(j)}(0)$ lie in the odd class. Thus, using Proposition 11.4.1, we get $\mathrm{res}_{x,l}\left(A^{(j)}(0)\right) = 0$. Hence, applying Theorem 11.4.2, $\mathrm{TR}_x(A)dx = \left(⨍_{T_x^*M} \mathrm{tr}_x\left(\sigma(A)(x,\xi)\right) đ\xi \right)dx$ defines a global density on M, so that $\mathrm{TR}(A) = \int_M \mathrm{TR}_x(A)dx$ is well defined, and the rest follows. □

Example 11.4.5 Let $A, Q \in C\ell(M, E)$ be odd-class operators such that Q is admissible with positive order q and Agmon angles α and $\alpha - q\pi$. The holomorphic family

$$A(z) = A\frac{Q_\alpha^z + Q_{\alpha-q\pi}^z}{2}$$

verify $A(0) = A$, and for $l > 1$ we have $A^{(l)}(0) = A(\log_\alpha^{\mathrm{sym}} Q)^l$. In odd dimension, $A(z)$ fulfills the theorem, so

$$\mathrm{fp}_{z=0}\mathrm{TR}\left(A\frac{Q_\alpha^z + Q_{\alpha-q\pi}^z}{2}\right) = \mathrm{TR}(A).$$

We recover the known fact [PS] that TR is cyclic on the algebra of odd-class log-polyhomogeneous operators.

Corollary 11.4.6 *Assume that the dimension of M is odd. For any odd-class operators* $A, B \in C\ell_{(-1)}^{\star,\star}(M, E)$,

$$\mathrm{TR}[A, B] = 0.$$

Proof This follows from applying Theorem 11.4.4 to the family $A(z) = [AQ_\theta^z, BQ_\theta^z]$ with Q an odd-class admissible operator with even order q and Agmon angle θ. □

11.5 The symmetrized canonical determinant

Inspired by Braverman [B], who introduced a symmetrized determinant using symmetrized regularized traces, we define a symmetrized determinant which involves the canonical trace and symmetrized logarithms.

Definition 11.5.1 Suppose that M is an odd-dimensional manifold. Let A be an odd-class admissible operator with nonnegative order a which admits Agmon angles θ and $\theta - a\pi$. A *determinant associated* to TR is defined by setting

$$\mathrm{DET}_\theta^{\mathrm{sym}}(A) := \exp\left(\mathrm{TR}(\log_\theta^{\mathrm{sym}} A)\right). \tag{11.6}$$

Remark 11.5.2 If A has even order, then $\mathrm{DET}_\theta^{\mathrm{sym}}$ coincides with the determinant defined in [PS], which in turn coincides with the ζ-determinant:

$$\log \mathrm{DET}_\theta^{\mathrm{sym}}(A) = \log \det_{\zeta,\theta}(A) = \mathrm{TR}(\log_\theta A).$$

Indeed, $\log_\theta^{\mathrm{sym}} A = \log_\theta A - ik\pi I$. Using Corollary 11.3.2 and Proposition 11.4.1, we get $\mathrm{tr}_\theta^A(\log_\theta^{\mathrm{sym}} A) = \mathrm{tr}_\theta^A(\log_\theta A)$. The result follows on applying Theorem 11.4.4.

Proposition 11.5.3 *Under the assumptions of Definition 11.5.1,* $\mathrm{DET}_\theta^{\mathrm{sym}}(A)$ *coincides with the symmetrized determinant introduced in* [B]*:*

$$\mathrm{Det}_\theta^{\mathrm{sym}} A := \exp\left(\frac{1}{2}\mathrm{Tr}^{\mathrm{sym}}\left(\log_\theta A + \log_{\theta-a\pi} A\right)\right) = \exp\left(\mathrm{Tr}^{\mathrm{sym}}\left(\log_\theta^{\mathrm{sym}} A\right)\right),$$

where $\mathrm{Tr}^{\mathrm{sym}} A := \mathrm{Tr}_\alpha^{Q,\mathrm{sym}} A = \frac{1}{2}\left(\mathrm{tr}_\alpha^Q A + \mathrm{tr}_{\alpha - q\pi}^Q A\right)$. *Here* Q *is any odd-class admissible operator with positive order* q *and Agmon angles* $\alpha,\ \alpha - q\pi$.

Proof It is easy to see that $\mathrm{TR}(\log_\theta^{\mathrm{sym}} A) = \mathrm{Tr}^{\mathrm{sym}}(\log_\theta^{\mathrm{sym}} A)$ by applying Theorem 11.4.4 to the family $A(z) = \frac{1}{2}(\log_\theta^{\mathrm{sym}} A)(A_\theta^z + A_{\theta - a\pi}^z)$. □

Remark 11.5.4 It is well known that the ζ-determinant generally depends on the choice of spectral cut. However, it is invariant under mild changes of spectral cut in the following sense: if $0 \leq \theta < \phi < 2\pi$ are two Agmon angles for A and if the cone $\Lambda_{\theta,\phi} := \{\rho\, e^{it},\ \infty > \rho \geq r,\quad \theta < t < \phi\}$ contains only a finite number of eigenvalues of A, then $\det_{\zeta,\theta}(A) = \det_{\zeta,\phi}(A)$. If the order of A is even, then by Remark 11.5.2, $\mathrm{DET}_\theta^{\mathrm{sym}}(A) = \det_{\zeta,\theta}(A)$, and hence the same property holds for the symmetrized determinant. As already remarked by Braverman, if the order of A is odd, that is no longer the case, for it might happen that there are infinitely many eigenvalues of A in the cone $\Lambda_{\theta - a\pi, \phi - a\pi}$. Nevertheless, we have the following proposition, proved in [B]:

Proposition 11.5.5 *Let* M *be an odd-dimensional manifold, and let* A *be an odd-class admissible operator with odd positive order* a *which admits Agmon angles* $\theta,\ \theta - \pi$ *and* $\phi,\ \phi - \pi$. *Suppose that* $0 \leq \phi - \theta < \pi$. *In the cases*

(i) only a finite number of eigenvalues of A *lie in* $\Lambda_{\theta,\phi} \cup \Lambda_{\theta-\pi,\phi-\pi}$,
(ii) all but finitely many eigenvalues of A *lie in* $\Lambda_{\theta,\phi} \cup \Lambda_{\theta-\pi,\phi-\pi}$,

we have

$$\mathrm{DET}_\theta^{\mathrm{sym}}(A) = \pm \mathrm{DET}_\phi^{\mathrm{sym}}(A).$$

From the traciality of TR on odd-class operator in odd dimensions, we infer the following multiplicative property.

Proposition 11.5.6 *Suppose that* M *is an odd-dimensional manifold. Let* A *be an odd-class admissible operator with positive order* a *and Agmon angles* θ *and* $\theta - a\pi$, *and let* B *be an odd-class admissible operator with positive order* b *and Agmon angles* ϕ *and* $\phi - b\pi$ *such that* AB *is also admissible with Agmon angles* ψ *and* $\psi - (a+b)\pi$. *If* $[A, B] = 0$ *then*

$$\mathrm{DET}_\psi^{\mathrm{sym}}(AB) = \mathrm{DET}_\theta^{\mathrm{sym}}(A)\,\mathrm{DET}_\phi^{\mathrm{sym}}(B).$$

This result generalizes to noncommuting operators, as was shown by Braverman [B] using, under suitable assumptions, the formula for the multiplicative anomaly established by Okikiolu [O]. For this, we first recall the following definition taken from [B].

Definition 11.5.7 Let θ be a principal angle for an operator $A \in C\ell_{(-1)}^a(M, E)$. An Agmon angle $\phi \geq \theta$ is *sufficiently close to* θ if there are no eigenvalues of A in the cones $\Lambda_{(\theta,\phi]}$ and $\Lambda_{(\theta-a\pi,\phi-a\pi]}$. We shall denote by $\log_{\widetilde{\theta}} A$, $\log^{\mathrm{sym}}_{\widetilde{\theta}} A$, $\mathrm{DET}^{\mathrm{sym}}_{\widetilde{\theta}} A$ the corresponding numbers obtained using an Agmon angle sufficiently close to θ. Clearly, those numbers are independent of the choice of $\widetilde{\theta}$.

Here, I give an alternative short proof of the multiplicativity of $\mathrm{DET}^{\mathrm{sym}}$ using the cyclicity of the canonical trace on odd-class operators in odd dimensions.

Proposition 11.5.8 *Let M be an odd-dimensional manifold. Suppose that A is an odd-class admissible operator with positive order a and Agmon angles θ and $\theta - a\pi$, and that B is an odd-class admissible operator with positive order b and Agmon angles ϕ and $\phi - b\pi$. Let us assume that for each $t \in [0, 1]$, $A^t_\theta B$ has principal angle $\psi(t)$, where $t \to \psi(t)$ is continuous. Set $\psi(0) = \phi$ and $\psi(1) = \psi$. Then for $\widetilde{\psi}$ an angle sufficiently close to ψ,*

$$\mathrm{DET}^{\mathrm{sym}}_{\widetilde{\psi}}(AB) = \mathrm{DET}^{\mathrm{sym}}_{\theta}(A)\,\mathrm{DET}^{\mathrm{sym}}_{\phi}(B).$$

Proof For a fixed t, the operator $A^t_\theta B$ is classical with order $at + b$, and it is easy to prove that $\log^{\mathrm{sym}}_{\psi(t)}(A^t_\theta B)$ is an odd-class operator. Let us set $\log \mathcal{M}^{\mathrm{sym}}(A^t_\theta, B) := \log \mathrm{DET}^{\mathrm{sym}}_{\psi(t)}(A^t_\theta B) - \log \mathrm{DET}^{\mathrm{sym}}_{\theta}(A^t_\theta) - \log \mathrm{DET}^{\mathrm{sym}}_{\phi}(B)$. Following arguments similar to Okiliolu's (see [O]), we build a finite partition $\bigcup_{k=1}^K J_k$ of $[0, 1]$ in such a way that we can choose on each of the intervals J_k a common fixed Agmon angle $\widetilde{\psi}_k$ of $A^t_\theta B$ sufficiently close to $\psi(t)$ when t varies in J_k, and we set $m = at + b$. We want to show that for all $t \in [0, 1]$, $\frac{d}{dt}\left(\log \mathcal{M}^{\mathrm{sym}}(A^t_\theta, B)\right) = 0$, i.e., for all $\tau \in [0, 1]$,

$$\frac{d}{dt}\Big|_{t=0}\left(\log \mathcal{M}^{\mathrm{sym}}(A^{t+\tau}_\theta, B)\right) = 0.$$

Let us start by proving the result at $\tau = 0$. In practice we work on each of the J_k with the Agmon angle $\widetilde{\psi}_k$; to simplify notation, we just write $\widetilde{\psi}$ instead of $\widetilde{\psi}_k$. We have

$$\begin{aligned}
&\frac{d}{dt}\Big|_{t=0}\left(\log \mathcal{M}^{\mathrm{sym}}(A^t_\theta, B)\right)\\
&= \frac{d}{dt}\Big|_{t=0}\mathrm{TR}\Big(\log^{\mathrm{sym}}_{\widetilde{\psi}}(A^t_\theta B) - \log^{\mathrm{sym}}_{\theta}(A^t_\theta) - \log^{\mathrm{sym}}_{\phi}(B)\Big)\\
&= \frac{1}{2}\mathrm{TR}\Big[\dot{\overline{(A^t_\theta B)}}_{\widetilde{\psi}}(A^t_\theta B)^{-1}_{\widetilde{\psi}} + \dot{\overline{(A^t_{\theta-a\pi} B)}}_{\widetilde{\psi}-m\pi}(A^t_{\theta-a\pi} B)^{-1}_{\widetilde{\psi}}\Big]_{|t=0}\\
&\quad - \frac{1}{2}\mathrm{TR}\Big[\dot{\overline{(A^t_\theta)}}_{\theta}(A^t_\theta)^{-1}_{\theta} + \dot{\overline{(A^t_{\theta-a\pi}})}_{\theta-a\pi}(A^t_{\theta-a\pi})^{-1}_{\theta-a\pi}\Big]_{|t=0} = 0,
\end{aligned}$$

where we have used the formula $\frac{d}{dt}\log C_t = \int_0^1 (C_t)^{-1-\lambda}(\dot{C}_t)(C_t)^\lambda\, d\lambda$ combined with the traciality of TR in the second identity.[1] Now, replacing B by $A^\tau_\theta B$

[1] Interchanging the trace TR and the differentiation is justified in [Ou].

yields $\frac{d}{dt}\big|_{t=0}\big(\log \mathcal{M}^{\mathrm{sym}}(A_\theta^t, A_\theta^\tau B)\big) = 0$. An easy computation shows that

$$\log \mathcal{M}^{\mathrm{sym}}(A_\theta^{t+\tau}, B) - \log \mathcal{M}^{\mathrm{sym}}(A_\theta^t, A_\theta^\tau B)$$
$$= \mathrm{TR}\Big(-\log_\theta^{\mathrm{sym}}(A_\theta^\tau) - \log_{\tilde{\psi}}^{\mathrm{sym}} B - \log_{\tilde{\psi}}^{\mathrm{sym}}(A_\theta^\tau B)\Big).$$

Because the r.h.s. of the previous equation in independent of t, it follows that for all $\tau \in [0, 1]$, $\frac{d}{dt}\big|_{t=0}\big(\log \mathcal{M}^{\mathrm{sym}}(A_\theta^{t+\tau}, B)\big) = 0$. □

Bibliography

[B] M. Braverman, *Symmetrized trace and symmetrized determinant of odd-class pseudo-differential operators*, arXiv:math-ph/0702060 (2007).

[D] C. Ducourtioux, *Weighted traces on pseudo-differential operators and associated determinants*, PhD thesis, Clermont-Ferrand (2001); *Multiplicative anomaly for the ζ-regularized determinant*, Geometric methods for quantum field theory (Villa de Leyva, 1999), 467–482, World Science Publ., River Edge, NJ, 2001.

[G] V. Guillemin, *A new proof of Weyl's formula on the asymptotic distribution of eigenvalues*, Adv. Math. **55** (1985), no. 2, 131–160; *Gauged Lagrangian distributions*, Adv. Math. **102** (1993), no. 2, 184–201.

[Gr] G. Grubb, *A resolvent approach to traces and zeta Laurent expansions*, Contemp. Math., **366**, Amer. Math. Soc., Providence, RI, 2005, 67–93.

[KV] M. Kontsevich, S. Vishik, *Determinants of elliptic pseudo-differential operators*, Max Planck Preprint (1994).

[K] Ch. Kassel, *Le résidu non commutatif [d'après Wodzicki]*, Sém. Bourbaki **708** (1989).

[L] M. Lesch, *On the noncommutative residue for pseudo-differential operators with log-polyhomogeneous symbols*, Ann. Global Anal. Geom. **17** (1998), 151–187.

[MSS] L. Mannicia, E. Schrohe, J. Seiler, *Uniqueness of the Kontsevich–Vishik trace*, arXiv:math.FA/0702250 (2007).

[O] K. Okikiolu, *The Campbell–Hausdorff theorem for elliptic operators and a related trace formula*, Duke. Math. J. **79** (1995), 687–722; *The multiplicative anomaly for determinants of elliptic operators*, Duke Math. J. **79** (1995), 722–749.

[Ou] M. F. Ouedraogo, Thesis in preparation.

[P] S. Paycha, *The noncommutative residue and the canonical trace in the light of Stokes' and continuity properties*, arXiv:math-OA/07062552 (2007); monograph in preparation.

[PS] S. Paycha, S. Scott, *A Laurent expansion for regularised integrals of holomorphic symbols*, Geom. Funct. Anal. **17** (2007), no. 2, 491–536.

[Se] R. T. Seeley, *Complex powers of an elliptic operator*, in Singular integrals, Proc. Symp. Pure Math., Chicago, Amer. Math. Soc., Providence (1966), 288–307.

[W] M. Wodzicki, *Noncommutative residue*, in Lecture Notes in Math. **1283**, Springer Verlag, 1987; *Spectral asymmetry and noncommutative residue* (in Russian), Thesis, (former) Steklov Institute, Sov. Acad. Sci., Moscow 1984.

12

Some remarks about cosymplectic metrics on maximal flag manifolds

MARLIO PAREDES AND SOFIA PINZÓN*

Abstract

In this chapter we provide some differential-type conditions for the Borel metrics on a maximal flag manifold to be cosymplectic. These conditions are obtained using the formula to calculate the codifferential of the Kähler form on a maximal flag manifold derived by the authors in a yet unpublished paper.

12.1 Introduction

This work deals with the maximal complex flag manifold

$$\mathbb{F}(n) = \frac{U(n)}{U(1) \times \cdots \times U(1)}.$$

The geometry of this manifold has been studied in several papers. Burstall and Salamon [5] showed the existence of a bijective relation between almost complex structures on $\mathbb{F}(n)$ and tournaments with n vertices. Mo and Negreiros [16], by using moving frames and tournaments, showed explicitly the existence of an n-dimensional family of invariant $(1, 2)$-symplectic metrics on $\mathbb{F}(n)$. In [19], the first author proved the existence of several families of $(1, 2)$-symplectic metrics on $\mathbb{F}(n)$; see too [18] and [20].

The main motivation to study $(1, 2)$-symplectic and cosymplectic metrics is the construction of harmonic maps using a known result due to Lichnerowicz [13]: let (M, g, J_1) and (N, h, J_2) be almost Hermitian manifolds with M cosymplectic and N $(1, 2)$-symplectic; then any $\pm$-holomorphic map $\phi : (M, g, J_1) \to (N, h, J_2)$ is harmonic.

* We would like to thank Professor David Colón Arroyo for his help reviewing the English language in this work.

Cohen, Negreiros and San Martin in [9] proved that $\mathbb{F}(n)$ admits a $(1, 2)$-symplectic invariant metric if and only if the tournament associated to the almost complex structure is cone-free. This result was improved in [10] using local transitivity of the associated tournament.

Mo and Negreiros [16], using the moving frames technique, found a formula to calculate the differential of the Kähler form. In [21], the authors used a similar method to obtain a formula to calculate the codifferential of the Kähler form.

In the present work, we use the formula for the codifferential of the Kähler form to show that the Borel metrics (see [1–3]) on flag manifolds are cosymplectic if and only if they satisfy a system of partial differential equations.

12.2 Maximal flag manifolds

The classical maximal flag manifold is defined by

$$\mathbb{F}(n) = \{(L_1, \ldots, L_n) : L_i \text{ is a subspace of } \mathbb{C}^n, \dim_{\mathbb{C}} L_i = 1, L_i \perp L_j\}. \quad (12.1)$$

The unitary group $U(n)$ acts transitively on $\mathbb{F}(n)$, turning this manifold into the homogeneous space

$$\mathbb{F}(n) = \frac{U(n)}{U(1) \times U(1) \times \cdots \times U(1)} = \frac{U(n)}{T}, \quad (12.2)$$

where $T = U(1) \times U(1) \times \cdots \times U(1)$ is any maximal torus of $U(n)$.

Let $\mathfrak{p}$ be the tangent space of $\mathbb{F}(n)$ at the point (T). It is known that $\mathfrak{u}(n)$, the Lie algebra of skew-Hermitian matrices, decomposes as

$$\mathfrak{u}(n) = \mathfrak{p} \oplus \mathfrak{u}(1) \oplus \cdots \oplus \mathfrak{u}(1),$$

where $\mathfrak{p} \subset \mathfrak{u}(n)$ is the subspace of zero-diagonal matrices.

In order to define any tensor on $\mathbb{F}(n)$ it is sufficient to give it on $\mathfrak{p}$, because the action of $U(n)$ on $\mathbb{F}(n)$ is transitive. An invariant almost complex structure on $\mathbb{F}(n)$ is determined by a linear map $J : \mathfrak{p} \to \mathfrak{p}$ such that $J^2 = -I$, and commutes with the adjoint representation of the torus T on $\mathfrak{p}$.

For each almost complex structure we assign a directed graph. A tournament, or n-tournament, $\mathcal{T}$ consists of a finite set $T = \{p_1, \ldots, p_n\}$ of n players together with a dominance relation, $\to$, which assigns to every pair of players a winner, that is, $p_i \to p_j$ or $p_j \to p_i$. A tournament $\mathcal{T}$ can be represented by a directed graph in which T is the set of vertices and any two vertices are joined by an oriented edge. If the dominance relation is transitive, then the tournament is called transitive. For a complete reference on tournaments see [15].

Given an invariant complex structure J, we define the associated tournament $\mathcal{T}(J)$ in the following way: if $J(a_{ij}) = (a'_{ij})$, then $\mathcal{T}(J)$ is such that for $i < j$

$$\left(i \to j \Leftrightarrow a'_{ij} = \sqrt{-1}\, a_{ij}\right) \quad \text{or} \quad \left(i \leftarrow j \Leftrightarrow a'_{ij} = -\sqrt{-1}\, a_{ij}\right);$$

see [16].

We consider $\mathbb{C}^n$ equipped with the standard Hermitian inner product, that is, for $V = (v_1, \dots, v_n)$ and $W = (w_1, \dots, w_n)$ in $\mathbb{C}^n$, we have $\langle V, W \rangle = \sum_{i=1}^n v_i \overline{w_i}$. We use the convention $v_{\bar{i}} = \overline{v_i}$ and $f_{\bar{i}j} = \overline{f_{i\bar{j}}}$.

A *frame* consists of an ordered set of n vectors $(Z_1, \dots, Z_n)$, such that $Z_1 \wedge \dots \wedge Z_n \neq 0$, and it is called unitary if $\langle Z_i, Z_j \rangle = \delta_{i\bar{j}}$. The set of unitary frames can be identified with the unitary group $U(n)$.

If we write $\mathrm{d}Z_i = \sum_j \omega_{i\bar{j}} Z_j$, the coefficients $\omega_{i\bar{j}}$ are the Maurer–Cartan forms of the unitary group $U(n)$. They are skew-Hermitian, that is, $\omega_{i\bar{j}} + \omega_{\bar{j}i} = 0$. For more details see [7].

We may define all left-invariant metrics on $(\mathbb{F}(n), J)$ by (see [1] or [17])

$$\mathrm{d}s^2_\Lambda = \sum_{i,j} \lambda_{ij} \omega_{i\bar{j}} \otimes \omega_{\bar{i}j}, \tag{12.3}$$

where $\Lambda = (\lambda_{ij})$ is a symmetric real matrix such that

$$\begin{cases} \lambda_{ij} > 0 & \text{if} \quad i \neq j, \\ \lambda_{ij} = 0 & \text{if} \quad i = j, \end{cases} \tag{12.4}$$

and the Maurer–Cartan forms $\omega_{i\bar{j}}$ are such that

$$\omega_{i\bar{j}} \in \mathbb{C}^{1,0} \text{ (forms of type (1,0))} \quad \Longleftrightarrow \quad i \overset{\mathcal{T}(J)}{\longrightarrow} j. \tag{12.5}$$

The metrics (12.3) are called Borel type, and they are almost Hermitian for every invariant almost complex structure J, that is, $\mathrm{d}s^2_\Lambda(JX, JY) = \mathrm{d}s^2_\Lambda(X, Y)$ for all tangent vectors X, Y. When J is integrable, $\mathrm{d}s^2_\Lambda$ is said to be Hermitian.

Let J be an invariant almost complex structure on $\mathbb{F}(n)$, $\mathcal{T}(J)$ the associated tournament, and $\mathrm{d}s^2_\Lambda$ an invariant metric. The Kähler form with respect to J and $\mathrm{d}s^2_\Lambda$ is defined by

$$\Omega(X, Y) = \mathrm{d}s^2_\Lambda(X, JY), \tag{12.6}$$

for any tangent vectors X, Y. For each permutation σ of n elements, the Kähler form can be written as follows (see [16]):

$$\Omega = -2\sqrt{-1} \sum_{i<j} \mu_{\sigma(i)\sigma(j)} \omega_{\sigma(i)\overline{\sigma(j)}} \wedge \omega_{\overline{\sigma(i)}\sigma(j)}, \tag{12.7}$$

where $\mu_{\sigma(i)\sigma(j)} = \varepsilon_{\sigma(i)\sigma(j)}\lambda_{\sigma(i)\sigma(j)}$ and

$$\varepsilon_{ij} = \begin{cases} 1 & \text{if} \quad \sigma(i) \to \sigma(j), \\ -1 & \text{if} \quad \sigma(j) \to \sigma(i), \\ 0 & \text{if} \quad \sigma(i) = \sigma(j). \end{cases}$$

$\mathbb{F}(n)$ is said to be almost Kähler if and only if Ω is closed, that is, $\mathrm{d}\Omega = 0$. If J is integrable and Ω is closed, then $\mathbb{F}(n)$ is said to be a Kähler manifold.

Mo and Negreiros proved in [16] that

$$\mathrm{d}\Omega = 4 \sum_{i<j<k} C_{\sigma(i)\sigma(j)\sigma(k)} \Psi_{\sigma(i)\sigma(j)\sigma(k)}, \tag{12.8}$$

where

$$C_{ijk} = \mu_{ij} - \mu_{ik} + \mu_{jk} \tag{12.9}$$

and

$$\Psi_{ijk} = \mathrm{Im}(\omega_{i\bar{j}} \wedge \omega_{\bar{i}k} \wedge \omega_{j\bar{k}}). \tag{12.10}$$

We denote by $\mathbb{C}^{p,q}$ the space of forms of type (p, q) on $\mathbb{F}(n)$. Then, for any i, j, k, we have either $\Psi_{ijk} \in \mathbb{C}^{0,3} \oplus \mathbb{C}^{3,0}$ or $\Psi_{ijk} \in \mathbb{C}^{1,2} \oplus \mathbb{C}^{2,1}$. An invariant almost Hermitian metric $\mathrm{d}s^2_\Lambda$ is said to be (1, 2)-symplectic if and only if $(\mathrm{d}\Omega)^{1,2} = 0$. If $\delta\Omega = 0$ (the codifferential of the Kähler form is zero), then the metric is said to be cosymplectic.

Given a tournament $\mathcal{T}$, we define a winner in $\mathcal{T}$ as the unique vertex that wins against all the other vertices, and a loser in $\mathcal{T}$ as the unique vertex that loses to all the other vertices. Not every tournament has a winner or a loser; however, when a winner or a loser exists, it is not part of any cycle.

Up to isomorphism, there are four distinct 4-tournaments. The two that contain a single 3-cycle are called coned 3-cycles. Each of them contains a cycle and a winner or a loser. An n-tournament $\mathcal{T}$ is called cone-free if each 4-subtournament of $\mathcal{T}$ is not a coned 3-cycle.

The following theorem, due to Cohen, Negreiros and San Martin [9], classifies the $(1, 2)$-symplectic metrics on $\mathbb{F}(n)$.

Theorem 12.2.1 *Let* $(\mathbb{F}(n), J, \mathrm{d}s^2_\Lambda)$ *be the maximal flag manifold. The metric* $\mathrm{d}s^2_\Lambda$ *is* $(1, 2)$*-symplectic if and only if the associated tournament* $\mathcal{T}(J)$ *is cone-free.*

Another way to see the cone-free tournaments is their local transitivity. Given a tournament $\mathcal{T}$ and a vertex $v \in \mathcal{T}$, we define the following subtournaments:

$$\mathcal{T}^-(v) = \{x \in \mathcal{T} : x \to v\} \quad \text{and} \quad \mathcal{T}^+(v) = \{x \in \mathcal{T} : v \to x\}, \tag{12.11}$$

which are called the in-neighbor and the out-neighbor of v, respectively. $\mathcal{T}$ is called locally transitive if and only if the subtournaments $\mathcal{T}^-(v)$ and $\mathcal{T}^+(v)$ are transitive for each vertex v (see [4]). The following proposition was proved in [10].

Proposition 12.2.2 *A tournament $\mathcal{T}$ is cone-free if and only if it is locally transitive.*

This proposition implies the following theorem, which is equivalent to Theorem 12.2.1.

Theorem 12.2.3 *Let $(\mathbb{F}(n), J, ds^2_\Lambda)$ be the maximal flag manifold. The metric ds^2_Λ is $(1,2)$-symplectic if and only if the associated tournament $\mathcal{T}(J)$ is locally transitive.*

The proof of this theorem is more direct than that of the last one (Theorem 12.2.1), because the local transitivity concept is more natural than the cone-free concept. In addition, the local transitivity concept can be generalized to directed graphs associated to f-structures (see [22, 24]). In [8], Theorem 12.2.3 was generalized to f-structures using local transitivity. References [6] and [14] are important for future developments on the subject of this chapter.

12.3 Cosymplectic metrics on $\mathbb{F}(n)$

We are interested in studying cosymplectic metrics on $\mathbb{F}(n)$. San Martin and Negreiros [23, Proposition 7.3] proved that the metrics ds^2_Λ in the last theorem are also cosymplectic. In fact, they proved that every invariant metric ds^2_Λ on $\mathbb{F}(n)$ is cosymplectic.

The condition for ds^2_Λ to be cosymplectic is that the codifferential of the Kähler form $\delta\Omega$ is zero. However, San Martin and Negreiros did not calculate this codifferential, because they used another equivalent condition due to Gray and Hervella [11]. Here we use an explicit formula to calculate this codifferential; see Theorem 12.3.1.

Remember that, up to isomorphisms, the Kähler form can be written in the following way:

$$\Omega = -2\sqrt{-1}\sum_{i<j}\mu_{ij}\omega_{i\bar{j}} \wedge \omega_{\bar{i}j},$$

where $\mu_{ij} = \varepsilon_{ij}\lambda_{ij}$ and

$$\varepsilon_{ij} = \begin{cases} 1 & \text{if} \quad i \to j\,, \\ -1 & \text{if} \quad j \to i\,, \\ 0 & \text{if} \quad i = j\,. \end{cases}$$

Following the book by Griffiths and Harris [12], because $\omega_{i\bar{j}} \in \mathbb{C}^{1,0}$ we can write

$$\omega_{i\bar{j}} = \sum_k f_k^{ij} \mathrm{d}z_k\,, \tag{12.12}$$

where f_k^{ij} are complex functions. Then, applying conjugation, we have

$$\omega_{\bar{i}j} = \overline{\omega_{i\bar{j}}} = \sum_k \overline{f_k^{ij}} \mathrm{d}\bar{z}_k$$

and

$$\omega_{i\bar{j}} \wedge \omega_{\bar{i}j} = \sum_{k,l} f_k^{ij} \overline{f_l^{ij}} \mathrm{d}z_k \wedge \mathrm{d}\bar{z}_l\,.$$

So the Kähler form can be written in the following way:

$$\Omega = -2\sqrt{-1} \sum_{i<j} \mu_{ij} \left(\sum_{k,l} f_k^{ij} \overline{f_l^{ij}} \mathrm{d}z_k \wedge \mathrm{d}\bar{z}_l \right). \tag{12.13}$$

Using the definition of the codifferential and the definition of the Hodge star operator in [12], we can prove the following result:

Theorem 12.3.1 *The codifferential of the Kähler form is given by*

$$\delta\Omega = -2^{2+2N}\sqrt{-1} \sum_{i<j} \mu_{ij} \left\{ \sum_{k,l} \frac{\partial \left(f_k^{ij} \overline{f_l^{ij}} \right)}{\partial z_l} \mathrm{d}z_k - \frac{\partial \left(f_k^{ij} \overline{f_l^{ij}} \right)}{\partial \bar{z}_k} \mathrm{d}\bar{z}_l \right\}. \tag{12.14}$$

The proof of this theorem is included in the unpublished paper cited in the introduction. This theorem provides us the following result

Proposition 12.3.2 *A metric on $(\mathbb{F}(n), J)$ is cosymplectic if and only if the functions f_k^{ij} in the Kähler form satisfy the equation*

$$\sum_{i<j} \mu_{ij} \left\{ \sum_{k,l} \frac{\partial \left(f_k^{ij} \overline{f_l^{ij}} \right)}{\partial z_l} \mathrm{d}z_k - \frac{\partial \left(f_k^{ij} \overline{f_l^{ij}} \right)}{\partial \bar{z}_k} \mathrm{d}\bar{z}_l \right\} = 0. \tag{12.15}$$

Expanding the sums over k and l in (12.14) and reordering, we obtain

$$
\begin{aligned}
\delta\Omega &= -2^{2+2N}\sqrt{-1}\sum_{i<j}\mu_{ij}\left\{\sum_{k,l}\frac{\partial\left(f_k^{ij}\overline{f_l^{ij}}\right)}{\partial z_l}\mathrm{d}z_k - \frac{\partial\left(f_k^{ij}\overline{f_l^{ij}}\right)}{\partial \overline{z}_k}\mathrm{d}\overline{z}_l\right\} \\
&= -2^{2+2N}\sqrt{-1}\left[\sum_{i<j}\mu_{ij}\left\{\left(\frac{\partial\left(f_1^{ij}\overline{f_1^{ij}}\right)}{\partial z_1}+\cdots+\frac{\partial\left(f_1^{ij}\overline{f_N^{ij}}\right)}{\partial z_N}\right)\mathrm{d}z_1\right.\right. \\
&\quad + \left(\frac{\partial\left(f_2^{ij}\overline{f_1^{ij}}\right)}{\partial z_1}+\cdots+\frac{\partial\left(f_2^{ij}\overline{f_N^{ij}}\right)}{\partial z_N}\right)\mathrm{d}z_2+\cdots \\
&\quad \left. + \left(\frac{\partial\left(f_N^{ij}\overline{f_1^{ij}}\right)}{\partial z_1}+\cdots+\frac{\partial\left(f_N^{ij}\overline{f_N^{ij}}\right)}{\partial z_N}\right)\mathrm{d}z_N\right\} \\
&\quad - \sum_{i<j}\mu_{ij}\left\{\left(\frac{\partial\left(\overline{f_1^{ij}}f_1^{ij}\right)}{\partial \overline{z_1}}+\cdots+\frac{\partial\left(\overline{f_1^{ij}}f_N^{ij}\right)}{\partial \overline{z_N}}\right)\mathrm{d}\overline{z_1}\right. \\
&\quad + \left(\frac{\partial\left(\overline{f_2^{ij}}f_1^{ij}\right)}{\partial \overline{z_1}}+\cdots+\frac{\partial\left(\overline{f_2^{ij}}f_N^{ij}\right)}{\partial \overline{z_N}}\right)\mathrm{d}\overline{z_2}+\cdots \\
&\quad \left.\left. + \left(\frac{\partial\left(\overline{f_N^{ij}}f_1^{ij}\right)}{\partial \overline{z_1}}+\cdots+\frac{\partial\left(\overline{f_N^{ij}}f_N^{ij}\right)}{\partial \overline{z_N}}\right)\mathrm{d}\overline{z_N}\right\}\right].
\end{aligned}
$$

Then a metric on $(\mathbb{F}(n), J)$ is cosymplectic if and only if the functions f_k^{ij} in the Kähler form satisfy the following system of partial differential equations:

$$
\sum_{i<j}\mu_{ij}\left(\frac{\partial\left(f_k^{ij}\overline{f_1^{ij}}\right)}{\partial z_1}+\cdots+\frac{\partial\left(f_k^{ij}\overline{f_N^{ij}}\right)}{\partial z_N}\right)=0, \qquad k=1,\ldots,N, \quad (12.16)
$$

$$
\sum_{i<j}\mu_{ij}\left(\frac{\partial\left(\overline{f_1^{ij}}f_1^{ij}\right)}{\partial \overline{z_1}}+\cdots+\frac{\partial\left(\overline{f_1^{ij}}f_N^{ij}\right)}{\partial \overline{z_N}}\right)=0, \qquad k=1,\ldots,N. \quad (12.17)
$$

Actually, the equation (12.17) is the conjugate of the equation (12.16); thus we have the following result.

Proposition 12.3.3 *A metric on $(\mathbb{F}(n), J)$ is cosymplectic if and only if the functions f_k^{ij} in the Kähler form satisfy the system of partial differential* *equations*

$$\sum_{i<j} \mu_{ij} \left(\frac{\partial \left(f_k^{ij} \overline{f_1^{ij}} \right)}{\partial z_1} + \cdots + \frac{\partial \left(f_k^{ij} \overline{f_N^{ij}} \right)}{\partial z_N} \right) = 0, \qquad k = 1, \ldots, N. \quad (12.18)$$

Now, we can write the equation (12.14) in the following way:

$$\begin{aligned} \delta\Omega = & -2^{2+2N}\sqrt{-1} \sum_{i<j} \mu_{ij} \left\{ \sum_{k=1}^{N} \left(\frac{\partial \left(f_k^{ij} \overline{f_1^{ij}} \right)}{\partial z_1} + \cdots + \frac{\partial \left(f_k^{ij} \overline{f_N^{ij}} \right)}{\partial z_N} \right) \mathrm{d}z_k \right. \\ & \left. - \sum_{k=1}^{N} \left(\frac{\partial \left(\overline{f_1^{ij}} f_1^{ij} \right)}{\partial \overline{z_1}} + \cdots + \frac{\partial \left(\overline{f_1^{ij}} f_N^{ij} \right)}{\partial \overline{z_N}} \right) \mathrm{d}\overline{z_k} \right\} \\ = & -2^{2+2N}\sqrt{-1} \sum_{i<j} \mu_{ij} \sum_{k=1}^{N} \left\{ \left(\sum_{l=1}^{N} \frac{\partial}{\partial z_l} \left(f_k^{ij} \overline{f_l^{ij}} \right) \right) \mathrm{d}z_k - \left(\sum_{l=1}^{N} \frac{\partial}{\partial \overline{z_l}} \left(\overline{f_k^{ij}} f_l^{ij} \right) \mathrm{d}\overline{z_k} \right) \right\}. \end{aligned}$$

Because $z - \overline{z} = 2\sqrt{-1}\,\mathrm{Im}z$ for every complex number z, then

$$\begin{aligned} \delta\Omega & = -2^{2+2N}\sqrt{-1} \sum_{i<j} \mu_{ij} \sum_{k=1}^{N} \left\{ 2\sqrt{-1}\,\mathrm{Im} \left(\sum_{l=1}^{N} \frac{\partial}{\partial z_l} \left(f_k^{ij} \overline{f_l^{ij}} \right) \mathrm{d}z_k \right) \right\} \\ & = 2^{3+2N} \sum_{i<j} \mu_{ij} \left\{ \sum_{k,l=1}^{N} \mathrm{Im} \left(\frac{\partial}{\partial z_l} \left(f_k^{ij} \overline{f_l^{ij}} \right) \mathrm{d}z_k \right) \right\} \\ & = 2^{3+2N} \mathrm{Im} \left\{ \sum_{i<j} \mu_{ij} \left(\sum_{k,l=1}^{N} \left(\frac{\partial}{\partial z_l} \left(f_k^{ij} \overline{f_l^{ij}} \right) \mathrm{d}z_k \right) \right) \right\}. \end{aligned}$$

Then, we have the following proposition, equivalent to Propositions 12.3.2 and 12.3.3.

Proposition 12.3.4 *A metric on $(\mathbb{F}(n), J)$ is cosymplectic if and only if the functions f_k^{ij} in the Kähler form satisfy the equation*

$$\mathrm{Im} \left\{ \sum_{i<j} \mu_{ij} \left(\sum_{k,l=1}^{N} \left(\frac{\partial}{\partial z_l} \left(f_k^{ij} \overline{f_l^{ij}} \right) \mathrm{d}z_k \right) \right) \right\} = 0. \quad (12.19)$$

We can write this equation in real coordinates using the complex operators

$$\frac{\partial}{\partial z_i} = \frac{1}{2} \left(\frac{\partial}{\partial x_i} - \sqrt{-1} \frac{\partial}{\partial y_i} \right), \qquad \frac{\partial}{\partial \overline{z_i}} = \frac{1}{2} \left(\frac{\partial}{\partial x_i} + \sqrt{-1} \frac{\partial}{\partial y_i} \right)$$

and the complex differential forms

$$\mathrm{d}z_i = \mathrm{d}x_i + \sqrt{-1}\,\mathrm{d}y_i, \qquad \mathrm{d}\overline{z}_i = \mathrm{d}x_i - \sqrt{-1}\,\mathrm{d}y_i.$$

We have

$$\begin{aligned}
&\frac{\partial\left(f_k^{ij}\overline{f_l^{ij}}\right)}{\partial z_l}\mathrm{d}z_k - \frac{\partial\left(f_k^{ij}\overline{f_l^{ij}}\right)}{\partial \overline{z}_k}\mathrm{d}\overline{z}_l \\
&= \frac{1}{2}\left(\frac{\partial\left(f_k^{ij}\overline{f_l^{ij}}\right)}{\partial x_l} - \sqrt{-1}\frac{\partial\left(f_k^{ij}\overline{f_l^{ij}}\right)}{\partial y_l}\right)(\mathrm{d}x_k + \sqrt{-1}\,\mathrm{d}y_k) \\
&\quad -\frac{1}{2}\left(\frac{\partial\left(f_k^{ij}\overline{f_l^{ij}}\right)}{\partial x_k} + \sqrt{-1}\frac{\partial\left(f_k^{ij}\overline{f_l^{ij}}\right)}{\partial y_k}\right)(\mathrm{d}x_l - \sqrt{-1}\,\mathrm{d}y_l), \\
&= \frac{1}{2}\left\{\left(\frac{\partial\left(f_k^{ij}\overline{f_l^{ij}}\right)}{\partial x_l}\mathrm{d}x_k + \frac{\partial\left(f_k^{ij}\overline{f_l^{ij}}\right)}{\partial y_l}\mathrm{d}y_k\right) + \sqrt{-1}\left(\frac{\partial\left(f_k^{ij}\overline{f_l^{ij}}\right)}{\partial x_l}\mathrm{d}y_k - \frac{\partial\left(f_k^{ij}\overline{f_l^{ij}}\right)}{\partial y_l}\mathrm{d}x_k\right)\right\} \\
&\quad -\frac{1}{2}\left\{\left(\frac{\partial\left(f_k^{ij}\overline{f_l^{ij}}\right)}{\partial x_k}\mathrm{d}x_l + \frac{\partial\left(f_k^{ij}\overline{f_l^{ij}}\right)}{\partial y_k}\mathrm{d}y_l\right) + \sqrt{-1}\left(\frac{\partial\left(f_k^{ij}\overline{f_l^{ij}}\right)}{\partial y_k}\mathrm{d}x_l - \frac{\partial\left(f_k^{ij}\overline{f_l^{ij}}\right)}{\partial x_k}\mathrm{d}y_l\right)\right\} \\
&= \frac{1}{2}\left\{\left(\frac{\partial\left(f_k^{ij}\overline{f_l^{ij}}\right)}{\partial x_l}\mathrm{d}x_k - \frac{\partial\left(f_k^{ij}\overline{f_l^{ij}}\right)}{\partial x_k}\mathrm{d}x_l\right) + \left(\frac{\partial\left(f_k^{ij}\overline{f_l^{ij}}\right)}{\partial y_l}\mathrm{d}y_k - \frac{\partial\left(f_k^{ij}\overline{f_l^{ij}}\right)}{\partial y_k}\mathrm{d}y_l\right)\right\} \\
&\quad +\frac{\sqrt{-1}}{2}\left\{\left(\frac{\partial\left(f_k^{ij}\overline{f_l^{ij}}\right)}{\partial x_l}\mathrm{d}y_k - \frac{\partial\left(f_k^{ij}\overline{f_l^{ij}}\right)}{\partial y_k}\mathrm{d}x_l\right) + \left(\frac{\partial\left(f_k^{ij}\overline{f_l^{ij}}\right)}{\partial x_k}\mathrm{d}y_l - \frac{\partial\left(f_k^{ij}\overline{f_l^{ij}}\right)}{\partial y_l}\mathrm{d}x_k\right)\right\}.
\end{aligned}$$

Thus a metric on $(\mathbb{F}(n), J)$ is cosymplectic if and only if the functions f_k^{ij} in the Kähler form satisfy the following equations:

$$\sum_{i<j}\mu_{ij}\left\{\sum_{k,l}\left(\frac{\partial\left(f_k^{ij}\overline{f_l^{ij}}\right)}{\partial x_l}\mathrm{d}x_k - \frac{\partial\left(f_k^{ij}\overline{f_l^{ij}}\right)}{\partial x_k}\mathrm{d}x_l\right) + \left(\frac{\partial\left(f_k^{ij}\overline{f_l^{ij}}\right)}{\partial y_l}\mathrm{d}y_k - \frac{\partial\left(f_k^{ij}\overline{f_l^{ij}}\right)}{\partial y_k}\mathrm{d}y_l\right)\right\} = 0, \tag{12.20}$$

$$\sum_{i<j}\mu_{ij}\left\{\sum_{k,l}\left(\frac{\partial\left(f_k^{ij}\overline{f_l^{ij}}\right)}{\partial x_l}\mathrm{d}y_k - \frac{\partial\left(f_k^{ij}\overline{f_l^{ij}}\right)}{\partial y_k}\mathrm{d}x_l\right) + \left(\frac{\partial\left(f_k^{ij}\overline{f_l^{ij}}\right)}{\partial x_k}\mathrm{d}y_l - \frac{\partial\left(f_k^{ij}\overline{f_l^{ij}}\right)}{\partial y_l}\mathrm{d}x_k\right)\right\} = 0. \tag{12.21}$$

These sums are calculated over all k and l; therefore the left side of the equation (12.20) vanishes. So, we have the following result.

Proposition 12.3.5 *A metric on* $(\mathbb{F}(n), J)$ *is cosymplectic if and only if the functions* f_k^{ij} *in the Kähler form satisfy the following equation:*

$$\sum_{i<j} \mu_{ij} \left\{ \sum_{k,l} \left(\frac{\partial \left(f_k^{ij} \overline{f_l^{ij}} \right)}{\partial x_l} \mathrm{d}y_k - \frac{\partial \left(f_k^{ij} \overline{f_l^{ij}} \right)}{\partial y_k} \mathrm{d}x_l \right) + \left(\frac{\partial \left(f_k^{ij} \overline{f_l^{ij}} \right)}{\partial x_k} \mathrm{d}y_l - \frac{\partial \left(f_k^{ij} \overline{f_l^{ij}} \right)}{\partial y_l} \mathrm{d}x_k \right) \right\} = 0.$$

Bibliography

[1] M. Black. *Harmonic maps into homogeneous spaces* (Essex: Pitman Research Notes in Mathematics Series 255, Longman Scientific and Technical, 1991).

[2] A. Borel. Kählerian coset spaces of semi-simple Lie groups. *Proc. Natl. Acad. Sci. USA*, **40** (1954), 1147–1151.

[3] A. Borel and F. Hirzebruch. Characteristic classes and homogeneous spaces I. *Amer. J. Math.*, **80** (1958), 458–538.

[4] A. E. Brouwer. The enumeration of locally transitive tournaments. *Afdeling Zuivere Wiskunde [Department of Pure Mathematics]*, **138** (1980), Mathematisch Centrum, Amsterdam.

[5] F. E. Burstall and S. Salamon. Tournaments, flags and harmonic maps. *Math. Ann.*, **277** (1987), 249–265.

[6] G. Caristi and M. Ferrara. On a class of almost cosymplectic manifolds. *Differential Geometry – Dynamical Systems*, **4** (2002), 1–4.

[7] S. S. Chern and J. G. Wolfson. Harmonic maps of the two-sphere into a complex Grassmann manifold II. *Ann. of Math.*, **125** (1987), 301–335.

[8] N. Cohen, C. J. C. Negreiros, M. Paredes, S. Pinzón and L. A. B. San Martin. f-Structures on the classical flag manifold which admit (1, 2)-symplectic metrics. *Tohoku Math. J. (2)*, **57** (2005), 261–271.

[9] N. Cohen, C. J. C. Negreiros and L. A. B. San Martin. (1, 2)-symplectic metrics, flag manifolds and tournaments. *Bull. London Math. Soc.*, **34** (2002), 641–649.

[10] N. Cohen, M. Paredes and S. Pinzón. Locally transitive tournaments and the classification of (1, 2)-symplectic metrics on maximal flag manifolds. *Illinois J. Math.*, **48** (2004), 1405–1415.

[11] A. Gray and L. M. Hervella. The sixteen classes of almost Hermitian manifolds and their linear invariants. *Ann. Mat. Pura Appl. (4)*, **123** (1980), 35–58.

[12] P. Griffiths and J. Harris, *Principles of algebraic geometry* (New York: John Wiley & Sons, 1978).

[13] A. Lichnerowicz. Applications harmoniques et variétés kählériennes. *Sympos. Math.* **3** (1970), 341–402.

[14] I. Mihai, R. Rosca and L. Verstraelen. Some aspects of differential geometry of vector fields. In *Proceedings of PADGE*, vol. 2, K. U. Leuven and K. U. Brussel (1996).

[15] J. W. Moon, *Topics on tournaments* (New York: Holt, Rinehart and Winston, 1968).

[16] X. Mo and C. J. C. Negreiros. (1, 2)-Symplectic structures on flag manifolds. *Tohoku Math. J. (2)*, **52** (2000), 271–282.

[17] C. J. C. Negreiros. Some remarks about harmonic maps into flag manifolds. *Indiana Univ. Math. J.*, **37** (1988), 617–636.

[18] M. Paredes. Aspectos da geometria complexa das variedades bandeira. Ph.D. thesis, State University of Campinas, Brazil (2000).

[19] M. Paredes. Some results on the geometry of full flag manifolds and harmonic maps. *Rev. Colombiana Mat.*, **34** (2000), 57–89.
[20] M. Paredes. Families of (1, 2)-symplectic metrics on full flag manifolds. *Int. J. Math. Math. Sci.*, **29** (2002), 651–664.
[21] M. Paredes and S. Pinzón. (In Press) On the codifferential of the Kähler form and cosymplectic metrics on maximal flag manifolds. *Turkish J. Math.*
[22] M. Paredes and S. Pinzón. Geometry of flag manifolds. *Rev. Acad. Colombiana Cienc. Exact. Fís. Natur.*, **28** (2004), 123–134.
[23] L. A. B. San Martin and C. J. C. Negreiros. Invariant almost Hermitian structures on flag manifolds. *Adv. Math.*, **178** (2003), 277–310.
[24] K. Yano. On a structure defined by a tensor field of type (1, 1) satisfying $\mathcal{F}^3 + \mathcal{F} = 0$. *Tensor*, **14** (1963), 99–109.

13

Heisenberg modules over real multiplication noncommutative tori and related algebraic structures

JORGE PLAZAS*

Abstract

I review some aspects of the theory of noncommutative two-tori with real multiplication, focusing on the role played by Heisenberg groups in the definition of algebraic structures associated to these noncommutative spaces.

13.1 Introduction

Noncommutative tori have played a central role in noncommutative geometry since the early stages of the theory. They arise naturally in various contexts and have provided a good testing ground for many of the techniques from which noncommutative geometry has developed [1, 14]. Noncommutative tori are defined in terms of their algebras of functions. The study of projective modules over these algebras and the corresponding theory of Morita equivalences leads to the existence of a class of noncommutative tori related to real quadratic extensions of $\mathbb{Q}$. These real multiplication noncommutative tori are conjectured to provide the correct geometric setting under which to attack the explicit class field theory problem for real quadratic fields [7]. The right understanding of the algebraic structures underlying these spaces is important for these applications.

The study of connections on vector bundles over noncommutative tori gives rise to a rich theory, which has been recast recently in the context of complex algebraic geometry [1, 3, 5, 11, 12, 16]. The study of categories of holomorphic bundles has thrown light on some algebraic structures related to real multiplication noncommutative tori [9, 10, 18]. Some of these results arise in a natural way from the interplay between Heisenberg groups and noncommutative tori.

* I want to thank the organizers of the 2007 summer school "Geometric and topological methods for quantum field theory" and I.H.E.S. for their support and hospitality. This work was supported in part by ANR-Galois grant NT05-2 44266.

13.2 Noncommutative tori and their morphisms

In many situations arising in various geometric settings it is possible to characterize spaces and some of their structural properties in terms of appropriate rings of functions. One instance of this duality is provided by Gelfand's theorem, which identifies the category of locally compact Hausdorff topological spaces with the category commutative C^*-algebras. This correspondence assigns to a space X the commutative algebra $C_0(X)$ consisting of complex-valued continuous functions on X vanishing at infinity. Topological invariants of the space X can be obtained by the corresponding invariants of $C_0(X)$ defined in the context of C^*-algebras. If X is a smooth manifold, the smooth structure on X singles out the $*$-subalgebra $C_0^\infty(X)$ consisting of smooth elements of $C_0(X)$. Considering the space X in the framework of differential topology leads to structures defined in terms of the algebra $C_0^\infty(X)$.

Various geometric notions which can be defined in terms of rings of functions on a space do not depend on the fact that the rings under consideration are commutative, and can therefore be extended in order to consider noncommutative rings and algebras. In noncommutative geometry, spaces are defined in terms of their rings of functions, which are noncommutative analogs of commutative rings of functions. This passage is far from being just a translation of classical ideas to a noncommutative setting. Many extremely rich new phenomena arise in this context (see [2]). The noncommutative setting also enriches the classical picture in that noncommutative rings may arise in a natural way from classical geometric considerations.

We will be considering noncommutative analogs of the two-torus $\mathbb{T}^2 = S^1 \times S^1$. The reader may consult [1, 2, 4, 6, 7, 14, 15] for the proofs of the results on noncommutative tori mentioned in this section.

13.2.1 The C^*-algebra A_θ

Under the Gelfand correspondence, compact spaces correspond to commutative unital C^*-algebras and $\mathbb{T}^2$ is dual to $C(\mathbb{T}^2)$. At a topological level a noncommutative two-torus is defined in terms of a unital noncommutative C^*-algebra, which plays the role of its algebra of continuous functions.

Given $\theta \in \mathbb{R}$, let A_θ be the universal C^*-algebra generated by two unitaries U and V subject to the relation

$$UV = e^{2\pi\imath\theta} VU.$$

Then:

(i) If $\theta \in \mathbb{Z}$, the algebra A_θ is isomorphic to $C(\mathbb{T}^2)$.
(ii) If $\theta \in \mathbb{Q}$, the algebra A_θ is isomorphic to the algebra of global sections of the endomorphism bundle of a complex vector bundle over $\mathbb{T}^2$.
(iii) If $\theta \in \mathbb{R} \setminus \mathbb{Q}$, the algebra A_θ is a simple C^*-algebra.

For irrational values of θ we will refer to A_θ as *the algebra of continuous functions on the noncommutative torus* $\mathbb{T}^2_\theta$. Thus, as a topological space the noncommutative torus $\mathbb{T}^2_\theta$ is defined as the dual object of $C(\mathbb{T}^2_\theta) := A_\theta$.

There is a natural continuous action of the compact group $\mathbb{T}^2$ on the algebra A_θ. This action can be given in terms of the generators U and V by

$$\alpha_\varphi(U) = e^{2\pi\imath\varphi_1}U,$$
$$\alpha_\varphi(V) = e^{2\pi\imath\varphi_2}V,$$

where $\varphi = (\varphi_1, \varphi_2) \in \mathbb{T}^2$.

One of the main structural properties of the algebra A_θ is the existence of a canonical trace whose value at each element is given by the average over $\mathbb{T}^2$ of that action.

Theorem 13.2.1 *Let θ be an irrational number. Then there exist a unique normalized trace*

$$\chi : A_\theta \to \mathbb{C}$$

invariant under the action of $\mathbb{T}^2$.

For the remaining part of the chapter θ will denote an irrational number. Also, for any complex number $z \in \mathbb{C}$ we will use the notation

$$\mathrm{e}(z) = \exp(2\pi\imath z), \qquad \bar{\mathrm{e}}(z) = \exp(-2\pi\imath z).$$

13.2.2 Smooth elements

The action of $\mathbb{T}^2$ on A_θ induces a smooth structure on the noncommutative torus $\mathbb{T}^2_\theta$. An element $a \in A_\theta$ is called smooth if the map

$$\mathbb{T}^2 \longrightarrow A_\theta,$$
$$\varphi \longmapsto \alpha_\varphi(a)$$

is smooth. The set of smooth elements of A_θ is a dense $*$-subalgebra, which we denote by $\mathcal{A}_\theta$. Elements in this subalgebra should be thought of as smooth functions on the noncommutative torus $\mathbb{T}^2_\theta$; thus we take $C^\infty(\mathbb{T}^2_\theta) := \mathcal{A}_\theta$. The algebra $\mathcal{A}_\theta$ can be characterized in the following way:

$$\mathcal{A}_\theta = \left\{ \sum_{n,m\in\mathbb{Z}} a_{n,m} U^n V^m \in A_\theta \mid \{a_{n,m}\} \in \mathcal{S}(\mathbb{Z}^2) \right\},$$

where $\mathcal{S}(\mathbb{Z}^2)$ denotes the space of sequences of rapid decay in $\mathbb{Z}^2$. In the algebra $\mathcal{A}_\theta$ the trace χ is given by

$$\chi\left(\sum a_{n,m}U^nV^m\right) = a_{0,0}.$$

The Lie algebra $L = \mathbb{R}^2$ of $\mathbb{T}^2$ acts on $\mathcal{A}_\theta$ by derivations. A basis for this action is given by the derivations

$$\delta_1(U) = 2\pi\imath U, \qquad \delta_1(V) = 0,$$
$$\delta_2(U) = 0, \qquad \delta_2(V) = 2\pi\imath V.$$

A complex parameter $\tau \in \mathbb{C} \setminus \mathbb{R}$ induces a complex structure on $L = \mathbb{R}^2$ given by the isomorphism

$$\mathbb{R}^2 \longrightarrow \mathbb{C},$$
$$x = (x_1, x_2) \longmapsto \tilde{x} = \tau x_1 + x_2.$$

The corresponding complex structure on $\mathcal{A}_\theta$ is given by the derivation

$$\delta_\tau = \tau\delta_1 + \delta_2.$$

13.2.3 Vector bundles and K-theory

If X is a smooth compact manifold, the space of smooth sections of a vector bundle over X is a finite-type projective module over $C^\infty(X)$, and any such module arises in this way. In our setting, finite-type projective right $\mathcal{A}_\theta$-modules will play the role of vector bundles over the noncommutative torus $\mathbb{T}^2_\theta$.

As before, we denote by $L = \mathbb{R}^2$ the Lie algebra of $\mathbb{T}^2$ acting as an algebra of derivations on $\mathcal{A}_\theta$. If P is a finite-type projective right $\mathcal{A}_\theta$-module, a connection on P is given by an operator

$$\nabla : P \to P \otimes L^*$$

such that

$$\nabla_X(\xi a) = \nabla_X(\xi)a + \xi\delta_X a$$

for all $X \in L, \xi \in P$ and $a \in \mathcal{A}_\theta$. The connection ∇ is determined the operators

$$\nabla_i : P \to P, \qquad i = 1, 2,$$

giving its values on the basis elements δ_1, δ_2 of L.

The K_0 group of $\mathcal{A}_\theta$ is by definition the enveloping group of the abelian semigroup given by isomorphism classes of finite-type projective right $\mathcal{A}_\theta$-modules together with the direct sum. The trace χ extends to an injective morphism

$$\mathrm{rk} : K_0(\mathcal{A}_\theta) \to \mathbb{R}$$

whose image is

$$\Gamma_\theta = \mathbb{Z} \oplus \theta\mathbb{Z}.$$

13.2.4 Morphisms of noncommutative tori

Because noncommutative tori are defined in terms of their function algebras, one should expect a morphism $\mathbb{T}^2_\theta \to \mathbb{T}^2_{\theta'}$ between two noncommutative tori $\mathbb{T}^2_\theta$ and $\mathbb{T}^2_{\theta'}$ to be given by a morphism $\mathcal{A}_{\theta'} \to \mathcal{A}_\theta$ of the corresponding algebras of functions. It turns out that algebra morphisms are in general insufficient to describe the type of situations arising in noncommutative geometry. The right notion of morphisms in our setting is given by Morita equivalences. A Morita equivalence between $\mathcal{A}_{\theta'}$ and $\mathcal{A}_\theta$ is given by the isomorphism class of an $\mathcal{A}_{\theta'}$–$\mathcal{A}_\theta$ bimodule E which is projective and of finite type both as a left $\mathcal{A}_{\theta'}$-module and as a right $\mathcal{A}_\theta$-module. If such bimodule exists, we say that $\mathcal{A}_{\theta'}$ and $\mathcal{A}_\theta$ are Morita equivalent. We can consider a Morita equivalence between $\mathcal{A}_{\theta'}$ and $\mathcal{A}_\theta$ as a morphism between $\mathcal{A}_{\theta'}$ and $\mathcal{A}_\theta$ inducing a morphism between $\mathbb{T}^2_\theta$ and $\mathbb{T}^2_{\theta'}$. Composition of morphisms is provided by the tensor product of modules.

Let $SL_2(\mathbb{Z})$ act on $\mathbb{R} \setminus \mathbb{Q}$ by fractional linear transformations, i.e., given

$$g = \begin{pmatrix} a & b \\ c & d \end{pmatrix} \in SL_2(\mathbb{Z}), \qquad \theta \in \mathbb{R} \setminus \mathbb{Q},$$

we take

$$g\theta = \frac{a\theta + b}{c\theta + d}.$$

Morita equivalences between noncommutative tori are characterized by the following result:

Theorem 13.2.2 (Rieffel [13]) *Let θ', $\theta \in \mathbb{R} \setminus \mathbb{Q}$. Then the algebras $\mathcal{A}_{\theta'}$ and $\mathcal{A}_\theta$ are Morita equivalent if and only if there exists a matrix $g \in SL_2(\mathbb{Z})$ such that $\theta' = g\theta$.*

In Section 13.4 we will construct explicit bimodules realizing this equivalences. In what follows, whenever we refer to a right $\mathcal{A}_\theta$-module (left $\mathcal{A}_{\theta'}$-module, $\mathcal{A}_{\theta'}$–$\mathcal{A}_\theta$ bimodule), we mean a projective and finite-type right $\mathcal{A}_\theta$-module (left $\mathcal{A}_{\theta'}$-module, $\mathcal{A}_{\theta'}$–$\mathcal{A}_\theta$ bimodule)

Given a irrational number θ, an $\mathcal{A}_\theta$–$\mathcal{A}_\theta$ bimodule E induces Morita self-equivalences of $\mathcal{A}_\theta$. We denote by $End_{Morita}(\mathcal{A}_\theta)$ the group of Morita self-equivalences of $\mathcal{A}_\theta$. For example, given any positive integer n, the free bimodule

$\mathcal{A}_\theta^n$ induces a Morita self-equivalence of $\mathcal{A}_\theta$. A Morita self-equivalence defined via a free module is called a trivial Morita self-equivalence.

A Morita self-equivalence of $\mathcal{A}_\theta$ given by an $\mathcal{A}_\theta$–$\mathcal{A}_\theta$ bimodule E defines an endomorphism of $K_0(\mathcal{A}_\theta)$ via

$$\phi_E : [P] \mapsto [P \otimes_{\mathcal{A}_\theta} E]$$

for P a right projective finite-rank $\mathcal{A}_\theta$ module and $[P] \in K_0(\mathcal{A}_\theta)$ its K-theory class.

Via the map rk, the endomorphism ϕ_M becomes multiplication by a real number. Thus we get a map

$$\phi : End_{Morita}(\mathcal{A}_\theta) \to \{\alpha \in \mathbb{R} \mid \alpha\Gamma_\theta \subset \Gamma_\theta\}.$$

This map turns out to be surjective.

We can summarize the situation as follows (see [7]):

Theorem 13.2.3 *Let $\theta \in \mathbb{R}$ be irrational. The following conditions are equivalent:*

- *$\mathcal{A}_\theta$ has nontrivial Morita autoequivalences.*
- *$\phi(End_{Morita}(\mathcal{A}_\theta)) \neq \mathbb{Z}$.*
- *There exists a matrix $g \in SL_2(\mathbb{Z})$ such that*

$$\theta = g\theta.$$

- *θ is a real quadratic irrationality:*

$$[\mathbb{Q}(\theta) : \mathbb{Q}] = 2.$$

If any of these equivalent conditions holds, we say that the noncommutative torus $\mathbb{T}_\theta^2$ with algebra of smooth functions $\mathcal{A}_\theta$ is a *real multiplication noncommutative torus*. If $\mathbb{T}_\theta^2$ is a real multiplication noncommutative torus, then

$$\begin{aligned} \phi(End_{Mor}(\mathcal{A}_\theta)) &= \{\alpha \in \mathbb{R} \mid \alpha\Gamma_\theta \subset \Gamma_\theta\} \\ &= \mathbb{Z} + f\mathcal{O}_k, \end{aligned}$$

where $f \geq 1$ is an integer and $\mathcal{O}_k$ is the ring of integers of the real quadratic field $k = \mathbb{Q}(\theta)$.

These results should be compared with the analogous results for elliptic curves leading to the theory of complex multiplication. The strong analogy suggests that noncommutative tori may play a role in number theory similar to the role played by elliptic curves. In particular, noncommutative tori with real multiplication could give the right geometric framework to attack the explicit class field theory problem for real quadratic fields (see [7]).

Let $\tau \in \mathbb{C} \setminus \mathbb{R}$. Consider the lattice $\Lambda_\tau = \mathbb{Z} \oplus \tau\mathbb{Z}$ and the elliptic curve

$$X_\tau = \mathbb{C}/\Lambda_\tau.$$

The following conditions are equivalent:

- $End(X_\tau) \neq \mathbb{Z}$.
- τ generates a quadratic extension:

$$[\mathbb{Q}(\tau) : \mathbb{Q}] = 2.$$

In this case we have

$$\begin{aligned} End(X_\tau) &= \{\alpha \in \mathbb{C} \mid \alpha\Lambda_\tau \subset \Lambda_\tau\} \\ &= \mathbb{Z} + f\mathcal{O}_k, \end{aligned}$$

where $f \geq 1$ is an integer and $\mathcal{O}_k$ is the ring of integers of the imaginary quadratic field $k = \mathbb{Q}(\tau)$.

13.3 Heisenberg groups and their representations

Various aspects of the theory of representations of Heisenberg groups arise naturally when considering geometric constructions associated to noncommutative tori. This fact gives relations between noncommutative tori and elliptic curves through theta functions and plays a useful role in the study of the arithmetic nature of related algebraic structures. In this section I sketch the parts of the theory of Heisenberg groups that are relevant in order to describe these results. I follow Mumford's Tata lectures [8], which I also recommend as a reference for the material in this section.

Let G be a locally compact group lying in a central extension:

$$1 \to \mathbb{C}_1^* \to G \to K \to 0,$$

where $\mathbb{C}_1^*$ is the group of complex numbers of modulus 1, and K is a locally compact abelian group. Assume that the exact sequence splits, so as a set $G = \mathbb{C}_1^* \times K$ and the group structure is given by

$$(\lambda, x)(\mu, y) = (\lambda\mu\psi(x, y), x + y),$$

where $\psi : K \times K \to \mathbb{C}_1^*$ is a two-cocycle in K with values in $\mathbb{C}_1^*$. The cocycle ψ induces a skew multiplicative pairing

$$\begin{aligned} e : \ K \times K &\longrightarrow \mathbb{C}_1^*, \\ (x, y) &\longmapsto \frac{\psi(x, y)}{\psi(y, x)}. \end{aligned}$$

This pairing defines a group morphism $\varphi : K \to \widehat{K}$ from K to its Pontrjagin dual given by $\phi(x)(y) = e(x, y)$.

Definition 13.3.1 If φ is a isomorphism, we say that G is a *Heisenberg group.*

For a Heisenberg group G lying in a central extension we use the notation $G = \mathrm{Heis}(K)$. The main theorem about the representation of Heisenberg groups states that groups of this kind admit a unique normalized irreducible representation which can be realized in terms of a maximal isotropic subgroup of K. A subgroup H of K is called *isotropic* if $e|_{H\times H} \equiv 1$; this is equivalent to the existence of a section of G over H:

$$\sigma : K \longrightarrow G,$$
$$x \longmapsto (\alpha(x), x).$$

We say that a subgroup H of K is *maximal isotropic* if it is maximal with this property. A subgroup H of K is maximal isotropic if and only if $H = H^{\perp}$, where for $S \subset H$ we have

$$S^{\perp} = \{x \in K \mid e(x, y) = 1 \text{ for all } y \in S\}.$$

Theorem 13.3.2 (Stone, von Neumann, Mackey) *Let G be a Heisenberg group. Then:*

- *G has a unique irreducible unitary representation in which $\mathbb{C}_1^*$ acts by multiples of the identity.*
- *Given a maximal isotropic subgroup $H \subset K$ and a splitting σ as described in the preceding, let $\mathcal{H} = \mathcal{H}_H$ be the space of measurable functions $f : K \to \mathbb{C}$ satisfying*
  (i) $f(x + h) = \alpha(h)\psi(h, x)^{-1} f(x)$ for all $h \in H$,
  (ii) $\int_{K/H} |f(x)|^2 dx < \infty$.

Then G acts on $\mathcal{H}$ by

$$U_{(\lambda,y)} f(x) = \lambda\psi(x, y) f(x + y),$$

and $\mathcal{H}$ is an irreducible unitary representation of G.

We call such representation a *Heisenberg representation of G*. The following theorem will be useful later:

Theorem 13.3.3 *Given two Heisenberg groups*

$$1 \to \mathbb{C}_1^* \to G_i \to K_i \to 0, \qquad i = 1, 2,$$

with Heisenberg representations $\mathcal{H}_1$ and $\mathcal{H}_2$, then

$$1 \to \mathbb{C}_1^* \to G_1 \times G_2/\{(\lambda, \lambda^{-1}) | \lambda \in \mathbb{C}_1^*\} \to K_1 \times K_2 \to 0$$

is a Heisenberg group, and its Heisenberg representation is $\mathcal{H}_1 \hat{\otimes} \mathcal{H}_2$.

13.3.1 Real Heisenberg groups

Let $K = \mathbb{R}^2$, and let ε be a positive real number. We endow $G = K \times \mathbb{C}_1^*$ with the structure of a Heisenberg group defined by the cocycle ψ and the pairing e given by

$$\psi(x, y) = \mathrm{e}\left(\frac{1}{\varepsilon}\frac{(x_1 y_2 - y_1 x_2)}{2}\right),$$

$$e(x, y) = \mathrm{e}\left(\frac{1}{\varepsilon}(x_1 y_2 - y_1 x_2)\right),$$

where $x = (x_1, x_2),\ y = (y_1, y_2) \in K$.

If we choose as maximal isotropic subgroup $H = \{x = (x_1, x_2) \in K \mid x_2 = 0\}$, then the values of the functions in the corresponding Heisenberg representation (Theorem 13.3.2) are determined by their values on $\{x = (x_1, x_2) \in K \mid x_1 = 0\}$, and we may identify the space $\mathcal{H}_H$ with $L^2(\mathbb{R})$. The action of G is given by

$$U_{(\lambda, y)} f(x) = \lambda \mathrm{e}\left(\frac{1}{\varepsilon}\left(x y_2 + \frac{y_1 y_2}{2}\right)\right) f(x + y_1)$$

for $(\lambda, y) = (\lambda, (y_1, y_2)) \in G$ and $f \in L^2(\mathbb{R})$. In particular we have

$$U_{(1,(y_1,0))} f(x) = f(x + y_1),$$

$$U_{(1,(0,y_2))} f(x) = \mathrm{e}\left(\frac{1}{\varepsilon} x y_2\right) f(x).$$

We will denote this Heisenberg representation by $\mathcal{H}_\varepsilon$.

For any $X \in Lie(G)$ and any Heisenberg representation $\mathcal{H}$ there is a dense subset of elements $f \in \mathcal{H}$ for which the limit

$$\delta U_X(f) = \lim_{t \to 0} \frac{U_{\exp(tX)} f - f}{t}$$

exists. This formula for δU_X defines an unbounded operator on that set. An element $f \in \mathcal{H}$ is a smooth element for the representation $\mathcal{H}$ of G if

$$\delta U_{X_1} \delta U_{X_2} \cdots \delta U_{X_n}(f)$$

is well defined for any n and any $X_1, X_2, \ldots, X_n \in Lie(G)$. The set of smooth elements of $\mathcal{H}$ is denoted by $\mathcal{H}_\infty$. We may realize $Lie(G)$ as an algebra of operators on $\mathcal{H}_\infty$. If choose a basis $\{A, B, C\}$ for the Lie algebra $Lie(G)$ such that

$$\exp(tA) = (1, (t, 0)), \quad \exp(tB) = (1, (0, t)), \quad \exp(tC) = (\mathrm{e}(t), (0, 0)),$$

then a complex number $\tau \in \mathbb{C}$ with nonzero imaginary part gives a decomposition of $Lie(G) \otimes \mathbb{C}$ into conjugate abelian complex subalgebras:

$$W_\tau = \langle \delta U_A - \tau \delta U_B \rangle,$$
$$W_{\bar{\tau}} = \langle \delta U_A - \bar{\tau} \delta U_B \rangle.$$

Theorem 13.3.4 *Fix $\tau \in \mathbb{C}$ with $\mathrm{Im}(\tau) > 0$. Then in any Heisenberg representation of G there exists an element f_τ, unique up to a scalar, such that $\delta U_X(f_\tau)$ is defined and equal to 0 for all $X \in W_\tau$.*

In the Heisenberg representation $\mathcal{H}_\varepsilon$ we have

$$\delta U_A f(x) = \frac{d}{dx} f(x),$$
$$\delta U_B f(x) = \frac{2\pi \imath x}{\varepsilon} f(x),$$
$$\delta U_C f(x) = 2\pi \imath \, f(x),$$

and $\mathcal{H}_{\varepsilon,\infty}$ is the Schwartz space $\mathcal{S}(\mathbb{R})$. The element f_τ in Theorem 13.3.4 is given by

$$f_\tau = \mathrm{e}\left(\frac{1}{2\varepsilon} \tau x^2\right).$$

The complex parameter τ induces a complex structure on $\mathbb{R}^2$ given by the isomorphism

$$\mathbb{R}^2 \longrightarrow \mathbb{C},$$
$$x = (x_1, x_2) \longmapsto \tilde{x} = \tau x_1 + x_2.$$

We can realize a representation of G which is canonically dual to the Heisenberg representation $\mathcal{H}_1$ in terms of this complex structure.

Theorem 13.3.5 *Let $\tau \in \mathbb{C}$ with $\mathrm{Im}(\tau) > 0$, and let $\mathcal{H}_\tau$ be the Hilbert space of holomorphic functions h on $\mathbb{C}$ with*

$$\int_{\mathbb{C}} |h(\tilde{x})|^2 e^{-2\pi \mathrm{Im}(\tau) x^2} dx_1 dx_2 < \infty.$$

Then

$$U_{(\lambda,y)} h(\tilde{x}) = \lambda^{-1} \mathrm{e}\left(\frac{1}{\varepsilon}\left(y_1 \tilde{x} + \frac{y_1 \tilde{y}}{2}\right)\right) h(\tilde{x} + \tilde{y})$$

defines an irreducible unitary representation of G.

We call this representation the *Fock representation of G*.

13.3.2 Heis($(\mathbb{Z}/c\mathbb{Z})^2$)

Let c be a positive integer, and let $K = (\mathbb{Z}/c\mathbb{Z})^2$. We endow $G = K \times \mathbb{C}_1^*$ with the structure of a Heisenberg group defined by the cocycle ψ and the pairing e given by

$$\psi(([n_1], [n_2]), ([m_1], [m_2])) = \mathrm{e}\left(\frac{1}{2c}(n_1 m_2 - m_1 n_2)\right),$$

$$e(([n_1], [n_2]), ([m_1], [m_2])) = \mathrm{e}\left(\frac{1}{c}(n_1 m_2 - m_1 n_2)\right),$$

where $([n_1], [n_2]), ([m_1], [m_2]) \in K$.

If we choose as maximal isotropic subgroup $H = \{([n_1], [n_2]) \in K \mid [n_2] = 0\}$, we may realize the Heisenberg representation as the action of G on $C(\mathbb{Z}/c\mathbb{Z})$ given by

$$U_{(\lambda,([m_1],[m_2]))}\phi([n]) = \lambda \mathrm{e}\left(\frac{1}{c}\left(nm_2 + \frac{m_1 m_2}{2}\right)\right)\phi([n + m_1])$$

for $(\lambda, ([m_1], [m_2])) \in G$ and $\phi \in C(\mathbb{Z}/c\mathbb{Z})$. In particular we have

$$U_{(1,([m_1],0))}\phi([n]) = \phi([n + m_2]),$$

$$U_{(1,(0,[m_2]))}\phi([n]) = \mathrm{e}\left(\frac{1}{c}nm_2\right) f([n]).$$

Remark 13.3.6 This type of Heisenberg groups is related to *algebraic Heisenberg groups* or, more generally, *Heisenberg group schemes*. These are given by central extensions of the form

$$1 \to \mathbb{G}_m \to \mathcal{G} \to \mathcal{K} \to 0,$$

where $\mathcal{K}$ is a finite abelian group scheme aver a base field k. These groups arise in a natural way on considering ample line bundles on abelian varieties over the base field k. The corresponding Heisenberg representations can be realized as canonical actions of $\mathcal{G}$ on the spaces of sections of these bundles. The action of $Gal(\bar{k}/k)$ on the geometric points of $\mathcal{G}$ implies important algebraicity results about these representations. The abelian varieties that play a role in the constructions that follow are the elliptic curves whose period lattice is spanned by the parameter τ which defines the complex structure on the noncommutative torus.

The relevant finite groups that arise in this context are given by the automorphisms of the induced by a translation on the elliptic curve, and the space of sections on which the Heisenberg group scheme is represented is given by the elements in the Fock representation which are invariant under the action of the lattice.

13.4 Heisenberg modules over noncommutative tori with real multiplication

Let $\theta \in \mathbb{R}$ be a quadratic irrationality, and let

$$g = \begin{pmatrix} a & b \\ c & d \end{pmatrix} \in SL_2(\mathbb{Z})$$

be a matrix fixing θ. In this section I describe the construction of a $\mathcal{A}_\theta$–$\mathcal{A}_\theta$ bimodule E_g whose isomorphism class gives a Morita self-equivalence of $\mathcal{A}_\theta$. In what follows we assume that c and $c\theta + d$ are positive. Let

$$\varepsilon = \frac{c\theta + d}{c},$$

and consider the following operators on the Schwartz space $\mathcal{S}(\mathbb{R})$:

$$(\check{U} f)(x) = f(x - \varepsilon),$$
$$(\check{V} f)(x) = \mathrm{e}(x) f(x),$$
$$(\hat{U} f)(x) = f\left(x - \frac{1}{c}\right),$$
$$(\hat{V} f)(x) = \mathrm{e}\left(\frac{x}{c\varepsilon}\right) f(x).$$

Note that each pair of operators corresponds to the Heisenberg group action of two generators of $\mathbb{R}^2$, where, as before, we identify the Schwartz space $\mathcal{S}(\mathbb{R})$ with the set of smooth elements of the Heisenberg representation $\mathcal{H}_\varepsilon$ of $\mathrm{Heis}(\mathbb{R}^2)$.

We consider also the following operators on $C(\mathbb{Z}/c\mathbb{Z})$:

$$(\check{u}\phi)([n]) = \phi([n-1]),$$
$$(\check{v}\phi)([n]) = \bar{\mathrm{e}}\left(\frac{dn}{c}\right) \phi([n]),$$
$$(\hat{u}\phi)([n]) = \phi([n-a]),$$
$$(\hat{v}\phi)([n]) = \bar{\mathrm{e}}\left(\frac{n}{c}\right) \phi([n]).$$

Because both a and d are prime relative to c, each pair of operators corresponds to the Heisenberg group action of two generators of $(\mathbb{Z}/c\mathbb{Z})^2$ on the Heisenberg representation $C(\mathbb{Z}/c\mathbb{Z})$ of $\mathrm{Heis}((\mathbb{Z}/c\mathbb{Z})^2)$.

Taking into account the commutation relations satisfied between each of the preceding pairs of operators and the fact that $g\theta = \theta$, we see that the space

$$E_g = \mathcal{S}(\mathbb{R}) \otimes C(\mathbb{Z}/c\mathbb{Z})$$

becomes an $\mathcal{A}_\theta$–$\mathcal{A}_\theta$ bimodule on defining

$$(f \otimes \phi)U = (\check{U} \otimes \check{u})(f \otimes \phi),$$
$$(f \otimes \phi)V = (\check{V} \otimes \check{v})(f \otimes \phi),$$
$$U(f \otimes \phi) = (\hat{U} \otimes \hat{u})(f \otimes \phi),$$
$$V(f \otimes \phi) = (\hat{V} \otimes \hat{v})(f \otimes \phi),$$

where $f \in \mathcal{S}(\mathbb{R})$ and $\phi \in C(\mathbb{Z}/c\mathbb{Z})$.

Theorem 13.4.1 (Connes [1]) *With the preceding bimodule structure, E_g is finite-type and projective both as a right $\mathcal{A}_\theta$-module and as a left $\mathcal{A}_\theta$-module. Considering it as a right module, we have* $\mathrm{rk}(E_g) = c\theta + d$*. The left action of $\mathcal{A}_\theta$ gives an identification*

$$\mathrm{End}_{\mathcal{A}_\theta}(E_g) \simeq \mathcal{A}_\theta.$$

We refer to this kind of modules as *Heisenberg modules*.

Taking into account the $\mathcal{A}_\theta$–$\mathcal{A}_\theta$ bimodule structure of E_g, we may consider the tensor product $E_g \otimes_{\mathcal{A}_\theta} E_g$. This is one of the main consequences of the real multiplication condition. There is a natural identification (see [5, 12]):

$$E_g \otimes_{\mathcal{A}_\theta} E_g \simeq E_{g^2}.$$

To see this consider first the completed tensor product over $\mathbb{C}$ of the space E_g with itself:

$$\begin{aligned} E_g \hat{\otimes} E_g &= [\mathcal{S}(\mathbb{R}) \otimes C(\mathbb{Z}/c\mathbb{Z})] \,\hat{\otimes}\, [\mathcal{S}(\mathbb{R}) \otimes C(\mathbb{Z}/c\mathbb{Z})] \\ &= \left[\mathcal{S}(\mathbb{R}) \;\hat{\otimes}\; \mathcal{S}(\mathbb{R})\right] \otimes [C(\mathbb{Z}/c\mathbb{Z}) \otimes C(\mathbb{Z}/c\mathbb{Z})] \\ &= [\mathcal{S}(\mathbb{R} \times \mathbb{R})] \otimes [C(\mathbb{Z}/c\mathbb{Z} \times \mathbb{Z}/c\mathbb{Z})]\,. \end{aligned}$$

The space $\mathcal{S}(\mathbb{R} \times \mathbb{R})$ is the space of smooth elements of the Heisenberg representation of $\mathrm{Heis}(\mathbb{R}^4)$ obtained as a product of the Heisenberg representations of $\mathrm{Heis}(\mathbb{R}^2)$. Likewise, $C(\mathbb{Z}/c\mathbb{Z} \times \mathbb{Z}/c\mathbb{Z})$ is the Heisenberg representation of $\mathrm{Heis}((\mathbb{Z}/c\mathbb{Z})^4)$ obtained as a product of the Heisenberg representations of $\mathrm{Heis}((\mathbb{Z}/c\mathbb{Z})^2)$.

To pass from $E_g \;\hat{\otimes}\; E_g$ to $E_g \otimes_{\mathcal{A}_\theta} E_g$ we have to quotient $E_g \;\hat{\otimes}\; E_g$ by the space spanned by the relations

$$[(f \otimes \phi)U] \,\hat{\otimes}\, [g \otimes \omega]\,, = [f \otimes \phi] \,\hat{\otimes}\, [U(g \otimes \omega)]\,,$$
$$[(f \otimes \phi)V] \,\hat{\otimes}\, [g \otimes \omega] = [f \otimes \phi] \,\hat{\otimes}\, [V(g \otimes \omega)]\,,$$

where $f, g \in \mathcal{S}(\mathbb{R})$ and $\phi, \omega \in C(\mathbb{Z}/c\mathbb{Z})$. At the level of the Heisenberg representations involved, this amounts to restricting to the subspaces of $\mathcal{S}(\mathbb{R} \times \mathbb{R})$ and

$C(\mathbb{Z}/c\mathbb{Z} \times \mathbb{Z}/c\mathbb{Z})$ which are invariant under the action of the subgroups of $\text{Heis}(\mathbb{R}^4)$ and $\text{Heis}((\mathbb{Z}/c\mathbb{Z})^4)$ generated by the elements giving these relations.

The corresponding space of invariant elements in $\mathcal{S}(\mathbb{R} \times \mathbb{R})$ is canonically isomorphic to the space $\mathcal{S}(\mathbb{R})$ of smooth elements of the Heisenberg representation of $\text{Heis}(\mathbb{R}^2)$ with $c\varepsilon^2/(a+d)$ playing the role of ε. In $C((\mathbb{Z}/c\mathbb{Z}) \times (\mathbb{Z}/c\mathbb{Z}))$ the corresponding invariant subspace is canonically isomorphic to the Heisenberg representation $C(\mathbb{Z}/c(a+d)\mathbb{Z})$ of $\text{Heis}((\mathbb{Z}/c(a+d)\mathbb{Z})^2)$. Thus we get

$$\begin{aligned} E_g \otimes_{\mathcal{A}_\theta} E_g &\simeq \mathcal{S}(\mathbb{R}) \otimes C(\mathbb{Z}/c(a+d)\mathbb{Z}) \\ &= E_{g^2}. \end{aligned}$$

The compatibility of the module structures in this isomorphism is implied by the the compatibility of the Heisenberg representations involved.

In a similar manner one may obtain isomorphisms

$$\underbrace{E_g \otimes_{\mathcal{A}_\theta} \cdots \otimes_{\mathcal{A}_\theta} E_g}_{n} \simeq E_{g^n}.$$

13.5 Some rings associated to noncommutative tori with real multiplication

Noncommutative tori may be considered as noncommutative projective varieties. In noncommutative algebraic geometry, varieties are defined in terms of categories, which play the role of appropriate categories of sheaves on them (see [17]). In [10] Polishchuk analyzed real multiplication noncommutative tori from this point of view. Given a real quadratic irrationality θ and a complex structure δ_τ on $\mathcal{A}_\theta$, Polishchuk constructed a homogeneous coordinate ring associated with $\mathbb{T}^2_\theta$ and δ_τ.

During the rest of this section g will denote a matrix in $SL_2(\mathbb{Z})$ fixing a quadratic irrationality θ. We will denote the elements of the powers of this matrix by

$$g^n = \begin{pmatrix} a_n & b_n \\ c_n & d_n \end{pmatrix}, \qquad n > 0.$$

Given a Heisenberg $\mathcal{A}_\theta$–$\mathcal{A}_\theta$ bimodule E_g, we use the complex structure δ_τ on $\mathcal{A}_\theta$ to single out a finite-dimensional subspace in each of the graded pieces of

$$\begin{aligned} \mathcal{E}_g &= \bigoplus_{n \geq 0} \underbrace{E_g \otimes_{\mathcal{A}_\theta} \cdots \otimes_{\mathcal{A}_\theta} E_g}_{n} \\ &= \bigoplus_{n \geq 0} E_{g^n}. \end{aligned}$$

This should be done in a way compatible with the product structure of $\mathcal{E}_g$.

Given a Heisenberg $\mathcal{A}_\theta$–$\mathcal{A}_\theta$ bimodule E_g, we may define a connection on E_g by

$$(\nabla_1 f \otimes \phi)(x,[n]) = 2\pi\imath\left(\frac{x}{\varepsilon}\right)(f\otimes\phi)(x,[n]),$$

$$(\nabla_2 f \otimes \phi)(x,[n]) = \frac{d}{dx}(f\otimes\phi)(x,[n]).$$

Connections of this kind were studied in [3] in the context of Yang–Mills theory for noncommutative tori. Note that this connection corresponds to the action of the Lie algebra of Heis($\mathbb{R}^2$) on the left factor of $\mathcal{S}(\mathbb{R})\otimes C(\mathbb{Z}/c\mathbb{Z})$ given by the derivations δU_A and δU_B. Once we choose a complex parameter $\tau\in\mathbb{C}\setminus\mathbb{R}$ giving a complex structure on $\mathcal{A}_\theta$, the corresponding decomposition of the complexified Lie algebra singles out the element f_τ (Theorem 13.3.4). Thus it is natural to consider the spaces

$$\begin{aligned} R_g &= \{f_\tau\otimes\phi\in E_g \mid \phi\in C(\mathbb{Z}/c\mathbb{Z})\}\\ &= \left\{\mathrm{e}\left(\frac{1}{2\varepsilon}\tau x^2\right)\otimes\phi\in E_g \mid \phi\in C(\mathbb{Z}/c\mathbb{Z})\right\}. \end{aligned}$$

These are the the spaces of the holomorphic vectors considered in [5, 10, 12, 16]. We denote by $f_{\tau,n}$ the corresponding element on the left factor of E_{g^n}, and let

$$R_{g^n} = \left\{f_{\tau,n}\otimes\phi\in E_{g^n} \mid \phi\in C(\mathbb{Z}/c_n\mathbb{Z})\right\}.$$

Following [10], we define the *homogeneous coordinate ring for the noncommutative torus* $\mathbb{T}^2_\theta$ *with complex structure* δ_τ by

$$B_g(\theta,\tau) = \bigoplus_{n\geq 0} R_{g^n}.$$

The following result characterizes some structural properties of $B_g(\theta,\tau)$ in terms of the matrix elements of g:

Theorem 13.5.1 [10, Theorem 3.5] *Assume* $g\in SL_2(\mathbb{Z})$ *has positive real eigenvalues.*

(i) If $c\geq a+d$ *then* $B_g(\theta,\tau)$ *is generated over* $\mathbb{C}$ *by* R_g.
(ii) If $c\geq a+d+1$ *then* $B_g(\theta,\tau)$ *is a quadratic algebra.*
(iii) If $c\geq a+d+2$ *then* $B_g(\theta,\tau)$ *is a Koszul algebra.*

Let X_τ be the elliptic curve with complex points $\mathbb{C}/(\mathbb{Z}\oplus\tau\mathbb{Z})$. Taking into account the remarks at the end of Section 13.3, it is possible to realize each space R_{g^n} as the space of sections of a line bundle over X_τ. For this we consider the matrix coefficients obtained by pairing $f_{\tau,n}$ with functionals in the distribution completion of the Heisenberg representation which are invariant under the action of elements in

Heis($\mathbb{R}^2$) corresponding to a lattice in $\mathbb{R}^2$ associated to g^n. This matrix coefficients correspond to theta functions with rational characteristics which form a basis for the space of sections of the corresponding line bundle over X_τ. In these bases the structure constants for the product of $B_g(\theta, \tau)$ have the form

$$\vartheta_r(l\tau),$$

where $\ell \in \mathbb{Z}$, and $\vartheta_r(l\tau)$ is the theta constant with rational characteristic $r \in \mathbb{Q}$ defined by the series

$$\vartheta_r(l\tau) = \sum_{n\in\mathbb{Z}} \exp[\pi\imath(n+r)^2 l\tau].$$

This fact has the following consequence:

Theorem 13.5.2 [9] *Let $\theta \in \mathbb{R}$ be a quadratic irrationality fixed by a matrix $g \in SL_2(\mathbb{Z})$, and assume $c \geq a + d + 2$. Let k be the minimal field of definition of the elliptic curve X_τ. Then the algebra $B_g(\theta, \tau)$ admits a rational presentation over a finite algebraic extension of k.*

Remark 13.5.3 Analogous results hold for the rings of quantum theta functions considered in [18]. These rings correspond to Segre squares of the homogeneous coordinate rings $B_g(\theta, \tau)$ and can be analyzed in terms of the Heisenberg modules involved in their construction.

Bibliography

[1] A. Connes, *C^*-algèbres et géométrie différentielle.* C. R. Acad. Sci. Paris **290** (1980), 599–604.

[2] A. Connes, *Noncommutative geometry.* Academic Press (1994).

[3] A. Connes, M. Rieffel, *Yang–Mills for noncommutative two-tori. Operator algebras and mathematical physics.* Contemp. Math. **62**, Amer. Math. Soc. Providence, RI (1987), 237–266.

[4] K. R. Davidson, *C^*-algebras by example.* Fields Institute Monographs, 6. Amer. Math. Soc., Providence, RI (1996).

[5] M. Dieng, A. Schwarz, *Differential and complex geometry of two-dimensional noncommutative tori.* Lett. Math. Phys. **61** (2002), 263–270.

[6] E. G. Effros, F. Hahn, *Locally compact transformation groups and C^*-algebras.* Bull. Amer. Math. Soc. **73** (1967), 222–226.

[7] Y. Manin, *Real multiplication and noncommutative geometry.* In: "The legacy of Niels Henrik Abel." Springer Verlag, Berlin (2004).

[8] D. Mumford, *Tata Lectures on Theta III.* With the collaboration of M. Nori and P. Norman. Progr. Math. 97. Birkhäuser, Boston (1991).

[9] J. Plazas, *Arithmetic structures on noncommutative tori with real multiplication.* Preprint: arXiv Math. QA/0610127.

[10] A. Polishchuk, *Noncommutative two-tori with real multiplication as noncommutative projective varieties.* J. Geom. and Phys. **50** (2004), 162–187.

[11] A. Polishchuk, *Classification of holomorphic vector bundles on noncommutative two-tori.* Doc. Math. **9** (2004), 163–181.

[12] A. Polishchuk, A. Schwarz. *Categories of holomorphic bundles on noncommutative two-tori*. Comm. Math. Phys. **236** (2003), 135–159.

[13] M. A. Rieffel, *C^*-algebras associated with irrational rotations.* Pacific J. Math. **93** (1981), 415–429.

[14] M. A. Rieffel, *Noncommutative tori – a case study of noncommutative differentiable manifolds.* In: "Geometric and topological invariants of elliptic operators," Contemp. Math., 105, Amer. Math. Soc., Providence, RI (1990), 191–211.

[15] M. A. Rieffel, *The cancellation theorem for projective modules over irrational rotation C^*-algebras.* Proc. London Math. Soc. (3) **47** (1983), 285–302.

[16] A. Schwarz, *Theta functions on noncommutative tori.* Lett. Math. Phys. **58** (2001), 81–90.

[17] J. T. Stafford, M. van den Bergh, *Noncommutative curves and noncommutative surfaces.* Bull. Amer. Math. Soc. (N.S.) **38** (2001), 171–216.

[18] M. Vlasenko, *The graded ring of quantum theta functions for noncommutative torus with real multiplication*. Int. Math. Res. Not. Art. ID 15825 (2006).